Novel Advancements in Electrical Power Planning and Performance

Smita Shandilya
Sagar Institute of Research, Technology and Science, India

Shishir Kumar Shandilya
Vellore Institute of Technology, India

Tripta Thakur
Maulana Azad National Institute of Technology, India

Atulya K. Nagar
Liverpool Hope University, UK

A volume in the Advances in Environmental Engineering and Green Technologies (AEEGT) Book Series

IGI Global
DISSEMINATOR OF KNOWLEDGE

Published in the United States of America by
IGI Global
Engineering Science Reference (an imprint of IGI Global)
701 E. Chocolate Avenue
Hershey PA, USA 17033
Tel: 717-533-8845
Fax: 717-533-8661
E-mail: cust@igi-global.com
Web site: http://www.igi-global.com

Library of Congress Cataloging-in-Publication Data

Names: Shandilya, Smita, 1982- editor.
Title: Novel advancements in electrical power planning and
performance / Smita Shandilya, Shishir Kumar Shandilya, Tripta Thakur, and
Atulya Kumar Nagar, editors.
Description: Hershey, PA : Engineering Science Reference, 2020. | Includes
bibliographical references and index.
Identifiers: LCCN 2018055859| ISBN 9781522585510 (hardcover) | ISBN
9781522585534 (ebook) | ISBN 978-1-5225-8552-7 (softcover)
Subjects: LCSH: Electric power systems--Planning.
Classification: LCC TK1001 .H3525 2020 | DDC 333.793/20684--dc23 LC record available at https://lccn.loc.gov/2018055859

This book is published in the IGI Global book series Advances in Environmental Engineering and Green Technologies (AEEGT) (ISSN: 2326-9162; eISSN: 2326-9170)

British Cataloguing in Publication Data
A Cataloguing in Publication record for this book is available from the British Library.

Advances in Environmental Engineering and Green Technologies (AEEGT) Book Series

Sang-Bing Tsai
University of Electronic Science and Technology of China Zhongshan Institute, China
Ming-Lang Tseng
Lunghwa University of Science and Technology, Taiwan
Yuchi Wang
University of Electronic Science and Technology of China Zhongshan Institute, China

ISSN:2326-9162
EISSN:2326-9170

Mission

Growing awareness and an increased focus on environmental issues such as climate change, energy use, and loss of non-renewable resources have brought about a greater need for research that provides potential solutions to these problems. Research in environmental science and engineering continues to play a vital role in uncovering new opportunities for a "green" future.

The **Advances in Environmental Engineering and Green Technologies (AEEGT)** book series is a mouthpiece for research in all aspects of environmental science, earth science, and green initiatives. This series supports the ongoing research in this field through publishing books that discuss topics within environmental engineering or that deal with the interdisciplinary field of green technologies.

Coverage

- Renewable Energy
- Contaminated Site Remediation
- Biofilters and Biofiltration
- Green Technology
- Cleantech
- Alternative Power Sources
- Air Quality
- Green Transportation
- Radioactive Waste Treatment
- Pollution Management

IGI Global is currently accepting manuscripts for publication within this series. To submit a proposal for a volume in this series, please contact our Acquisition Editors at Acquisitions@igi-global.com or visit: http://www.igi-global.com/publish/.

The Advances in Environmental Engineering and Green Technologies (AEEGT) Book Series (ISSN 2326-9162) is published by IGI Global, 701 E. Chocolate Avenue, Hershey, PA 17033-1240, USA, www.igi-global.com. This series is composed of titles available for purchase individually; each title is edited to be contextually exclusive from any other title within the series. For pricing and ordering information please visit http://www.igi-global.com/book-series/advances-environmental-engineering-green-technologies/73679. Postmaster: Send all address changes to above address.

Titles in this Series

For a list of additional titles in this series, please visit: www.igi-global.com/book-series

Handbook of Research on Energy-Saving Technologies for Environmentally-Friendly Agricultural Development
Valeriy Kharchenko (Federal Scientific Agroengineering Center VIM, Russia) and Pandian Vasant (Universiti Teknologi Petronas, Malaysia)
Engineering Science Reference • ©2020 • 554pp • H/C (ISBN: 9781522594208) • US $295.00

Hydrology and Water Resources Management in Arid, Semi-Arid, and Tropical Regions
Christopher Misati Ondieki (Kenyatta University, Kenya) and Johnson Utu Kitheka (South Eastern Kenya University, Kenya)
Engineering Science Reference • ©2019 • 356pp • H/C (ISBN: 9781799801634) • US $215.00

Global Perspectives on Air Pollution Prevention and Control System Design
G. Venkatesan (Anna University, India) and Jaganthan Thirumal (Anna University, India)
Engineering Science Reference • ©2019 • 345pp • H/C (ISBN: 9781522572893) • US $195.00

Retrofitting for Optimal Energy Performance
Adrian Tantau (Bucharest University of Economic Studies, Romania)
Engineering Science Reference • ©2019 • 339pp • H/C (ISBN: 9781522591047) • US $245.00

Advanced Design of Wastewater Treatment Plants Emerging Research and Opportunities
Athar Hussain (Ch. Brahm Prakash Government Engineering College, India) and Ayushman Bhattacharya (Ch. Brahm Prakash Government Engineering College, India)
Engineering Science Reference • ©2019 • 350pp • H/C (ISBN: 9781522594413) • US $195.00

Advanced Multi-Criteria Decision Making for Addressing Complex Sustainability Issues
Prasenjit Chatterjee (MCKV Institute of Engineering, India) Morteza Yazdani (Universidad Loyola Andalucía, Spain) Shankar Chakraborty (Jadavpur University, India) Dilbagh Panchal (Dr. B. R. Ambedkar National Institute of Technology (NIT) Jalandhar, India) and Siddhartha Bhattacharyya (RCC Institute of Information Technology Kolkata, India)
Engineering Science Reference • ©2019 • 360pp • H/C (ISBN: 9781522585794) • US $195.00

Amelioration Technology for Soil Sustainability
Ashok K. Rathoure (Biohm Consultare Pvt Ltd, India)
Engineering Science Reference • ©2019 • 280pp • H/C (ISBN: 9781522579403) • US $185.00

IGI Global
DISSEMINATOR OF KNOWLEDGE

701 East Chocolate Avenue, Hershey, PA 17033, USA
Tel: 717-533-8845 x100 • Fax: 717-533-8661
E-Mail: cust@igi-global.com • www.igi-global.com

Table of Contents

Detailed Table of Contents

This chapter presents a generalized day-ahead combined dynamic economic emission dispatch (DEED) problem incorporating demand response (DR) strategy for power system networks with mutual communication between electricity customers and power utility. A nonconvex mixed binary integer programming technique is used to solve the demand response optimization problem. Fixed and flexible home appliances connected as load to the power system network are considered in the demand response strategy. The optimization of the DEED problem is done using particle swarm optimization (PSO) technique. The proposed PSO algorithm takes into account thermal power generation unit ramp rates and their power generation constraints.

Renewable energy sources always drag the attention of researchers as alternate sources of power generation. These sources are inexhaustible and free of cost, which makes them very important for fulfilling electrical load demand. Due to stochastic nature of these sources as these are nature dependent, power generation from these sources varies. In order to mitigate this issue, these sources are integrated with distributed generation along with energy storage system so as to maintain the system stability. This chapter focuses on diminishing the frequency variation of microgrid incorporated hybrid power system. A hybrid system consisting of solar, wind, diesel along with a controller and superconducting magnetic energy storage unit is simulated. Whenever load demand of the system increases, frequency falls as a result deviation occurred in the system. This is overcome by the automatic generation control mechanism. Superconducting magnetic energy storage unit absorbs the excessive power available during offload condition and injects the same during peak load condition.

Moth-flame optimization algorithm (MFOA) based on the navigation strategy of moths in universe is a novel bio-inspired optimization technique and has been exerted for determining the maximum loading limit of power system. This process is highly effective for traversing long distances following a straight path. As a matter of fact, moths follow a deadly spiral path as artificial lights tend to confuse them. Exploration and exploitation are two vital aspects of the algorithm, used in tuning of the parameters. The algorithm is verified on MATPOWER case30 and case118 systems. Comparison of the performance of MFOA has been done with other evolutionary algorithms such as multi-agent hybrid PSO (MAHPSO), differential evolution (DA), hybridized DE, and PSO (DEPSO). The performance of MFOA in determining maximum loading limit is verified from the results. In much reduced time, MFO algorithm also gives high maximum loading point (MLP).

This chapter presents the different dynamics in power system reliability as a result of the intrinsic behavior of distributed renewable energy sources. The output power of distributed renewable energy sources depends on the amount of available respective resource at any given time. This output power generally experiences fluctuations when compared with the output of conventional power generation units. The phenomenon is not usually included in traditional reliability worth evaluation methods for power system networks with distributed generation. In this chapter, a reliability worth evaluation model for power system networks with time-dependent distributed renewable generation resources is presented and analyzed. Time sequential Monte Carlo simulation technique is used, and the operational efficiency of the distributed generation unit is measured using the primary reliability worth index, ECOST. The derived index is fitted to a beta distribution function to show the inherent skewness of the supply reliability worth index.

Widespread adoption of electric vehicles would bring a paradigm shift in the way distribution infrastructure is planned and electricity markets operate. Electric vehicle adoption could help in meeting the worldwide targets for greenhouse gas emissions. Moreover, the health benefits for the public would be immense as the source of emissions would be far away from the massively populated areas. For electricity markets, electric vehicles can serve as a distributed plug in facility of energy storage at low cost requiring minimal capital investment from grid utilities. However, widespread electric vehicle adoption faces a number of hurdles such as limited range in comparison to Internal combustion engines, but from the grid perspective, it faces issues such as limitations of available charging infrastructure to charge large number of electric vehicles and longer charging time currently as compared to refueling fuel driven vehicles. This chapter explores such issues and their remedies in the current literature.

In recent times, most of the power systems are made to operate close to their operating limits owing to various reasons like slow pace of transmission line expansion, environmental constraints, deregulated electricity market, etc. Therefore, the issue of maintaining the system stability has become the primary objective of the utility companies. The recent development and integration of renewable energy sources have further pushed the modern power systems to system security risks. The voltage instability had been the major cause of recent blackouts around the world. The timely assessment of voltage stability can prevent the blackouts in the power systems. This chapter explores the classical as well as newly developed static voltage stability assessment techniques proposed by various researchers in recent years. Also, the chapter cater to the needs of undergraduate as well as graduate students, professional engineers, and researchers who all are working in the domain of power system voltage stability.

This chapter presents a new reactive power loss index for identification of weak buses in the system. This index can be used for identification of weak buses in the systems. The new reactive power loss index is illustrated on sample 5-bus system, and tested on sample 10-bus equivalent system and 72-bus equivalent system of Indian southern region power grid. The validation of the weak buses identification from the reactive power loss index with that from other existing methods in the literature is carried out to demonstrate the effectiveness of the index. Simulation results show that the identification of weak buses in the system from the new reactive power loss index is completely non-iterative, and thus requires minimal computational efforts as compared with other existing methods in the literature.

Energy is described as the amount of work that can be done by force. There are various forms of energy such as kinetic energy, potential energy, thermal energy, light energy, sound energy, and electromagnetic energy. As per the law of conservation of energy, it is neither created nor destroyed. In this modern era, energy became an integral part of our life. The life without energy is not at all possible nowadays. The

Preface

As the demand for efficient energy sources continues to grow, electrical systems are becoming more essential to meet these increased needs. Electrical generation and transmission plans must remain cost-effective, reliable, and flexible for further future expansion. As these systems are being utilized more frequently, it becomes imperative to find ways of optimizing their overall function. Novel Advancements in Electrical Power Planning and Performance is an essential reference source that provides vital research on the specific challenges, issues, strategies, and solutions that are associated with electrical transmission and distribution systems and features emergent methods and research in the systemic and strategic planning of energy usage. Featuring research on topics such as probabilistic modeling, voltage stability, and radial distribution, this book is ideally designed for electrical engineers, practitioners, power plant managers, investors, industry professionals, researchers, academicians, and students seeking coverage on the methods and profitability of electrical expansion planning.

Highlighting theoretical perspectives and empirical research, it is assumed that this reference book will prove to be a comprehensive reference source for researchers, practitioners, students, and professionals interested in the current advancements and efficient use in electrical systems.

This reference book is a catalogue of twelve chapters presented by various authors across the globe. All the issues related to Electrical Power Planning and Performance are covered in following chapters as:

Chapter 1 presents a generalized day-ahead combined dynamic economic emission dispatch (DEED) problem incorporating demand response (DR) strategy for power system networks with mutual communication between electricity customers and power utility. The optimization of the DEED problem is done using particle swarm optimization (PSO) technique.

Chapter 2 focuses on diminishing the frequency variation of micro-grid incorporated hybrid power system. A hybrid system consists of solar, wind, diesel system along with a controller and superconducting magnetic energy storage unit is simulated.

In Chapter 3, Moth-Flame optimization algorithm (MFOA) based on the navigation strategy of moths in universe, is novel bio-inspired optimization technique and has been exerted for determining the maximum loading limit of power system. This process is highly effective for traversing long distances following a straight path. In this chapter, the algorithm is verified on MATPOWER case30 and case118 systems. The performance of MFOA in determining maximum loading limit is verified from the results.

Chapter 4 presents the different dynamics in power system reliability as a result of the intrinsic behaviour of distributed renewable energy sources. The output power of distributed renewable energy sources depends on the amount of available respective resource at any given time. This output power generally experiences fluctuations when compared with the output of conventional power generation units. The phenomenon is not usually included in traditional reliability worth evaluation methods for power system networks with distributed generation. In this chapter, an evaluation model for power system networks with time-dependent distributed renewable generation resources is presented and analysed.

Chapter 5 explores the issues and remedies in the current literature of Electric Vehicle Infrastructure Planning. Widespread adoption of Electric vehicles would bring a paradigm shift in the way distribution infrastructure is planned and electricity markets operate. Electric vehicle adoption could help in meeting the worldwide targets for greenhouse gas emissions. For electricity markets Electric vehicles can serve as a distributed plug in facility of energy storage at low cost requiring minimal capital investment from grid utilities.

Chapter 6: In recent times, most of the power-systems are made to operate close to their operating limits owing to various reasons like slow pace of transmission line expansion, environmental constraints, deregulated electricity market etc. Therefore, the issue of maintaining the system stability has become the primary objective of the utility companies. The recent development and integration of renewable energy sources have further pushed the modern power systems to system security risks. The voltage instability had been the major cause of recent blackouts around the world. The timely assessment of voltage stability can prevent the blackouts in the power systems. This chapter explores the classical as well as newly developed static voltage stability assessment techniques proposed by various researchers in recent years.

Chapter 7 presents a new reactive power loss index for identification of weak buses in the system. This index can be used for identification of weak buses in the systems. The new reactive power loss index is illustrated on sample 5-bus system and tested on sample 10-bus equivalent system and 72-bus equivalent system of Indian southern region power grid. The validation of the weak buses identification from the reactive power loss index with that from other existing methods in the literature is carried out to demonstrate the effectiveness of the index. Simulation results show that the identification of weak buses in the system from the new reactive power loss index is completely non-iterative, thus requires minimal computational efforts as compared with other existing methods in the literature.

Chapter 8 discusses the present scenario of energy while explaining the fundamental issues and future directions.

Chapter 9 describes the development of energy efficient and secure localization systems for wireless sensor networks (WSNs) based on anchor mobility control techniques. Mobility control schemes aim to explore few new nodes along with the existing ones in each subsequent anchor steps. Sensors would be able to localize themselves after receiving range (distance/angle) data from two distinct anchor positions.

Chapter 10: Understanding the voltage instability phenomenon and its effects in detail facilitates the research community to develop methodologies that can detect instability in a timely manner. This chapter discusses the various issues of voltage stability which is affected by the load dynamics, voltage control devices like OLTC and hitting of over excitation limiters of the synchronous generators.

Chapter 11 focuses on load frequency control (LFC) problems in multi area power systems using nonlinear second order sliding mode control (NL-SOSMC) under load disturbances and parameter uncertainties. A sudden load disturbance can causes deviation in frequency and tie line power from their schedule value. The main objective of the chapter is to give knowledge about the application of robust control technique mainly sliding mode control (SMC) for load frequency problems. The designed controller ensures finite time convergence of frequency and tie line power deviations with chattering free control signal.

Chapter 12: The increasing greenhouse emissions have led to take advantage of renewable sources. The intermittency of these sources can be mitigated using energy storage systems. The work presented in this chapter shows 3 different strategies depending on the power management and other technical factors, such as energy quality, each one with a specific goal. An intelligent model approach is explained with the aim to predict the energy demand and generation.

We express our heartfelt gratitude to all the authors, reviewers and IGI-Global personnel, especially Ms. Lindsay Wertman, Mr. Joshua Witman, and Ms. Jan Travers, for their endless motivation and patience. Special thanks to Ms. Jordan Tepper for her constant support. We hope that this reference book will be beneficial to all the concerned readers.

Editors

Chapter 1
Load Management Using Swarm Intelligence:
Dynamic Economic Emission Dispatch Optimization

Oliver Dzobo
https://orcid.org/0000-0001-9602-6835
University of Johannesburg, South Africa

Yanxia Sun
University of Johannesburg, South Africa

ABSTRACT

This chapter presents a generalized day-ahead combined dynamic economic emission dispatch (DEED) problem incorporating demand response (DR) strategy for power system networks with mutual communication between electricity customers and power utility. A nonconvex mixed binary integer programming technique is used to solve the demand response optimization problem. Fixed and flexible home appliances connected as load to the power system network are considered in the demand response strategy. The optimization of the DEED problem is done using particle swarm optimization (PSO) technique. The proposed PSO algorithm takes into account thermal power generation unit ramp rates and their power generation constraints.

INTRODUCTION

The current planning, implementation and monitoring of power utility activities is now designed to influence electricity customer's use of energy in ways that will produce desired changes in the power system network load shape (Sethaolo, Xia, & Zhang, 2014; Morais, Faria, & Vale, 2014; Yoon, Bladick, & Novoselac, 2014; Parvania & Fotuhhi-Firuzabad, 2010). This is enabled by embedding intelligence into the power system network grid so that electricity customers and the power utility are able to mutually

DOI: 10.4018/978-1-5225-8551-0.ch001

communicate between each other (Khajavan, Monsef, & Abniki, 2010; Alami, Yousef, & Moghadam, 2010; Fahrioglu & Alvarado, 2001). Demand response and dynamic economic dispatch are two control strategies that are employed by electricity customers and power utility.

Demand response plays a very important role in facilitating the interaction between the electricity customers and the power utility. It has been estimated that it could translate into as much as US$ 59 billion in societal benefits. There are two basic demand response strategies employed by most power utilities (Nwulu, Xia, & Zhang, 2013; Fahrioglu & Alvarado, 2000). The first strategy is incentive based demand response where incentives options are given to electricity customers to reduce or curtail their loads when the power system network is stressed. The incentive can be in the form of rebates or lower electricity tariffs. The second strategy is price based DR strategy which uses time-of-use (TOU) or real time electricity tariffs to encourage electricity customers to curtail their loads during periods of higher electricity price and take advantage of favorable lower electricity prices. An example of a price based DR strategy is when the price of electricity is calculated using a three-tier electricity pricing; at peak, off-peak and standard times.

Dynamic economic dispatch (DED) results in great economic benefits in power system operation. The objective of the DED is to minimize the fuel consumption cost of committed thermal power generation units used to supply a given power system network load over a time horizon under ramp rate constraints and other constraints (Elaiw, Xia, & Shehata, 2012; Basu, 2006; Xia & Elaiw, 2010). Most recently pressure from environmental agencies have forced many power utilities to consider the amount of emissions from their thermal power generation units. Power utilities are requested to reduce emission of gaseous pollutants such as SO2; NOx; CO; and CO2 from fossil fuel fired thermal power generation units as they are hazardous to human health and the environment (Nwulu & Xia, 2015; Talaq, El-Hawary, & El-Hawary, 1994; Jeddiand & Vahidinasab, 2014; Xia & Elaiw, 2010; Basu, 2014). Commonly, the dynamic economic emission dispatch (DEED) mathematical problem is used to determine the optimal scheduling of the committed thermal power generation units output whilst supplying the power system network load over a scheduling period at minimum fuel operating cost and emission simultaneously under a set of constraints. In Xia & Elaiw (2010) and Basu (2014), a review of various dynamic economic dispatch (DED) mathematical formulations and solutions methods that have been applied to solve the problem are presented. In most cases, power utilities solve DEED with other associated tasks like unit commitment and also most recently with the incorporation of renewable energy sources like wind and solar energy (Osorio, Lujano-Rojas, Matias, & Catal, 2015; Aghaei, Niknam, Azizipanah-Abarghooee, & Arroyo, 2013).

Several research studies have considered DEED and DR optimization models separately when applying it to power system network management. The DEED optimization model is solely concerned with the supply end of the power system network while the DR optimization problem is applied to the demand side of the power system network. The shortfall of these approaches is that they are presented as a simplified problem without considering the cost saving opportunities that may exists between the simultaneous interaction of DEED and DR optimization models. Thus, they may forge some practical suboptimal electricity customer energy usage behavior as well as higher short-term marginal electricity production costs than they would otherwise be in an optimally efficient system. The disconnection between electricity production costs and overall electricity usage may lead to inefficient retail electricity price rates paid by electricity customers, higher fuel consumption cost and large amount of gaseous pollutants emission. This problem has resulted in an increasing need to convert electricity customers into active participants who engage with the power utility to balance electricity production costs and overall electricity usage. The engagement between electricity customers and power utility is achieved through the use of a combined DR-DEED optimization model with interactive control strategies (Dzobo & Xia, 2017). The combined DR-DEED optimization model seeks to minimize simultaneously the economic and emission of committed thermal power generation units; and DR costs over a scheduling horizon. The interactive control strategy between the power utility and the electricity customers will result in a stabilized and optimal power system scheduling plan to be reached. In Nwulu & Xia (2015), a computationally efficient scheduling model for power system scheduling problem of thermal generation units incorporating DR strategy was solved using CPLEX 12.6 solver. It has been mentioned that the approach resulted in a better quality of the optimal solutions. One limitation of the existing combined DR-DEED optimization models, however, is that they examine the value of DR assuming the electricity customer loads are operated using a deterministic load model. Electricity customer loads are composed of a variety of electrical equipment and appliances (e.g. heating, cooking, lighting, air conditioner etc) that sometimes rely on naturally occurring phenomena that are not perfectly predictable. For loads, uncertainties are caused by weather-related factors including temperature and precipitation, economic growth, new types of electronically-controlled loads, and variations in load power factors. Because real-time load variability is uncertain when commitment decisions are made, the set of thermal power generation units that are committed may be suboptimal if deterministic load models are used when making operational decisions day- and hour-ahead. Thus, for realistic optimal power system network scheduling plan representation with existing thermal power generation units, the inherently stochastic nature of the load must be included in the optimization problem.

The most common conventional methods used for solving DEED problems include lambda iteration method, gradient method, etc. These methods require monotonically increasing or piece-wise fuel cost and emission curves. The main disadvantage of the methods is that sometimes it converges to local optimal solution or diverge together. As different constraints have started to be included into the DEED problem formulation, this has added to the complexity of the DEED optimisation problem. As a result, more convenient numerical methods have been developed for solving the DEED problem. Artificial intelligence techniques such as Hopfield neural networks, genetic algorithms (GA) (Po-Hung & Hong-Chan, 1995), simulated annealing (SA) (Wong & Wong, 1994), particle swarm optimisation (PSO) (Sinha, Chakraborti, & Chattopadhyay, 2003), etc are popular heuristic techniques for solving complex DEED optimisation problems. In Po-Hung & Hong-Chan (1995), a GA method was presented for solving DEED problem of a large scale power system network and cost factors of thermal power generation units were considered. The PSO algorithm technique was introduced by Kennedy and Eberhart (Po-Hung & Hong-Chan, 1995) as one of the modern heuristic algorithms. It produces high quality solutions with time and stable convergence characteristics than other stochastic methods. In this chapter, a combine DEED incorporating demand response strategy is performed.

Flexible and fixed home appliance loads connected to the power system network are considered. The demand response optimisation problem is solved using a nonconvex mixed binary integer programming technique. The following section will introduce the formulation of the combined DR-DEED problem under a set of constraints; load demand constraint; generation limits and unit ramp rate constraints.

BACKGROUND

The power system network is one of the most complex networks built by human being. The main purpose of the power system network is to provide electricity whenever the connected consumers require it. As the connected customers become more and more dependent on it, the reliability of the power system has become very important. The power system network can be subdivided into three levels, namely; generation, transmission and distribution. Power system reliability can be performed at any of these levels or a combination of the levels. This section will provide highlights of the recent developments of power system that have taken place and ongoing.

Combined DR-DEED Problem Formulation

The objective of the combined DR-DEED optimisation problem is to reduce the peak power system network load thus reducing the fuel and emission cost of the thermal power generation units and in turn maximize the economic benefit by electricity customers through reducing their electricity bill/cost. To achieve this the demand side management controller (DSMC) designs an objective load curve according to the objective of the combined DR-DEED strategy. Figure 1 shows the proposed architecture for the day-ahead combined DR-DEED strategy. It shows the exchange of information between the DSMC and the electricity customers during the real-time operation. The modelling technique employed in the DR-DEED strategy evolves in two stages. The first stage involves the DR optimisation of the load demand from the power system network using a nonconvex mixed binary integer programming technique. The DR optimisation objective in this stage is to minimise the total daily electricity cost to the electricity consumers. The DR system receives the initial objective load curve and electricity pricing as an input, calculates the required load control actions in order to fulfil the desired load consumption and minimizing the electricity bill/cost.

The DR optimisation objective is carried out at the beginning of a predefined control period which is typically a day. The electricity bill/cost for the electricity customers is minimized under load demand constraints amidst other constraints. The optimised load demand is then used as an input to the PSO algorithm in the second stage. The DSMC receives the optimized load from the DR system and use this optimized load to perform the dynamic economic emission dispatch (DEED) optimization model. At

Figure 1. DEED-DR optimisation framework

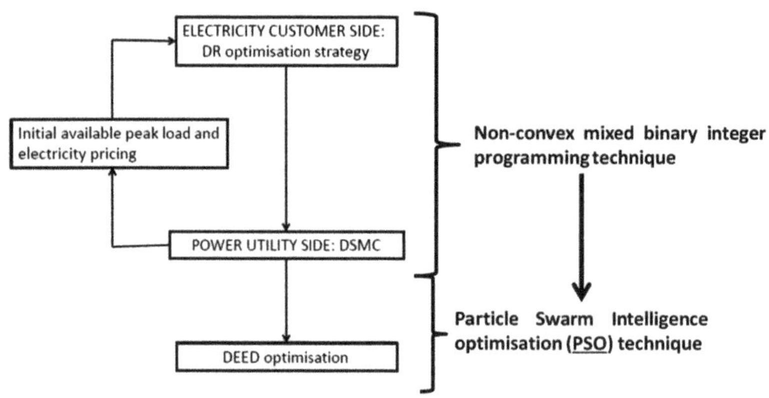

this stage the ramp rates, economic and emissions of the thermal power generation units are optimised. In this stage, a weighting factor μ is introduced to transform the multi-objective optimization problem into a weighted single objective optimization problem. This would allow the contradicting power generation fuel cost objective and emission objective functions to be simultaneously minimized. The weighting factor, μ, blends the power generation fuel cost and amount of emission of the thermal power generation units. The mathematical formulation of the weighted single objective optimization can be expressed as:

$$DR - DEED = \min\left[\mu C_{\cos t} + \left(1 - \mu\right) E_{\cos t} + DR_{\min}\right] \quad (1)$$

where: (i) $C_{\cos t}$ is the cost due to thermal power generation fuel consumption costs and (ii) $E_{\cos t}$ is the cost due to emission of thermal power generation units (iii) $DR_{\min}$ is the minimum electricity cost charged to the electricity customer.

Multi-Objective DEED Problem Formulation

Economic load dispatch in power system networks has been investigated extensively and several optimisation algorithms have been developed. The economic load dispatch problem optimizes the scheduling of thermal power generation units based on their power generation fuel consumption costs. Emission constraints of thermal power generation units were added to the economic load dispatch problem to formulate a multi-objective dynamic economic emission dispatch (DEED) problem.

Cost Objective Function

The objective function minimizes the power generation fuel consumption cost of thermal power generation units for a specific period of operation and in return satisfying the power system network load and operation constraints of the committed thermal generation units. The objective function can be formulated as:

$$C = \sum_{h=1}^{H}\sum_{i=1}^{N_g} C_i P_i^h = \sum_{h=1}^{H}\sum_{i=1}^{N_g} [a_i + b_i P_i^h + c_i (P_i^h)^2] \quad (2)$$

where N_g is the number of thermal power generation units, P_i^h is the active power generation of thermal power generation unit i at time h and a_i, b_i and c_i are the fuel consumption cost coefficients of the i_{th} thermal power generation unit. H is the scheduling horizon that indicates the number of hours ahead which are taken in account for decision making in the economic load dispatch. $C_i P_i^h$ is the fuel consumption cost of the thermal power generation unit i to producing power P_i^h. This cost objective function is normally obtained from the heat run test of thermal power generation units (Nwulu & Xia, 2015).

Emission Objective Function

The emission of pollutants from thermal power generation units affects not only human beings, but it also causes damage to materials and global warming. These effects can be interpreted as cost and can be formulated into an emission cost function (Talaq, El-Hawary, & El-Hawary, 1994). The main objective of this emission cost function is to distribute the power system network load to the scheduled thermal power generation units at minimum total emission. The emission cost function can be expressed as the sum of all types of emissions such as NO, SO2, particulate materials and thermal radiation with suitable pricing for each pollutant emitted. For SOx and CO2, it is generally acknowledged that emissions are proportional to the fuel consumptions of the thermal power generation units. The emission cost function for the production of NOx is less straightforward to represent since it is highly non-linear in the power output, and the slope of the curve is not always positive. However, in most cases, this emission is in proportion to the power output and it is represented by a combination of polynomial and exponential terms (Shehata & Xia, 2012). In this chapter, the emission cost functions for NO, SO2 and CO2 are modelled as quadratic cost function. The objective of emission dispatch is to minimize the total emission cost or the total pollutant emission due to the burning of thermal power generation unit fuels. The emission objective function can be expressed as:

$$E = \sum_{h=1}^{H} \sum_{i=1}^{N_g} E_i P_i^h = \sum_{h=1}^{H} \sum_{i=1}^{N_g} [d_i + e_i P_i^h + f_i (P_i^h)^2] \quad (3)$$

where $E_i P_i^h$ is the emission function of the thermal power generation unit i to producing power P_i^h. Constants d_i, e_i and f_i are the emission cost coefficients of the i_{th} thermal power generation unit.

Optimisation Constraints

Power Balance

This constraint ensures that at any given time, the total power generated by the committed thermal power generation units equals the respective power system network load. The constraint is based on the principle of equilibrium between total system generation, $\sum_i^{N_g} P_i^h$, and total system demand, P_D^h, and transmission power losses, $Loss^h$, that is:

$$\sum_i^{N_g} P_i^h = P_D^h + Loss^h h = 1,2,.,.,.,H \quad (4)$$

where, $Loss^h$, is obtained using Kron's loss formula, given by:

$$Loss_h = \sum_{i=1}^{N_g} \sum_{j=1}^{N_g} P_i^h B_{ij} P_j^h, h = 1,2,.,.,.,H \quad (5)$$

The coefficients B_{ij} are called transmission loss coefficients or B-coefficients. They are dependent on the impedance parameters of the transmission lines of the power system network [27], [28].

Generator Operation Constraints

This constraint ensures that the generator power output limits are not exceeded.

$$P_i^{\min} \le P_i^h \le P_i^{\max}, i = 1, 2, ., ., ., N_g; h = 1, 2, ., ., ., H \tag{6}$$

where $P_i^{\min}$ and $P_i^{\max}$ are lower and upper bounds for power outputs of the i_{th} generation unit, respectively.

Generation Unit Ramp Rates

This constraint ensures that the generation unit ramp rate limits are not violated.

$$DR_i \le P_i^{h+1} - P_i^h \le UR_i, i = 1, 2, ., ., ., N_g; h = 1, 2, ., ., ., H - 1 \tag{7}$$

where DR_i and UR_i are maximum down and up ramp rates of the i_{th} generation unit, respectively.

Optimization Problem

It can be clearly seen that the fuel consumption cost and the amount of emission conflict with each other. Minimization of fuel consumption cost maximizes the amount of emission and vice versa. Therefore, it is necessary to combine the fuel consumption cost and emission function objectives; and find out an operating point that strikes balance between them. This can be done by formulating the DEED problem which is a multi-objective optimization problem with two conflicting objectives, the fuel consumption cost and emission. According to the above objectives and constraints, the DEED problem can be mathematically formulated as a multi-objective optimization problem which can be converted into a single-objective optimization using the weighting method as follows:

$$\min\left[\mu C + (1-\mu)E\right] \tag{8}$$

where $\mu \in [0,1]$ is a weighting factor. μ is calculated with respect to the maximum and minimum power of the thermal power generation units.

$$\mu_{\max-\min} = \frac{a * P^2_{\max} + b * P_{\max} + c}{d * P^2_{\min} + e * P_{\min} + f}$$

$$\mu_{\max-\max} = \frac{a * P^2_{\max} + b * P_{\max} + c}{d * P^2_{\max} + e * P_{\max} + f}$$

$$\mu_{\min-\max} = \frac{a * P^2_{\min} + b * P_{\min} + c}{d * P^2_{\max} + e * P_{\max} + f}$$

$$\mu_{\min-\min} = \frac{a * P^2_{\min} + b * P_{\min} + c}{d * P^2_{\min} + e * P_{\min} + f}$$

Demand Response Problem Formulation

DSM programs can be categorized as demand response (DR) programs and energy efficiency programs. DR programs involves incentive options that are given to electricity customers in order to shift their loads during periods of peak demands (Alami, Yousef, & Moghadam, 2008; Malette & Venkatarmanan, 2010). This reduces the stress on the power system network and allows the power system network load to be supplied from less expensive base load generation (Sethaolo, Xia, & Zhang, 2014). Traditionally, DR programs are applied to large electricity consumers such as industrial or commercial buildings (Malette & Venkataramanan, 2010). The intensive use of electricity in these sectors allows for the cos-benefit of providing such control strategies affordable. DR control strategies are still very rare in residential customers where the implementation costs are considered to be relatively high. Traditionally, the current home appliance control activities by residential customers are mainly operated manually. The problem with this type of DR control strategy is that some residential customers may not have time to make such home appliance scheduling decisions and in addition, if electricity prices vary fast and frequently, home appliance scheduling may be too complex. Recent advances in development of smart home appliances have led to a reduction in cost of implementing DR control strategies in residential customers (Sethaolo, Xia, & Zhang, 2014; Alami, Yousef, & Moghadam, 2008). Most of the smart home appliances now have automated connectivity features that allow automated residential demand response control strategies (Erol-Kantarci & Mouftah, 2011).

Residential customers perform different daily activities during the day. These daily activities are characterized by different home appliances that are scheduled to operate at preferred time intervals (Song, Alvehag, Widen, & Parisio, 2014). Some of the home appliances are continuously connected throughout the whole day, e.g., refrigerator. These type of home appliances consume electricity throughout the day and have mechanisms to adjust their electricity consumption levels. In some cases, the home appliances are scheduled according to the user or customer preference constraints, e.g., a residential customer can choose the time interval to use a washing machine. Several research studies have modelled smart home appliances using different constraints (Sethaolo, Xia, & Zhang, 2014; Yoon, Bladick, Novoselac, 2014; Khahavi, Monsef, & Abniki, 2010; Nwulu, Xia, & Zhang, 2013; Dzobo & Xia, 2017). Commonly, average non-varying energy consumption level constraints for complete operating cycle of home appliances are assumed. However, some smart home appliances, e.g., dishwasher, washing machine etc have different operating cycles which consumes different energy levels. It is therefore not realistic to assume average values for the complete operating cycle of such home appliances, as this may lead to underestimation or overestimation of its energy consumption level. In some instances, the constraints considered ignore that home appliances may have to run sequentially without interruption. For example, when doing laundry, the washing machine will have to finish first before the dryer starts operating. It is therefore important that the resultant home appliance schedule achieved from the minimization of load and electricity cost is presented to reveal further if the reduction is not achieved as a result of inappropriate scheduling of the home appliances. The following sections outline how the smart home appliances were modelled for demand response optimization problem.

Home Appliance Load Modelling

The DR objective function minimizes the electricity bill of the electricity customers without affecting their daily energy requirements. Let A denote the set of home appliance load demand, then: L_a^h is the load demand by home appliance $a \in A$ at time slot h. The total daily energy requirement from home appliance a over one day can be defined as:

$$E_{DAILY,a} = \sum_{h=1}^{H} L_a^h \tag{9}$$

Fixed Home Appliance Modelling

Fixed home appliance has operating times that cannot be shifted. For example, refrigerator that operates throughout the whole day. If the home appliance is switched OFF then the electricity customer may incur more costs or losses than the cost of unsupplied energy. For fixed home appliances which cannot be shifted to any time slot, the total energy requirement for the whole operation period of the home appliance is given by:

$$\sum_{h=\alpha_a}^{\beta_a} L_a^h = E_a \forall a \in A \tag{10}$$

where E_a is the total energy requirement for the whole operation of home appliance, α_a is the beginning of acceptable operation time and β_a is the end of acceptable operation time. Equation 10 ensures that the operation period of the home appliance is finished before deadline and is equal to the total energy requirement of operation. It is also required that $L_a^h = 0 \forall h \prec \alpha_a \; and \; h \succ \beta_a$.

Example 1: A home appliance load 1 which has a power rating of 100W (i.e., $L_{load1}^h = 100W$) has two periods of operation, i.e., ($\alpha_1 = 12$; $\beta_1 = 16$) and ($\alpha_2 = 20$; $\beta_2 = 22$). The total daily energy requirement for home appliance load 1 is given as:

$$E_{DAILY,Load1}=\sum_{h=12}^{16}L_{Load1}^{h}+\sum_{h=20}^{22}L_{Load1}^{h}$$
$$=500+300$$
$$=800Wh$$

It is also required that

$$L_{Load1}^{h}=0\forall 1\leq h\leq 11;17\leq h\leq 19\,and\,23\leq h\leq 24$$

In case where the hourly home appliance load demand is known in terms of range of load levels, then L_a^h is treated as variable e.g., refrigerator. For home appliance load $a\in A$ which have a maximum hourly load level $\gamma_a^{\max}$ a and a minimum hourly load level $\gamma_a^{\min}$, the total daily energy requirement is given as:

$$\sum_{h=\alpha_a}^{\beta_a}L_a^h=E_{DAILY,a}\forall a\in A;h\in[\alpha_a,\beta_a];\gamma_a^{\min}\leq L_a^h\leq\gamma_a^{\max} \tag{11}$$

Example 2: If home appliance Load 2 operates for the whole 24 hours of the day, i.e., $\alpha=1$; $\beta=24$ and the energy requirement of the home appliance can take any value between $\gamma_{Load2}^{\min}=0$ and $\gamma_{Load2}^{\max}=50Wh$ at each hour h. If the total daily energy requirement for the home appliance is equal to 200MWh, i.e.,

$$\sum_{h=1}^{24}L_a^h=200Wh\forall Load2\in A;h\in[1,24];0\leq L_a^h\leq 50Wh$$

Flexible Home Appliance Modelling

Flexible home appliance can operate in many optional operational times and can be shifted according to customer preferences and/or different electricity cost rates during the day. Examples of such home appliances are water pumps, electric vehicle charging etc. Flexible home appliance can be arranged in several optional hours while ensuring the total energy requirement is supplied. Let $R(\subset A)$ denote the set of indexes of the flexible home appliance. For home appliance $a\in R$, if X_a denote the fundamental load demand pattern as ($\varepsilon_a^1,\varepsilon_a^2,.,.,.,\varepsilon_a^h,.,.,.,\varepsilon_a^H$) where $\varepsilon_a^h\geq 0$, the a_{th} flexible home appliance can have H possible patterns which are obtained by circular shifting the fundamental load demand pattern. In order to select one possible load demand pattern for optimization, a binary switching integer vector s_a is used. The binary switching integer vector s_a is defined as $s_a=(s_a^1,s_a^2,.,.,.,s_a^h,.,.,.,s_a^H)$ where $s_a^h\in(0,1)$. The position of a binary integer one (1) means the starting time at which the home appliance load is switched ON. The binary switching integer vector therefore has only one non-zero element equal to one (1) in order to ensure that each home appliance is switched only once per each operation. For home appliance $a \in R$, this constraint can be written as:

$$\sum_{h=1}^{H} s_a^h = 1 \tag{12}$$

By using s_a, the home appliance demand scheduling plan L_a can be written as:

$$L_a = s_a \times X_a^C \forall a \in R \tag{13}$$

where the columns of the 24 x 24 matrix X_a^C a is the circulant matrix of the fundamental load demand pattern, X_a, i.e.,

$$X_a^C = \begin{pmatrix} \varepsilon_a^1 & \varepsilon_a^2 & . & . & . & . & \varepsilon_a^2 \\ \varepsilon_a^2 & \varepsilon_a^1 & . & . & . & . & \varepsilon_a^3 \\ . & . & . & . & . & . & . \\ . & . & . & . & . & . & . \\ . & . & . & . & . & . & . \\ \varepsilon_a^{24} & \varepsilon_a^{23} & . & . & . & . & \varepsilon_a^1 \end{pmatrix}$$

Example 3: Let us take a 5 hour load schedule for home appliance a. If the fundamental load demand pattern is given as $X_a = [100 \quad 0 \quad 0 \quad 50 \quad 0]$ then:

$$X_a^C = \begin{pmatrix} 100 & 0 & 50 & 0 & 0 \\ 0 & 100 & 0 & 50 & 0 \\ 0 & 0 & 100 & 0 & 50 \\ 50 & 0 & 0 & 100 & 0 \\ 0 & 50 & 0 & 0 & 100 \end{pmatrix}$$

To choose one of the home appliance load demand pattern a binary switching vector s_a is used. For example, to choose the first column, the first element of s_a is set to binary integer 1, i.e., $X_a = [1 \quad 0 \quad 0 \quad 0 \quad 0]$. The home appliance load demand pattern for optimization is given as:

$$L_a^{\mathrm{T}} = X_a^C \times s_a^{\mathrm{T}}$$
$$= \begin{pmatrix} 100 & 0 & 50 & 0 & 0 \\ 0 & 100 & 0 & 50 & 0 \\ 0 & 0 & 100 & 0 & 50 \\ 50 & 0 & 0 & 100 & 0 \\ 0 & 50 & 0 & 0 & 100 \end{pmatrix} \begin{pmatrix} 1 \\ 0 \\ 0 \\ 0 \\ 0 \end{pmatrix}$$
$$\therefore L_a = \begin{bmatrix} 100 & 0 & 0 & 50 & 0 \end{bmatrix}$$

Similarly, to choose the second, third, fourth and fifth column of the circulant matrix X_a^C, the second, third, fourth, fifth element of s_a is set to binary integer 1 respectively.

If ε_a^h is variable and bounded by a minimum hourly load level $\gamma_a^{\min} \geq 0$ and a maximum hourly load level $\gamma_a^{\max}$, with a positive constant E_a denoting the total daily energy requirement limit for the a_{th} home appliance, the load demand scheduling plan L_a for the flexible home appliance can be written as:

$$L_a^{\mathrm{T}} = X_a^C \times s_a^{\mathrm{T}} \forall \sum_{h=\alpha}^{\beta} \varepsilon_a^h = E_a \gamma_a^{\min} \leq \varepsilon_a^h \leq \gamma_a^{\max}; a \in R \tag{16}$$

Demand Response Optimization Algorithm

The demand response strategy is to minimize the total daily electricity cost to the connected electricity consumers. This is achieved by shifting the operational times of the flexible home appliances. Assume the daily electricity cost rate vector is given as: $\rho = \left[\rho_1, \rho_2, ., ., ., \rho_H\right]$, the hourly electricity cost due to both flexible and fixed home appliances, $EL_C(h)$, is given as:

$$EL_c(h) = \rho(h)\left[\sum_{a \in A} E_a(h) + \sum_{b \in B} E_b(h)\right] \tag{17}$$

The total daily electricity cost, EL_{DC}, due to flexible and fixed loads is given by:

$$EL_{DC} = \sum_{h=1}^{24}\left[p_h\left(\sum_{a \in A} E_a(h) + \sum_{b \in B} E_b(h)\right)\right] \tag{18}$$

Therefore, the demand response optimization problem can be formulated as:

$$DR_{\min} = \min\left[\sum_{h=1}^{24}\left[p_h\left(\sum_{a \in A} E_a(h) + \sum_{b \in B} E_b(h)\right)\right]\right] \tag{19}$$

subject to constraints (9) - (16). Constraints (9) - (16) are all real constraints for both fixed and flexible loads used in the optimization problem. In general, the total daily electricity cost can be expressed as follows:

$$EH_{\cos t,h} = \rho_h \times L_{ET,h} \tag{20}$$

where

$$L_{ET,h} = L^h_{E\,fixed,a} + L^h_{E\,flexible,a} + L^h_{E\,nighttime,a}$$

i.e, sum of fixed, flexible and night-time home appliance loads at any given time t. The general DR optimization problem formulation for the whole day is given as:

$$DR_{\min} = \min\left[\sum_{h=1}^{H} \rho_h L_{ET,h}\right] \tag{21}$$

subject to: Flexible and non-flexible energy hub load constraints, i.e., Eqs 9 - 16.

The general formulation is expressed as Equation 22 below:

$$\begin{aligned} &\min_{x,z} f(x,z) = a^{\mathrm{T}}x + b^{\mathrm{T}}z \\ &subject\,to : G(x,z) = c; H(x,z) \le d \end{aligned} \tag{22}$$

where, $f(x,z)$ is the objective function presented as a linear combination of variables, x , real values and, z , binary values. The objective function defines the feasible solution to the variables that is optimal. $G(x,z)$ and $H(x,z)$ are the equalities and inequalities constraints respectively. The constraints are used to limit the optimal solution of the model in a feasible region.

The output optimal load demand pattern from the demand response optimization problem is used as an input to the PSO algorithm. The following subsection describes the steps used in the PSO algorithm to solve the dynamic combined economic and emission dispatch optimization problem.

Particle Swarm Optimization Algorithm

The PSO is also known as the swarm intelligence technique. It is a heuristic technique that mimics the behavior of swarm of birds looking for food. The swarm consists of a set of birds referred as 'particles' which are flying in different directions that are possible solutions to finding food. The position of each bird or particle in the food search space is referred as the 'personal best position' (P_{best}). The best position among the P_{best} positions that leads to finding the food is referred as to the 'global best position' (G_{best}). The position of each bird or particle is updated using its velocity. A new velocity of the particle is calculated using the current velocity and its distance from P_{best} and G_{best} positions. The velocity vector consists of three components that determines the next position. The first term is the previous velocity of the bird which is stored from the previous iteration and is used to prevent each particle from severe deviation of its direction in comparison from its previous direction. The social component is the term that is used to form the attractive force among the birds or particles to the best position G_{best} of the swarm. The third component of the velocity is the cognitive component which defines the attractive force of each bird or particle to its previous best position. To maintain a good balance between the individuality and sociality component, two constants are normally used, and these are made to be equal so that their effect is balanced. If the individuality (cognitive) component is weighed more that the sociality component it is possible that the resultant solution will be trapped in local solutions achieved by that individual particle or bird. Contrary, if the sociality component is weighed more than the individuality component then the resultant solution might fail to converge. To improve the convergence of the resultant best position G_{best} an inertia weighting parameter can be introduced in calculating the velocity. This would allow the velocity to start with very large values and then decreases as the iteration increases and thus limit very large deviation of particle directions towards the end of the optimization process. When a new P_{best} position is calculated it is compared with its previous P_{best} using a fitness function and this will give a new P_{best} position for the bird or particle. All the new P_{best} positions are compared to give a new G_{best} position. This process is iteratively repeated until a given criterion or number of iterations is achieved. The best optimization position for the optimization problem will be the resultant global best G_{best} position. Figure 2 below shows the positions that can be achieved by each bird or particle.

Advantages of Using PSO

1. The PSO is a non-gradient and non-derivative method which is able to deal with objective functions that are non-continuous, non-convex and non-differentiable.

Figure 2. Different positions that can be achieved by each bird

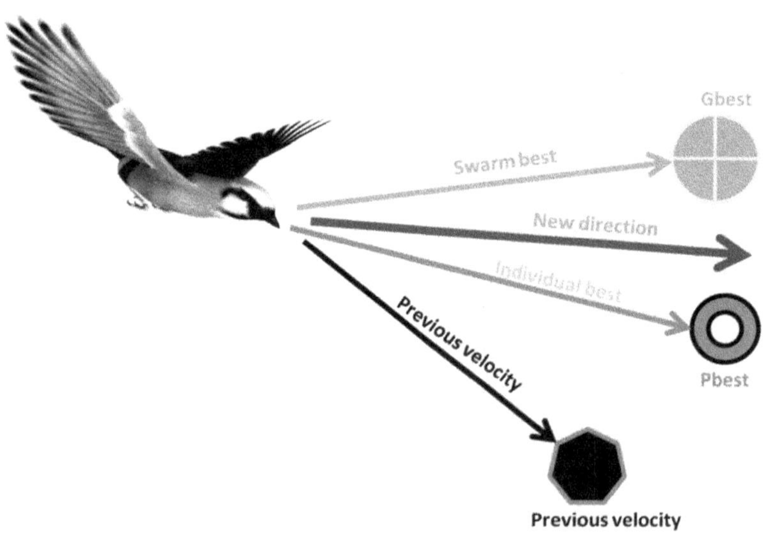

2. The PSO method utilizes the fitness function value to search for an optimal solution and this eliminates the approximation and assumption operations that are commonly performed in conventional optimization techniques on problem objectives and constraint functions.
3. The PSO method is a stochastic method which is able to handle objective functions that are stochastic and noisy in nature.
4. The resultant optimal solution from PSO is not dependent on the initial solution as in deterministic optimization techniques.
5. The PSO method evaluates several solutions in a single iteration which minimizes the chance of being trapped in a local minima.
6. The PSO method is vey flexible as it can be hybridized and integrated with other methods whether deterministic or heuristic

PSO Algorithm

The PSO technique described above is applied to a power system network with thermal power generation units. All simulations were performed using Matlab software. The general PSO algorithm that was applied is as outlined below.

Step 1: Initialize thermal power generation units data i.e. power generation constraints, fuel consumption coefficients, ramp rates limits, emission coefficients, number of thermal power generation units, starting reference point of thermal power generation units, load demand, inertia weights, acceleration constants, uniform random numbers, maximum number of iterations, etc

Step 2: Calculate the initial minimum power, $P_i^{\min}$, and maximum power, $P_i^{\max}$, using the starting refence point, P_i^0, maximum down ramp rate, DR_i and maximum up ramp rate, UR_i of each thermal power generation unit, i

$$(P_i^{\min})^0 = P_i^0 - DR_i$$

$$(P_i^{\max})^0 = P_i^0 - UR_i$$

Check the thermal power generation, unit constraints and adjust accordingly, i.e.

$$\begin{Bmatrix} (P_i^{\min})^1 = \max\left[P_i^{\min}, (P_i^{\min})^0\right] \\ (P_i^{\max})^1 = \min\left[P_i^{\max}, (P_i^{\max})^0\right] \end{Bmatrix}$$

Step 3: Calculate the velocity of all the random samples in the problem space using Equation 15 below

$$v_i^{\min} = -0.5 \times (P_i^{\min})^1$$

$$v_i^{\max} = 0.5 \times (P_i^{\max})^1$$

$$v_{i,j}^1 = v_i^{\min} + rand() \times (v_i^{\max} - v_i^{\min})$$

Equation 16 and 17 are used to calculate the minimum velocity, $v_i^{\min}$,and maximum velocity, $v_i^{\max}$, of each thermal power generation unit, i. Equation 18 determines the position of each random sample, j, in each thermal power generation unit, i.

Step 4: Determine the real power, $P_{i,j}$ of each thermal power generation unit, i, for each sample position, j, using results from *Step 2*.

$$P_{i,j} = (P_i^{\min})^1 + rand() \times \left[(P_i^{\min})^1 - (P_i^{\max})^1\right]$$

Check the quality below and adjust the thermal power generation units accordingly

$$0 = P_D^1 - \sum_{i=1}^{N} P_{i,j}$$

where, P_D^1, is the optimal load demand for $h = 1$, N is the maximum number of thermal power generation units in the power system network.

Step 5: Determine the objective or fitness function values based on the obtained power generation values of all thermal power generation units.

$$FC_{i,j} = a_i \times (P_{i,j})^2 + b_i \times P_{i,j} + c_i$$

$$ET_{i,j} = d_i \times (P_{i,j})^2 + eb_i \times P_{i,j} + f_i$$

$$TC_{i,j} = FC_{i,j} + \mu \times ET_{i,j}$$

Step 6: Select the initial best positions of thermal power generation units as follows:

i) The initial sample positions of each thermal power generation unit are considered as the best positions, P_i^{best}, of the respective thermal power generation units, i.e.,

$$P_i^{best} = \min(P_{i,j}^{best})$$

ii) Compare the P_i^{best} of all thermal power generation units to determine the global best position, G_i^{best}, i.e.,

$$G_i^{best} = \min(P_1^{best}, P_2^{best}, \ldots, P_N^{best})$$

Step 7: Increase iteration, $iter$, by 1 and update the velocity of each thermal power generation unit using Equation 23 below:

$$v_{i,j}^{new} = \omega_i \times v_{i,j}^{ite-1} + c_1 \times rand_1 \times \left[(P_i^{best})^{iter-1} - (P_i)^{iter-1}\right] + c_2 \times rand_2 \times \left[(G_i^{best})^{iter-1} - (P_i)^{iter-1}\right]$$

where, ω_i, is given as:

$$\omega_i = \omega_{\max} - \frac{i \times (\omega_{\max} - \omega_{\min})}{10}$$

Step 8: Check for violations of minimum and maximum velocities constraints, i.e.,

$$v_{i,j}^{new} = \begin{Bmatrix} v_i^{\min} & v_{i,j}^{new} \leq (v_{i,j}^{\min})^{iter-1} \\ v_{i,j}^{new} & v_i^{\min} \leq v_{i,j}^{new} \leq v_i^{\max} \\ v_i^{\max} & v_{i,j}^{new} \geq (v_{i,j}^{\min})^{iter-1} \end{Bmatrix}$$

Step 9: Check for violations of the following equality constraint and adjust accordingly,

$$0 = P_D^1 - \sum_{i=1}^{N} P_{i,j}^{new}$$

Step 10: Determine the new fitness function values based on the obtained new power generation values for each thermal power generation unit.

$$FC_{i,j}^{new} = a_i \times (P_{i,j}^{new})^2 + b_i \times P_{i,j}^{new} + c_i$$

$$ET_{i,j}^{new} = d_i \times (P_{i,j}^{new})^2 + e_i \times P_{i,j}^{new} + f_i$$

$$TC_{i,j}^{new} = FC_{i,j}^{new} + \mu \times ET_{i,j}^{new}$$

Step 11: Select the best sample position of each thermal power generation unit as follows:

i) The best position of the respective thermal power generation unit is given by Equation 25 below:

$$(P_i^{best})^{iter} = \min\left[(P_i^{best})^{iter-1}, (P_i^{best})^{new}\right]$$

ii) Compare the $(P_1^{best})^{iter}$ of all thermal power generation units to determine the global best, $(G_i^{best})^{iter}$, i.e.,

$$(G_i^{best})^{iter} = \min\left[(P_1^{best})^{iter}, (P_2^{best})^{iter}, ., ., (P_N^{best})^{iter}\right]$$

Step 12: Repeat *Step 6 – 11* until the maximum number of iterations or the criterion set is reached.
Step 13: Select the overall global best G_i^{best}, as the best optimal solution of the optimization problem.
Step 14: Increase time h by 1 and use the best sample positions of the thermal power generation units at $h-1$ as the initial power generation values of the respective thermal power generation units. Constrain (56) ensure that the thermal power generation unit ramp rate limits are not violated.

$$-DR_i \leq P_i^{h+1} - P_i^h \leq UR_i; i = 1, 2, ., ., ., N; h = 1, 2, ., ., ., H-1$$

where DR_i and UR_i are maximum down and up ramp rates of the ith thermal power generation unit, respectively.

Calculate the positions of all the thermal power generation units as in *Step 2*

Step 15: Do *Step 3 – 14* until the maximum time H is reached. For each step h store the overall global best positions of the thermal power generation units, total cost of fuel consumption, total

CASE STUDIES

To illustrate the effective use of PSO algorithm, the technique is applied to two different systems.

Case A: Power System Network With Six Thermal Power Generation Units

The test power system network consists of six generation units at the supply side and eight aggregated electricity customer loads at the customer side. At the supply side the goal is to obtain the optimal forecast demand in terms of the economic and emission costs, while at the electricity customer side, the main objective is to obtain the optimal electricity customer load schedule in view of the power utility electricity price and forecast demand. The initial TOU electricity tariffs are obtained from Eskom's 2015/16 Tariff book 1 and are shown in Table 1 below. The fixed and flexible electricity customer loads connected to the power system network are presented in Table 2 and 3 below. The fuel consumption cost coefficients

Table 1. TOU electricity tariffs for the six-unit system

Time (hr)	1	2	3	4	5	6	7	8	9	10	11	12
TOU Prices ($/ kWh)	0.4088	0.4088	0.4088	0.4088	0.4088	0.6413	0.6413	0.9293	0.9293	0.9293	0.6413	0.6413
Time (hr)	13	14	15	16	17	18	19	20	21	22	23	24
TOU Prices ($/ kWh)	0.6413	0.6413	0.6413	0.6413	0.6413	0.6413	0.9293	0.9293	0.6413	0.6413	0.4088	0.4088

Table 2. Fixed electricity customer loads

Time (hr)	1	2	3	4	5	6	7	8	9	10	11	12
Load (MW)	70	70	70	70	70	70	186	222	232	232	232	295
Time (hr)	13	14	15	16	17	18	19	20	21	22	23	24
Load (MW)	481	431	232	232	232	288	279	169	140	140	96	96

Table 3. Flexible electricity customer loads

<table>
<tr><th>Appliance</th><th>Type</th><th>E_a (MWh)</th><th>α_a(h)</th><th>β_a (h))</th><th>Z_a (h</th></tr>
<tr><td colspan="6">Flexible</td></tr>
<tr><td>Load 1</td><td>Time shiftable</td><td>1st hour – 800
2nd hour – 200</td><td>1</td><td>24</td><td>Once a day</td></tr>
<tr><td rowspan="2">Load 2</td><td rowspan="2">Time shiftable</td><td rowspan="2">100</td><td>10</td><td>12</td><td>2</td></tr>
<tr><td>20</td><td>22</td><td>2</td></tr>
<tr><td>Load 3</td><td>Time shiftable</td><td>Hourly consumption = 500</td><td>1</td><td>24</td><td>Once every 11 hours</td></tr>
<tr><td rowspan="2">Load 4</td><td rowspan="2">Non-shiftable (Customer preference)</td><td rowspan="2">Hourly consumption = 200</td><td>12</td><td>16</td><td>5</td></tr>
<tr><td>20</td><td>22</td><td>3</td></tr>
<tr><td rowspan="4">Load 5</td><td rowspan="4">Power shiftable (Customer preferences)</td><td>Hourly energy consumption = 0 – 1000</td><td colspan="3"></td></tr>
<tr><td>Total energy requirement = 1000</td><td>3</td><td>8</td><td></td></tr>
<tr><td>Total energy requirement = 1000</td><td>9</td><td>16</td><td></td></tr>
<tr><td>Total energy requirement = 1000</td><td>17</td><td>20</td><td></td></tr>
<tr><td rowspan="2">Load 6</td><td rowspan="2">Power shiftable (Customer preference)</td><td>Hourly consumption = 100 – 1200</td><td>21</td><td>9</td><td></td></tr>
<tr><td>Daily energy requirement = 3000</td><td colspan="3"></td></tr>
</table>

and the emission coefficients are obtained from Elaiw, Xia, & Shehata (2012) and are shown in Table 4 below. For the sake of simplicity, it is assumed that the electricity customer load classification does not change. The fixed electricity price is assumed to be \$0.6413/kW.

Case B: Power System Network With Ten Thermal Power Generation Units

The test power system network consists of ten generation units at the supply side and ten aggregated electricity customer loads at the electricity customer side. The initial TOU electricity tariffs are similarly obtained from the Eskom's 2015/16 Tariff book. Tables 6 and 7 show the fixed and flexible electricity customer loads connected to the power system network. The coefficients of the fuel cost function, emission cost function are taken from Basu (2008) and are given in Table 8. The objectives of both the power utility side and the electricity customer side is similar to that of the six-unit system. The fixed electricity price is also similar to that of the six-unit system.

Table 4. Power and emission data of the six-unit system

Gen No.	P_i^{max} MW	P_i^{min} MW	a_i \$/h	Bi \$/MWh	c_i \$/MW2h	d_i lb/h	e_i lb/MWh	f_i lb/MW2h	UR_i MW/h	DR_i MW/h
1	500	50	240	7.0	13.8593	0.00419	0.0070	0.32767	80	120
2	200	50	200	10.0	13.8593	0.32767	0.0095	0.00419	50	90
3	300	80	220	8.0	0.0090	40.2669	-0.54551	0.00683	65	100
4	150	50	200	11.0	0.0090	40.2669	-0.54551	0.00683	50	90
5	200	50	220	10.5	0.0080	42.8955	-0.51116	0.00461	50	90
6	120	50	190	12.0	0.0075	42.8955	-0.51116	0.00461	50	90

Figure 3. Optimal power generation by each thermal power generation unit for the day $\mu_{\max-\max}$

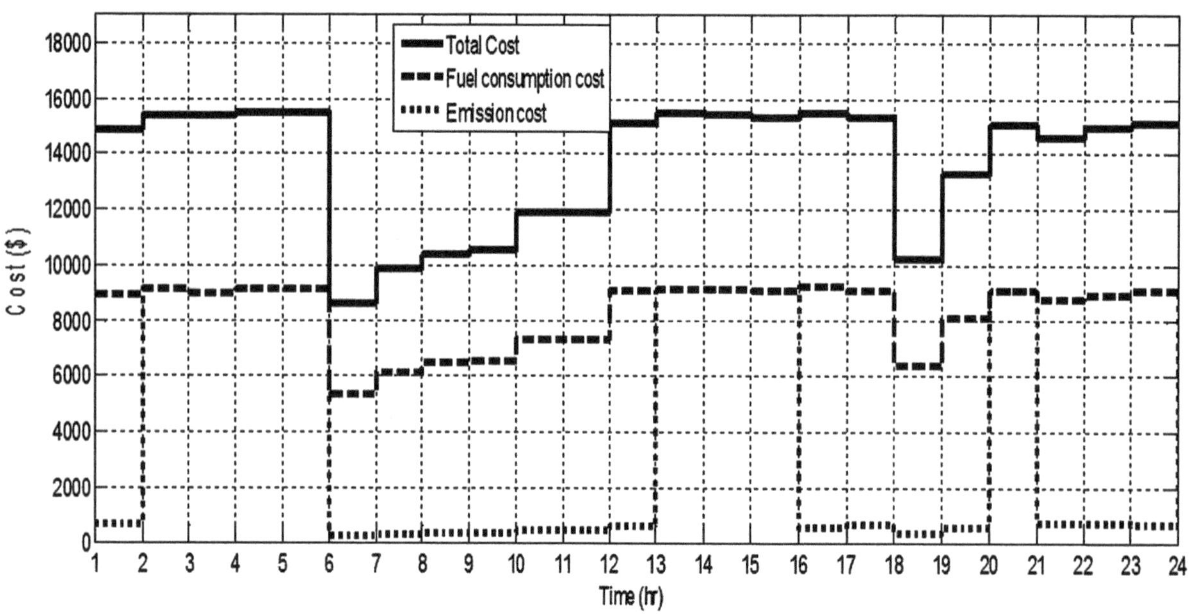

Figure 4. Cost of power generation $\mu_{\max-\max}$

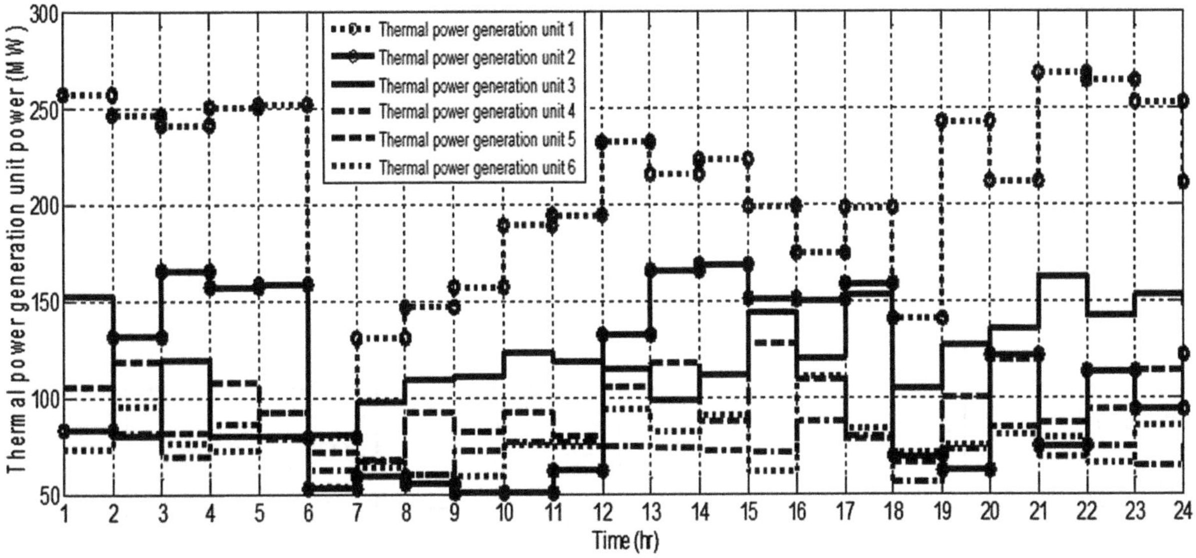

FUTURE RESEARCH DIRECTIONS

Future research work will include comparison of the proposed PSO algorithm with other artificial intelligence techniques such as GA, Simulated annealing, artificial neural networks and the inclusion of other power generation constraints such as transmission loss, prohibited zones etc.

CONCLUSION

As the competition amongst independent power utilities increases, it is important for the responsible energy regulator to develop proper policies that balance power system network load demand with fuel consumption costs and environmental preservation. The efficiency use of electricity by consumers should also be included in the analyses. Research studies have proved that demand response strategies can reduce the electricity consumers' electricity cost bill and in addition reduce the strain of load demand on the power system network. Development of proper policies that takes into account all practical constraints requires proper mathematical formulation of the CEED problem. In this paper, a PSO algorithm is presented to solve a dynamic CEED optimisation problem of a power system network with six thermal power generation units. The load demand input to the PSO algorithm is optimised using a nonconvex mixed binary integer programming technique. The demand response strategy takes into account consumer preferences and flexible loads. The proposed methodology optimally schedules thermal power generation units taking into account their fuel consumption cost and amount of emission. The problem is simulated for the whole day, thereby providing the power utility real-time analysis of the optimised power system network.

Table 5. Sensitivity analysis of penalty factor, μ

Time (hr)	1	2	3	4	5	6	7	8
$TC \times 10^4$ $\mu_{\max-\max}$	1.4868	1.5260	1.5288	1.5434	1.5221	0.8621	0.9825	1.0432
$TC \times 10^4$ $\mu_{\min-\min}$	2.0116	2.0025	2.0080	2.0089	1.9796	1.1279	1.2113	1.3005
$TC \times 10^4$ $\mu_{\max-\min}$	5.2308	4.9547	5.0207	4.9741	4.9779	2.8589	2.7931	3.0119
$TC \times 10^4$ $\mu_{\min-\max}$	1.0469	1.0807	1.0802	1.0781	1.0585	0.6535	0.7249	0.7658
Time (hr)	**9**	**10**	**11**	**12**	**13**	**14**	**15**	**16**
$TC \times 10^4$ $\mu_{\max-\max}$	1.0586	1.1897	1.1909	1.5087	1.5178	1.5369	1.4942	1.5293
$TC \times 10^4$ $\mu_{\min-\min}$	1.3212	1.5209	1.5211	1.9801	2.0161	2.0091	2.0008	2.0027
$TC \times 10^4$ $\mu_{\max-\min}$	3.0734	3.5710	3.6111	4.9543	4.9670	4.9906	4.9645	4.9750
$TC \times 10^4$ $\mu_{\min-\max}$	0.7746	0.8573	0.8556	1.0616	1.0862	1.0780	1.1003	1.0835
Time (hr)	**17**	**18**	**19**	**20**	**21**	**22**	**23**	**24**
$TC \times 10^4$ $\mu_{\max-\max}$	1.5436	1.0174	1.3292	1.5075	1.4570	1.4910	1.5100	1.5372
$TC \times 10^4$ $\mu_{\min-\min}$	1.9966	1.2674	1.7391	1.9847	1.9386	2.0016	1.9805	2.0039
$TC \times 10^4$ $\mu_{\max-\min}$	4.9560	2.9478	4.2517	5.0093	4.8412	4.9552	4.9550	4.9639
$TC \times 10^4$ $\mu_{\min-\max}$	1.0790	0.7483	0.9459	1.0614	1.0253	1.0670	1.0545	1.0749

Table 6. Fixed electricity customer loads

Time (hr)	1	2	3	4	5	6	7	8	9	10	11	12
Load Demand (MW)	1380	1380	1380	1380	1380	1380	1286	1122	1332	1332	1332	1295
TOU Prices ($/ kWh)	0.4088	0.4088	0.4088	0.4088	0.4088	0.6413	0.6413	0.9293	0.9293	0.9293	0.6413	0.6413
Time (hr)	13	14	15	16	17	18	19	20	21	22	23	24
Load Demand (MW)	1381	1104	1232	1232	1232	1128	1379	929	1140	1580	1016	1036
TOU Prices ($/ kWh)	0.6413	0.6413	0.6413	0.6413	0.6413	0.6413	0.9293	0.9293	0.6413	0.6413	0.4088	0.4088

Table 7. Flexible electricity customer loads

Appliance	**Type**	**E_a (MWh)**	**α_a(h)**	**β_a (h))**	**Z_a (h**
Flexible					
Load 1	Time shiftable	1st hour – 800 2nd hour – 200	1	24	Once a day
Load 2	Time shiftable	100	8	21	1
Load 3	Time shiftable	Hourly consumption = 500	1	24	Once every 11 hours
Load 4	Non-shiftable (Customer preference)	Hourly consumption = 200	12	16	5
			20	22	3
Load 5	Power shiftable (Customer preferences)	Hourly energy consumption = 0 – 1000			
		Total energy requirement = 1000	3	8	
		Total energy requirement = 1000	9	16	
		Total energy requirement = 1000	17	20	
Load 6	Power shiftable (Customer preference)	Hourly consumption = 100 – 1200	21	9	
		Daily energy requirement = 3000			

Table 8. Power and emission data of the ten-unit system

Gen No.	P_i^{max} MW	P_i^{min} MW	a_i \$/h	bi \$/ MWh	c_i \$/ MW2h	d_i lb/h	e_i lb/ MWh	f_i lb/ MW2h	$UR_i = DR_i$ MW/h
1	470	150	786.7988	38.5397	0.1524	103.3908	-2.4444	0.0312	80
2	470	135	451.3251	46.1591	0.1058	103.3908	-2.4444	0.0312	80
3	340	73	1049.9977	40.3965	0.0280	300.3910	-4.0695	0.0509	80
4	300	60	1243.5311	38.3055	0.0354	300.3910	-4.0695	0.0509	50
5	243	73	1658.5696	36.3278	0.0211	320.0006	-3.8132	0.0344	50
6	160	57	1356.6592	38.2704	0.0179	320.0006	-3.8132	0.0344	50
7	130	20	1450.7045	36.5104	0.0121	330.0056	-3.9023	0.0465	30
8	120	47	1450.7045	36.5104	0.0121	330.0056	-3.9023	0.0465	30
9	80	20	1455.6056	39.5804	0.1090	350.0056	-3.9524	0.0465	30
10	55	10	1469.4026	40.5407	0.1295	360.0012	-3.9524	0.0470	30

Table 9. Thermal power generation for DR load demand

Time (hr)	P1	P2	P3	P4	P5	P6	P7	P8	P9	P10
1	157.1334	320.0000	290.9118	250.7254	230.0574	150.0000	126.4253	120.0000	80.0000	55.0000
2	167.9854	240.0000	307.0342	277.2337	243.0000	160.0000	130.0000	120.0000	80.0000	55.0000
3	230.4050	316.5878	338.0792	300.0000	243.0000	160.0000	130.0000	120.0000	80.0000	55.0000
4	253.7397	291.3322	340.0000	300.0000	243.0000	160.0000	130.0000	120.0000	80.0000	55.0000
5	255.3394	289.7326	340.0000	300.0000	243.0000	160.0000	130.0000	120.0000	80.0000	55.0000
6	52.9677	220.1507	262.3640	250.3509	219.8043	130.9713	105.4732	104.1294	79.9195	53.8690
7	131.9793	159.4933	204.4907	200.3509	186.9409	136.0700	118.5241	120.0000	80.0000	48.1509
8	112.5576	145.8805	186.5139	199.9801	181.2473	151.2102	115.2963	99.4976	77.9178	51.8986
9	127.2464	146.6243	254.0642	249.9801	231.2473	137.8377	130.0000	120.0000	80.0000	55.0000
10	150.8453	161.8108	224.1367	244.3082	243.0000	160.0000	130.0000	120.0000	80.0000	53.1276
11	136.0603	145.6599	281.1790	256.9922	214.7690	154.9470	129.2472	116.4148	77.4618	54.4974
12	570.0000	149.8391	214.1591	209.0067	168.8033	127.7564	99.6947	90.0420	60.1851	41.8381
13	267.9246	229.8391	294.1591	259.0067	218.8033	160.0000	123.8043	119.8707	79.0444	55.0000
14	163.7486	182.1498	282.2757	209.0067	243.0000	160.0000	130.0000	120.0000	80.0000	55.0000
15	152.4650	207.1599	317.1105	259.0067	235.3768	160.0000	127.3758	116.5328	80.0000	54.1944
16	185.4159	158.3218	299.6671	277.8170	243.0000	160.0000	130.0000	120.0000	80.0000	55.0000
17	171.3315	199.9436	272.3353	300.0000	243.0000	160.0000	130.0000	120.0000	80.0000	55.0000
18	50.0000	135.0231	203.0711	250.1757	200.4588	127.0948	107.0001	108.3730	78.9126	49.5580
19	141.1376	156.4728	227.5155	200.1757	208.6984	160.0000	130.0000	120.0000	80.0000	55.0000
20	250.6095	236.4728	307.5155	250.1757	243.0000	160.0000	126.6161	120.0000	80.0000	55.0000
21	105.4664	156.4728	227.5155	200.1757	205.3696	160.0000	130.0000	120.0000	80.0000	55.0000
22	203.4894	236.4728	307.5155	250.1757	243.0000	160.0000	130.0000	120.0000	80.0000	55.0000
23	288.7887	316.4728	340.0000	300.0000	243.0000	160.0000	130.0000	120.0000	80.0000	55.0000
24	264.5816	300.7809	340.0000	300.0000	243.0000	160.0000	130.0000	120.0000	80.0000	55.0000

REFERENCES

Aalami, H., Yousef, G., & Moghadam, M. (2008, 21 - 24 April). Demand response model considering EDRP and TOU. In *IEEE PES*. Chicago, IL: Transmission and Distribution Conference and Exposition. doi:10.1109/TDC.2008.4517059

Aalami, H., Yousef, G. R., & Moghadam, M. P. (2010). Demand response modelling considering interruptible/curtailable loads and capacity market programs. *Applied Energy*, *87*(1), 243–250. doi:10.1016/j.apenergy.2009.05.041

Aghaei, J., Niknam, T., Azizipanah-Abarghooee, R., & Arroyo, J. M. (2013). Scenario-based dynamic economic emission dispatch considering load and wind power uncertainties. *International Journal of Electrical Power & Energy Systems*, *47*, 35167. doi:10.1016/j.ijepes.2012.10.069

Basu. (2008). Dynamic economic emission dispatch using non-dominated sorting genetic algorithm-ii. *Elect. Power Energy Syst*, *30*, 140-149.

Basu, M. (2006). Particle swarm optimization based goal-attainment method for dynamic economic emission dispatch. *Electric Power Components and Systems*, *34*(9), 1015–1025. doi:10.1080/15325000600596759

Basu, M. (2014). Teaching learning based optimization algorithm for multi-area economic dispatch. *Energy*, *68*, 21–28. doi:10.1016/j.energy.2014.02.064

Basu, M. (2007). Dynamic economic emission dispatch using evolutionary programming and fuzzy satisfied method. *Int J Emerg Electr Power Syst.*, *485*(8), 1-15.

Dzobo, O., & Xia, X. (2017). Optimal operation of smart multi-energy hub systems incorporating energy hub coordination and demand response strategy. *Journal of Renewable and Sustainable Energy*, *9*(4), 045501. doi:10.1063/1.4993046

Elaiw, Xia, & Shehata. (n.d.). *Minimization of fuel costs and gaseous emissions of electric power generation by model predictive control.* Math Prob Eng Article ID 906958.

Elaiw, A. M., Xia, X., & Shehata, A. M. (2012). Application of model predictive control to optimal dynamic dispatch of generation with emission limitations. *Electric Power Systems Research*, *84*(1), 31–44. doi:10.1016/j.epsr.2011.09.024

Erol-Kantarci, M., & Mouftah, H. (2011). Wireless sensor networks for cost efficient residential energy management in the smart grid. *IEEE Transactions on Smart Grid*, *1*, 320–325.

Fahrioglu, M., & Alvarado, F. L. (2000). Designing incentive compatible contracts for effective demand management. *IEEE Transactions on Power Systems*, *15*(4), 1255–1260. doi:10.1109/59.898098

Fahrioglu, M., & Alvarado, F. L. (2001). Using utility information to calibrate customer demand management behaviour models. *IEEE Transactions on Power Systems*, *450*(16), 317–322. doi:10.1109/59.918305

Jeddiand, B., & Vahidinasab, V. (2014). A modified harmony search method for environmental/economic load dispatch of real-world power systems. *Energy Conversion and Management*, *78*, 66175.

Khajavi, P., Monsef, H., & Abniki, H. (2010). Load profile reformation through demand response programs using smart grid. *Proc. Of the international symposium on modern electric power systems.*

Malette, M., & Venkataramanan, G. (2010). *Financial incentives to encourage demand response participation by plug-in hybrid electric vehicle owners. In IEEE Energy Conversation Congress and Exposition* (pp. 4278–4284). ECCE.

Morais, H., Faria, P., & Vale, Z. (2014). Demand response design and use based on network locational marginal prices. *International Journal of Electrical Power & Energy Systems*, *61*, 18091. doi:10.1016/j.ijepes.2014.03.024

Nwulu, N. I., & Xia, X. (2015). Multi-objective dynamic economic emission dispatch of electric power generation integrated with game theory based demand response programs. *Energy Conversion and Management*, *89*, 963–974. doi:10.1016/j.enconman.2014.11.001

Nwulu, N. I., & Xia, X. (2015). Implementing a model predictive control strategy on the dynamic economic emission dispatch problem with game theory based demand response programs. *Energy*, *91*, 404419. doi:10.1016/j.energy.2015.08.042

Nwulu, N. I., Xia, X., & Zhang, J. (2013). Determining the optimal incentive and number of retrofits for a demand response program in South Africa. *Proc. Of the 5th international conference on applied energy.*

Osorio, G. J., Lujano-Rojas, J. M., Matias, J. C. O., & Catal, J. P. S. (2015). A probabilistic approach to solve the economic dispatch problem with intermittent renewable energy sources. *Energy*, *82*, 949–959. doi:10.1016/j.energy.2015.01.104

Parvania, M., & Fotuhi-Firuzabad, M. (2010). Demand response scheduling by stochastic SCUC. *IEEE Transactions on Smart Grid*, *1*(1), 8998. doi:10.1109/TSG.2010.2046430

Po-Hung, C., & Hong-Chan, C. (1995). Large-scale economic dispatch by genetic algorithm. *IEEE Transactions on Power Systems*, *12*(2), 128–149.

Setlhaolo, D., Xia, X., & Zhang, J. (2014). Optimal scheduling of household appliances for demand response. *Electric Power Systems Research*, *116*, 248. doi:10.1016/j.epsr.2014.04.012

Shehata, A. E. A., & Xia, X. (2012). Application of model predictive control to optimal dynamic dispatch of generation with emission limitations. *Electric Power Systems Research*, *84*(1), 31–44. doi:10.1016/j.epsr.2011.09.024

Sinha, N., Chakraborti, R., & Chattopadhyay, P. (2003). Evolutionary programming techniques for economic load dispatch. *IEEE Transactions on Evolutionary Computation*, *7*(1), 83–94. doi:10.1109/TEVC.2002.806788

Song, M., Alvehag, K., Widen, J., & Parisio, A. (2014). Estimating the impacts of demand response by simulating household behaviours under price and CO2 signals. *Electric Power Systems Research*, *111*, 103–114. doi:10.1016/j.epsr.2014.02.016

Talaq, J. H., El-Hawary, F., & El-Hawary, M. E. (1994). A summary of environmental/economic dispatch algorithms. *IEEE Transactions on Power Systems*, *9*(3), 150816. doi:10.1109/59.336110

Wong, K., & Wong, Y. (1994). Genetic and genetic/simulated-annealing approaches to economic dispatch. *IEEE Proc Generation Transmission Distribution, 141*(5), 507–13. 10.1049/ip-gtd:19941354

Xia, X., & Elaiw, A. M. (2010). Optimal dynamic economic dispatch of generation: A review. *Electric Power Systems Research, 80*(8), 975–986. doi:10.1016/j.epsr.2009.12.012

Xia, X., & Elaiw, A. M. (2010). Optimal dynamic economic dispatch of generation: A review. *Electric Power Systems Research, 80*(8), 975–986. doi:10.1016/j.epsr.2009.12.012

Yoon, J. H., Bladick, R., & Novoselac, A. (2014). Demand response for residential buildings based on dynamic price of electricity. *Energy and Building*, *80*, 53141. doi:10.1016/j.enbuild.2014.05.002

Chapter 2
Optimized Hybrid Power System Using Superconducting Magnetic Energy Storage System:
Hybrid Power System Using SMES

Sandeep Bhongade
https://orcid.org/0000-0003-0575-5093
Shri G. S. Institute of Technology and Science, India

Ritu Verma
Shri G. S. Institute of Technology and Science, India

ABSTRACT

Renewable energy sources always drag the attention of researchers as alternate sources of power generation. These sources are inexhaustible and free of cost, which makes them very important for fulfilling electrical load demand. Due to stochastic nature of these sources as these are nature dependent, power generation from these sources varies. In order to mitigate this issue, these sources are integrated with distributed generation along with energy storage system so as to maintain the system stability. This chapter focuses on diminishing the frequency variation of microgrid incorporated hybrid power system. A hybrid system consisting of solar, wind, diesel along with a controller and superconducting magnetic energy storage unit is simulated. Whenever load demand of the system increases, frequency falls as a result deviation occurred in the system. This is overcome by the automatic generation control mechanism. Superconducting magnetic energy storage unit absorbs the excessive power available during offload condition and injects the same during peak load condition.

DOI: 10.4018/978-1-5225-8551-0.ch002

GENERAL

In a power system, whenever load disturbance takes place response of synchronous generators is not quick enough to maintain the stable system. In order to maintain system stability, FACTs devices are used for high-speed reactive power control. FACTS devices sometimes also control real power with the circulation of power in the converter. In order to achieve a reliable system and good power quality, energy storage systems are used. Due to advancements in power electronics technologies, the energy storage system plays a vital role in the field of power applications. These technologies have various advantages such as leveling of load at large scale, maintaining system stability, power quality, power transfer, etc.

According to a principle "energy can neither be created nor it can be destroyed but can transform from one form to another" i.e.

Total electrical energy input = mechanical energy output +energy stored in total + total energy dissipation

In an AC system, electricity is converted and can be stored in the form of potential as well as kinetic energy, electromagnetically as well as electrochemically. The amount and the rate of energy at which electrical energy is stored and transferred into the system or out of the system depend on the factors characterizing their suitability and these factors are peak power rating and response rate of the device as discussed by Singh et al (2014). These devices inject and absorb active and reactive power into or out of the system to solve power system failures like voltage dip, load leveling, quality of power, etc.

ENERGY STORAGE SYSTEMS

Energy storage systems classified into three main categories on the basis of a specific principle are:

1. Physical Energy Storage Systems
2. Electrochemical ESS
3. Electromagnetic ESS

This ESS possesses several advantages and disadvantages with its applications in the field of the power system. These technologies have a strength of maintaining system stability, large storage capacity, improved dynamic response, etc. A brief explanation of the above ESS are showed in Xun et al (2012)

Physical ESS

At present these energy storage systems are quite practical and mature storage systems. Due to the limitation of geographical conditions, these technologies do not possess large scale promotion. These technologies include pumped hydroelectric storage, compressed air energy storage and flywheel storage. CAES is mainly used in load-leveling with energy conversion efficiency less than 70%. On the other hand, the pumped storage system has a large unit volume and various environmental issues. Flywheel coupled with electrical machines stores energy in the power system. All these physical ESS have lower energy conversion efficiency and various limitations limit their use in the system.

Electrochemical ESS

These technologies are moving faster includes Li-ion battery, sodium Sulphur, and fluid flow batteries. BESS is commonly used in several industrial applications but has several disadvantages such as limitations in voltage, current and life cycle and various environmental hazards. These technologies have increased the conversion efficiency of about 80%. In these technologies battery stored energy electrochemically and is cost-effective. Whenever an internal chemical reaction takes place due to applied potential, batteries are charged and get discharge when the reverse phenomenon occurs. Since DC is stored by BESS a power convertor unit is required for interfacing with an AC system. Several advantages of BESS have increased energy density, energy, and cycling capability, initial cost, etc.

Electromagnetic ESS

This ESS involves superconducting magnetic energy storage and a supercapacitor energy storage system. These systems possess higher conversion efficiency from the above-mentioned system. SMES stores energy in the magnetic field created by the dc current flowing in the coil. SMES unit has several advantages such as fast response, active and reactive power injection simultaneously, load leveling, voltage dip, power quality, etc. These systems are used to mitigate the deviation that occurred in the system due to the disturbance occurred between load and supply. It provides stable power system operation when employed with microgrids also. These technologies have high cost and complex structure but due to its fast response and large storage capacity made it very useful in power systems. SMES is also incorporated with FACTS devices in the transmission system.

A comparative analysis of energy storage systems with various parameters are given below which is showed by Ali M H et al (2010).

These classifications of storage technologies are shown in figure 1.

Figure 1. Classification of storage technologies

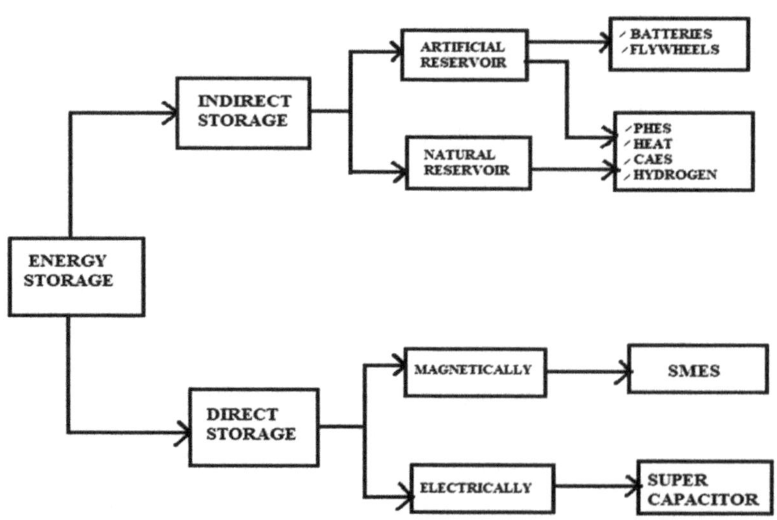

SMES SYSTEM

SMES stands for Superconducting Magnetic Energy Storage System and acts as an energy storage device in power systems. SMES plays an important role in the field of power and energy system as it injects and absorbs power into or from the system. SMES unit possesses fast response and higher energy conversion efficiency makes it an important energy storage device. SMES always drags its attention to electrical utilities and military areas Zargar et al (2017). In comparison with other ESS, SMES devices are costly but this limitation is eliminated by integrating SMES coil with the FACTS devices as the coil is the major costly element of the device. SMES stores energy in the magnetic field created by DC current flowing through the coil which is superconducting in nature and cooled at cryogenic temperature. SMES stores energy by converting electrical into magnetic energy in the magnetic field. SMES stored power by flowing a DC circulating current into the coil in absence of mechanical linkage. By reversing the power, SMES also pumps out power as explained by Xun et al (2012).

Energy stored in joules is given as

$$E = \frac{1}{2} L_{sm} I^2_{coil} \tag{1}$$

And rated power is given as

$$P = \frac{dE}{dt} = L_{sm} I_{sm} \frac{dI_{sm}}{dt} = VI \tag{2}$$

Where,

E = energy stored in the coil
L_{sm} = inductance of the coil
I_{sm} = DC current flowing through the coil
V = voltage across the coil

Energy stored as a circulating current drawn from the system with an instantaneous response of energy stored or injected over a fraction of second. SMES unit is commercially used for improving power quality at manufacturing plants which requires clean power for fabrication purposes. SMES also provides stability of the grid in the distribution system and used in utility applications. It is also used for enhancing the transmission loop stability. SMES is based on three concepts that are not applied to other energy systems are material carrying current with no resistive loss, the magnetic field produced by current and magnetic energy is a pure form of energy which can be stored.

CONFIGURATION OF SMES UNIT

SMES is an energy storage device with higher conversion efficiency and stores energy in the magnetic field. DC current flowing through the conductor created a magnetic field and the coil is cooled at cryogenic temperature. A typical SMES unit is categorized into three parts as explained by Ali M H et al (2010):

1. Superconducting coil
2. Power conditioning system
3. Cryogenic refrigerator

A description of SMES components is given below:

Superconducting Coil

SMES device stores energy in the magnetic field produced by a DC current flowing through a coil which is cryogenically cooled below a critical temperature. SMES comprises of a high conductance coil act as a constant current source. Whenever disturbance is occurred in the system than due to the coil's superconducting nature, energy will be stored for a long duration and also injects power to the grid. When the superconducting coil is charged, magnetic energy is stored due to the stoppage in the decay of current. The superconducting coil is highly efficient in storing electrical energy due to three main reasons viz. no resistive loss, magnetic field and energy storage. Due to the continuous flow of circulating current which produces energy makes it attractive in the group of energy storage devices. When poured in DC current SMES adopts DC charging system due to a negligible resistive loss in the coil and also it transforms AC to DC or vice versa. Due to this reason no thermodynamic losses occur. Niobium-titanium is used for molding superconductor and it is 12.5H inductance and 100 MJ of storage. SMES coil is wrapped in a cylindrical type double pancake structure. While designing the coil self and mutual inductance are taken into consideration and configured as toroid type winding and solenoid type winding. Solenoid type is quite popular because of its low cost, simple, also it reduces stray field involving more conductors in a single solenoid model. This limitation is overcome by the toroid type structure by reducing stray field effects to a large extent. Some important factors are the configuration of the coil, energy capability, operating temperature, structure, least cost, energy or mass ratio, etc. Also stray magnetic field, low loss, etc. for better reliability and stability of SMES. Due to the high cost of the superconducting coil, SMES is used for a short time period during storage of energy; hence it is more prominent in power quality improvement.

Power Conditioning System

A power conditioning unit transfers energy from SMES into the grid. It comprises of a chopper (dc-dc based) and a 3 phase VSC. Active and reactive power is easily controlled by the voltage angle control method. Constant current in the coil of SMES is maintained by the chopper and power transfer to VSC through the dc-link capacitor. In order to inject active or reactive power, the capacitor acts as a temporary dc voltage source to VSC. On the basis of the control loop and switching characteristics, two types of coupling models are involved i.e. VSC and CSC with the help of these models SMES co-operate with the transmission system. Due to the high capacity power devices of the chopper, SMES coil charged and discharged quickly. Energy is stored by charging the coil and can be released back to the network by discharging the coil. PCS uses the inverter and rectifier circuit for converting AC to DC power or vice versa. Due to power electronic equipment, SMES is highly efficient with low losses. Voltage and current rating of SMES are determined by its maximum power injection and absorption in the system and these ratings are application dependent. A low-temperature superconductor is used whereas high temperature super conducts are in the development stage.

Cryogenic Refrigerator

A superconducting coil of the system is cooled at a cryogenic temperature below the superconducting critical temperature. Cryogenic refrigeration is the most important part of the SMES system. Superconductors are cooled in liquid nitrogen and helium for heat removal using an external refrigerator. The cryogenic temperature of cooling is maintained by a cryostat or dewar containing helium and nitrogen liquid vessels. For maintaining a superconducting nature and strong magnetic field created by high current, this temperature is ranging between 4 to 10° K. as this temperature is realized by liquid helium, it acts as the heart of the cryogenic system. When the coil of SMES is carrying current greater than its rated value, heat dissipation results in the breach of cryostat due to the breakdown of the cooling system. A bypass switch is used during standby for reducing energy loss also it provides bypassing of coil current, converter servicing period protects coil when cooling is lost. A schematic diagram of SMES unit is shown in figure 2.

MODELING OF SMES UNIT

Under the specific range of values, DC voltage generated is constantly regulating in nature. But it can be regulated by varying firing angles of the convertor. On the application of small positive voltage inductor current reaches its rated value. In order to maintain the constant current, when the current reaches its rated value voltage across the inductor should be reduced to zero as derived by Singh et al (2017).

DC voltage of SMES system is given as-

Figure 2. Schematic diagram of SMES unit
(Ali & Abd-Elazim, 2013)

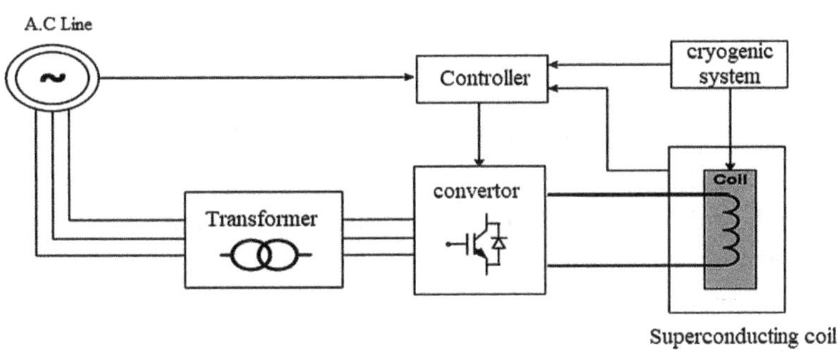

$$E_d = 2V_d \cos\alpha - 2I_d RC . \tag{3}$$

Where E_d is the DC voltage applied to an inductor, α is controlled charging and discharging of the superconducting coil, I_d is the current flowing through the inductor, R_C is an equivalent communicating resistance, V_{do} is the maximum circuit voltage.

1. $\alpha < 90°$ then converter mode
2. $\alpha > 90°$ then inverter mode

Whenever frequency falls below the nominal value, power fed back to the system and direction of current does not changes so the control voltage is negative in nature. Incremental change in the voltage applied to the inductor is expressed as derived by Bhatti et al (1997),:

$$\Delta E_d = \left[\frac{K_{SMES}}{1+sT_{dc}}\right]\Delta F . \tag{4}$$

Where T_{dc} is a converter time delay, K_{SMES} is the control loop gain, ΔF is the actuating signal to control unit of SMES.

The inductor current deviation (ΔI_d). is calculated by the following:

$$\Delta I_d = \frac{\Delta E_d}{sL} . \tag{5}$$

An overall change in SMES power flow due to system frequency change is,

$$\Delta P_{SMES} = \Delta E_d \cdot I_d \tag{6}$$

$$I_d = I_{do} + \Delta I_d \tag{7}$$

where I_d is the net inductor current. The mathematical modeling of SMES unit is shown in figure 3.

APPLICATION

Since the energy content of SMES is very small hence some methods are implemented to increase its storage capacity for large scale operation. For superconducting applications, cryogenics is required and on the other hand, a simple and robust mechanical structure helps in containing Lorentz forces produced from magnetic coils. SMES possess several advantages over other ESS such as short time delay between charging and discharging, instantaneous power supply, low loss due to negligible resistance. Also, SMES main parts are motionless results in higher reliability and greater efficiency. The main advantage of this

Figure 3. Modeling of SMES unit
(Singh, Joshi, Chanana, & Verma, 2014)

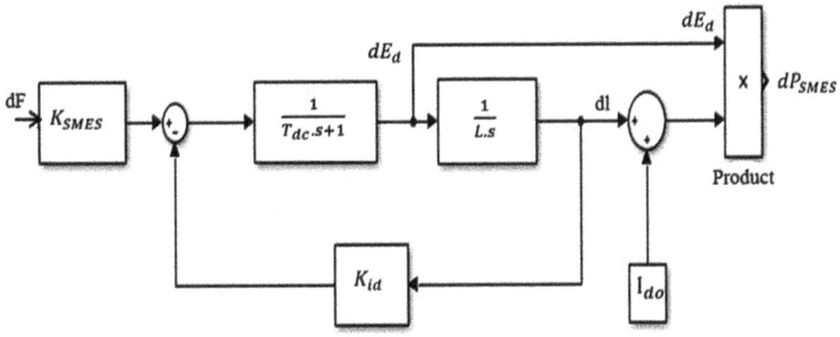

system is that it can discharge the surplus amount of power during a short interval of time. On the basis of specific principles, superconducting coil stores energy efficiently. It is mainly used for large scale load leveling, but due to its fast discharging ability, it is used in pumped power system stability applications. This work utilizes its fast response of charging and discharging of power so as to damp out the frequency fluctuation. Following are the several advantages of SMES in power systems as detailed explained by Ali et al (2010), Ramakrishna.et al (2010), Singh et al (2013; 2014):

- Energy storage
- Load following
- System stability
- Automatic Generation Control
- Spinning Reserve
- Reactive volt-ampere (VAR) control and power factor correction
- Bulk energy management
- Transient voltage dip improvement
- Backup power supply
- Lower use of oil and gas
- Increased efficiency and reduced maintenance of generating units
- Deferral of new conventional capacity
- Deferral of new transmission capacity
- Increased availability of generating units
- Environmentally sound etc.
- Power quality improvement

HYBRID POWER SYSTEM (HPS)

Renewable energy sources attract the energy sectors for large scale power generation. Solar and wind power sources are efficiently used as power generating sources because of its ease of availability and environmental advantages in local power generation. Due to the stochastic nature of wind and solar energy, their use as power generation sources is somehow limited on a larger scale. Due to climatic change and unpredictable nature, they do not match the time distribution of demand. But with a higher cost, these energy systems are oversized for becoming completely reliable. Also, it is clear that neither of these solar or wind sources, continuously supply power during standalone condition due to climatic effects. These problems are overcome partially or completely on the integration of more than one renewable source properly, utilizing the strength of one source over the drawback of the other one. Researchers always attracted to these renewable energy sources for power generation with increased power demand due to its clean energy, low cost, etc. factors. Solar and wind energy sources are promising renewable sources with energy storage devices involvement in the development stage as discussed by Singh & Shakya (2017). A hybrid power system is defined as a combination of two or more nonconventional energy sources for power generation. PV and wind are primary generation sources and HPS involving these two sources is highly reliable and requires a low running cost. MPPT techniques are employed in the system for extracting maximum power and providing it to the grid. Maximum power is obtained from the PV panel by using a control strategy irrespective of irradiance and temperature. Similarly neglecting the wind speed, maximum power is tracked from small wind turbines. A combination of solar and wind sources along with ESS results in reliable, clean energy with minimum maintenance costs. MPPT techniques are employed for achieving higher efficiency, voltage based MPPT is used for wind and solar panels due to its fast-tracking and simple operation. In order to minimize the power variation and increase the power output, the hybrid power system consists of PV and wind is used. A schematic diagram of the considered hybrid power system (HPS) is shown in figure 4.

OFF-GRID VS ON-GRID HPS

A hybrid power system is of two types depending on the basis of availability and they are standalone and grid-connected systems. Stand-alone systems required a proper storage facility so as to handle power

Figure 4. Schematic diagram of HPS
(Shayanfar, Shayeghi, & Younesi, 2015)

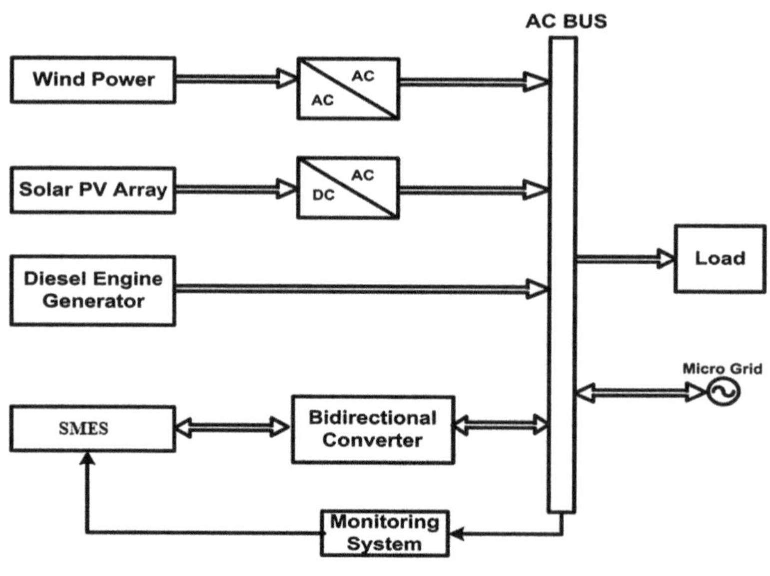

fluctuation due to variable power generation. Such a type of system is taken as a microgrid, whereas on the other hand grid-connected systems, microgrid sources supply power to the local load as well as to the grid. Also, these distributed generations provide reactive power or voltage regulation to the grid. The storage capacity required for grid-connected systems is small due to the system backup. This system results in voltage, frequency, and harmonic distortion while operating with the connection of the grid.

ADVANTAGES OF HPS

Hybrid power systems are commonly used in a remote location where connection to the utility grid is required. They stabilize the grid by improving the power quality and power handling capacity in certain areas. Following are the benefits of using HPS as showed by Pandey et al (2014):

- The possible combination of renewable energy sources on the basis of local demand.
- The clean environment due to reduced emission of gases
- Low running costs such as solar radiation for solar power panels and wind speed for wind power sources considering the cost of fossils and nuclear.
- Energy diversity with secured supply
- Fuel is available in the excess amount due to renewable nature, hence free and inexhaustible

MICROGRID

Renewable energy sources produce instability in power generation due to several environmental conditions, which are quite unfavorable for standalone or on-grid power systems. But these sources are economically useful, and hence play a vital role in power generation. A microgrid is an integration of various distributed generation and energy storage systems for improving power quality and supplying power to the local utility. The power generated through various sources such as solar and wind is collected in DC bus via various power conversion systems. Microgrids are used to supply power to various utility points using various energy storage systems as discussed by Ayyarao et al (2017).

In order to maintain system stability, improving efficiency using distributed generations, ESS, power control method is employed in microgrid during optimization. Power generation and consumption from microgrid are not usually limited in the case of on-grid systems, similarly, the grid provides voltage and frequency regulation. In some cases power exchange between grid and MG is specified otherwise microgrid stability will be lost. Microgrids can operate either autonomous or with the utility grid. Hence possess the advantage of disconnecting itself from the grid and act standalone for power supply to local loads during a fault condition. A utility grid-connected HPS is shown in figure 5.

Different modes of operation for microgrids are interconnected, islanded, the transition from one grid to off-grid. Heat generation from various micro power sources is used for meeting local heat demand under any operating mode. Microgrids can operate either with off-grid or with medium voltage distribution systems. Microgrids enhances the reliability of the system, power quality, reduces emissions, voltage dip and low energy cost by providing thermal and electrical needs. DG reduces the power distribution and transmission demand from the utility point. Microgrids also reduce the effect of climatic fluctuations, emissions, loss reductions and act as a substitute for network assets.

Figure 5. Utility grid connected HPS
(Singh et al., 2014)

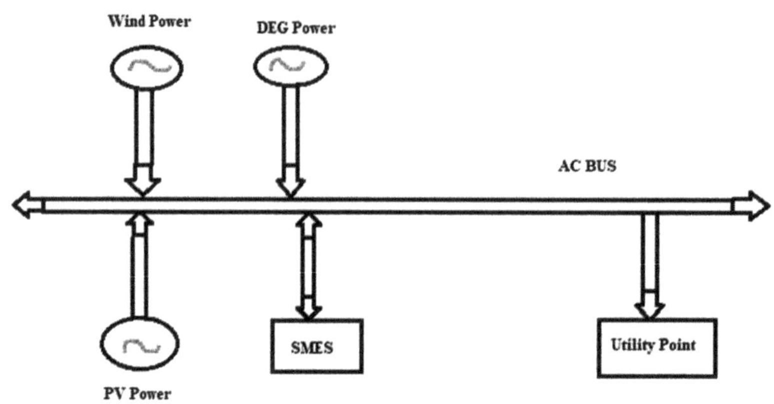

For maintaining stable operation of the power system, power stability, power quality requires the development of control strategies of microgrids in islanding mode so as to provide frequency and voltage stability during load change conditions. When microgrid is connected parallel with the main grid, its voltage and frequency are controlled by the utility grid. According to the loading conditions of the main grid, it supplies or absorbs power and becomes controllable load and source. Whenever a fault occurs in the grid, microgrids ability of disconnection plays an important role. Due to this ability, power quality is improved by improving voltage control. Microgrids also adjust their voltage and frequency by adjusting active and reactive power. Microgrid supplies power to the local load during the islanded mode of operation and these loads are from villages, universities, buildings, etc. The purpose of a microgrid is not completed until the load is properly served with voltage and frequency stability.

MODELING OF SOLAR PHOTOVOLTAIC CELL

Solar PV cells are the combination of PV arrays in connection with protective parts. PV arrays are a combination of current sources in parallel with diode and its output is proportional to the light fallen on the cells. The equation of ideal solar cell that symbolizes the solar cell model is explained by Singh & Singh (2014),

$$I = I_L - I_R \left[\exp\left(\frac{V}{A V_i} \right) - 1 \right] \tag{8}$$

Where,

I_L is photocurrent (A)
I_R is reverse saturation current (A)
V is diode voltage (V)
V_i is thermal voltage

A is an ideality factor

An equivalent circuit of the solar power model used in this work is shown in figure 6 and solar photovoltaic arrangement is shown in figure 7.

The photocurrent depends on the irradiation and temperature of the sun and is given as

$$I_L = \frac{1}{I_{ref}}\left[I_{scref} + \mu_{scref}\left(T - T_{ref}\right)\right]. \tag{9}$$

Where,

I_L = photocurrent
I_{scref} = short circuit solar cell condition at reference
μ_{scref} = temperature coefficient of short circuit solar cell

When solar cells are integrated with protection devices in series and parallel combinations, it is called a solar module. A characteristic equation of series and parallel combination PV module is given as:

Figure 6. Equivalent circuit of solar power model

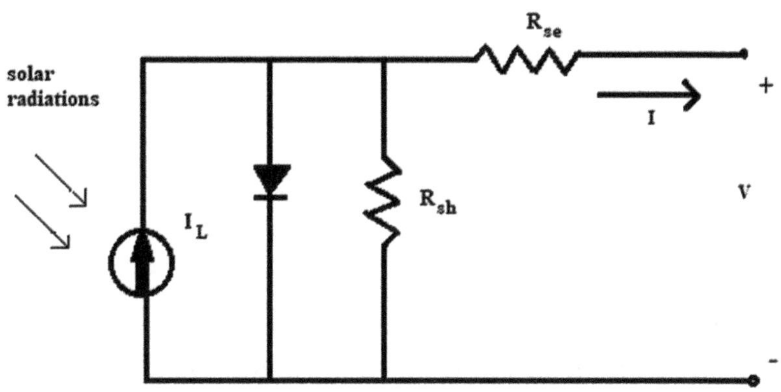

Figure 7. Solar photovoltaic arrangement

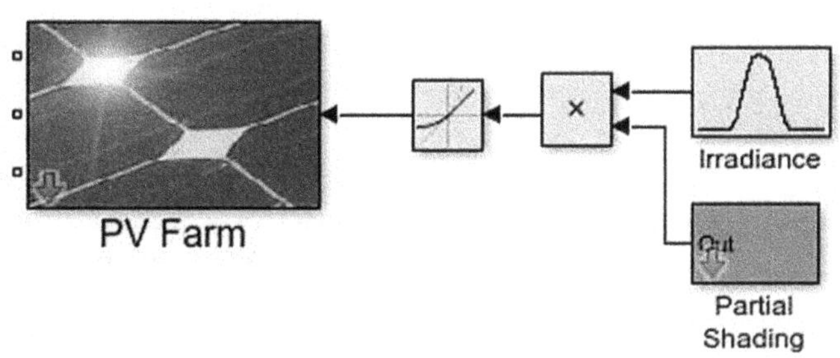

$$I^M = N_P I_L - N_P I_R \left[\exp\left(\frac{\frac{V^M}{N_S} + \frac{I^M}{N_S}}{AV_i} \right) \right] - \frac{\left(N_P / N_S \right) V^M + I^M R_{se}}{R_{sh}} \tag{10}$$

Where,

N_p = parallel number of cells
N_S = series cells

MODELING OF WIND POWER CONVERSION MODEL

Wind speed to power conversion model is used to present a wind power model. Wind power is obtained by the wind flow from wind turbines into mechanical power generators for the production of electrical power. Wind velocity is a combination of various wind speeds as explained by Singh & Shakya (2017). A simple wind conversion model is shown in figure 8.

The resultant wind velocity is given as,

$$v_w(t) = v_{wa}(t) + v_{wr}(t) + v_{wg}(t) + v_{wt}(t) \tag{11}$$

Where,

V_{wa} = average speed
V_{wr} = wind speed ramp
W_{wg} = wind gust
V_{wt} = wind turbulence

A wind profile used by the wind farm is shown in figure 9.

Due to the stochastic nature of wind speed, fluctuation produced will be diminished by employing a low pass filter in the conversion model. Power generation from a wind turbine is showed by Singh et al (2014):

Figure 8. Wind conversion model
(Singh, Verma, & Shakya, 2017)

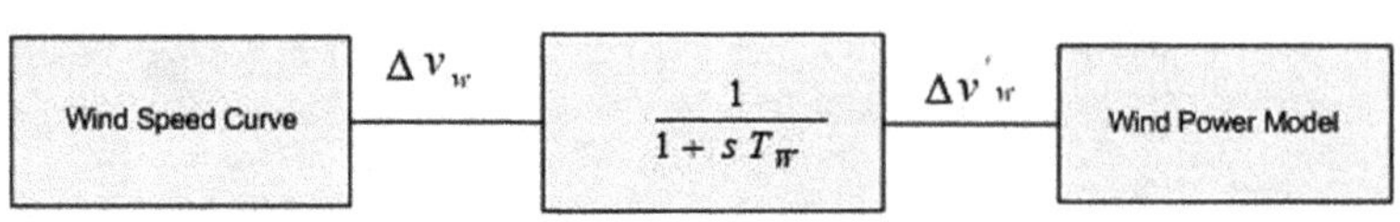

Figure 9. Wind power arrangement

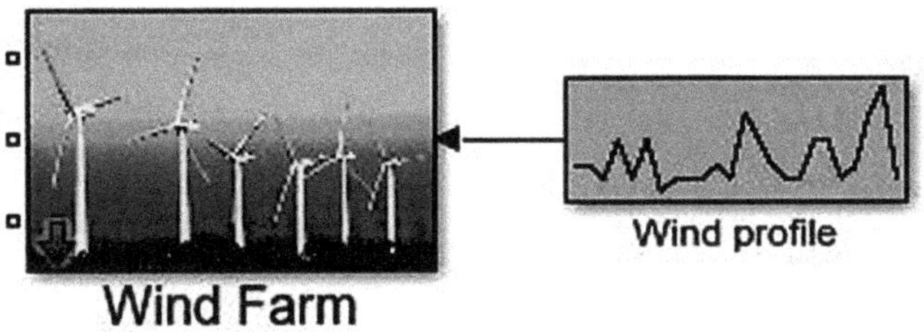

$$P_w = c_p(\lambda, \beta)\frac{\sigma}{2}r^2 v_w^{\ 3} \tag{4.5}$$

Where,

c_p = wind turbine power coefficient
λ = tip speed ratio and λ is defined as

$$\lambda = \frac{\omega_T \cdot r}{V_w} \tag{12}$$

DIESEL ENGINE GENERATOR (DEG) MODEL

Diesel engine generator consists of a diesel engine, a generator along with various ancillary devices like a control system, circuit breakers, etc. This diesel engine is modeled using a model of first-order transfer function with a minimum time delay between injecting oil and mechanical torque production with T_D time constant showed by Singh et al (2017).

DEG power generation is governed by the following equation given by Ayyarao et al (2017):

$$\Delta P_d = -\left(\frac{1}{R} + \frac{K_I}{s}\right)\left(\frac{1}{1+sT_{sg}}\right)\left(\frac{K_{DEG}}{1+sT_{DEG}}\right)\Delta f_e \tag{13}$$

Where R is the speed regulation, T_{DEG} is DEG time constant, K_{DEG} is DEG gain. Modeling of DEG has been derived by Ayyarao et al (2017) and shown in figure 10.

Figure 10. Modelling of DEG
(Ayyarao et al., 2017)

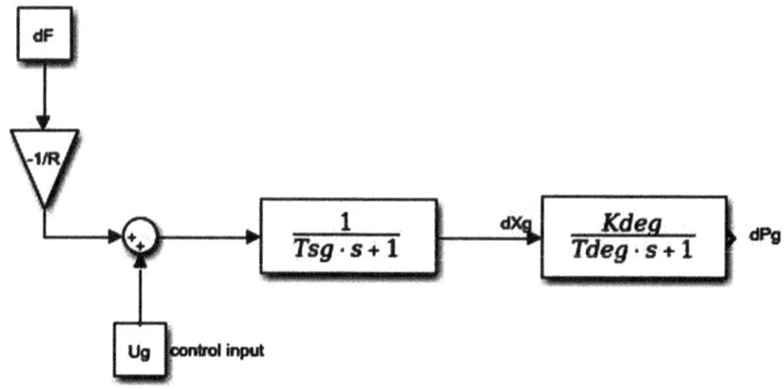

MODELING OF HPS

The mathematical model of HPS is considered for modeling. As PCS is working in its defined position properly. For smooth operation of the power system, demand and supply should be maintained. A modelling of HPS is shown in figure 11.

The control strategy is determined by the controlling error which is the difference between the change in load demand (P_D) and change in net power production (P_T) as given by Singh et al (2017)

$$\Delta P_T = \Delta P_{PV} + \Delta P_W \pm \Delta P_{DEG} \pm \Delta P_{SMES}. \tag{14}$$

Net controlling power error,

$$\Delta P_S = \Delta P_T - \Delta P_D. \tag{15}$$

Since an inherent time delay exists between system frequencies. The transfer function for system frequency variation to per unit power deviation is expressed in the below equation

Figure 11. modelling of HPS
(Singh, Verma, & Shakya, 2017)

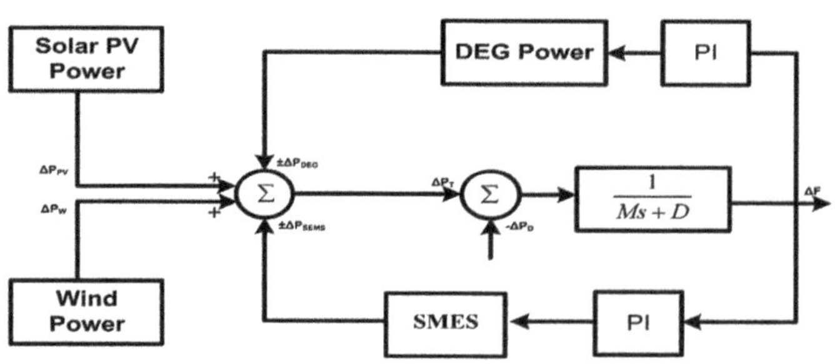

$$\Delta F = \frac{1}{D + sM} \cdot \Delta P_s \,. \tag{16}$$

$$\mathrm{D} = \frac{1}{K_{PS}} \ \& \ \mathrm{M} = \frac{T_{PS}}{K_{PS}}$$

Where M is inertia constant & D is damping constant.

OPERATION STRATEGY

This section represents the HPS operating strategy shown in figure 12 . Frequency of the system is load change sensitive i.e. a sudden change in load demand results in frequency variation. Whenever a load is increased in the system, the frequency falls below a specified limit hence to maintain the demand-supply ration constant SMES comes into the picture and injects power into the system. Similarly, when load demand decreases frequency increases, as a result, excessive power is available in the system is absorbed by the SMES unit. For proper operation of the system, the frequency must lie in the specified range and in order to neutralize the fluctuations power supply should be maintained accordingly.

When frequency fluctuation of the system is zero or change in frequency in the grid is zero, SMES is neither in charging nor in discharging mode. It remains in standby mode and acts only under variation in operating conditions.

OVERVIEW OF GREY WOLF OPTIMIZER ALGORITHM

Grey Wolf Optimization (GWO)

In order to find solutions for steady and transient environments, several operational problems are formulated. Thus, different optimization algorithms have been adopted for finding optimal settings in the output of operational units. Meta heuristics optimization algorithm is becoming very popular over the last few decades, due to some of its main advantages viz. fairly simple, flexible for various problems, derivation free mechanism due to their stochastic nature and having superior abilities for local minima avoidance.

These algorithms are basically inspired by nature, activities or behavior of animals, etc. These biologically inspired algorithms always develop an interest of researchers to find solutions for various operational problems. These algorithms are classified into two: single solution and population-based. Population-based module possesses several advantages over a single solution. Single candidate solutions are improved over some course of iterations whereas population-based solutions perform optimization using a set of solutions. Population-based metaheuristics techniques are based on the behavior of swarm intelligence as discussed by Mirjalili et al (2014). The most common features of swarm intelligence techniques are exploration and exploitation phases which are generally used in finding a solution showed by Guha et al (2016). When the implementation of mathematical methods for finding accuracy is difficult, these techniques bring control parameters to the edge in nonlinear problems. Thus, some effective optimization

Figure 12. Operating strategy of HPS
(Singh et al., 2014)

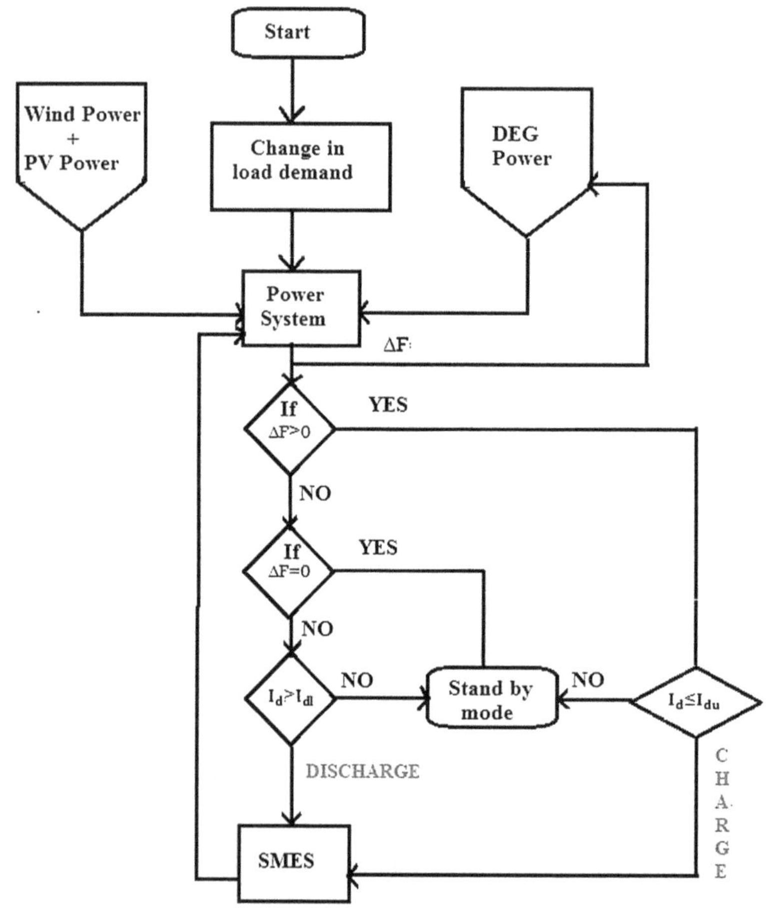

techniques are required for nonlinear problems. A worth mentioning theorem, No Free Lunch theorem logically proves that no meta-heuristic techniques solve all optimization problems, but these techniques are showing very good results in some set of problems but advantages in a different set of problems.

A grey wolf optimizer algorithm was proposed by Mirjalili et al (2013). This algorithm is based on the biological activities of grey wolves. It is a population-based optimization technique having advantages of minimum controlling parameters, the ability of strong global convergence, and ease of implementation. This technique is good at exploration but has some disadvantages in exploitation which is further been improved. These techniques are very popular due to its simple implementation, flexible in application, derivative-free mechanism, and local minima avoidance.

This algorithm is inspired by the hunting technique of the grey wolves also known as Canis lupus, the most successful predators in the food chain. This technique mimics the leadership skills and hunting method of grey wolves. These wolves possess a strong dominating hierarchy. These wolves are generally moved in the group of 10 to 12. The group of these wolves is known as a pack. A pack of wolves consists of four types of wolves which are classified on the basis of their skills. They are alpha, beta, delta, and omega wolves as explained by Ayyarao et al (2017). A Hierarchy of Grey wolf is shown in figure 13.

Figure 13. Hierarchy of Grey wolves
(Guha, Roy, & Banerjee, 2016)

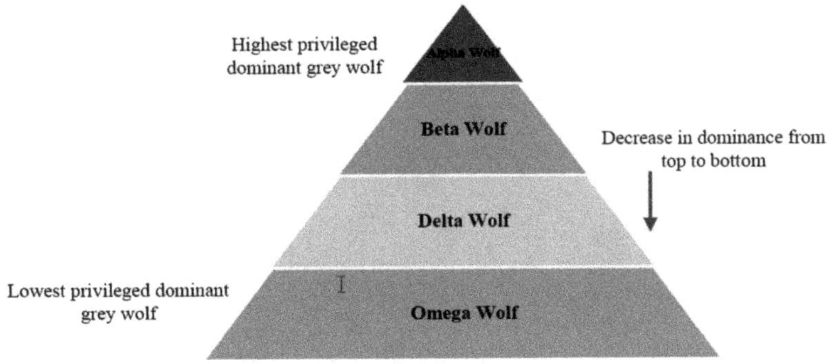

This pack of wolves is led by leader i.e. alpha wolves. Alpha wolves possess the ability to lead the group making decisions of hunting and various other activities. Alpha dictates its decision to the pack. Second level wolves are beta group, they are the subordinates of alpha wolf and after any incident with alpha, and they become an alpha wolf. These wolves help alpha in making decisions. Lower level wolves are an omega, they follow the order of other wolves. One category of wolves is delta wolves; they also subordinate alpha and beta wolves but dominate omega wolves as given by Guha et al (2016).

The hunting mechanism of the grey wolf is adopted by GWO to find the solution in search space or hunting prey. The main phase of hunting phenomenon is given as –

1. Tracking (following) and chasing (dashing) so as to approach the prey.
2. Encircling a prey and annoying him till it halts.
3. Attacking the prey.

GWO attacking the prey is shown in figure 14 showed by Ayyarao et al (2017).

Mathematical Modeling of GWO

In this section, the social behavior, chasing, encircling, searching, hunting and attacking phases are modeled for grey wolf optimizer as showed by Mirjalili et al (2014).

Social Hierarchy

During the designing stage of GWO, the best fittest solution is considered i.e. alpha wolves (α). Other subordinate wolves are termed as beta (β) and delta wolves (δ) and lower-level wolves are given as omega wolves (ω).

Surrounding a Prey

Encircling behavior of grey wolves during hunting is mathematically modeled as

Figure 14. GWO attacking the prey
(Ayyarao et al., 2017)

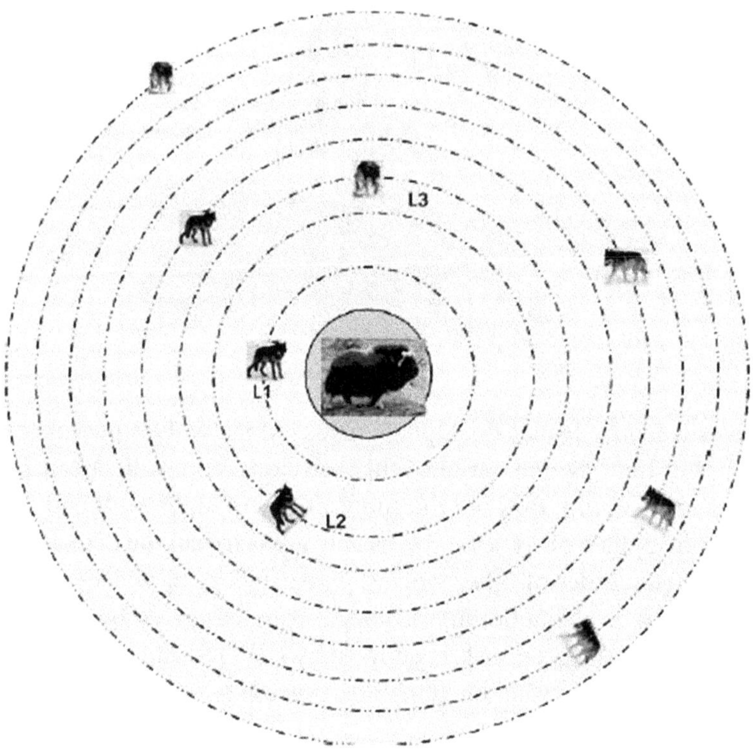

$$\bar{D} = \left| \bar{C} \cdot \overline{X_P}(t) - \bar{X}(t) \right| \tag{17}$$

$$\bar{X}(t+1) = \overline{X_P}(t) - \bar{A}.\bar{D} \tag{18}$$

Where t is the current iteration, $\bar{A}$ and $\bar{C}$ represent coefficient vectors. X_P shows position vectors of prey and X is the position vector of the grey wolf.

$\bar{A}$ and $\bar{C}$ vectors are evaluated as

$$\bar{A} = 2\bar{a} \cdot \overline{r_1} - \bar{a} \tag{19}$$

$$\bar{C} = 2 \cdot \overline{r_2} \tag{20}$$

Where $\bar{a}$ are decreasing linearly from 2 to 0 over a course of an iteration. r_1, r_2 random vectors in the interval [0,1].

Hunting a Prey

Since the wolves have the capability of finding the location of prey and hunting it. The Hunt process is led by the alpha wolves and subordinate wolves help the leader. In a search space, the idea of the location of prey is not present. Thus, to simulate the behavior, three best fitness solutions are obtained while finding solutions in search space and given as

$$\overline{D_\alpha} = |\overline{C_1}.\overline{X_\alpha} - \overline{X}| \tag{21}$$

$$\overline{D_\beta} = |\overline{C_2}.\overline{X_\beta} - \overline{X}| \tag{22}$$

$$\overline{D_\delta} = |\overline{C_3}.\overline{X_\delta} - \overline{X}| \tag{23}$$

Also,

$$\overline{X_1} = \overline{X_\propto} . \overline{A_1} .(\overline{D_\propto}) \tag{24}$$

$$\overline{X_2} = \overline{X_\beta} . \overline{A_2} .(\overline{D_\beta}) \tag{25}$$

$$\overline{X_3} = \overline{X_\delta} . \overline{A_3} .(\overline{D_\delta}) \tag{26}$$

Now, alpha, beta, and delta evaluate the position of prey and start updating their positions around prey. Updated position of these wolves is given as

$$\bar{X}(t+1) = \frac{\overline{X_1} + \overline{X_2} + \overline{X_3}}{3} \tag{27}$$

A 2-D and 3-D position vector with their updated position is shown in figure 15 and 16.

Attack Towards the Prey

Hunting process is finished by attacking the prey, as it halts. Mathematical modeling of this stage is shown by the decreasing of $\bar{a}.\overline{A}$ is also fluctuating in the range of decreasing $\bar{a}$ from 2 to 0 over a certain course of iteration. This phase of algorithm allows the search agent of search space to update its position accordingly and attacks the prey.

Figure 15. 2D position vector with their new update positions
(Guha et al., 20116)

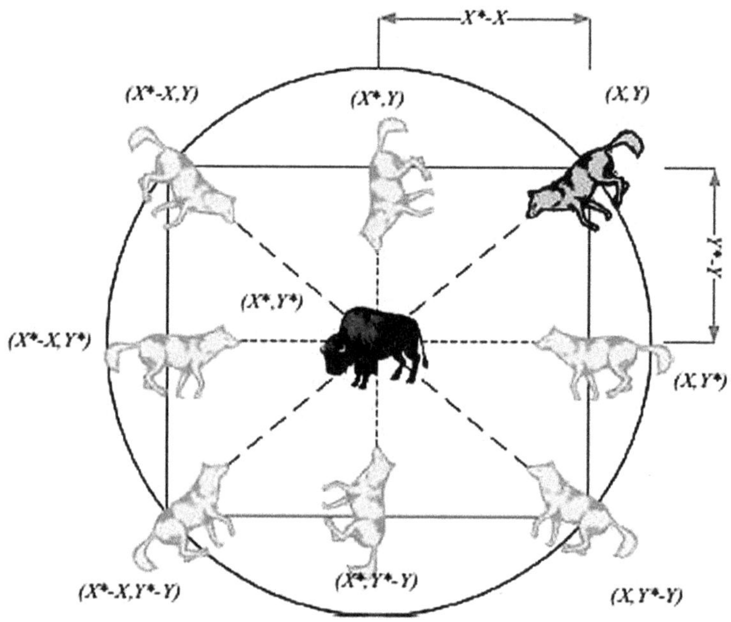

Figure 16. 3D position vector with their updated position
(Guha et al., 2016)

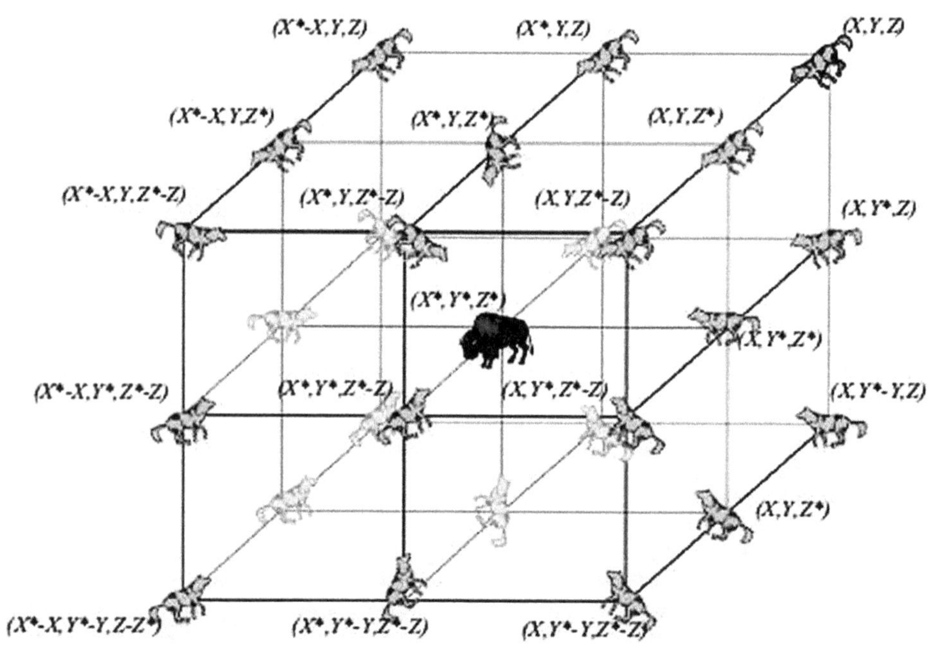

Searching a Prey

This phase of algorithm enables the exploration process in the search space. After updating their positions, grey wolves diverge from their location and in order to converge them $\bar{A}$ plays a significant role. Figure 17 shows a Position updating mechanism of search agents and effects of *A*.

When $|\bar{A}|>1$, grey wolves diverges from the prey to find the fittest prey. This implies exploration and searches globally.

$|\bar{A}|<1$, grey wolves converges to attack the prey and Algorithm terminated with the satisfactory end criterion.

Implementation of GWO Algorithm

Grey wolf optimization is implemented in the following steps as demonstrated by Padhy et al (2017):

1. Social hierarchy: this phase is used for the designing stage.
2. Searching the prey: this phase enables the algorithm to explore the search process.
3. Encircling prey: in this phase grey wolves encircles the prey and their position are updated
4. Hunting: The optimum position is obtained by saving three best positions corresponding to alpha, beta and delta wolves and remaining solutions including omega have competed

Figure 17. Position updating mechanism of search agents and effects of A on it
(Guha et al., 2016)

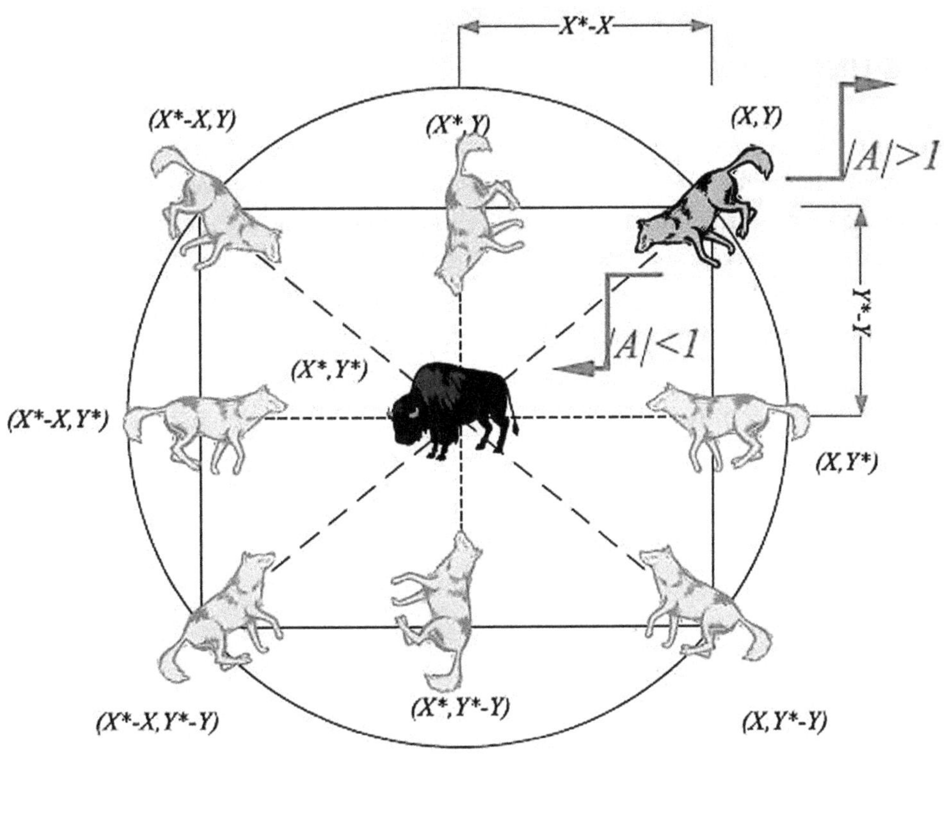

5. Attacking prey: this phase enables the algorithm to exploit the search process.

The maximum exploration of search space results in a lower probability of getting trapped in local optima. Also, too much exploration provides randomness and difficulty in finding the optimal solution. Similarly, more exploitation results in less randomness and the probability of reaching global optima is also less.

Flow Chart of GWO

In order to explain the algorithm, a flow chart is drawn which shows complete steps of optimization.

This flow chart explains the basic code of the algorithm and can be explained as:

1. First of all the population of the pack is initialized, iterations should be fed and dimensions of the problem are set. The lower and upper limit of these parameters is also set initially and leads are initialized
2. Then the main loop starts working, as when iteration < maximum iteration
3. Positions are generated and their fitness is evaluated also the random position of grey wolves along with its fitness value is also generated.
4. Position of leads are updated
5. The position of grey wolves is also updated with respect to the leads.

Increment in iteration and then the process terminated.

A step by the step flow chart of an implementation of GWO is shown in figure 18.

OBJECTIVE FUNCTION

In order to minimize the deviation, controller parameters tuned with GWO using an objective function. For designing a controller an optimization method is employed, this objective function is specified, generally depends on some performance criteria such as Integral of Time Absolute Error (ITAE), Integral Time Square Error (ITSE), Integral of Absolute Error (IAE), Integral Square Error (ISE). Depending on the system performance ITSE is used for formulating objective function.

The tuning of the PI controller depends upon two parameters (K_P & K_I) to regulate the change in frequency error (Δf). PI controller is selected for minimizing the deviation in frequency. PI controller is given by its weighted sum of two functions of time. These distinct weights are K_p, determines the present error value influence on the control mechanism. K_I, decides the reaction on the basis of area under the error time curve till the present point as derived by Ayyarao et al (2017).

In general PI control is given as

$$G(s) = K_p + \frac{K_I}{S} \tag{27}$$

Figure 18. Flow chart of GWO

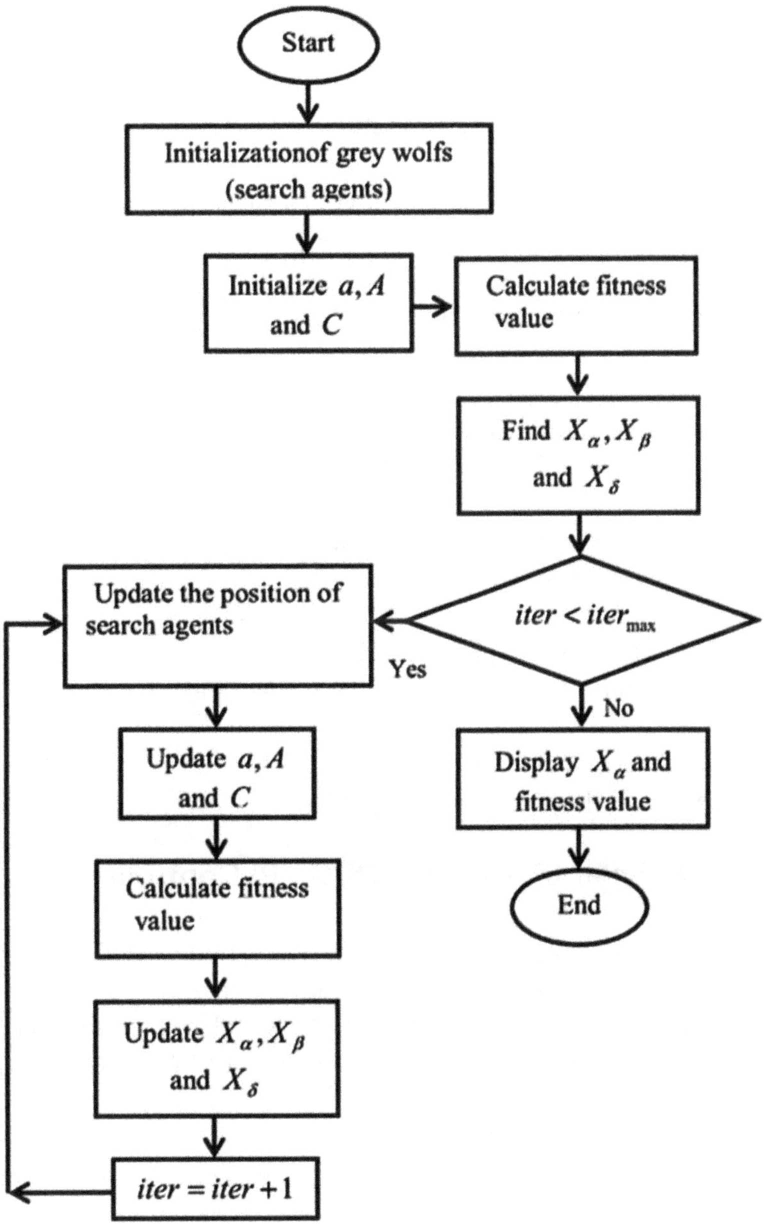

In order to achieve optimum system response, the integral time square error (ITSE) method is used. The objective function is given as *(Ayyarao et al., 2017):*

$$F(X) = ITES = \int_{0}^{t_{sim}} t * (\Delta f)^2 dt \tag{28}$$

Where X is the variable in terms of the value of K_p & K_i, which are simulated in the MATLAB environment. The optimal values of the PI controller are determined with the Grey Wolf optimization technique.

SIMULATION RESULTS AND DISCUSSION

This section presents the simulation model along with the results of the microgrid incorporated power system. A 2 MW (base) system has been considered for study under different loading conditions. In order to analyze the system different scenarios of the hybrid power system is considered. HPS is very much efficient with renewable energy sources. This scenario is simulated in two-phase as presented by Singh et al (2017):

1. The steady-state stability analysis using the step load response of the system model.
2. The dynamic stability analysis with 24 hours of simulations with random variation in load demand.

Hybrid power model involves various generation sources, energy storage devices etc. following cases are simulated under two different scenarios:

1. HPS with Solar PV+ Wind + DEG +PI Controller
2. HPS with Solar PV+ Wind + DEG +PI Controller + SMES unit

A detailed description of these scenarios along with their cases is explained in this chapter. Parameters of the controller and effect of SMES have been analyzed.

Scenario (I) HPS With Solar PV+ Wind + DEG +PI Controller

In this scenario PV, wind, DEG incorporated a hybrid power system is studied. A PI controller is also connected in the system so as to minimize the frequency deviation or to perform load frequency control. In order to find out the parameters of these controllers, the Grey wolf optimization technique is employed so as to tune the controller using ITSE. In this case, the effect of SMES for meeting the required load demand is not considered.

A simulation model of HPS without SMES is shown in figure 19. Each block of the model has its own performance parameters. PI controller is tuned using GWO to helps in minimizing the deviation that occurred in the system due to variation in load. This power system is more efficient when the power generation from renewable sources is more and consumption from DEG is minimum. The system is simulated for different time duration under both the cases and the effect of the controller in mitigating the deviation is seen. Under this scenario, steady-state analysis and transient analysis has done under different loading and operating conditions. The algorithm is analyzed in the MATLAB environment and the objective function is evaluated for optimization.

Figure 19. Simulation model of HPS without SMES

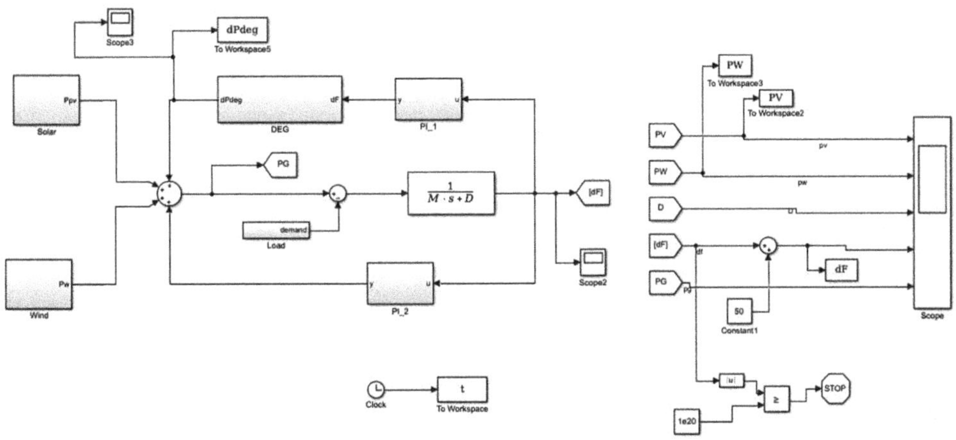

Steady-State Analysis

In this section, a simulation has been done during sudden load changes in the hybrid power system. The steady-state stability analysis of the hybrid power system is carried out through step load change during the maximum and minimum power consumption period. The system response is analyzed with the load change in the absence of a superconducting magnetic energy storage system. The Simulink model is simulated for the time period of 30 seconds. Under this analysis two cases has been studied-

1. Peak Load Condition
2. Off-Peak Load Condition

Both cases have been studied in the MATLAB environment under different operating conditions. Frequency deviation under both conditions is obtained and the result obtained is analyzed.

Case 1: During Peak Load

In this case, wind and PV power variations are assumed to be absent. When the load on the system changes, the system frequency deviation will take place. A step load change of 0.1 p.u. occurred as peak load is considered. When a sudden change in load demand i.e. as the peak load change occurs, system frequency falls below a predefined limit which affects the system performance. In order to maintain the system frequency within the desired limit, the energy storage system plays an important role. The effect in frequency response without SMES in HPS but using a Grey wolf optimized PI controller is shown.

Power generation from solar is shown in figure 20 and from wind in figure 21. Generation from these sources is considered to be invariable, thus power from these two sources remain constant.

Figure 22 represents the load demand profile, as sudden load change occurs in the system during peak hours. Total power generation from all the generation sources is shown in figure 23 as when the load demand increases, generation should also need to be increased, thus the involvement of DEG starts for the fulfillment of required demand.

Figure 20. PV Power profile

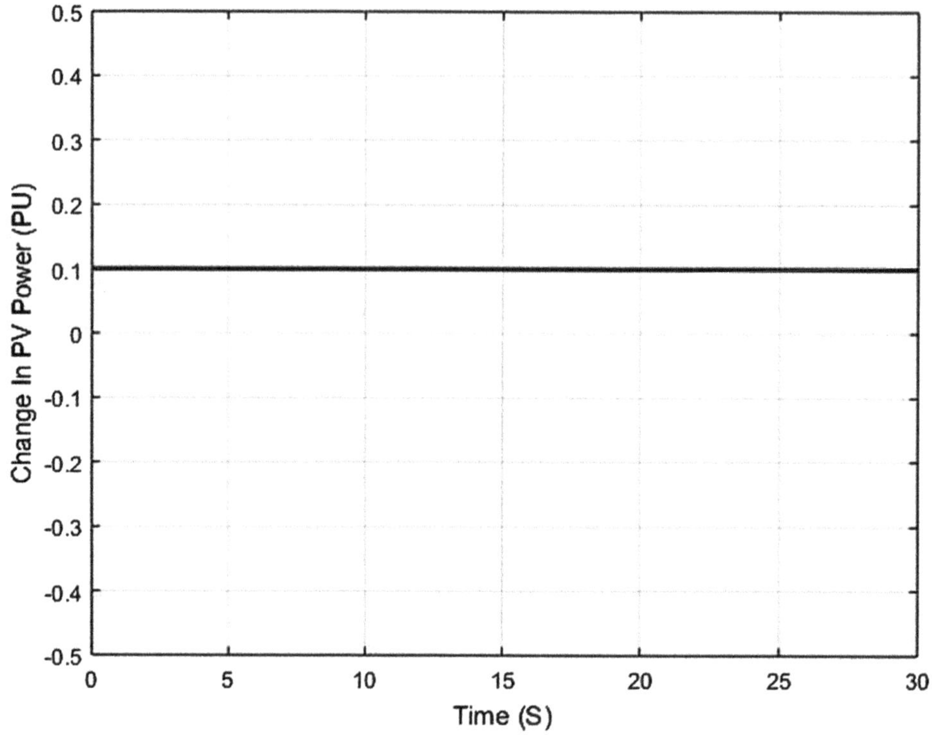

Figure 21. Wind power profile

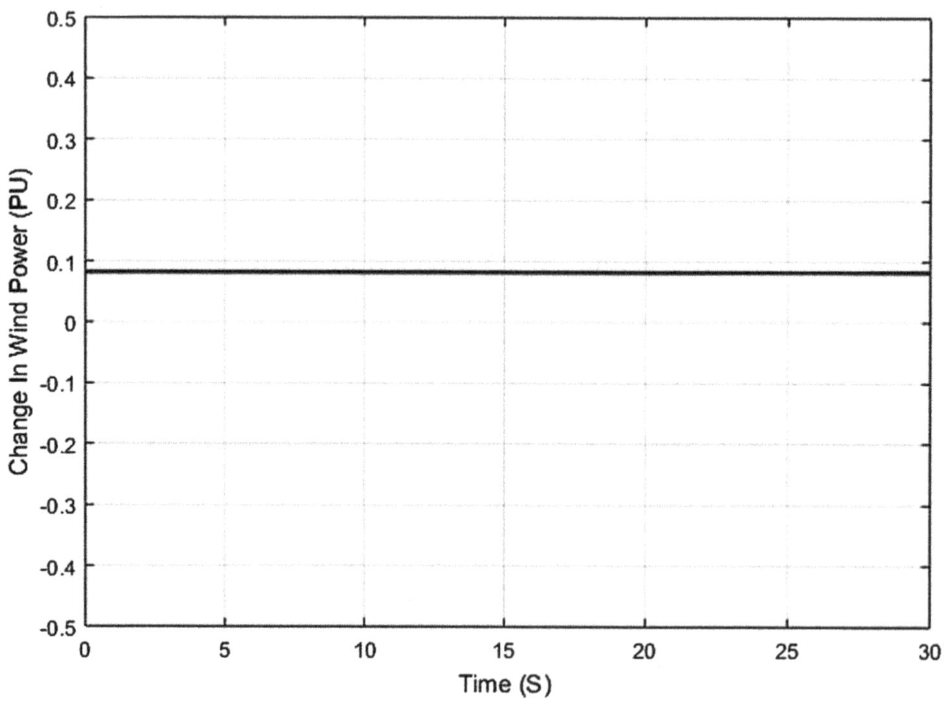

Figure 22. Load demand profile

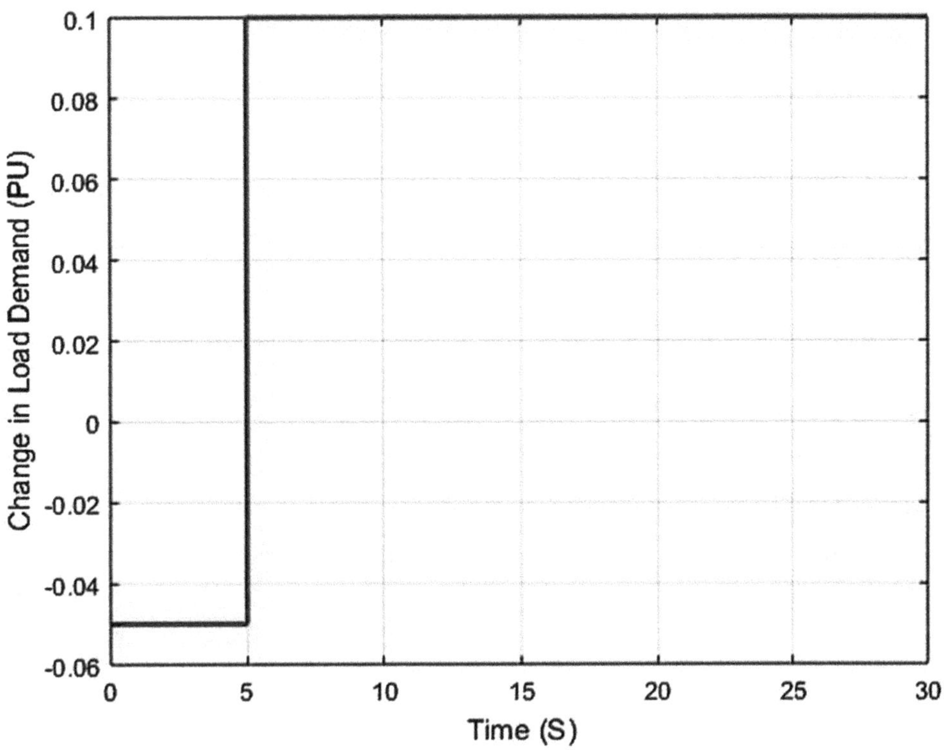

Figure 23 shows the power generation from solar, wind and DEG. At t=5 seconds when sudden load change occurs in the system frequency deviation occurs which is minimized by grey wolf optimized PI controller shown in figure 25 and best fitness value obtained with respect to iteration for the optimized controller is sown in figure 24.

Case 2: During the Off-Peak Load

Similar conditions as above are applied here but during offload condition. A step load change of 0.05p.u. is occurred and the system response without considering SMES is shown. When the load demand decreases, the frequency of the system increases above the nominal value. in order to mitigate this issue, a controller is optimized using the GWO algorithm and their parameters are tuned accordingly. The GWO algorithm finds the best fitness function to optimize the parameters of the controller. In this case, wind and PV power variations are assumed to be absent. The results are shown below.

Figure 20 and 21 shows the power generation from solar and wind unit. These figure shows that the constant power from both the units is considered.

Figure 26 represents the load demand profile, as at t=5 sec, sudden load changes in the system. Similarly, total power generation from all the generation sources is shown as when the load demand increases, generation should also need to be increased, thus the involvement of DEG starts for the fulfillment of required demand.

Figure 23. Total power generation

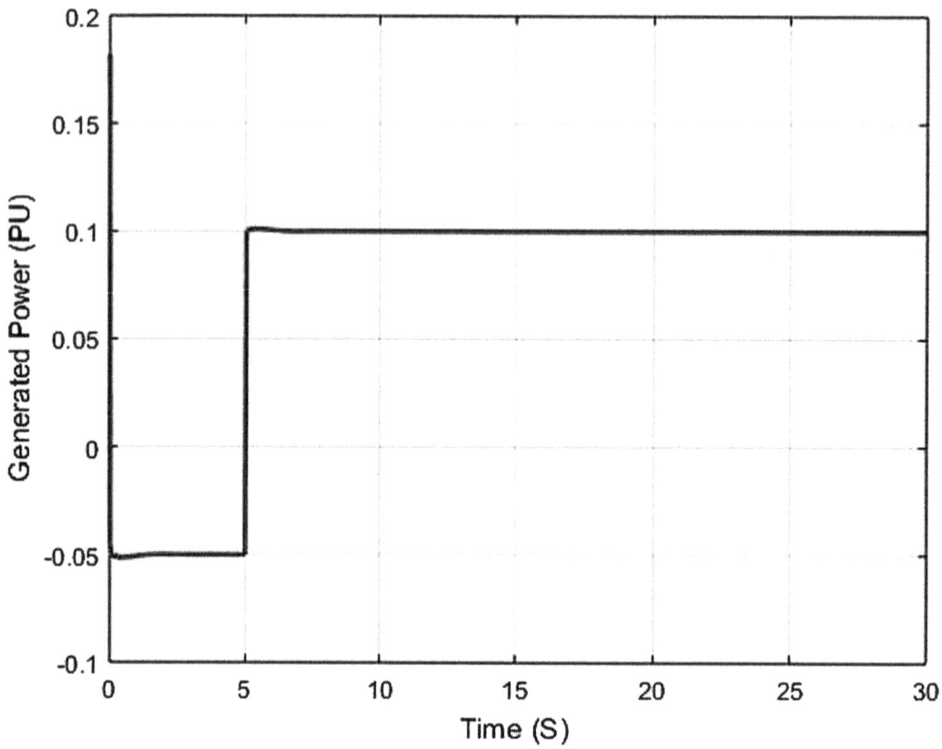

Figure 24. Best fitness obtained with respect to iteration

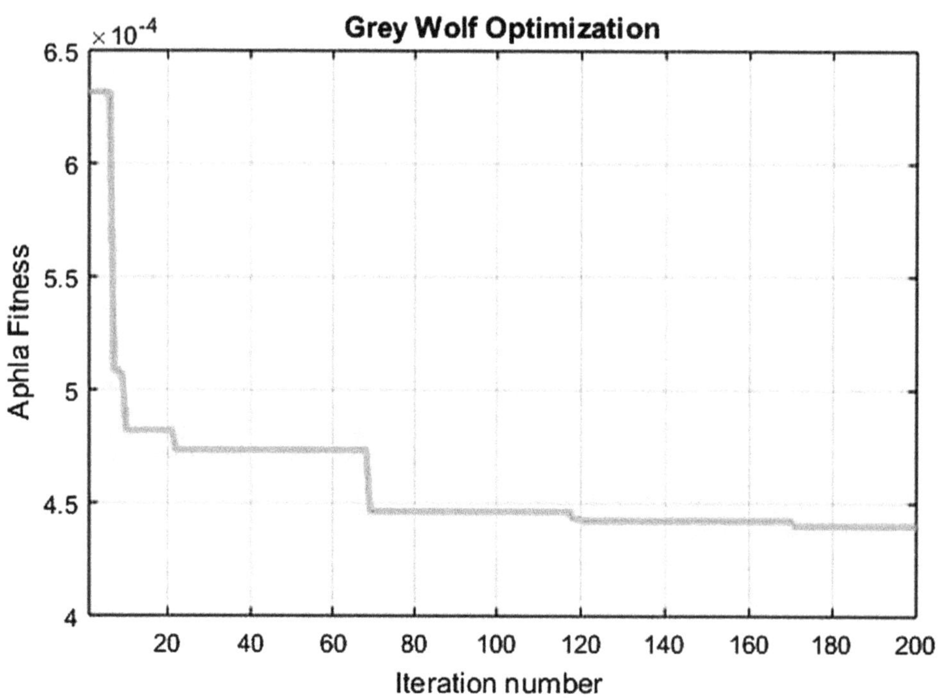

Figure 25. Change in frequency without SMES

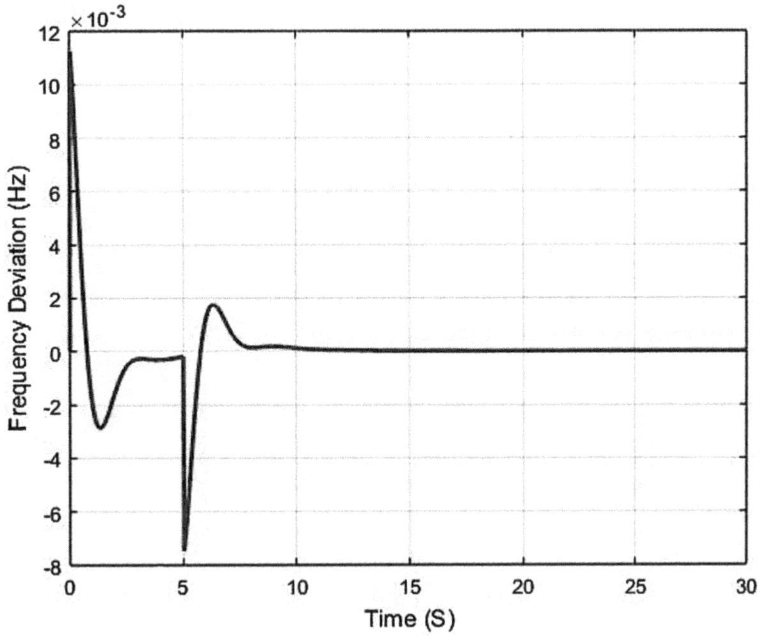

Figure 26. Load demand profile

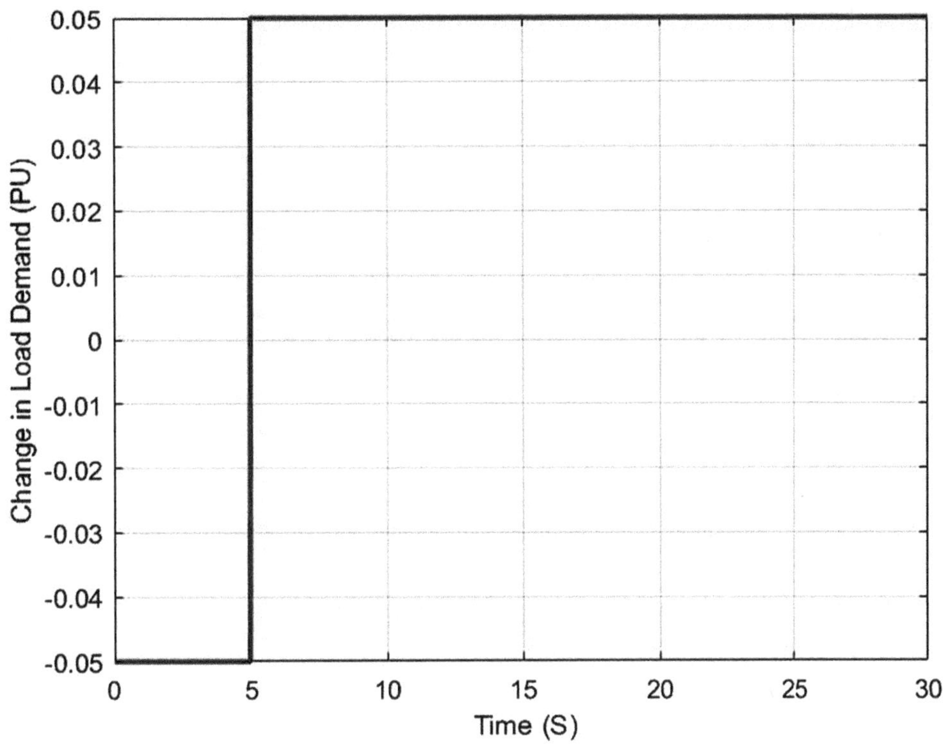

Case (2) studies show the mitigation in frequency deviation figure 29 due to the tuned PI controller whereas figure 27 shows the total power generation from various power generation sources along with its best fitness value for the optimized PI controller in figure 28.

Dynamic Stability Analysis

In this analysis, a model is simulated for 24 hours and their transient performance is analyzed. This analysis is carried out by varying the load demand as well as the generation from various sources throughout a complete day. With the increase in load, system frequency falls below a predefined limit. In order to meet the required power demand, the generation should be increased to maintain the proper equilibrium. When the dip occurred in the system frequency, power through DEG source should be enlarged up to its rated capacity to maintain the generation in the system

Similarly, when the load decreases, system frequency increases, as a result, the surplus amount of power is available in the system as the generation is not varying. DEG is not involved in the system, so as to reduce consumption. The effect in frequency response due to the optimized PI controller without using SMES is shown.

A solar model of PV power profile for a hybrid power system throughout a day is considered which shown in figure 30. Maximum power is extracted somewhere throughout the day.

Figure 27. Total power generation

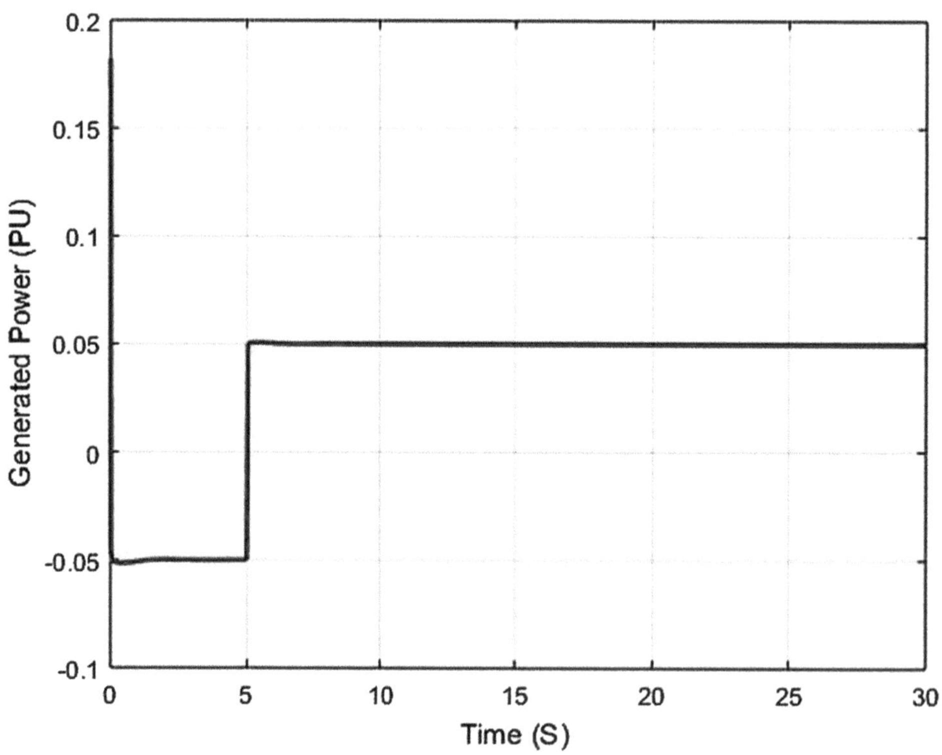

Figure 28. Best fitness obtained so far

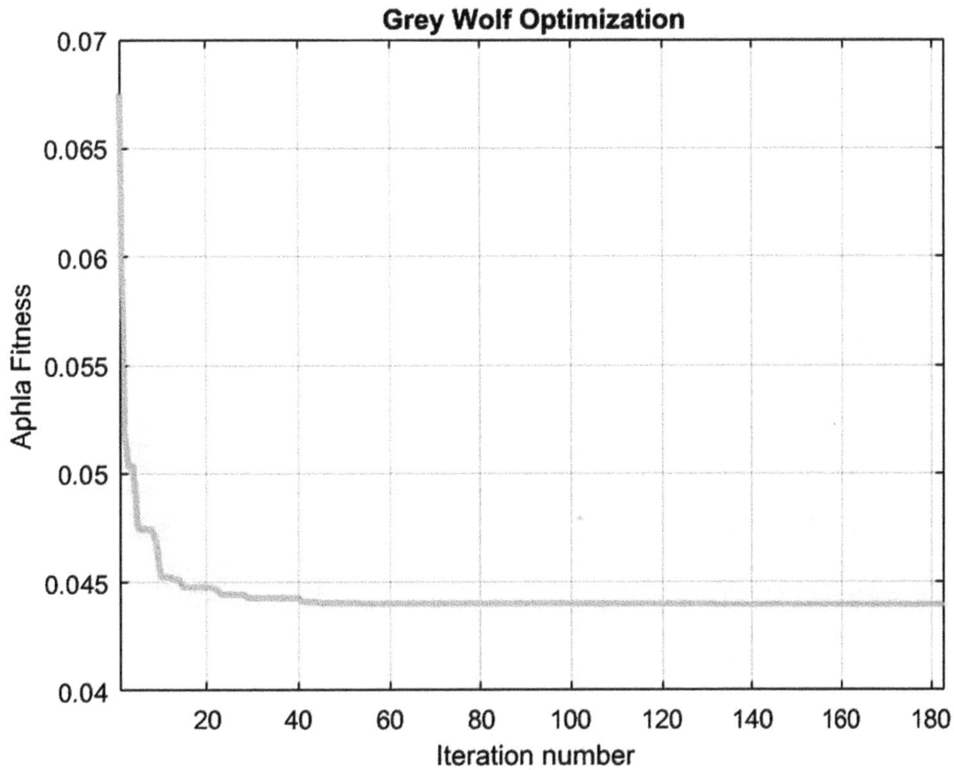

Similarly, wind power profile is also considered for 24 hours as shown in figure 31. Maximum wind power is also obtained, whereas a low pass filter is used to convert wind speed curve into the wind model. The change in power from the diesel engine generator is also shown in figure 32.

The above-shown figure shows the transient analysis of microgrid incorporated hybrid power systems. In this analysis, both the load demand profile shown in figure33 and power generation are variable in nature.

The system frequency response with both an increase and decrease of load demand without SMES is analyzed. Frequency deviation occurred in the system is shown in figure 35 along with its best fitness value in figure 34.

Optimized Controller Parameters

Since the controller is tuned in the system using the GWO technique. The controller parameter is optimized using Integral Time Square Error function. Since two controllers are incorporated in the system. The first controller, for the area, consists of DEG and another one in a different area. The obtained parameters for the above-studied cases are given below.

1. Steady-state stability analysis- parameters of controllers are further tuned for different loading conditions, i.e. during peak loading or off-peak loading.

Figure 29. Frequency deviation in the HPS

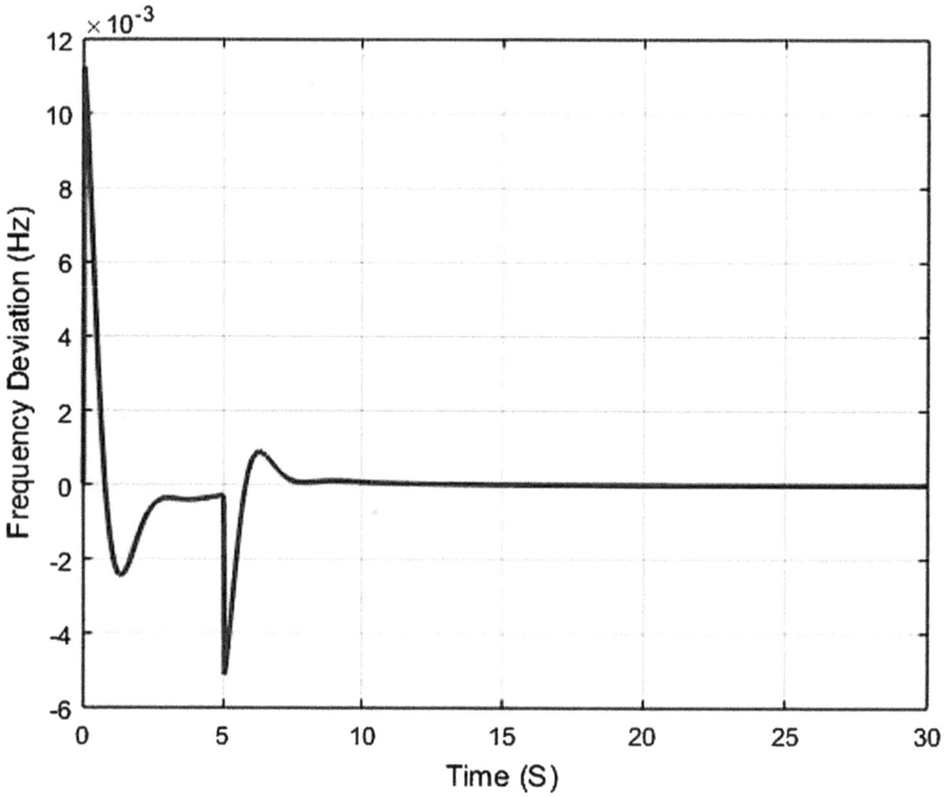

Figure 30. PV power profile for a typical day

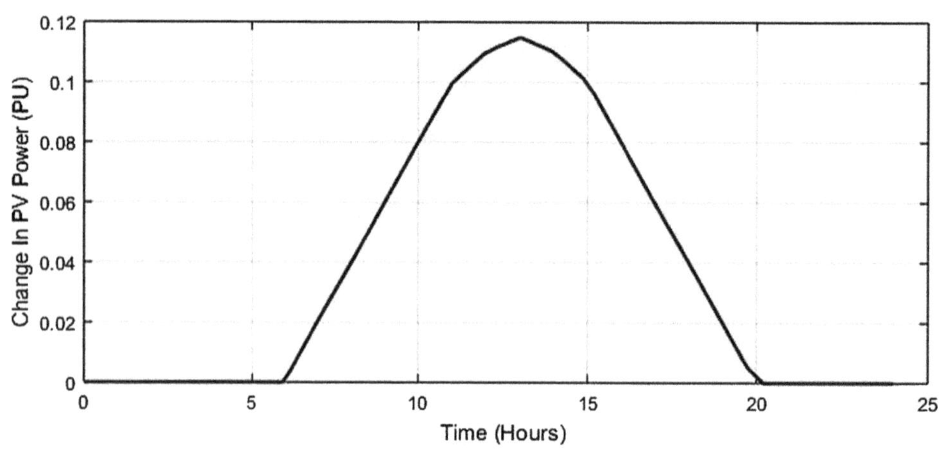

Figure 31. Wind power profile throughout a day

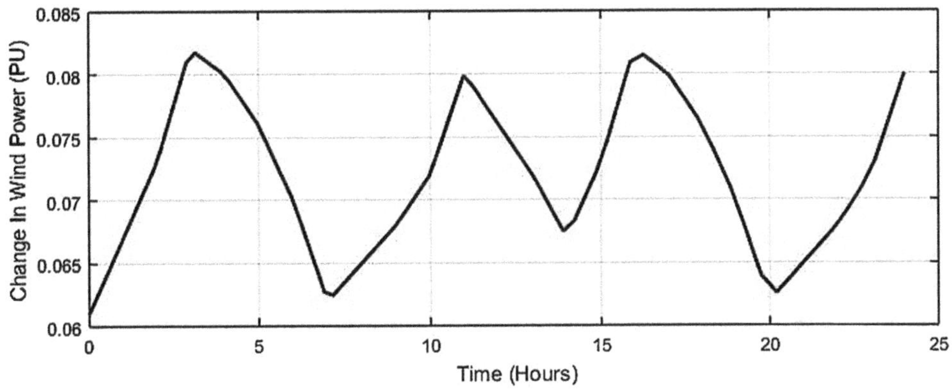

Figure 32. Change in DEG power

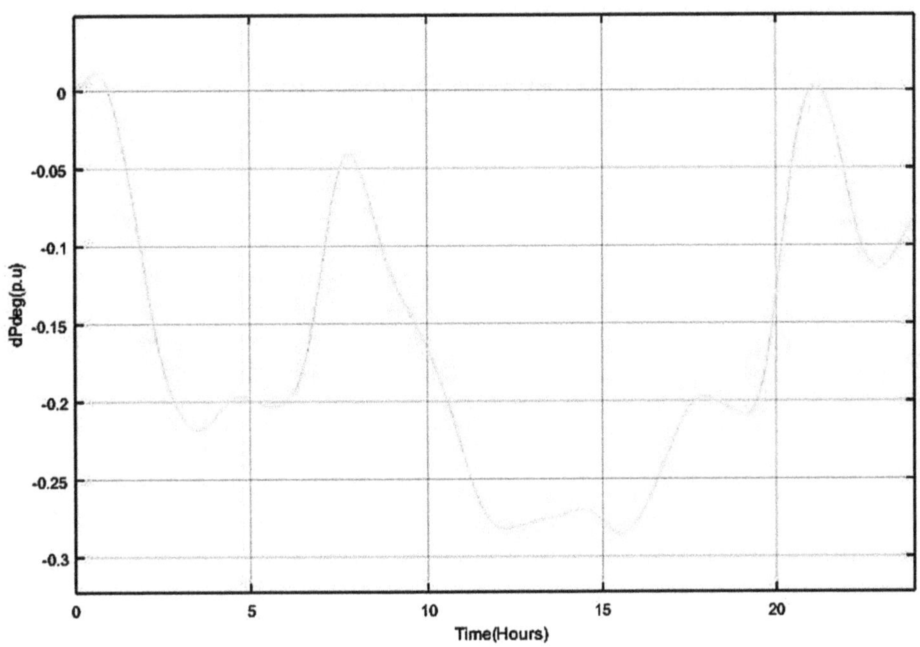

2. Dynamic stability analysis- parameters are tuned for random load change and power generation throughout a typical day.

Under steady-state stability analysis of microgrid containing HPS without considering SMES, controller parameters for different cases are given below. Since four parameters are simulated under this operating condition as the effect of SMES is not given, it's again parameter is also not seen in the simulation

1. During Peak load (without SMES)
2. During Off-Peak load (without SMES)

Figure 33. Load demand profile

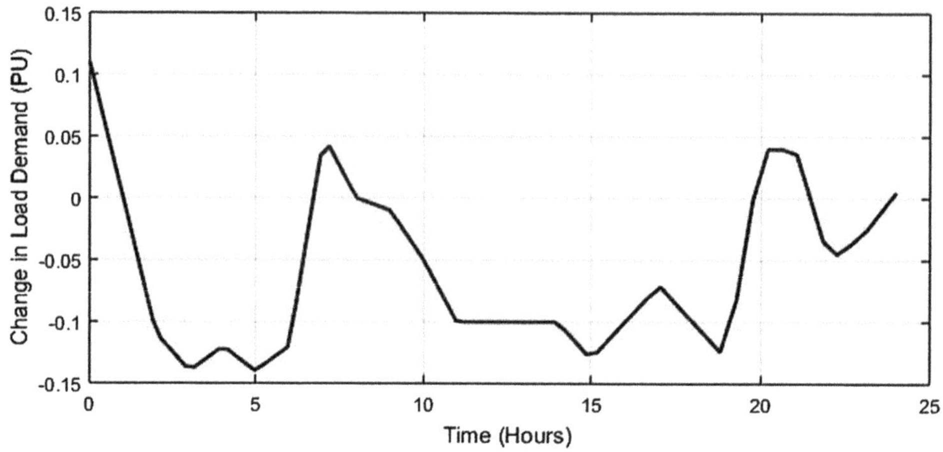

Figure 34. Best fitness obtained w.r.t iteration

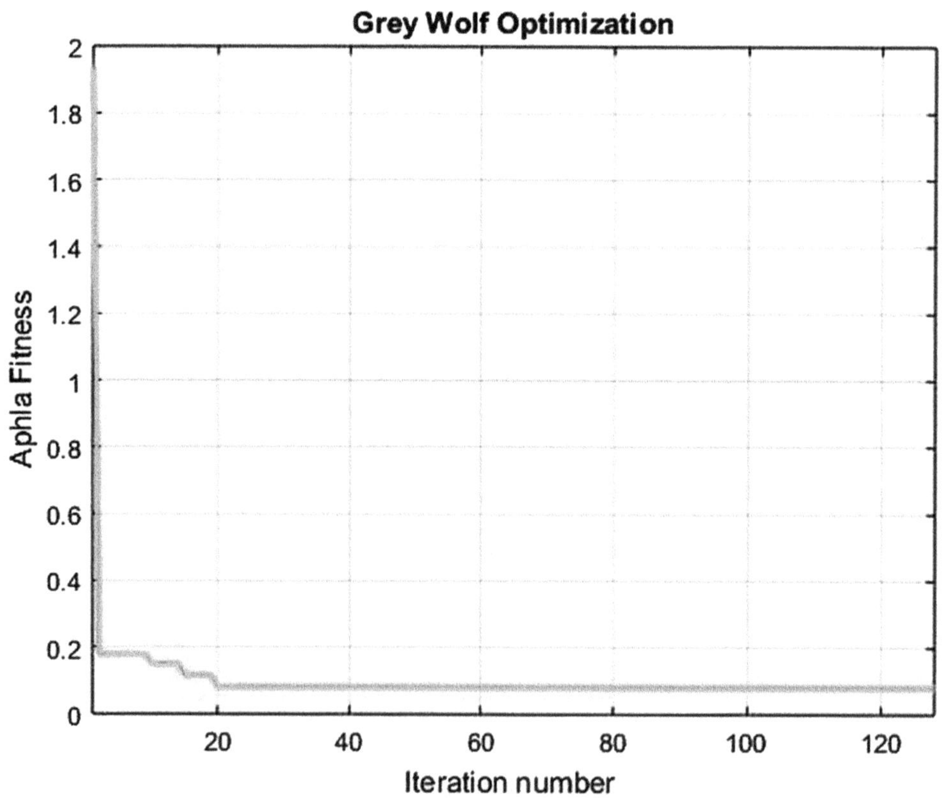

Figure 35. Frequency deviation

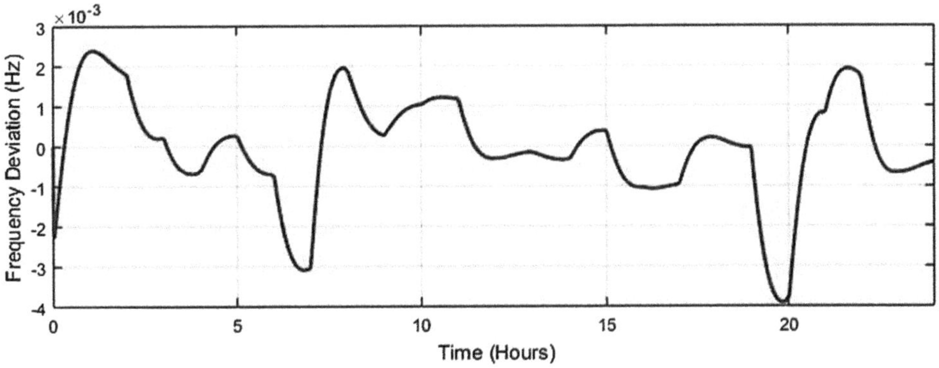

Dynamic stability analysis of hybrid power systems involving solar, wind and DEG without considering SMES is shown.

Tuned controller parameters during peak load and off-peak load without SMES are given in Tables 1 and 2. Similarly, Table 3 gives tuned controller parameters during dynamic stability analysis of the hybrid power system without SMES.

Table 1. controller parameters at steady state during 0.1p.u. load without SMES

Parameters	Values
K_{P1}	20
K_{P2}	-20
K_{I1}	7.6587
K_{I2}	-20

Table 2. controller parameters during 0.05p.u. load without SMES

Parameters	Values
K_{P1}	20
K_{P2}	-20
K_{I1}	9.7826
K_{I2}	-20

Table 3. controller parameters during dynamic simulation without SMES

Parameters	Values
K_{P1}	20
K_{P2}	-20
K_{I1}	-0.8519
K_{I2}	-0.0937

Scenario (II) HPS With Solar PV+ Wind + DEG +PI Controller + SMES.

In this scenario PV, wind, DEG incorporated a hybrid power system using SMES is studied. A PI controller is also connected in association with DEG or SMES unit in the system so as to minimize the frequency deviation or to perform load frequency control. In order to find out the parameters of these controllers, the Grey wolf optimization technique is employed so as to tune the controller using ITSE. In this case, the effect of SMES for meeting the required load demand is considered.

A SMES based model of HPS is shown in figure 11. Each block of the model has its own performance parameters. PI controller is tuned using GWO to helps in minimizing the deviation that occurred in the system due to variation in load. The hybrid power system is more efficient when the power generation from renewable sources is more and consumption from DEG is minimum. The system is simulated for different time duration under both the cases and the effect of the controller in mitigating the deviation is seen using SMES. The impact of SMES is widely seen in the system response due to its fast-acting nature and power quality improvement characteristics. Under this scenario, steady-state analysis and transient analysis has done under different loading and operating conditions. GWO Algorithm is analyzed in the MATLAB environment and the objective function is evaluated for optimization. The simulation model of this case employing SMES unit is shown in figure 11.

Steady-State Analysis

This analysis has been done during a sudden load change in the hybrid power system. The steady-state stability analysis of the hybrid power system employed with SMES is carried out through step load change during the maximum and minimum power consumption. The system response is analyzed with the load change in the presence of superconducting magnetic energy storage system or and the effect of SMES under different operating condition. The Simulink model is simulated for the time period of 30 seconds. Under this analysis, two cases have also been studied-

1. Peak Load Condition
2. Off-Peak Load Condition

Both cases have been studied in the MATLAB environment under different operating conditions. Frequency deviation under both conditions is obtained and the result obtained is analyzed.

Case 1: During Peak Load

A step load change of 0.1p.u. occurs as peak load is considered. When a sudden change in load demand i.e. as the peak load change occurs, system frequency falls below a pre-defined limit which affects the system performance. In order to maintain the system frequency within the desired limit, the energy storage system plays an important role. The effect in the frequency response of HPS using SMES with PI controllers whose values are optimized using GWO is shown.

Power generation from solar and wind is shown in figure 36 and 37.

Generation from these sources is considered to be invariable, thus power from these two sources remain constant. A load demand profile is shown in figure 38.

The total power generation of various sources along with the SMES is shown in figure 39. When the load of the system changes, the frequency deviation takes place. As sudden load change occurred in the system and generation is quite insufficient, SMES starts acting in discharging mode and supplying a sufficient amount of power so as to maintain the frequency regulation.

Analysis: When sudden load change occurs in the system SMES starts acting in inverter mode i.e starts injecting power to the system so as to maintain the required demand is shown and frequency de-

Figure 36. PV power profile

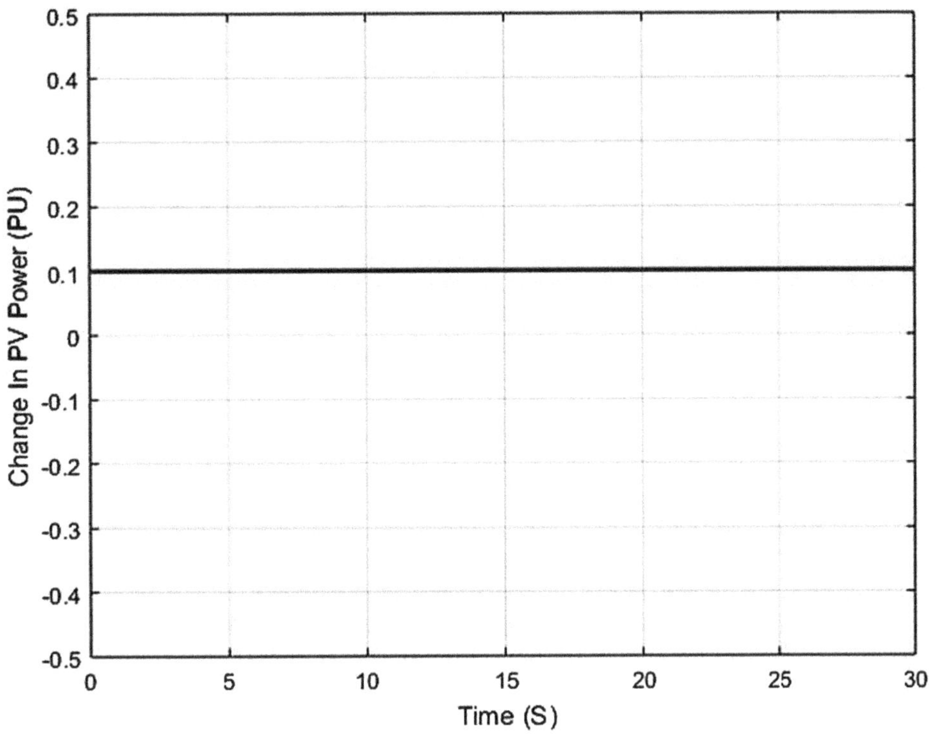

Figure 37. Wind power profile

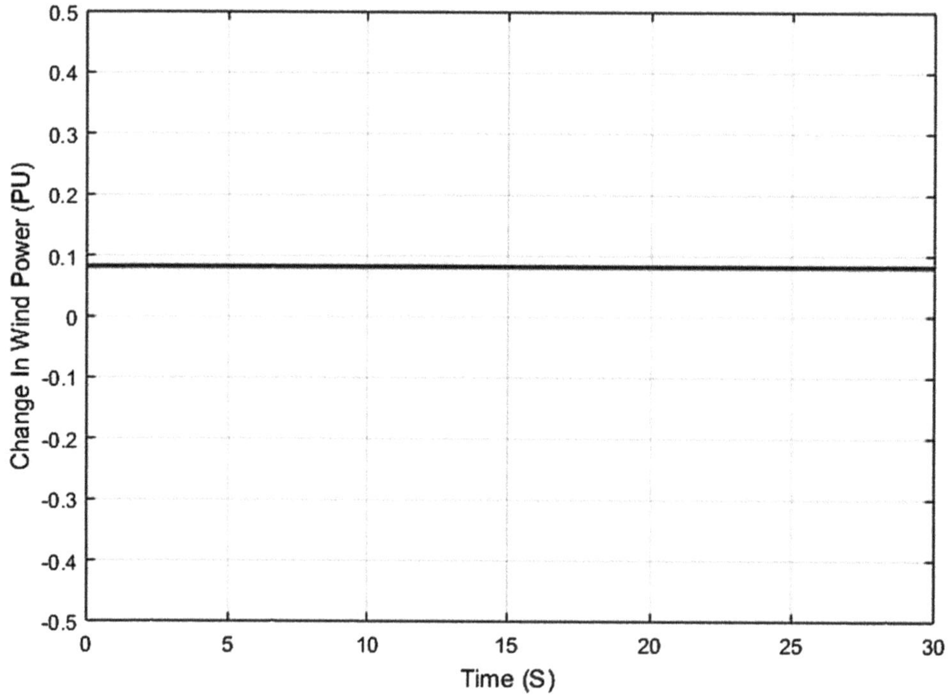

Figure 38. Load demand profile

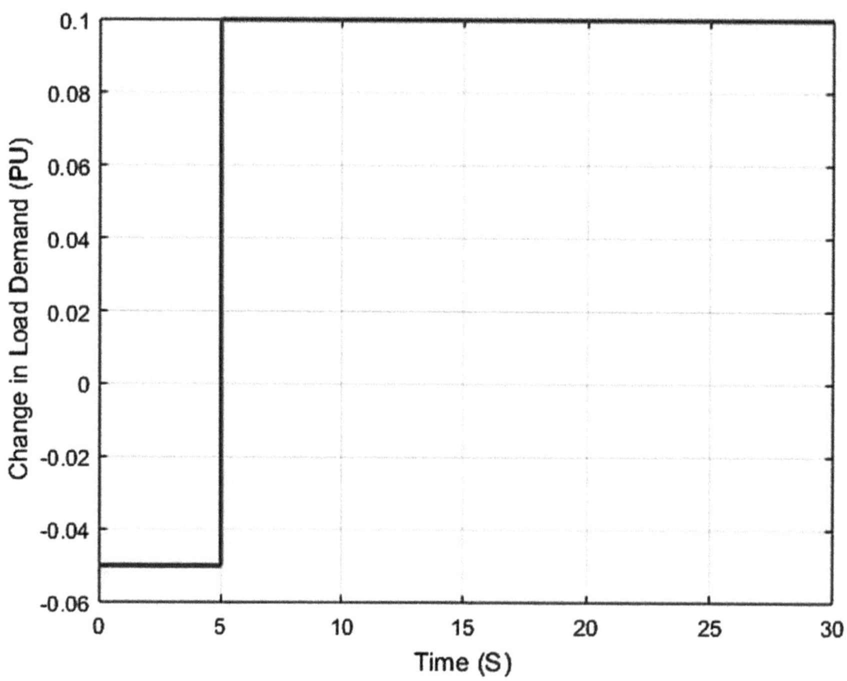

Figure 39. Total power generation considering the effect of SMES

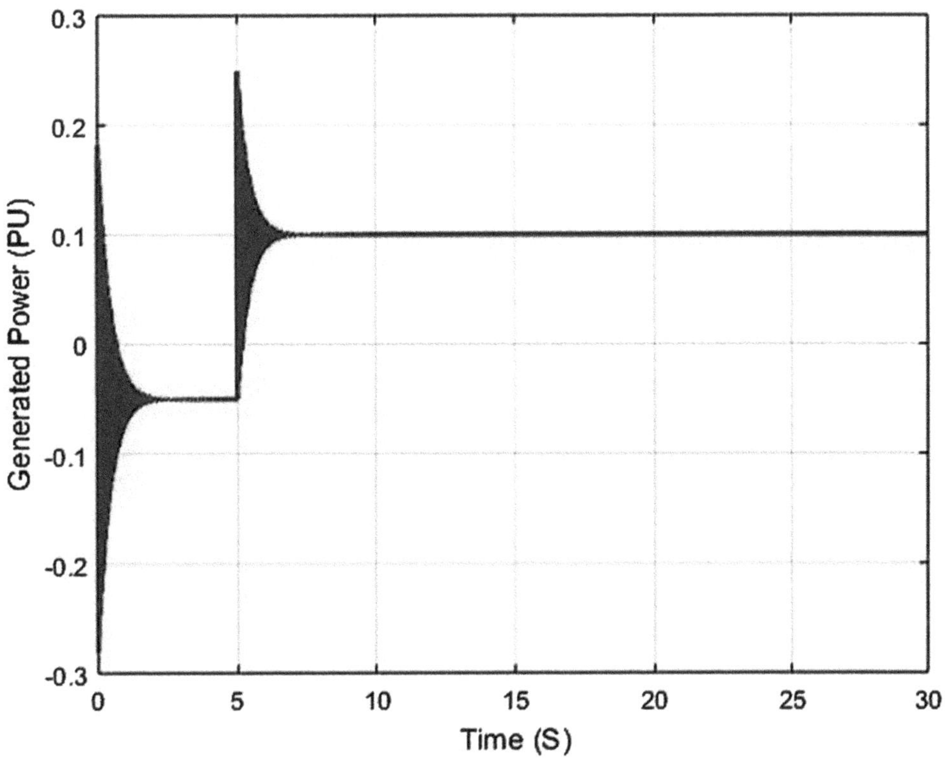

viation with respect to this load change is shown in figure 41 and best fitness value for grey wolf tune PI controller is given in figure 40.

Case 2: During the Off-Peak Load

Similar conditions as above are applied here but during offload condition along with SMES unit. A step load change of 0.05p.u. is occurred and the system response with considering SMES is shown. When the load demand decreases, the frequency of the system increases above the nominal value. The results are shown below.

Load change at t= 5 seconds is shown in the figure 42 and solar and wind power generation, which is assumed to be constant throughout the simulation period is shown in Figures 36 and 37.

The best fitness value of an optimized PI controller is shown in figure 43 and the total power generation considering the impact of SMES is seen in figure 44. Change in frequency is shown in figure 45 below.

Dynamic Stability Analysis

In this analysis, the hybrid model is simulated for 24 hours and their transient performance is analyzed. This analysis is carried out by varying the load demand as well as the generation from various sources throughout a complete day using the impact of SMES technology. With the increase in load, a sudden dip occurs in the system frequency falls below a predefined limit. In order to meet the required power

Figure 40. Best fitness value w.r.t iteration

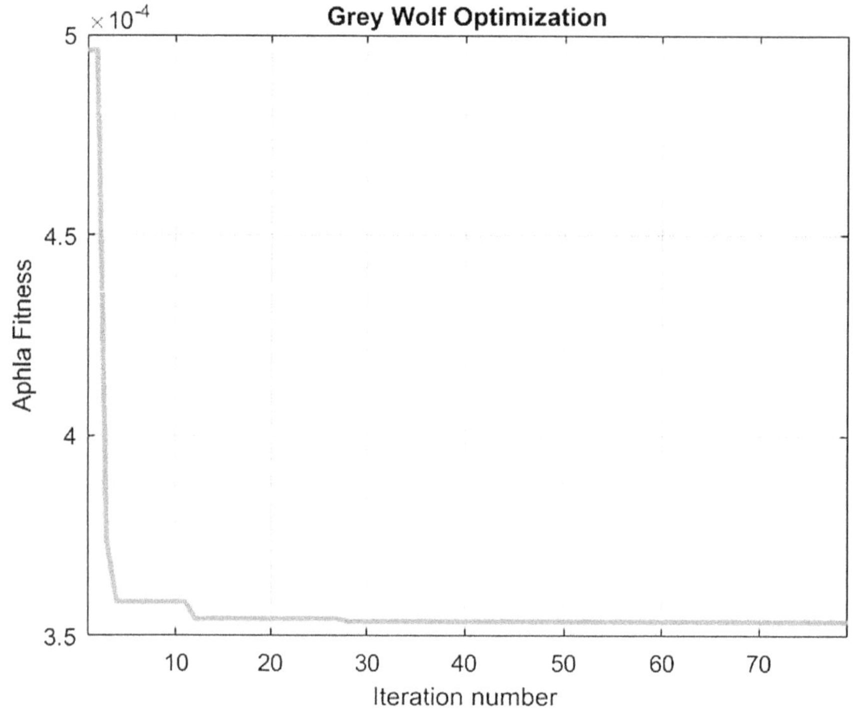

Figure 41. Change in frequency

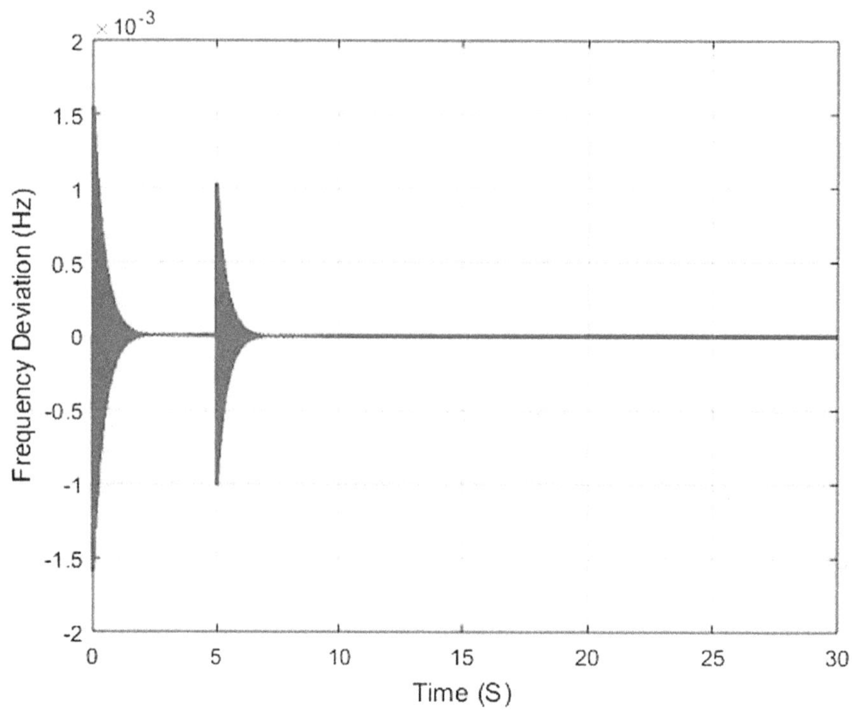

Figure 42. Load demand profile

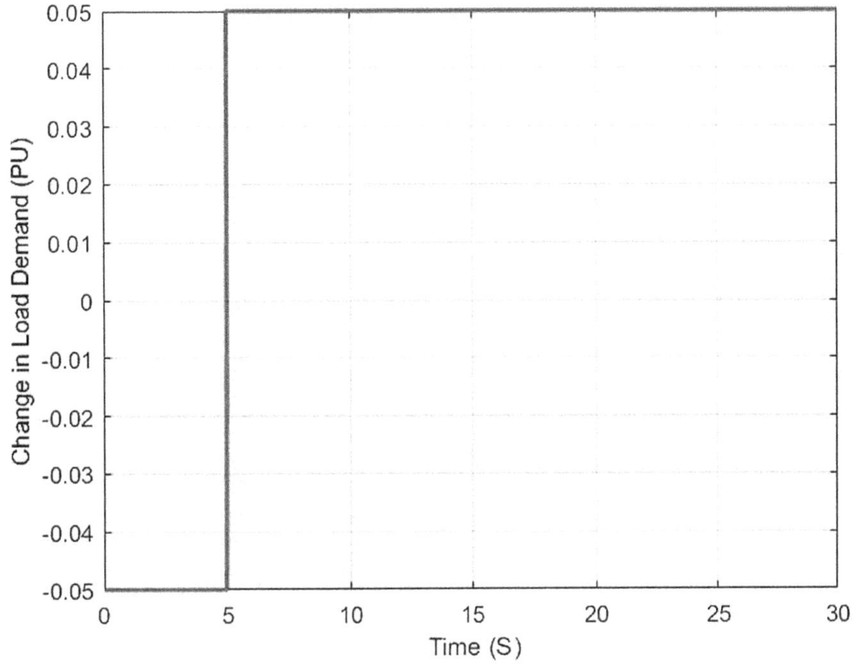

Figure 43. Best fitness with respect to iteration

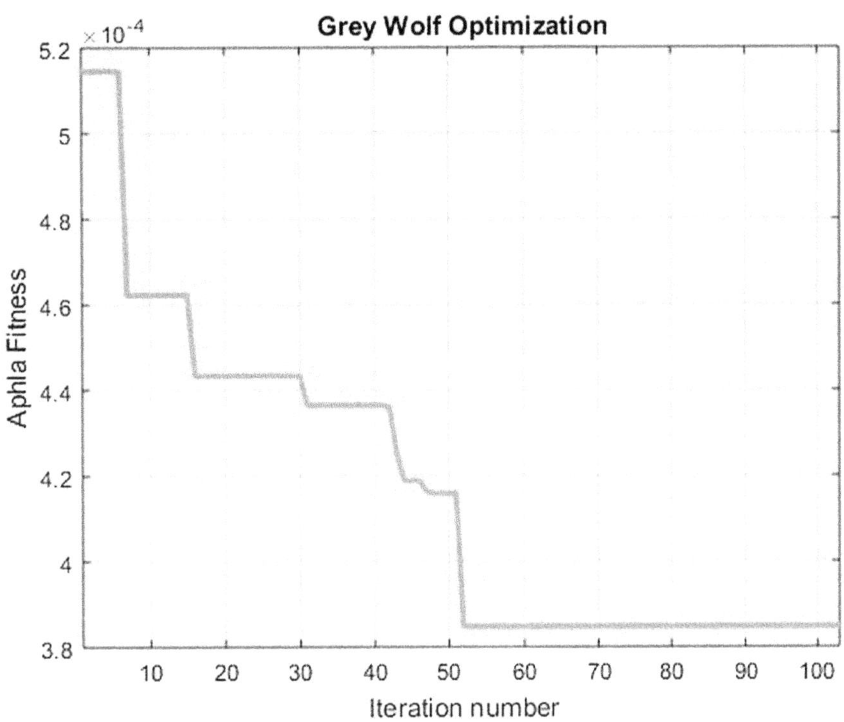

Figure 44. Total power generation

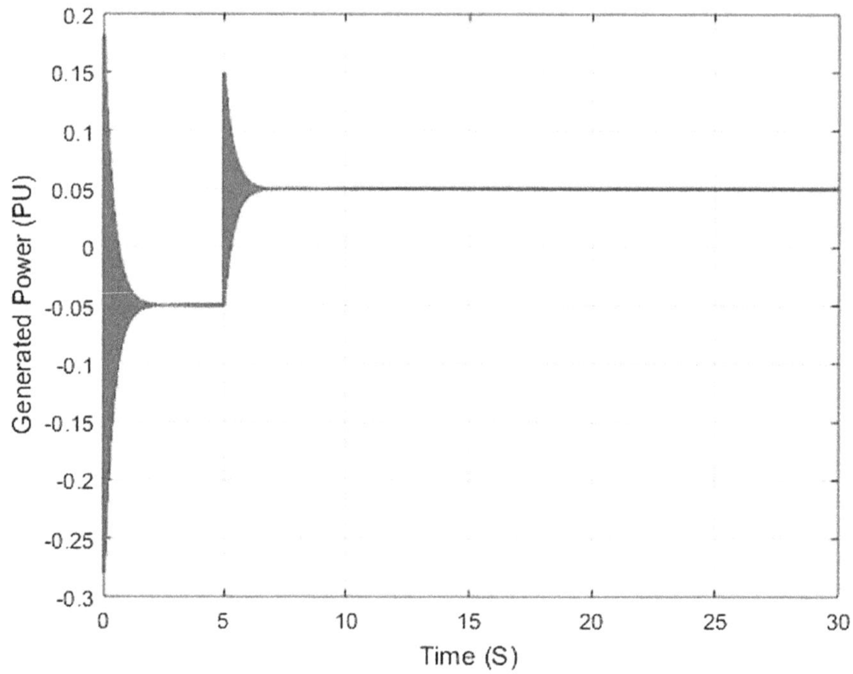

Figure 45. Change in frequency

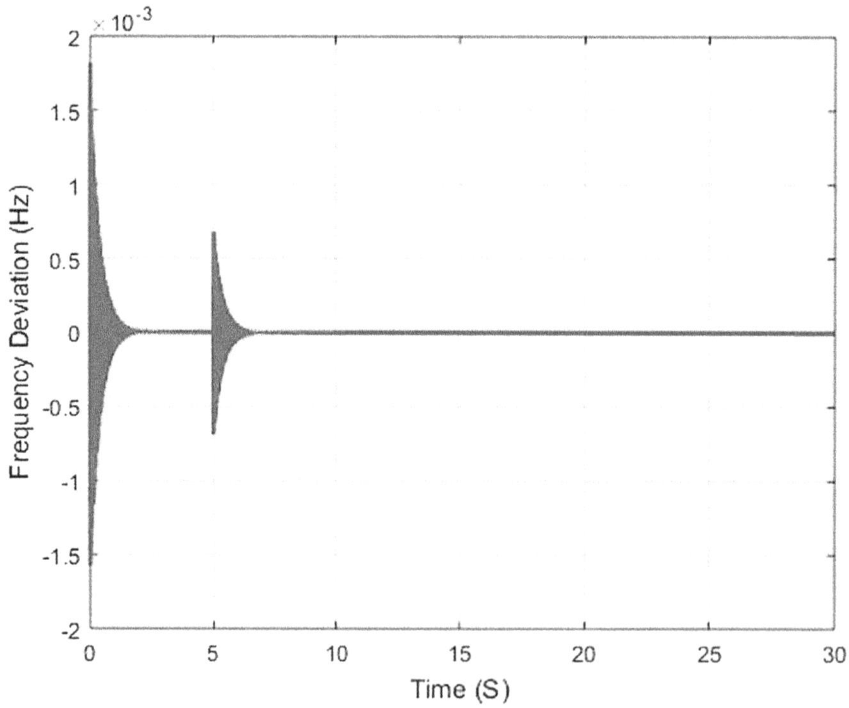

demand, the generation should be increased to maintain the proper equilibrium, thus the SMES starts acting and provide the necessary power required. When the dip occurred in the system frequency, power through DEG source should be enlarged up to its rated capacity along with SMES to maintain the generation in the system.

Similarly, when the load decreases, system frequency increases, as a result, the surplus amount of power is available in the system as the generation is not varying, SMES starts absorbing the power for load compensation. DEG is not involved in the system, so as to reduce consumption. The effect in frequency response due to the optimized PI controller using SMES is shown.

A PV power profile and a wind power profile for a hybrid power system throughout a day is considered and shown in figure 46 and figure 47. Maximum power is extracted somewhere throughout the day. Similarly, wind power profile is also considered for 24 hours.

Maximum wind power is also obtained, whereas a low pass filter is used to convert wind speed curve into the wind model. Power from diesel engine generators is also shown. The effect of SMES throughout a day with random variation in load or power generation is seen and a load profile throughout a day is

Figure 46. PV power profile for a typical day

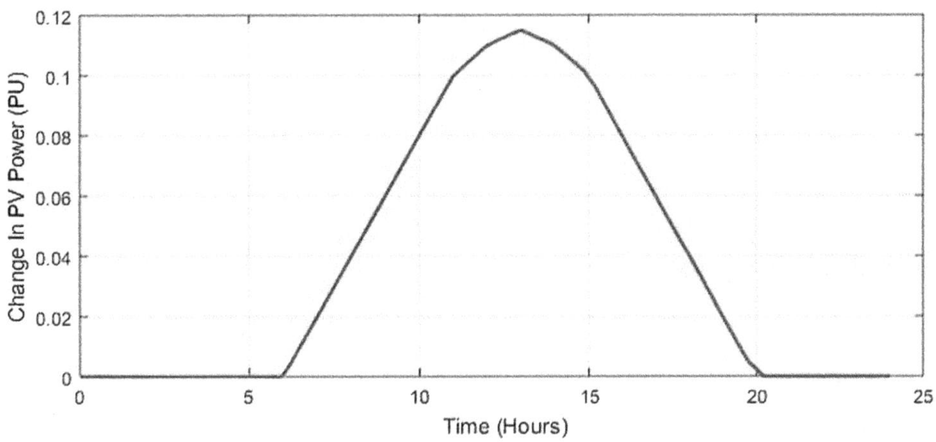

Figure 47. Wind power profile through a day

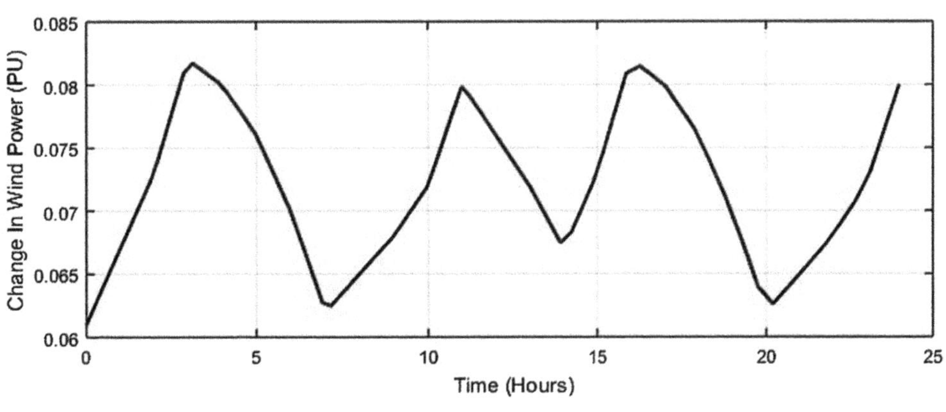

in figure 48. SMES acts in both charging and discharging mode so as to mitigate the fluctuations and change in SMES power is shown in figure 49.

As the random change in load and variable generation from different power sources occurred in the system, frequency deviation is mitigated with the effect of SMES whose input is frequency deviation occurred in the system.

Change in frequency with the effect of SMES is shown in figure 51 and its best fitness value is obtained in figure 50.

Figure 48. Load profile throughout a day

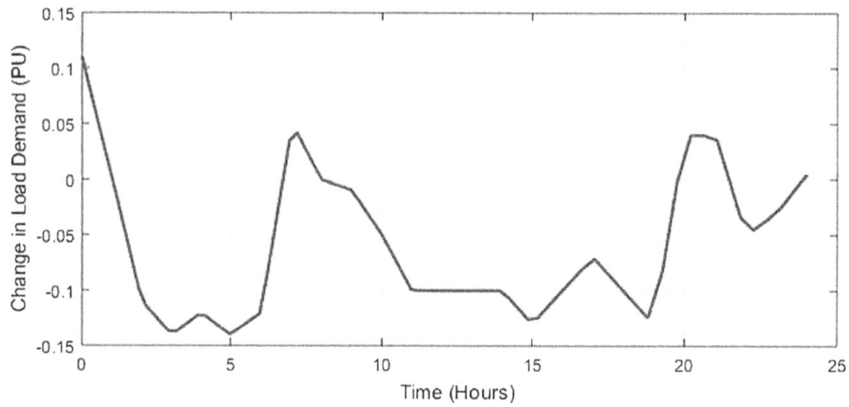

Figure 49. Change in SMES power

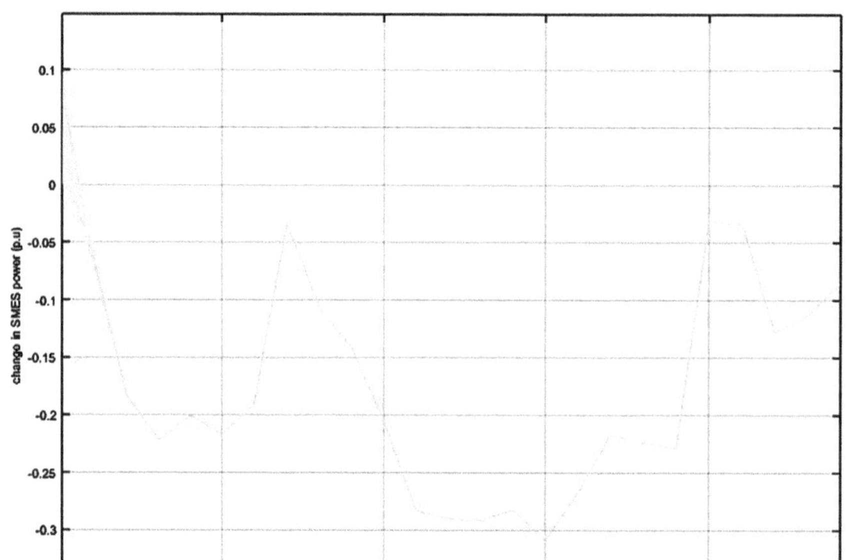

Figure 50. Best fitness value

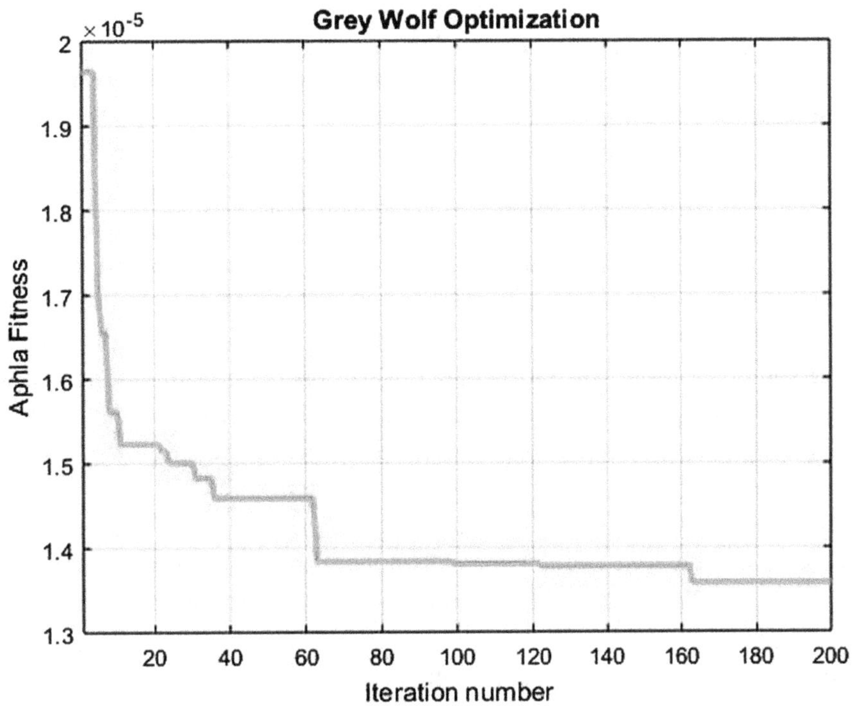

Figure 51. Change in frequency with the effect of SMES

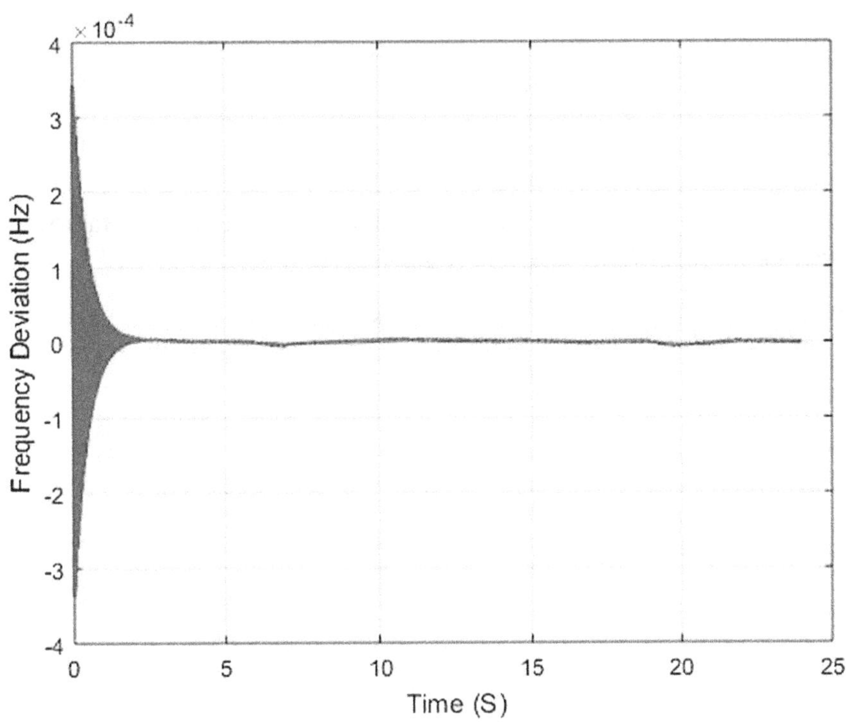

CONCLUSION

The main purpose of this chapter is to mitigate the deviation in frequency occurred in the system, during the variation in load demand. So as to fulfill this objective a grey wolf optimized PI controller is designed for a microgrid connected hybrid power system. Due to random power generation from PV and wind sources, energy management between load and demand side is distorted. In order to minimize this distortion energy storage device, SMES is employed in the system. SMES plays an important role in meeting the required load demand when the generation is quite insufficient. A DEG provides supplementary control action to limit the fluctuation in frequency occurred during a sudden load change. The hybrid power system involves microgrid, as renewable energy sources are very popular in the field of the power system. With increased development these conventional energy sources, provides various benefits such as an alternate source of power generation, reducing the energy consumption of DEG, clean energy, etc. Microgrid incorporated hybrid power model is simulated under various loading conditions. On the basis of obtained simulation results, some important points are concluded:

1. Microgrid involved a hybrid power system possesses more reliability as they involved various generation sources. When one of the sources is not capable of fulfilling required power demand, due to the various atmospheric conditions, other sources operate and start sharing the required load demand.
2. SMES as an energy storage system also acts either as a source of power or a sink of power. Due to its fast response and load sharing capacity, deviation occurred in the system is limited. The main importance of SMES is that it injects active and reactive power simultaneously.
3. In order to damp out power system oscillations, SMES can be used as a power quality improvement device.

REFERENCES

Adefarati, Bansal, & Jackson. (2017). Reliability and economic evaluation of a microgrid power system. *Proceedings of 9th International Conference on Applied Energy.*

Ali, E. S., & Abd-Elazim, S. M. (2013). BFOA based design of PID controller for two area Load Frequency Control with nonlinearities. *Electrical Power and Energy Systems, 51*(1), 224–231. doi:10.1016/j.ijepes.2013.02.030

Ali, M. H., Wu, B., & Dougal, R. A. (2010). An overview of SMES applications in power and energy systems. *IEEE Trans Sustain Energy, 1*(1), 38–47. doi:10.1109/TSTE.2010.2044901

Asma, A., Shafiullah, G. M., Amanullah, M. T. O., & Alex, S. (2010). Automatic Generation Control in Wind Integrated Power System: New Perspectives and Challenges. *Int. J. of Thermal & Environmental Engineering, 12*(1), 27–38.

Ayyarao, Tummala, Alluri, & Ramanarao. (2017). Optimal load frequency control of electric vehicle-based micro-grid using Grey Wolf Optimization. *Proceedings of International Conference on Computing of Power, Energy & Communication, 6*, 11-17.

Bhatti, T. S., Al-Ademi, A. A. F., & Bansal, N. K. (1997). Load Frequency Control of Isolated Wind Diesel Hybrid Power Systems. *Energy Conservation and Mgmr, 38*(9), 829–837. doi:10.1016/S0196-8904(96)00108-2

Bhongade, S., Tyagi, B., & Gupta, H. O. (2011). Effect of SMES unit in load following contract in a restructured power system. *International Journal of Engineering Science and Technology, 3*(1), 255–264. doi:10.4314/ijest.v3i1.67652

Chaiyatham, Ngamroo, Pothiya, & Vachirasricirikul. (2009). Design of Optimal Fuzzy Logic-PID Controller using Bee Colony Optimization for Frequency Control in An Isolated Wind-Diesel System. *Proceedings of IEEE Transmission & Distribution Conference & Exposition: Asia and Pacific,* 1-4.

Dipayan Guha, P. K. (2016). Load frequency control of interconnected power system using grey wolf optimization. *Swarm and Evolutionary Computation, 27*(2), 97–115. doi:10.1016/j.swevo.2015.10.004

Elgerd, O. I. (1971). *Electrical Energy systems theory: an Introduction* (2nd ed.). McGraw Hill.

Elgerd, O. I., & Fosha, C. E. (1970). Optimum megawatt-frequency control of diverse-area electric energy systems. *IEEE Transactions on Power Apparatus and Systems, PAS-89*(4), 556–563. doi:10.1109/TPAS.1970.292602

Farrokhabadi, M., Ca˜nizares, C., & Bhattacharya, K. (2016). Unit Commitment For Isolated Microgrids Considering Frequency Control. *IEEE Transactions on Smart Grid, 9*(4), 3270–3280. doi:10.1109/TSG.2016.2629982

Guha, Roy, & Banerjee. (2016). Load frequency control of large scale power system using quasi-oppositional grey wolf optimization algorithm. *Engineering Science and Technology, an International Journal, 19*(1), 1693–1713.

Hassan, H. A., & Zellagui, M. (2017). Application of Grey Wolf Optimizer Algorithm for Optimal Power Flow Of Two-Terminal HVDC Transmission System. *Advances in Electrical And Electronic Engineering - Power Engineering And Electrical Engineering, 15*(5), 701–712.

Jin, X., & Chen, X. Y. (2012). Study on the SMES Application Solutions for Smart Grid. *Physics Procedia, 36*, 902–907. doi:10.1016/j.phpro.2012.06.227

Kanellos, F. D., Tsouchnikas, A. I., & Hatziargyriou, N. D. (2005). Micro-Grid Simulation during Grid-Connected andIslanded Modes of Operation. *Proceedings of International Conference on Power Systems Transients (IPST'05).*

Khamis, Ghani, Kim, Kamarudin, & Aras. (2017). Voltage and Frequency Control of Microgrid Systems with Demand Response. *International Journal of Simulation: Systems. Science & Technology, 25*(1), 1–6.

Lazarov, Notton, Zarkov, & Bochev. (2005). Hybrid Power Systems with Renewable Energy Sources – Types, Structures, Trends For Research And Development. *Proceedings of Eleventh International Conference On Electrical Machines, Drives And Power Systems.*

Lee, D. J., & Wang, L. (2008). Small-signal stability analysis of an autonomous hybrid renewable energy power generation/energy storage system part I: Time-domain simulations. *IEEE Transactions on Energy Conversion, 23*(1), 311–320. doi:10.1109/TEC.2007.914309

Lin, X., & Lei, Y. (2017). *Coordinated Control Strategies for SMES-Battery Hybrid Energy Storage Systems* (Vol. 5). IEE Access.

Long, W., Jiao, J., Liang, X., & Tang, M. (2018). An exploration-enhanced grey wolf optimizer to solve high-dimensional numerical optimization. *Engineering Applications of Artificial Intelligence, 68*(2), 63–80. doi:10.1016/j.engappai.2017.10.024

Maiti, Acharya, Chakraborty, Konar, & Janarthana. (2008). Tuning PID and PIλDδ Controllers using the Integral Time Absolute Error Criterion. *Proceedings of 4th IEEE International Conference on Information and Automation for Sustainability, 516.*

Mirjalili, S., Mirjalili, S. M., & Lewis, A. (2014). Grey Wolf Optimizer. *Advances in Engineering Software, 69*(1), 46–61. doi:10.1016/j.advengsoft.2013.12.007

Padhy, Panda, & Mahapatra. (2017). A modified GWO technique based cascade PI-PD controller for AGC of power systems in presence of Plug in Electric Vehicles. *Engineering Science and Technology, an International Journal, 20*(2), 427-442.

Pandey, Irfaan, & Bhardwaj. (2014). Automatic Generation Control of Multi-Area Power System with Wind Power Generation. *International Journal of Science. Engineering and Technology, 2*(6), 1364–1369.

Parise, G., Martirano, L., & Kermani, M. (2017). Designing a Power Control Strategy in a Microgrid Using PID / Fuzzy Controller Based on Battery Energy Storage. *Proceedings of 2017 IEEE International Conference on Environment and Electrical Engineering.*

Prathibha, M., & Bhavani, M. (2014). Automatic Generation Control in Restructured Power System with Wind Integrated System. *Proceedings of International Conference on Innovations in Engineering and Technology (ICIET'14), Engineering and Technology.*

Rahimian, M. A., & Tavazoei, M. S. (2013). Improving integral square error performance with implementable fractional-order PI controllers. *Optimal Control Applications And Methods Wiley Online Library, 35*(3), 303–313. doi:10.1002/oca.2069

Ramakrishna, K. S. S., Sharma, P., & Bhatti, T. (2010). Automatic generation control of interconnected power systems with diverse sources of power generation. *International Journal of Engineering Science and Technology, 2*(5), 51–65. doi:10.4314/ijest.v2i5.60102

Rao & Mishra. (2014). Comparative study of P, PI and PID controller for speed control of VSI-fed induction motor. *International Journal of Engineering Development and Research, 2*(2), 2740–2744.

Sabah Nimma, K., Al-Falahi, M. D. A., Nguyen, H. D., Jayasinghe, S. D. G., & Mahmoud, T. S. (2018). Grey Wolf Optimization-Based Optimum Energy-Management and Battery-Sizing Method for Grid-Connected Microgrids. *Energies, 11*(4), 847. doi:10.3390/en11040847

Shayanfar, H. A., Shayeghi, H., & Younesi, A. (2015). Optimal PID Controller Design Using Krill Herd Algorithm for Frequency Stabilizing in an Isolated Wind-Diesel System. *Proceedings of Int'l Conf. Artificial Intelligence.*

Shen, J., Jiang, C., & Li, B. (2015). Controllable Load Management Approaches in Smart Grids. *Energies, 8*(1), 11187–11202. doi:10.3390/en81011187

Singh, S., Joshi, H., Chanana, S., & Verma, R. K. (2014). Impact of superconducting magnetic energy storage on the frequency stability of an isolated hybrid power system. *Proceedings of International Conference on Computing for sustainable global development,* 141-145.

Singh, S., Singh, M., Chanana, S., & Raveendhra, D. (2013). Operation and control of a hybrid wind-diesel-battery energy system connected to micro-grid. *Proceedings of International conference on control and automation, robotics and embedded system, 1*(6). 10.1109/CARE.2013.6733758

Singh, S., Singh, M., Chanana, S., & Semwal, S. (2015). Frequency regulation of isolated hybrid Wind/ Diesel, power generation with fuel cell system. In Lecture Notes in Electrical Engineering, (pp. 853–862). Springer Verlag Publication.

Singh, S., Singh, S. K., Chanana, S., & Singh, Y. P. (2014). Frequency regulation of an isolated hybrid power system with a battery energy storage system. In *Proceedings of Power and Energy Systems: towards Sustainable Energy (PESTSE), Amrita Vishwa Vidyapeetham Bengaluru Campus*. Bengaluru, India: Amrita School of Engineering. doi:10.1109/PESTSE.2014.6805286

Singh, S., Verma, R. K., Shakya, A. K., & Pratap Singh, S. (2017). Frequency Regulation of Microgrid Connected Hybrid Power System with SMES. *Technology and Economics Smart Grids Sustain Energy, 2*(1), 2–13. doi:10.100740866-017-0028-3

Teeparthi, K., & Vinod Kumar, D. M. (2016). Grey wolf optimization algorithm based dynamic security constrained optimal power flow. *Proceedings of National Power System Conference 2016*. 10.1109/ NPSC.2016.7858858

Tungadio, D. H., & Bansal, R. C. (2017). Active power reserve estimation of two interconnected Microgrids. *Energy Procedia, 105*, 3909–3914. doi:10.1016/j.egypro.2017.03.807

Uehara, T., Ludin, G. A., Datta, M., & Senjyu, T. (2015). Frequency Control using Real-Time Pricing for Isolated Power Systems. *Proceedings of 9th International Conference on Power Electronics-ECCE Asia*. 10.1109/ICPE.2015.7167929

Zargar, M. Y., Mufti, M. D., & Ahmed Lone, S. (2017). Voltage and frequency control of a hybrid Wind-Diesel system using SVC and predictively controlled SMES. *Proceedings of 6th International Conference on Computer Applications in Electrical Engineering-Recent Advances (CERA).*

Zhou, X., Xiao, Y. C., & Jian, X. J. (2011). Development of SMES Technology and Its Applications in Power Grid. *Proceedings of IEEE International Conference on Applied Superconductivity and Electromagnetic Devices, 260-269.*

Chapter 3
Application of Moth-Flame Optimization Algorithm for the Determination of Maximum Loading Limit of Power System:
Application of MFO for Maximum Loading Limit

Suvabrata Mukherjee
NSHM Durgapur, India

Provas Kumar Roy
https://orcid.org/0000-0002-3433-5808
Kalyani Government Engineering College, India

ABSTRACT

Moth-flame optimization algorithm (MFOA) based on the navigation strategy of moths in universe is a novel bio-inspired optimization technique and has been exerted for determining the maximum loading limit of power system. This process is highly effective for traversing long distances following a straight path. As a matter of fact, moths follow a deadly spiral path as artificial lights tend to confuse them. Exploration and exploitation are two vital aspects of the algorithm, used in tuning of the parameters. The algorithm is verified on MATPOWER case30 and case118 systems. Comparison of the performance of MFOA has been done with other evolutionary algorithms such as multi-agent hybrid PSO (MAHPSO), differential evolution (DA), hybridized DE, and PSO (DEPSO). The performance of MFOA in determining maximum loading limit is verified from the results. In much reduced time, MFO algorithm also gives high maximum loading point (MLP).

DOI: 10.4018/978-1-5225-8551-0.ch003

1. INTRODUCTION

Being considered as the biggest man-made gadget in the world, electric power grid comprises of transmission lines transformers, synchronous generators, active and reactive components, switches, relays, and loads. Being nonlinear and complex in nature, power systems are prone to faults and disturbances. Increased loading causes heavy stress on the power systems that may cause voltage instability problems. Numerous studies in recent times are being carried out for detection of loading limits of power systems related to security monitoring and voltage stability. Maximum loading limit (MLL) can be defined as the periphery between the maximum loading point (MLP) and operating point of the system (Ajjarapu and Christy, 1992; Irisarri et al., 1997).Various control parameters operation in such a system requires providing a solution to an optimization problem (Sato, 2002; Karysinos et al, 2005).Among some of the methods; the continuation power flow (CPF) is a widely used approach. However, with increase in step length, the method is inefficient in providing an accurate result. Another effective approach used for solving the maximum loading problem is the Interior point method (IP). However, there are limitations in initial and concluding conditions. In sequential quadratic programming (SQP), differentiation of the constraints is an important aspect. But calculation of large number of matrices during the iteration process makes the method complex and sluggish. An efficient approach that has been used for determination of maximum loading limit is the Fuzzy logic technique (Babulal et al., 2008). However, the results highlight its inefficiency to provide global optima.

Recently, many researchers have used evolutionary algorithms for determining the MLL. Not heavily afflicted by nonlinearity of the problem, the Particle swarm optimization (PSO) is an efficient computational intelligence-based approach. Because of these advantages, for most problems, PSO can converge to the optimal solution (del Valle et al, 2008). By fusing breeding of genetic algorithm, the performance of PSO is enhanced (EL-Dib Amgadet al, 2006) which has been again used for determining the MLL (EL-Dib Amgadet al, 2006). The article, however, has not considered the voltage limit. Also the hybrid optimization technique is unsuitable for large systems. Determining the optimum generator allocation at the MLP, MLL has been determined using the Multi agent hybrid PSO (MAHPSO) (Shunmugalatha et al., 2008) where till the voltage limits are breached, the load is increased uniformly in all the load buses. MLL is also determined using co-ordinated aggregation based particle swarm optimization (CAPSO) (Arya et al, 2010) and genetic algorithm (GA) (Acharjee, 2012). In this case the voltage stability limit where the voltage magnitude is much lower than the lower limit is assumed to be the maximum limit. MLL is also examined considering the dynamic security and steady state constraints (Gan et al., 2003).

A very successful evolutionary computation algorithm is the differential evolution algorithm (DEA). DEA has excellent convergence characteristics and is easy to understand and implement because of the fewer number of control parameters to be provided by the users (Shunmugalatha et al., 2008; Acharjee, 2012; Arya et al., 2010). Although DEA is studied to be fast, robust and precise, the faster convergence property sometimes causes premature convergence. Also there is a high probability in searching tendency in the direction of local optima.

In recent times, the economic load dispatch problem (ELD) and reactive power management problem respectively, has been solved using hybrid differential evolution algorithm (HDE) (Bhattacharya and Chattopadhyay, 2010; Sakr et al., 2017). In HDE, the search space is effectively explored and exploited by combining a hybrid migration operator with crossover mutation and selection operators of DE. However the results indicate that HDE suffers from heavy computational burden in case of large-scale problems.

Also most of the aforesaid methods are unable to satisfactorily explore the search space with increased system complexity because of increase in number of unknown variables.

In order to mitigate aforesaid drawbacks, MLL using moth flame optimization algorithm (MFOA) method, based on the navigation strategy of moths in universe, has been introduced in this article (Mirjalili, 2015). Obtaining global solution consistently highlights the usefulness of MFOA technique. Other positive features of this algorithm are the use of roulette wheel selection for achieving faster convergence and the ability of evaluation of continuous and discrete optimization issues. However, to the knowledge of the present authors, there is no previous study of maximum loadability using MFO in the optimization literature. The authors in the present paper have tried to develop MFOA technique that is reliable, robust, and versatile enough to determine the maximum loading limit of power system.

The remaining paper is organized as follows: Framework of MLL as an optimization problem is presented in Section 2. MFO is explained. MFOA in determining the maximum loading limit is presented in section 4. In section 5, the results on applying MFOA to different systems, is reviewed in order to verify the robustness of the algorithm. The conclusion is finally discussed in section 6.

2. FRAMEWORK OF MAXIMUM LOADING LIMIT AS AN OPTIMIZATION PROBLEM

Maximum loading limit in power system when structured as an optimization problem, determination of maximum load increase without violating the voltage limits is the prime objective. This can be represented as follows:

Maximize $f = \sum_i \lambda, i \varepsilon N_{PQ}$

Where the number of load buses is represented by N_{PQ}, subjected to following constraints:

1. 1. Equality Constraints:

$$0 = P_{Gi} - P_{Di} - V_i \Sigma V_j Y_{ij} \cos\left(\theta_{ij} + \delta_j - \delta_i\right); i \in N_0 \quad (1)$$

$$0 = Q_{Gi} - Q_{Di} + V_i \Sigma V_j Y_{ij} \sin\left(\theta_{ij} + \delta_j - \delta_i\right); i \in N_{PQ} \quad (2)$$

$$N_0 = N_{PQ} + N_{PV} \quad (3)$$

$$For\, i \in N_{PQ},\ P_{Di} = P_{Do} + \lambda \quad (4)$$

2. Inequality Constraints:

All PQ and PV bus voltage limit

$$V_j^{\min} \leq V_j^{\max}; j \in nb \tag{5}$$

PV bus real and reactive power limits

$$P_{gi}^{\min} \leq P_{gi} \leq P_{gi}^{\max} \tag{6}$$

$$Q_{gi}^{\min} \leq Q_{gi} \leq Q_{gi}^{\max} \tag{7}$$

In equations (1)-(7), N_{PV} denotes the number of PV buses, voltage magnitude at bus i is given by V_i, voltage angle at bus I is given by δ_i. Y_{ij} gives the magnitude of *i*th and *j*th element in bus admittance matrix. Matrix angle of element Y_{ij} in bus admittance matrix is given by θ_{ij}.Real power demand and real power generation at bus i is denoted by P_{Di} and P_{Gi} respectively. Reactive power generation and reactive power demand at bus i is given by Q_{Di} and Q_{Gi} respectively.

For all load buses, at any given generation, the real power demand is increased randomly. For this loading condition the load flow equations represented by equations(2) and (3),are determined by Newton Raphson approach. Increase in load is considered if voltage limits are maintained, else increase in load is 0.

3. MOTH-FLAME OPTIMIZATION ALGORITHM

Transverse orientation is a special movement mechanism employed by moths during their movement at night. As per moon's position during the night, moths fly by keeping a fixed angle accordingly. This technique proves to be highly conducive for the moths as it allows them to maintain a rectilinear motion particularly for remote source of light. The moths try to fly spirally around a nearer light source and finally converge towards the source after a few improvements. Primary constituents in the MFO algorithm being moths and flames are each conceived as results. However, they differ, during each iteration, in their updation process. In this algorithm, factual search agents are the moths moving in search space, whilst the best current search agent position is given by the flames. Therefore, by exploring around a flame every single moth is able to update itself to avail a superior solution. Hence, a moth is always at its finest location (Mirjalili, 2015). The matrix representation of a set of moths is represented by equation (8):

$$M = \begin{bmatrix} m_{11} & m_{12} \cdots m_{1(d-1)} & m_{1d} \\ m_{21} & m_{22} \cdots m_{2(d-1)} & m_{2d} \\ & \vdots & \\ & \vdots & \\ m_{n1} & m_{n2} \cdots m_{n(d-1)} & m_{nd} \end{bmatrix} \quad (8)$$

Where d gives the total variables (dimension) and total moths is given by n. The corresponding fitness values for all the moths are stored in an array, and is represented by:

$$OM = \begin{bmatrix} OM_1 \\ OM_2 \\ \vdots \\ OM_n \\ \end{bmatrix} \quad (9)$$

where the number of moths is given by n.

Similarly, the matrix representation of the flames is given by equation (10):

$$G = \begin{bmatrix} G_{11} & G_{12} \cdots G_{1(d-1)} & G_{1d} \\ G_{21} & G_{22} \cdots G_{2(d-1)} & G_{2d} \\ & \vdots & \\ & \vdots & \\ G_{n1} & G_{n2} \cdots G_{n(d-1)} & G_{nd} \end{bmatrix} \quad (10)$$

Similarly corresponding fitness values are stored in an array, given by:

$$OG = \begin{bmatrix} OG_1 \\ OG_2 \\ \vdots \\ OG_n \\ \end{bmatrix} \quad (11)$$

Equation (12) expresses the general design of MFOA including three-tuple estimation function:

$$MFO = (I, P, T) \quad (12)$$

I give the random population of moths. The complementary fitness value of moths is represented by Equation (13):

$$M(i,j) = [ul(i) - ll(i) * rand(i) + ll(i)] \quad (13)$$

Where ul denotes upper limit of the variables and the lower limit of variables is denoted by ll.

$$om = fitness\, function(m) \quad (14)$$

The function *P*, because of moths move around the search space, is run iteratively after initialization, till *T* (termination function)is reached. As the algorithm is established on the transverse orientation of the moths in space, the logarithmic spiral function is considered as the outline for the updated position of every moth with respect to the flame. With respect to a flame, the updated position of each moth is represented by Equation (15):

$$M_i = S(M_i, G_j) \quad (15)$$

Equation (16) represents the logarithmic spiral:

$$S(M_i, G_j) = D_i.e^{bt}.\cos(2\pi t) + F_j \quad (16)$$

Where the *i*th moth is indicated by M_i, F_j indicates the *j*th flame, and *i*th moth distance from *j*th flame is indicated by D_i. The adaptive convergence constant given by*r*, is used to accelerate the convergence around the flames representing, and is linearly decreased from -1 to -2 over the course of iterations. Smaller the value of *t*, lesser is separation from flame. Figure 4 in (Mirjalili, 2015) provides the updated positions of moths around the flame.

In final iterative steps, with respect to the best flame, position of the moths is updated. A few observations which are noteworthy in MFO algorithm are:

- If every moth is designated a flame with flame sequence being updated over the iterations, then increase in analysis of search space can be obtained. This assists in minimizing sluggishness of local optima.
- Convergence is assured as the moths never get trapped in sluggishness since the current available best solution provides a guiding path for them.
- Exploration and exploitation, by using an adaptive number of flames, can be balanced.

4. PROPOSED MFOA IMPLEMENTATION IN MAXIMUM LOADABILITY PROBLEM

MFOA based approach to maximum loadability problem, takes the procedure as follows:

Step 1: Test system data is interpreted.

Step 2: A set of random values for λ is initialize; $X_i = \left[_{ij}\right]$ for $i = 1$ to $N_p; j \in N_{PQ}$

Step 3: Set number of search agents, maximum iterations, maximum generation,, andmaximum and minimum boundary limits of the control variables.

Step 4: All the control variables are randomly initialized within some effective maximum and minimum control variable bounds.

Step 5: Iteration count is set at one (Itr = 1).

Step 6: Using (1) the objective function is calculated [$f\left(X_i^{Itr}\right)$]. NR power flow is run for at least five iterations checking the voltage limits for all PQ buses. If voltage limit violation is observed for any of the bus voltages, then $f\left(X_i^{Itr}\right) = 0$.

Step 7: Based on the $objective\,function$ value, sort the population and identify elite solutions. Elite term here is used for the solution sets, that provide best $objective\,function$ results. Without making any changes, the best solution sets are retained after individual iteration.

Step 8: Tuning process on control variables of non-elite solutions is carried out plausibly

Using Equations (8-12) and modification of bus voltages of the load buses along with voltage phase angle of all the buses except the slack buses are done using Equations (10-11).

Step 9: After each controlled variable is tuned, the feasibleness as a problem solution needs to be established. All feasible solutions are separated from those that are infeasible. This approach co-evolves the population of infeasible solutions until they become feasible.

Step 10: If maximum specified number of iterations reaches, show results; otherwise return to step 4.

5. RESULTS

MFOA has been implemented for solving the proposed maximum loading limit problem. Considering a flat start, MATPOWER 30 bus and MATPOWER 57 bus are two test systems on which the algorithms have been applied. From literature survey it is observed that compared to other evolutionary algorithms such asMAPSO, HPSO, MAHPSO gives superior results. Hence, the results attained using MFOA has been compared with results attained using DE and MAHPSO. A detailed study of DE and PSO is presented by (HaoZhi-Feng, 2007). For carrying out the simulation, both maximum iteration and population size are chosen as 50. The simulation studies have been carried out in Pentium Dual core, 1800 MHz system using MATLAB.

5.1 MATPOWER Case30 System (Zimmerman & Gan, 1997)

Initially, MFO algorithm is tested on MATPOWERcase30 system. The system comprises of 24 load buses, 6 generator buses and 41 branches of which 4 branches have tap setting transformers and 9 buses have shunt capacitors connected to it. 0.9 p.u.-1.05 p.uare the operating limits assumed for voltage magnitudes at PQ buses. The test data is given in Table A1 in appendix.

The maximum load of all the buses obtained using DEPSO (Gnanambal and Babulal, 2012) and MFO for the system are given in Table 1. For this case a population size of 30 is assumed. Results of (Gnanambal and Babulal, 2012) show the superiority of DEPSO as compared, PSO, MAPSO, HPSO and MAHPSO algorithms. Thus result for maximum loading at each bus has been correlated to results of DEPSO and is presented in Table 1. From the results it is also observed that the maximum loading in case of MFOA is greater than that of DEPSO. Depicted in Table 2, are the bus voltages along with phase angles obtained at this maximum loading condition for all buses.

Table 1. Maximum loading (PDmax) for case30 system using MFOA

Bus Number	Real Power Demand in p.u		Bus Number	Real Power Demand in p.u		Bus Number	Real Power Demand in p.u	
	DEPSO	MFO		DEPSO	MFO		DEPSO	MFO
2	0.2170	0.2956	11	0.0905	0.0250	21	0.2381	0.2625
3	0.1500	0.0120	12	0.1977	0.1629	22	0	0.0250
4	0.1199	0.1140	13	0	0.0250	23	0.0320	0.0480
5	0.1175	0.0250	14	0.0620	0.0930	24	0.1046	0.1305
6	0.0658	0.0250	15	0.2086	0.0500	25	0.0057	0.0250
7	0.2280	0.3420	16	0.0406	0.0501	26	0.0518	0.0450
8	0.3042	0.3601	17	0.0900	0.0766	27	0	0.0250
9	0.0007	0.0250	18	0.0459	0.0224	28	0.0044	0.0250
10	0.0619	0.0468	19	0.0950	0.1170	29	0.0240	0.0242
			20	0.0224	0.0123	30	0.1178	0.1574
Total maximum loadability				DEPSO		WOA		
				2.6974		2.7448		

The simulation is carried out for 50 iterations. Considering the best solution, convergence characteristics is obtained as shown in Figure 1. From the convergence characteristics curve it is observed that after 31 iterations there is a minimal transition in fitness values. A comparison of total maximum loading obtained by MAHPSO, DE, DEPSO and MFO is given in Table 3.

Table 2. Bus voltages and phase angles at maximum loading point for case30 system using MFOA

Bus Number	Bus Voltage in p.u	Phase Angle in deg	Bus Number	Bus Voltage in p.u	Phase Angle in deg	Bus Number	Bus Voltage in p.u	Phase Angle in deg
1	1.0000	0	11	0.9768	-0.1549	21	0.9928	-0.1735
2	1.0000	-0.0432	12	0.9847	-0.1251	22	1.0000	-0.1724
3	0.9755	-0.0767	13	1.0000	-0.0760	23	1.0000	-0.1406
4	0.9716	-0.0928	14	0.9739	-0.1445	24	0.9855	-0.1656
5	0.9756	-0.0932	15	0.9803	-0.1429	25	0.9865	-0.1546
6	0.9635	-0.1110	16	0.9762	-0.1485	26	0.9657	-0.1665
7	0.9562	-0.1195	17	0.9767	-0.1623	27	1.0000	-0.1355
8	0.9501	-0.1227	18	0.9684	-0.1640	28	0.9641	-0.1186
9	0.9768	-0.1494	19	0.9646	-0.1742	29	0.9738	-0.1674
10	0.9843	-0.1634	20	0.9688	-0.1720	30	0.9565	-0.1937

Figure 1.

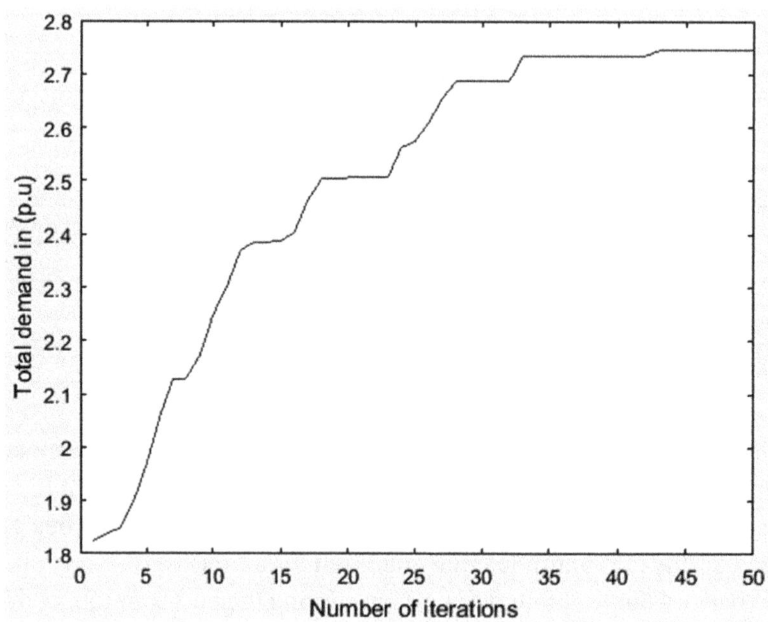

Table 3. Comparison of maximum loading limits $\sum P_{D_{\max}}$ in p.u

System	DE	MAHPSO	DEPSO	MFO
30 bus	2.6709	2.608080	2.6974	2.7448
118 bus	56.5430	56.450	57.0156	60.3075

5.2 IEEE 118 Bus System

Prospective MFOA is applied to MATPOWER 118-bus system so as to determine the optimization capability in solving bigger systems. The system comprises of54 generator buses,64 load buses, and 186 branches out of which there are9 branches are under load tap setting transformer along with 14 buses connected to shunt capacitors.0.9 p.u.-1.05 p.u are the operating limits assumed for voltage magnitudes at PQ buses. The test data is given in Table A2 in appendix. With an increased population size of 50, simulation is carried over variable ranges of 20independent trials considering 50 iterations for each trial. As mentioned in section 5.1, the performance of MFO has been compared with some other evolutionary algorithms (Gnanambal and Babulal, 2012). Table 3 the comparison between DEPSO and MFOis given. Maximum loading of 60.3075 p.u is attained by application of MFOA, whereas by the application of DEPSO, maximum loading obtained is 57.0156p.u. This clearly suggests of superiority of the proposed MFOA. Table 4 below gives the maximum loading for all buses using MFOA.

The characteristic curve is given in Figure 2. As observed from the figure, using MFOA the variation reduces significantly after 32 iterations and converges smoothly in the 41 iterations. The characteristic curve proves the robustness of proposed MFOA method for bigger systems. As convergence is achieved in reduced number of iterations therefore total computational time also reduces. Hence as compared to DE and MAHPSO, proposed MFOA is able to provide enhanced results for the maximum loading problem.

Table 4. The maximum loading (PDmax) for case118 system using MFOA

Sl No.	*PD*(p.u)	Sl No	*PD*(p.u)	Sl No	*PD*(p.u)	Sl No	*PD*(p.u)	Sl No.	*PD*(p.u)	Sl No	*PD*(p.u)	Sl No	*PD*(p.u)	Sl No.	*PD*(p.u)
1	0.7650	16	0.1554	31	0.6450	46	0.4200	61	0.0750	76	1.0200	91	0.1500	106	0.6450
2	0.3000	17	0.1650	32	0.5916	47	0.4909	62	1.1550	77	0.9150	92	0.9750	107	0.7500
3	0.5850	18	0.9000	33	0.1297	48	0.2580	63	0.0551	78	1.0650	93	0.1540	108	0.0300
4	0.5631	19	0.6497	34	0.8850	49	1.3050	64	0.0750	79	0.5850	94	0.4500	109	0.1200
5	0.0750	20	0.2700	35	0.4950	50	0.2550	65	0.0714	80	1.9500	95	0.6300	110	0.5045
6	0.7800	21	0.2100	36	0.4476	51	0.2550	66	0.5850	81	0.0750	96	0.5700	111	0.0715
7	0.2850	22	0.1500	37	0.0749	52	0.2700	67	0.4200	82	0.8100	97	0.2250	112	1.0200
8	0.4200	23	0.1050	38	0.0750	53	0.1428	68	0.0750	83	0.3000	98	0.5100	113	0.0900
9	0.0716	24	0.1950	39	0.4050	54	1.6950	69	0	84	0.1650	99	0.6300	114	0.1200
10	0.0250	25	0.0750	40	0.4164	55	0.9450	70	0.9194	85	0.1903	100	0.5550	115	0.3006
11	0.5055	26	0.0749	41	0.5550	56	1.2600	71	0.0750	86	0.3150	101	0.3300	116	2.7600
12	0.7050	27	0.7082	42	1.4400	57	0.1800	72	0.1800	87	0.0573	102	0.0645	117	0.2888
13	0.5100	28	0.2455	43	0.2477	58	0.1800	73	0.0792	88	0.6101	103	0.3450	118	0.4950
14	0.2100	29	0.3436	44	0.2010	59	4.1550	74	1.0200	89	0.0750	104	0.2746		
15	0.6664	30	0.0715	45	0.7950	60	1.1700	75	0.2504	90	2.4450	105	0.4650		
Total maximum loadability													60.3075		

**PD* – *Real power demand*

6. CONCLUSION

In the present article, a new population-based algorithm, MFOA, has been applied for determination of maximum loading limit of medium and large-scale power systems. For verifying the effectiveness and robustness of MFOA, the algorithm is applied on MATPOWER30-bus system and MATPOWER 118-bus system. Obtained results using MFOA has been compared with results obtained using DE, DEPSO and MAHPSO (Gnanambal and Babulal, 2012). It is observed from the results that MFOA delivers higher maximum loading limit with more accuracy compared to the other mentioned algorithms. Also with lesser number of iterations higher accuracy is achieved. The effectiveness and robustness of MFOA when applied to large systems is well observed from the characteristic curves in figure 2.

Figure 2.

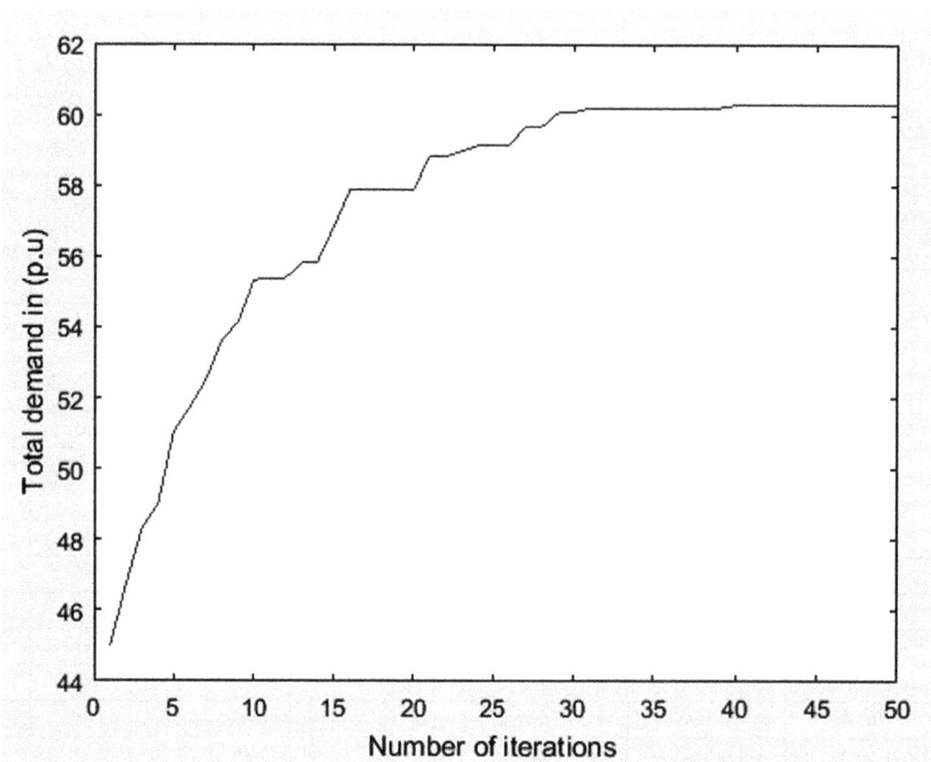

REFERENCES

Acharjee, P. (2012). Identification of maximum loadability limit and weak buses using security constraintgenetic algorithm. *International Journal of Electrical Power and Energy Systems*, *36*(1), 40–50. doi.org/10.1016/j.ijepes.2011.10.021

Ajjarapu, V., & Christy, C. (1992). The continuation power flow: A tool for steady state voltage stability analysis. *IEEE Transactions on Power Systems*, *7*(10), 304–311. doi:10.1109/59.141737

Anita, J. M., Babulal, C. K., Kannan, P. S., & Venkatesh, B. (2008). Determination of voltage Stability analysis limit using fuzzy logic. *International Journal of Power Energy Systems*, *28*(1), 3651–3692. doi:10.2316/Journal.203.2008.1.203-3651

Arya, L. D., Choube, S. C., Kothari, D. P., & Shrivastava, M. (2010). Loadability margin enhancement using co-ordinated aggregation based particle swarm optimization (CAPSO). *International Journal of Electrical Power & Energy Systems*, *32*(9), 975–984. doi:10.1016/j.ijepes.2010.02.007

Azmy, El-Sehiemy, & Sakr. (2017). Adaptive differential evolution algorithm for efficient reactive power management. *Applied Soft Computing*, *59*, 336–351. doi:10.1016/j.asoc.2017.01.004

Bhattacharya, A., & Chattopadhyay, P. K. (2010). Solving economic emission load dispatch problems using hybrid differential evolution. *Electrical Power and Energy Systems*, *11*(2). doi.org/10.1016/j.asoc.2010.09.008

Bhesdadiya, Parmar, Trivedi, & Jangir, Bhoye, & Jangir. (2016). Optimal Active and Reactive Power Dispatch Problem Solution using Whale Optimization Algorithm. *Indian Journal of Science and Technology*, *9*(1), 1–6. doi:10.17485/ijst/2016/v9iS1/101941

Bijwe, P. R., & Kelapure, S. M. (2003). Nondivergent fast decoupled power flow methods. *IEEE Transactions on Power Systems*, *18*(2), 633–688. doi:10.1109/TPWRS.2003.810905

Bouktir, T., Herbadji, O., & Slimani, L. (2013). BiogeographyBased Optimization Approach for Solving Optimal Power Flow Problem. *International Journal of Hybrid Information Technology*, *6*(5), 183–196. doi:10.14257/ijhit.2013.6.5.17

Del Valle, Y., Harley, R. G., Hernandez, J. C., Mohagheghi, S., & Venayagamoorthy, G. K. (2008). Particle swarm optimization: Basic concepts, variants and applications in power systems. *IEEE Transactions on Evolutionary Computation*, *12*(2), 171–195. doi:10.1109/TEVC.2007.896686

Duman, S., Güvenç, U., Sönmez, Y., & Yörükeren, N. (2012). Optimal power flow using gravitational search algorithm. *Energy Conversion and Management*, *59*, 86–95. doi:10.1016/j.enconman.2012.02.024

Dwivedi, S. K. (n.d.). *Load Modelling In Optimal Power Flow Studies*. Retrieved from http://ethesis@nitrkl.ac.in

El-Dib, El-Metwally, Hosam, Osrnan, & Youssef. (2004). Load flow solution using hybrid particle swarm optimization. *ICEEC 2004,International conference electrical electronic and computer engineering*, 742–746.

El-Dib, El-Metwally, Hosam, Osrnan, & Youssef. (2006). Maximum loadability of power systems using hybrid particle swarm optimization. *Electric Power Systems Research, 76*, 485–492.

Fukuyama, Y., & Yoshida, H. (2001). A particle swarm optimization for reactive power and voltage control in electrical power systems. *Proceedings of 2001 congress on evolutionary computation, 1*, 87–93. 10.1109/CEC.2001.934375

Gan, D., Qu, Z., & Wu, X. (2003). Loadability of power systems with steady-state and dynamic securityconstraints. *International Journal of Electrical Power & Energy Systems, 25*(2), 91–96. doi:10.1016/S0142-0615(02)00030-3

Irisarri, G. D., Mokhtari, S., Tong, J., & Wang, X. (1997). Maximum loadability of power systems Using interior point non-linear optimization methods. *IEEE Transactions on Power Systems, 12*(1), 162–172. doi:10.1109/59.574936

Karysinos, M. E., Nikolaidis, V. C., & Vorunas, C. D. (2005). Loadability limits and emergency Counter measures against voltage collapse. *Proceeding of the 15th Power system computation conference*, 22–26.

Lewis, A., Mirjalili, S., & Mirjalili, S. M. (2014). Grey wolf optimizer. *Advances in Engineering Software, 69*, 46–61. doi:10.1016/j.advengsoft.2013.12.007

Li, A., Law, M. Y., & Wong, K. P. (1997). Development of constrained genetic algorithm load flow method. *IEE Proceedings. Generation, Transmission and Distribution, 144*(2), 91–99. doi:10.1049/ip-gtd:19970847

Mary, R. (2008). Optimum cost of generation for Maximum loadability limit of power system using hybrid particle swarm optimization. *International Journal of Electrical Power & Energy Systems, 30*(8), 486–490. doi:10.1016/j.ijepes.2008.04.001

Mirjalili,S. (2015).Moth-flame optimization algorithm: A novel nature-inspired heuristic paradigm. *Knowledge-Based System*, 89, 228-49.doi.org/10.1016/j.knosys.2015.07.006.

Sato, H. (2002). Computation of power system loadability limits. *IEEE Conference on Transmission and Distribution, 3*(2), 1707–1011. 10.1109/TDC.2002.1177711

Slochanal & Shunmugalatha. (2008). Maximum loadability of a powersystem using multi agent-based hybrid particle swarm optimization. *Electric Power Components and Systems, 36*(6), 575–86. doi.org/10.1080/15325000701801579

Urdaneta, A. J., Gomez, J. F., Sorrentino, E., Flores, L., & Diaz, R. (1999). A hybrid genetic algorithm for optimal reactive power planning based upon successive linear programming. *IEEE Transactions on Power Systems, 14*(4), 1292–1298. doi:10.1109/59.801887

Venkatesh, P., Gnanadas, R., & Padhy, N. P. (2003). Comparison and application of evolutionary programming techniques to combined economic emission dispatch with line flow constrained. *IEEE Transactions on Power Systems, 18*(2), 688–697. doi:10.1109/TPWRS.2003.811008

APPENDIX

Table 5. Test data for Matpower case30 system

From Bus	To Bus	Line Impedance		Half Line Charging B in p.u.	From Bus	To Bus	Line Impedance		Half Line Charging B in p.u.
		R in p.u.	X in p.u				R in p.u.	X in p.u	
1	2	0.0200	0.0600	0.0300	15	18	0.1100	0.2200	0
1	3	0.0500	0.1900	0.0200	18	19	0.0600	0.1300	0
2	4	0.0600	0.1700	0.0200	19	20	0.0300	0.0700	0
3	4	0.0100	0.0400	0	10	20	0.0900	0.2100	0
2	5	0.0500	0.2000	0.0200	10	17	0.0300	0.0800	0
2	6	0.0600	0.1800	0.0200	10	21	0.0300	0.0700	0
4	6	0.0100	0.0400	0	10	22	0.0700	0.1500	0
5	7	0.0500	0.1200	0.0100	21	22	0.0100	0.0200	0
6	7	0.0300	0.0800	0.0100	15	23	0.1000	0.2000	0
6	8	0.0100	0.0400	0	22	24	0.1200	0.1800	0
6	9	0	0.2100	0	23	24	0.1300	0.2700	0
6	10	0	0.5600	0	24	25	0.1900	0.3300	0
9	11	0	0.2100	0	25	26	0.2500	0.3800	0
9	10	0	0.1100	0	25	27	0.1100	0.2100	0
4	12	0	0.2600	0	28	27	0	0.4000	0
12	13	0	0.1400	0	27	29	0.2200	0.4200	0
12	14	0.1200	0.2600	0	27	30	0.3200	0.6000	0
12	15	0.0700	0.1300	0	29	30	0.2400	0.4500	0
12	16	0.0900	0.2000	0	8	28	0.0600	0.2000	0.0200
14	15	0.2200	0.2000	0	6	28	0.0200	0.0600	0.0100
16	17	0.0800	0.1900	0					

Table 6. MATPOWER 118 bus-system test data

From Bus	To Bus	Line Impedance		Half Line Charging B in p.u.	From Bus	To Bus	Line Impedance		Half Line Charging B in p.u.	From Bus	To Bus	Line Impedance		Half Line Charging B in p.u.
		R in p.u.	X in p.u.				R in p.u.	X in p.u.				R in p.u.	X in p.u.	
1	2	0.0303	0.0999	0.0254	46	47	0.0380	0.1270	0.0316	79	80	0.0156	0.0704	0.0187
1	3	0.0129	0.0424	0.0108	46	48	0.0601	0.1890	0.0472	68	81	0.0018	0.0202	0.8080
4	5	0.0018	0.0080	0.0021	47	49	0.0191	0.0625	0.0160	81	80	0	0.0370	0
3	5	0.0241	0.1080	0.0284	42	49	0.0715	0.3230	0.0860	77	82	0.0298	0.0853	0.0817
5	6	0.0119	0.0540	0.0143	42	49	0.0715	0.3230	0.0860	82	83	0.0112	0.0367	0.0380
6	7	0.0046	0.0208	0.0055	45	49	0.0684	0.1860	0.0444	83	84	0.0625	0.1320	0.0258
8	9	0.0024	0.0305	1.1620	48	49	0.0179	0.0505	0.0126	83	85	0.0430	0.1480	0.0348
8	5	0	0.0267	0	49	50	0.0267	0.0752	0.0187	84	85	0.0302	0.0641	0.0123
9	10	0.0026	0.0322	1.2300	49	51	0.0486	0.1370	0.0342	85	86	0.0350	0.1230	0.0276
4	11	0.0209	0.0688	0.0175	51	52	0.0203	0.0588	0.0140	86	87	0.0283	0.2074	0.0445
5	11	0.0203	0.0682	0.0174	52	53	0.0405	0.1635	0.0406	85	88	0.0200	0.1020	0.0276
11	12	0.0060	0.0196	0.0050	53	54	0.0263	0.1220	0.0310	85	89	0.0239	0.1730	0.0470
2	12	0.0187	0.0616	0.0157	49	54	0.0730	0.2890	0.0738	88	89	0.0139	0.0712	0.0193
3	12	0.0484	0.1600	0.0406	49	54	0.0869	0.2910	0.0730	89	90	0.0518	0.1880	0.0528
7	12	0.0086	0.0340	0.0087	54	55	0.0169	0.0707	0.0202	89	90	0.0238	0.0997	0.1060
11	13	0.0222	0.0731	0.0188	54	56	0.0027	0.0095	0.0073	90	91	0.0254	0.0836	0.0214
12	14	0.0215	0.0707	0.0182	55	56	0.0049	0.0151	0.0037	89	92	0.0099	0.0505	0.0548
13	15	0.0744	0.2444	0.0627	56	57	0.0343	0.0966	0.0242	89	92	0.0393	0.1581	0.0414
14	15	0.0595	0.1950	0.0502	50	57	0.0474	0.1340	0.0332	91	92	0.0387	0.1272	0.0327
12	16	0.0212	0.0834	0.0214	56	58	0.0343	0.0966	0.0242	92	93	0.0258	0.0848	0.0218
15	17	0.0132	0.0437	0.0444	51	58	0.0255	0.0719	0.0179	92	94	0.0481	0.1580	0.0406
16	17	0.0454	0.1801	0.0466	54	59	0.0503	0.2293	0.0598	93	94	0.0223	0.0732	0.0188
17	18	0.0123	0.0505	0.0130	56	59	0.0825	0.2510	0.0569	94	95	0.0132	0.0434	0.0111
18	19	0.0112	0.0493	0.0114	56	59	0.0803	0.2390	0.0536	80	96	0.0356	0.1820	0.0494
19	20	0.0252	0.1170	0.0298	55	59	0.0474	0.2158	0.0565	82	96	0.0162	0.0530	0.0544
15	19	0.0120	0.0394	0.0101	59	60	0.0317	0.1450	0.0376	94	96	0.0269	0.0869	0.0230
20	21	0.0183	0.0849	0.0216	59	61	0.0328	0.1500	0.0388	80	97	0.0183	0.0934	0.0254
21	22	0.0209	0.0970	0.0246	60	61	0.0026	0.0135	0.0146	80	98	0.0238	0.1080	0.0286
22	23	0.0342	0.1590	0.0404	60	62	0.0123	0.0561	0.0147	80	99	0.0454	0.2060	0.0546
23	24	0.0135	0.0492	0.0498	61	62	0.0082	0.0376	0.0098	92	100	0.0648	0.2950	0.0472
23	25	0.0156	0.0800	0.0864	63	59	0	0.0386	0	94	100	0.0178	0.0580	0.0604
26	25	0	0.0382	0	63	64	0.0017	0.0200	0.2160	95	96	0.0171	0.0547	0.0147
25	27	0.0318	0.1630	0.1764	64	61	0	0.0268	0	96	97	0.0173	0.0885	0.0240
27	28	0.0191	0.0855	0.0216	38	65	0.0090	0.0986	1.0460	98	100	0.0397	0.1790	0.0476
28	29	0.0237	0.0943	0.0238	64	65	0.0027	0.0302	0.3800	99	100	0.0180	0.0813	0.0216
30	17	0	0.0388	0	49	66	0.0180	0.0919	0.0248	100	101	0.0277	0.1262	0.0328
8	30	0.0043	0.0504	0.5140	49	66	0.0180	0.0919	0.0248	92	102	0.0123	0.0559	0.0146
26	30	0.0080	0.0860	0.9080	62	66	0.0482	0.2180	0.0578	101	102	0.0246	0.1120	0.0294
17	31	0.0474	0.1563	0.0399	62	67	0.0258	0.1170	0.0310	100	103	0.0160	0.0525	0.0536
29	31	0.0108	0.0331	0.0083	65	66	0	0.0370	0	100	104	0.0451	0.2040	0.0541

continued on following page

Table 6. Continued

From Bus	To Bus	Line Impedance		Half Line Charging B in p.u.	From Bus	To Bus	Line Impedance		Half Line Charging B in p.u.	From Bus	To Bus	Line Impedance		Half Line Charging B in p.u.
		R in p.u.	X in p.u.				R in p.u.	X in p.u.				R in p.u.	X in p.u.	
23	32	0.0317	0.1153	0.1173	66	67	0.0224	0.1015	0.0268	103	104	0.0466	0.1584	0.0407
31	32	0.0298	0.0985	0.0251	65	68	0.0014	0.0160	0.6380	103	105	0.0535	0.1625	0.0408
27	32	0.0229	0.0755	0.0193	47	69	0.0844	0.2778	0.0709	100	106	0.0605	0.2290	0.0620
15	33	0.0380	0.1244	0.0319	49	69	0.0985	0.3240	0.0828	104	105	0.0099	0.0378	0.0099
19	34	0.0752	0.2470	0.0632	68	69	0	0.0370	0	105	106	0.0140	0.0547	0.0143
35	36	0.0022	0.0102	0.0027	69	70	0.0300	0.1270	0.1220	105	107	0.0530	0.1830	0.0472
35	37	0.0110	0.0497	0.0132	24	70	0.0022	0.4115	0.1020	105	108	0.0261	0.0703	0.0184
33	37	0.0415	0.1420	0.0366	70	71	0.0088	0.0355	0.0088	106	107	0.0530	0.1830	0.0472
34	36	0.0087	0.0268	0.0057	24	72	0.0488	0.1960	0.0488	108	109	0.0105	0.0288	0.0076
34	37	0.0026	0.0094	0.0098	71	72	0.0446	0.1800	0.0444	103	110	0.0391	0.1813	0.0461
38	37	0	0.0375	0	71	73	0.0087	0.0454	0.0118	109	110	0.0278	0.0762	0.0202
37	39	0.0321	0.1060	0.0270	70	74	0.0401	0.1323	0.0337	110	111	0.0220	0.0755	0.0200
37	40	0.0593	0.1680	0.0420	70	75	0.0428	0.1410	0.0360	110	112	0.0247	0.0640	0.0620
30	38	0.0046	0.0540	0.4220	69	75	0.0405	0.1220	0.1240	17	113	0.0091	0.0301	0.0077
39	40	0.0184	0.0605	0.0155	74	75	0.0123	0.0406	0.0103	32	113	0.0615	0.2030	0.0518
40	41	0.0145	0.0487	0.0122	76	77	0.0444	0.1480	0.0368	32	114	0.0135	0.0612	0.0163
40	42	0.0555	0.1830	0.0466	69	77	0.0309	0.1010	0.1038	27	115	0.0164	0.0741	0.0197
41	42	0.0410	0.1350	0.0344	75	77	0.0601	0.1999	0.0498	114	115	0.0023	0.0104	0.0028
43	44	0.0608	0.2454	0.0607	77	78	0.0038	0.0124	0.0126	68	116	0.0003	0.0040	0.1640
34	43	0.0413	0.1681	0.0423	78	79	0.0055	0.0244	0.0065	12	117	0.0329	0.1400	0.0358
44	45	0.0224	0.0901	0.0224	77	80	0.0170	0.0485	0.0472	75	118	0.0145	0.0481	0.0120
45	46	0.0400	0.1356	0.0332	77	80	0.0294	0.1050	0.0228	76	118	0.0164	0.0544	0.0136

Chapter 4
Probabilistic Power System Reliability Assessment:
Distributed Renewable Energy Sources

Oliver Dzobo
https://orcid.org/0000-0001-9602-6835
University of Johannesburg, South Africa

Kehinde O. Awodele
University of Cape Town, South Africa

ABSTRACT

This chapter presents the different dynamics in power system reliability as a result of the intrinsic behavior of distributed renewable energy sources. The output power of distributed renewable energy sources depends on the amount of available respective resource at any given time. This output power generally experiences fluctuations when compared with the output of conventional power generation units. The phenomenon is not usually included in traditional reliability worth evaluation methods for power system networks with distributed generation. In this chapter, a reliability worth evaluation model for power system networks with time-dependent distributed renewable generation resources is presented and analyzed. Time sequential Monte Carlo simulation technique is used, and the operational efficiency of the distributed generation unit is measured using the primary reliability worth index, ECOST. The derived index is fitted to a beta distribution function to show the inherent skewness of the supply reliability worth index.

INTRODUCTION

Power system reliability means the ability of a power system network to generate adequate amount of electricity for an intended period of time under encountered operating conditions to meet the electricity demand. Recently, there has been a growing dependence on safe and reliable electricity supply among electricity consumers. This has been coupled with the growing need to reduce carbon emissions and

DOI: 10.4018/978-1-5225-8551-0.ch004

effective asset utilization in power system management. The reduction of carbon emissions is achieved through the use of renewable energy sources (Tazvinga & Dzobo, 2018). Development of new commercial renewable energy technologies has reduced the cost of installing renewable energy sources compared to the conventional power grid. In most cases, renewable energy sources are implemented in remote areas where grid extension is nearly impossible (Tazvinga & Dzobo, 2018; Tazvinga, Xia, & Zhang, 2013). The increasing interest in the deployment of distributed renewable energy sources has therefore provided alternative energy-based solutions which are believed to play a vital role in smart grid and modern utility. However, large scale implementation of distributed renewable energy sources into the conventional power grid will pose great challenges to the design, operation and management of the new-look power grid as it changes from a 'passive' power system network to an automated 'active' power system network. Consequently, there is need for new approaches to the planning and operation of the power system network that enables greater amount of priority to be given to the challenges of an automated power system network.

This chapter presents a supply reliability worth evaluation methodology that realistically represents time-dependent distributed renewable generation (DRG) resources and explicitly examines the influence of supply-to-load correlation on supply reliability. Probability distributions are used to model component failure and repair in time-sequential Monte Carlo simulation, while historical data is used to model the time-dependent DRG source. A probabilistic approach is employed to evaluate the supply reliability worth by simulating hourly generation availability of a solar photovoltaic (PV) source, load profiles, transmission availability, and other factors to estimate a standard supply reliability worth index, ECOST. The impact of large scale deployment of distributed generation sources on the grid reliability, performance, and operation is also considered. The chapter also describes the distributed generation technology landscape, design challenges and a vision of the modern power utility.

BACKGROUND

The power system network is one of the most complex networks built by human beings. The main purpose of the power system network is to provide electricity whenever the connected consumers require it. As the connected customers have become more and more dependent on it, the reliability of the power system has become very important. The power system network can be subdivided into three levels, namely; generation, transmission and distribution. Power system reliability can be performed at any of these levels or a combination of the levels. This section will provide highlights of the recent developments of power systems that have taken place and are ongoing.

Power System Deregulation

Traditionally the power system network has been managed through a vertical organizational hierarchy with a central monopoly. As more energy producing players came into effect, there has been pressure for the power system network to be deregulated so as to allow competition. The decentralized power system network was perceived as being more efficient because of the competitive electricity markets that energy producing players were creating. This has led to many countries shifting from the centralized monopolistic policies to a more liberalized competitive market. As a result, in most countries the power generation and distribution system network have been privatized. Consequently, electricity has become cheaper as power generation companies compete, and electricity consumers are provided with options to buy electricity from cheaper sources. However, complications remain as the transmission sector is still regulated and relatively monopolized in order to protect the electricity consumers from paying exorbitant network operation costs. The energy regulators still have control over the investments made by the transmission sectors to avoid unnecessary investments in the power system network. During the last decade, energy players from distributed renewable power generation have increased and the energy sector has started to accept reasonable connections to its transmission and distribution system networks. Research studies have predicted that DRG being installed would contribute about 20% of power generation in the next decade (European Commission, 2003; Winkler, 2005). Carbon emission reduction, economic considerations and technological advancements are some of the accelerating factors for DRG inclusion. Governments have also come up with different programs that include increased incentives and/or rebates for generation of power from renewable energy sources that seek to relieve the power system network from peak load demands and more importantly reduce greenhouse gas emission. Supply reliability worth evaluation has become an integral part of the planning and operation of this new look power system network.

Distributed Generation

Traditionally, the conventional power system network has been very dependent on fossil fuels for its power generation. This is because fossil fuels can be easily stored and would allow for a stable and predictable generation of electricity. However, recent technological developments in renewable energy sources and pressure from environmental activists to reduce greenhouse gas emission have accelerated the gradual replacement of conventional power generators with renewable energy sources. Recently, a binding agreement was made among 37 industrialized nations around the world under the Kyoto protocol to limit the level of greenhouse gas emission. Table 1 below shows some European Union countries commitment for the reduction in greenhouse gases with a 1990 baseline level (European Commission, 2003). Developing countries such as South Africa have also started to implement different government programs in order to reduce the greenhouse gases (Department of Minerals and Energy Republic of South Africa, 2003; Winkler, 2005). The South African government predicted that there would be around 10 GW of power generation from renewable energy by 2030 (Department of Minerals and Energy Republic of South Africa, 2003). The large contributors of renewable power generation are going to come from wind and solar energy.

Technology advancement in renewable energy has allowed renewable power generation to be deployed across a wider geographical space. Distributed generation (DG) of power from renewable energy has been supported as it can be connected close to the electricity consumer and thus reduces the long distance transfer of power and thereby reducing the transmission power losses/costs. This is one of the economic benefits of DG which makes electricity price more affordable. Electricity consumers are able to produce their own electricity and can export excess power back to the transmission grid and sell it to system operators. This has changed the operational landscape in the power system network to be bi-directional. Electricity prosumers are more likely to participate in real-time electricity market due to their micro-power generation ability. This scenario has introduced more challenges to the transmission grid as it needs to be more flexible and reliable to accommodate the participation of the new players in the electricity market.

Table 1. Kyoto protocol commitments from EU member states

Kyoto Protocol – EU Member State Commitments	
Country	**Internal Commitment (% Change From 1990 Levels)**
Austria	-13
Belgium	-7.5
Denmark	-21
Finland	0
France	0
Germany	-25
Greece	+25
Ireland	+13
Italy	-6.5
Luxembourg	-28
Netherlands	-6
Portugal	+27
Spain	+15
Sweden	+4
United Kingdom	-12.5

(European Commission, 2003)

Much attention has been given to microgrids as an alternative or complement to large centralized power system networks (Cabellero, Sauma, & Yanine, 2013; Mohamed & Kovio, 2010; Papaefthymiou, Schavemaker, van der Sluis, Kling, Kurowicka, & Cooke, 2006; Sun, Bollen, & Ault, 2006; Tan, Li, Wang, 2013). A microgrid can be defined as a section of a distribution system network that contains DRG and can be isolated from the rest of the power system network in the event of an upstream fault. Its ability to increase supply reliability for electricity customers depends on the sufficiency of the DRG resources and whether or not the local conventional generation units can quickly recover or even ride-through a nearby fault (Bae, Kim, Kim, & Singh, 2004; Barker & De Mellow, 2007; Hegazy, Salama, & Chikhani, 2003). Extensive literature on supply reliability worth evaluation of power system networks with DRG can be found although the information has not yet been formalized or standardized. Previous studies have illustrated substantial supply reliability worth benefits of DRGs connected to the power system network (Bae, Kim, & Kim, 2007; Hegazy, Salama, & Chikhani, 2003; Moradi, & Abedini, 2012; Niknam, Golestaneh, & Malekpour, 2012). The DRGs are usually considered to be dispatchable resources or stochastic but with sufficient energy storage to make it essentially dispatchable. These studies have focused primarily on supply reliability impacts as a function of the DRG size, location, and operating strategy (Bae, Kim, Kim, & Singh, 2004; Bae, Kim, & Kim, 2007; Barker & De Mellow, 2000; Caballero, Sauma, & Yanine, 2013; Morado & Abedini, 2012; Niknam, Golestaneh, & Malekpour, 2012; Zeinalzadeh, Mohammadi, & Moradi, 2015). There have been relatively fewer studies that have examined microgrids with purely time-dependent DRG, such as solar, even though this type of time-dependent DRG may become the most prevalent and practical type of DRG in power system networks (Bae, Kim, Kim, & Singh, 2004; Yang, Lu, & Burnett, 2003). The fact that the hourly solar irradiation incident on PV module at a given location is a function of time of day and time of the year means that the fraction of the load supplied by the PV module is not constant. The available power at any point in time may therefore not satisfy all loads. The dynamic reconfiguration of the microgrid network during these times must therefore be included in the supply reliability worth evaluation. In this chapter, a reliability evaluation model which considers prioritized load restoration with dynamic reconfiguration is examined. The time dependent DRG considered is capacity limited and energy storage is not considered. To measure the benefits of supply reliability worth due to inclusion of DRG in the power system networks, several indices have been proposed e.g. ECOST, IEAR etc. (Billinton & Allan, 1996; Cross, Herman, & Gaunt, 2006; Dzobo, Gaunt, & Herman, 2011). The selection and definition of the indices is dependent on the methodology and purpose of the study (Billinton & Allan, 1996; Cross, Herman, & Gaunt, 2006; Dzobo, Gaunt, & Herman, 2011; Edimu, Gaunt, & Herman, 2011). In this chapter, the supply reliability worth index, ECOST, is computed.

Several techniques have been developed for use in supply reliability worth evaluation of DRG application in power system networks. The techniques can be grouped as either deterministic or probabilistic (Bae, Kim, Kim, & Singh, 2004; Bae, Kim, & Kim, 2007; Barker & De Mello, 2000; Caballero, Sauma, & Yanine, 2013; Hegazy, Salama, & Chikhani, 2003; Moradi & Abedini, 2012; Niknam, Golestaneh, & Malekpour, 2012; apaefthymiou, Schavemaker, van der Sluis, Kling, Kurowicka & Cooke, 2006; Sun, Bollen, & Alut, 2006; Zeinalzadeh, Mohammadi, & Moradi, 2015). In standard deterministic techniques, the failure and repair of the various components are frequently considered independent of each other and of the load (Billinton & Allan, 1996; Cross, Herman, & Gaunt, 2006; Dzobo, Gaunt, & Herman, 2011; Edimu, Gaunt, & Herman, 2011). The techniques use mathematical models that use average input parameter values. These techniques are limited for the work presented in this chapter because it is almost impossible to apply them when there are multiple time-dependent processes and non-constant input parameters that need to be considered. Probabilistic techniques have the advantage that they can account for the stochastic nature of the power system network components (Cabral, Filho, Diniz, Martins, Toledo, & Neto, 2010; Cross, Herman, & Gaunt, 2006; Edimu, Gaunt, & Herman, 2011; Hegazy, Salama, & Chikhani, 2003; Papaefthymiou et al., 2006; Tina & Gagliano, 2011). The main probabilistic technique is simulation. The time sequential Monte Carlo simulation technique plays an important role in the work proposed in this chapter as it preserves the correlation between the time-dependent DRG power output and the load in a chronological order. Thus, at any point in time when the DRG may not satisfy all loads, a network reconfiguration which will alter the connectivity between the loads and the time-dependent DRG is effected according to some restoration priority. A load restoration order that is sensitive to the dynamic reconfiguration of the microgrid network is therefore specified and included in the supply reliability worth evaluation.

Monte Carlo Simulation Techniques

Monte Carlo simulation technique is a random sampling process which can be categorized into two, namely; non-sequential and time-sequential Monte Carlo simulation. The non-sequential MC simulates the basic intervals of the system lifetime by randomly choosing the intervals while time-sequential MC simulates the basic intervals of the system lifetime in a chronological order. Both non-sequential and time-sequential MC simulation techniques are dependent on the number of simulations conducted. For very small number of simulations both techniques produce poor estimate of probability, however, as the number of simulations increases the estimation of probability values converges toward true values.

Therefore, by increasing the number of simulations both techniques will eventually arrive at a point of convergence which indicates that most of the relevant system states have been sampled. The expected value of the sampled index will be obtained and thus provide an accurate picture of the index. The techniques will thus have the advantage of providing more useful information about the index as it will show the average and extreme values of the index. To capture this information usually a probability distribution function (PDF) is used.

Non-Sequential Monte Carlo Simulation Technique

The application of non-sequential Monte Carlo simulation technique in power system reliability analysis is very simple as it neglects the chronological order of events in the operation of the power system network. The statuses of the different components in the power system and its power supply status is discretized. The random combination of their status is simulated and an average value of the measured index is obtained. Commonly, a random probability value between 0 and 1 is sampled from a uniform distribution function for each power system network component. The unavailability U which is the failure probability of a component is also simulated using the assumed states of each component. For example, assume a two-state component (UP and DOWN states), the probability of failure is defined by Equation 1 below.

$$U = \frac{\lambda}{\lambda + \mu} \tag{1}$$

where λ and μ are the failure and repair rates of the component respectively. If the random probability value of the component is greater than its respective U, the component is considered to have failed. This is performed for all the components considered in the simulation and a combination of their resultant states is used to define the system state. The simulation is repeated for all possible combinations until the result converges and a reliability index value is evaluated. This simulation technique is a very simple process utilizing a uniform probability distribution, it has a shorter computational time and therefore requires smaller computer memory. The major drawback of this technique is that it does not take into account the impact of failure state transitions and the transition of a chronological load model. Each sampling of the states of components and load model is independent of each other and the transition of time among the samples is not considered. Hence, it is possible that crucial events such as when loss of load occurs are not sampled by the non-sequential Monte Carlo simulation technique.

Time-Sequential Monte Carlo Simulation Technique

In the time-sequential Monte Carlo simulation technique each component state is assumed to follow a probability distribution. The probability states of each component are derived from the respective probability distribution that represent the time-to-failure (TTF) and time-to-repair (TTR) of each component. The TTF is the duration of a component in the Up state before its transition into the failure state and TTR is the duration that is required to restore it to normal condition once its failure state is encountered. Each component can have as many states as possible and is dependent on the state duration distribution function assumed. Commonly the exponential probability distribution is assumed. However, probability distributions like Weibull, gamma and lognormal probability distribution function can also be used. The following are the steps that are considered when performing the time-sequential Monte Carlo simulation.

Step 1: All system components are initialized in the normal operating state.
Step 2: The duration of each system component to remain in its present state is sampled from the assumed respective probability distribution function. If an exponential distribution function is assumed for each component, the pdf of an exponential distribution given by Equation 2 below can be used to obtain an expression for the TTF as follows.

$$f(t) = \lambda e^{-\lambda t} \tag{2}$$

The cumulative probability distribution for the exponential distribution (Equation 2) is

$$U = F(t) = 1 - e^{-\lambda t} \tag{3}$$

where U is a uniformly distributed random number sampled from the uniform distributed probability distribution function.

Using the inverse transform method, the random TTF is given by:

$$TTF = -\frac{1}{\lambda} In(1 - U) \tag{4}$$

Since (1-U) distributes uniformly as U in the interval [0,1], Equation 4 can be simplified as:

$$TTF = -\frac{1}{\lambda} In\,(U) \tag{5}$$

λ is the failure rate of the component when the present state of the component is assumed to be UP state, or the contrary, if the *TTR* the of the component when the assumed state of the component is in the DOWN state is exponentially distributed, *TTF* can be replaced with *TTR* in equation (5) and λ replaced with the repair rate μ.

Step 3: The process in *Step 2* is repeated for all system components over the defined sampling period. The chronological state transitions of the system components are combined to obtain the resultant chronological state of the system.
Step 4: The system is analyzed for each system state to obtain the associated index.
Step 5: The expected index value is obtained by averaging the index values of all the system states. To get a full view of the index usually a probability distribution function is used to model the index from all the values obtained in *Step 4*.

As compared to the non-sequential Monte Carlo simulation technique, the time-sequential Monte Carlo simulation technique does not randomly choose the combination of state of the components to obtain the resultant system state. It considers all possible events from the beginning to the end in a chronological manner and thus requires less simulation iterations to converge. However, the technique requires more computational memory and time as it needs to store all the chronological behavior of the system components over a long period of time. Due to recent high technological advancement in computing technologies, this technique can be developed in high speed environments and results can be obtained much faster.

Rate of Convergence

Assume a vector of system state, $L(\psi)$, the expected reliability value, E(L), of this particular vector is expressed as in Equation 6 below.

$$E(L) = \sum_{\psi \in S} L(\psi) p(\psi) \tag{6}$$

where p is the probability of occurrence of the system state and S is the set of all possible system states. Using the Monte Carlo simulation technique, this expected reliability value can be estimated using the following formulation in Equation 7 below:

$$E(L) = \frac{1}{N} \sum_{\psi \in S} L(\psi) n(\psi) \tag{7}$$

where n is the number of occurrences of the system state and N is the total number of simulations needed to achieve the expected reliability value. From Equation7 it can be clearly seen that the rate and degree of confidence of convergence is dependent on the total number of system samples. E(L) from Equation 7 is a mean value obtained from random values and thus the statistical sample variance, V(X), is defined by Equation8 below.

$$V(X) = \frac{1}{N} \sum_{i=1}^{N} (X_i - E(L))^2 \tag{8}$$

where X_i is the recorded index value in a sampled state i.

The level of accuracy is determined by the standard error, η, which is defined by Equation9 below:

$$\eta = \sqrt{\frac{V(x)}{N}} = \frac{\delta}{\sqrt{N}} \tag{9}$$

δ is the standard deviation of the statistical sample. Convergence is attained when the standard error, η, drops below a specified fraction, ε, as in Equation10 below.

$$\eta \leq \varepsilon \tag{10}$$

This condition is used as the stopping criterion that dictates the end of the simulation once it is achieved. Commonly, the general consensus among the power system reliability study community is to use a convergence point of 5% or below as adequate in determining the true estimate values of power system reliability indices (Bae, Kim, Kim, & Singh, 2004; Zeinalzadeh, Mohammadi, & Moradi, 2015).

Application of PDFs in Power System Reliability Worth Assessment

The derived power system reliability index from Monte Carlo simulation techniques is a stochastic value which is dependent on the power system network topology, operating philosophy and conditions. Commonly, the resultant index is presented as an average value. The average value of the index shows how reliable the power system network is on average. However, power system operators and managers are more concerned with the extreme values at given risk levels as this would help them to prepare for any consequences that might occur in the system network. It is thus imperative to investigate the extreme values of the derived index. This can only be achieved by modeling the reliability index using a probability distribution function. In order to investigate the extreme values of the index, the shape and skewness of the distribution become very important and thus, the type of probability distribution function used to model the reliability index must be carefully selected. Different probability distribution functions have different properties when it comes to the modeling of different data sets. Most of the probability distributions are limited in the shapes they can exhibit, and therefore are only applied to specific data sets. It thus means that, certain probability distribution functions might be able to fit very well to a specific data set of the resultant reliability worth index and at the same time poorly fit to other data sets of the same reliability index. In order to avoid use of different PDFs in a single simulation of the power system, the PDF of choice must therefore be very versatile and should be able to fit well with a range of different data sets. Gamma and Weibull PDFs have these properties, however, they have a limitation that they do not have a finite range which is a property inherent in the beta PDF. The Beta PDF has been extensively used in power system reliability analysis and its appropriateness to model power system reliability indices has been confirmed. The Beta PDF is very versatile in the shapes it takes given different values of its two shape parameters, α and β. The two shape parameters are defined in Equations 11 and 12 below.

$$\alpha = \frac{C\mu - \mu^2 - \delta^2}{C\delta^2} \tag{11}$$

$$\beta = \frac{(C - \mu)(C\mu - \mu^2 - \delta^2)}{C\delta^2} \tag{12}$$

Figure 1 below shows some of the shapes that can be obtained by varying the values of α and β.

Figure 1. Different shapes shown by Beta PDF for different values of α and β

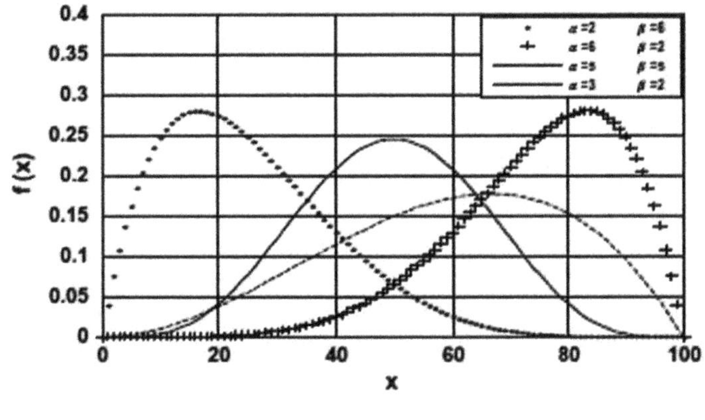

For more information on the beta distribution and how to derive the above equations the reader is referred to Cross, Herman, & Gaunt (2006). The beta PDFs can be used to provide the input data and output for reliability-worth analysis as illustrated in the following small case study.

TIME-DEPENDENT DISTRIBUTED RENEWABLE GENERATION MODEL

Solar PV generation is an established clean technology (Bizon, 2013; Cabral et al., 2010; Dursun & Kilic, 2012; Singh, 2013; Tina & Gagliano, 2011). On-grid solar PV-based generation units are being deployed globally to provide additional power for the relief of the peak demand during the day (Caballero, Sauma, & Yanine, 2013; Tan, Li, & Wang, 2013; Tazvinga, Xia, & Zhang, 2013; Yang, Lu, & Burnett, 2003). The solar PV array power output varies with time of the day, season and weather conditions (Collares-Pereira & Rabl, 1979; Hove, 2000; Tazvinga, Xia, & Zhang, 2013). Therefore, compared with conventional generation units, the power produced by the PV array may experience more fluctuations. The power output of a solar PV panel is obtained by utilizing the solar irradiance-power output characteristic of photovoltaic panels (Bizon, 2013; Hove, 2000; Singh, 2013; Tazvinga, Xia, & Zhang, 2013).

Solar Irradiance and Solar PV Panel Power Output Models

Solar irradiance can be modeled in different ways depending on the application and the frequency of observation (Hove, 2000). Extensive studies have been carried out to investigate the variability of solar irradiance (Bizon, 2013; Hove, 2000; Singh, 2013). Solar radiation has daily cycles and seasonal cycles due to the rotation and translation movement of the earth around the sun. The solar radiation for a specific location depends on the direct and scattered radiation as well as the clearness of the sky (Hove, 2000). The clearness is the degree of clouds blocking the direct radiation that can contribute to the scattered radiation. Hourly generation of solar irradiance depends on the cyclical periods of the solar irradiance. Figures 2 – 4 show the solar irradiance according to time of day and month for Cape Town (South Africa) location. The figures show that for each different month and time of day there is variation in the amount of solar irradiance received. The highest solar irradiance is received in summer between October – February. The highest solar irradiance of 1133W/m^2 was recorded in December.

Figures 2 and 3 shows solar irradiance for two different days in January. It can be clearly seen that there is significant difference in the solar irradiance received on the two days. The highest solar irradiance recorded on 04 January 2005 is 900W/m^2 at 1000hrs while a high of 850W/m^2 is recorded at 1300hrs on 07 January 2005. This implies that different times of the day receive different levels of solar irradiance and thus it is important to include this phenomenon in the solar PV panel power output model.

Lowest solar irradiance levels are experienced during the winter season between May – July. Figure 5 shows the hourly solar irradiance received on 04 June 2005. The maximum solar irradiance recorded is 598 W/m^2 at 1100hrs.

This variability has become an important issue in defining the operating policies to be adopted by power system operators and planners when accepting high solar PV penetration into the conventional power system network. The combination of solar panels can be represented as a single unit with single power output characteristic. The solar irradiance-power output characteristic can be represented by the expressions in Equation 13 shown below

$$PV_{output} = \gamma_{PV} I\, A_{PV} \tag{13}$$

where γ_{PV} is the efficiency of the PV array and corresponding converters, A_{PV} is the PV panel area (m^2), PV_{output} is the PV array output power (W), and I is the horizontal solar irradiance (W/m^2).

In this chapter, it is assumed that γ_{PV} is constant and equal to 12.3%. The power output is an hourly profile and is used as a distributed renewable generation point for supply reliability worth calculations. In this study, real time solar irradiance data from Solar Energy Service for Professionals (n.d.) have been used. The period of analysis is January - December 2005. Figure 6 shows the distribution of solar irradiance-power output calculated over all hours of the year 2005 for Cape Town (Western Cape, South

Figure 2. Variation of solar irradiance with time of day – 30 December 2005

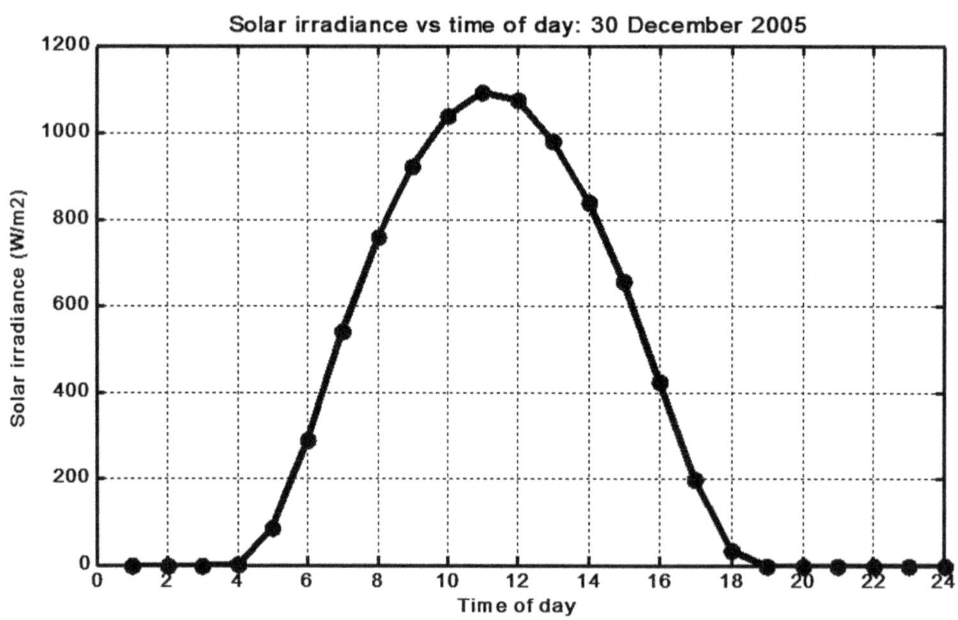

Figure 3. Variation of Solar irradiance with time of day – 04 January 2005

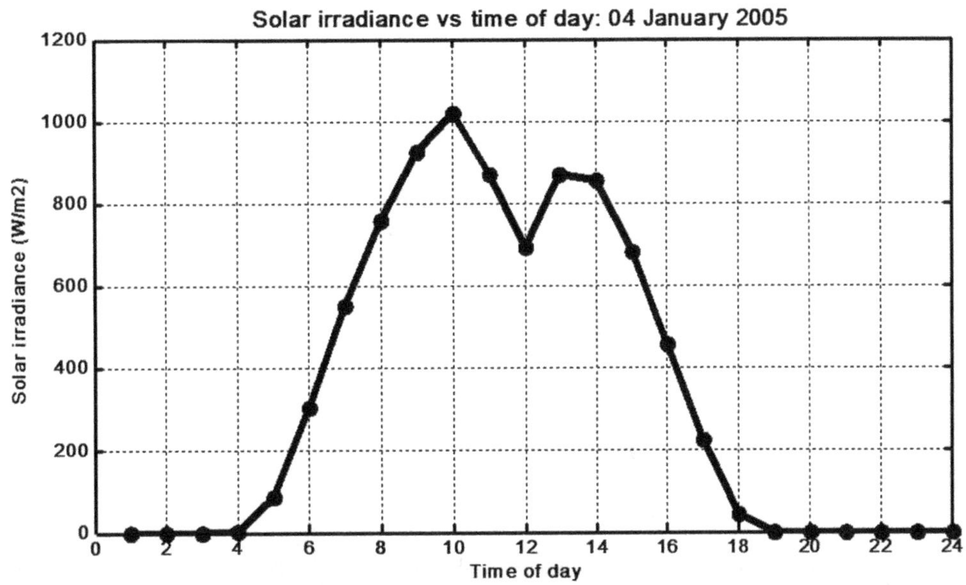

Figure 4. Variation of solar irradiance with time of day – 07 January 2005

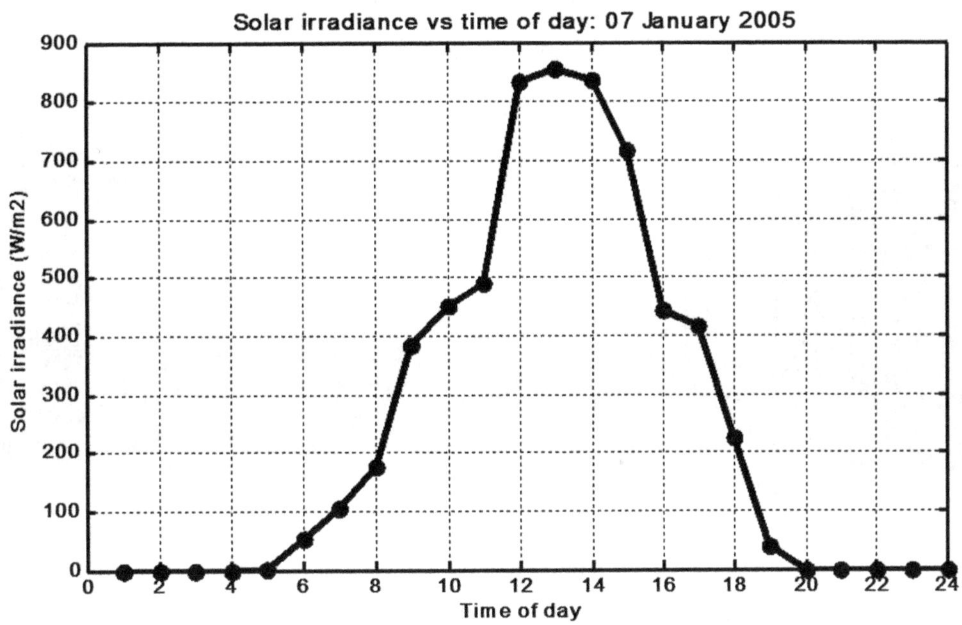

Africa, longitude 18.92, latitude -33.24) location. There are 8 760 dots in Figure 6, each representing the calculated value of solar irradiance-power output for an hour of the year.

Figure 5. Variation of solar irradiance with time of day – 04 June 2005

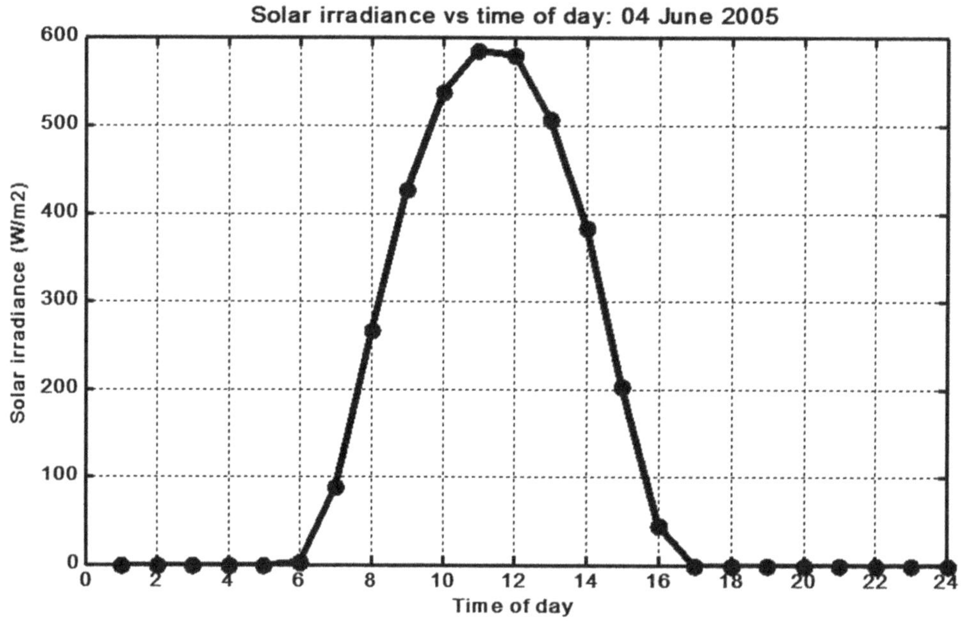

Figure 6. Variation of PV solar irradiance-power output for the whole year

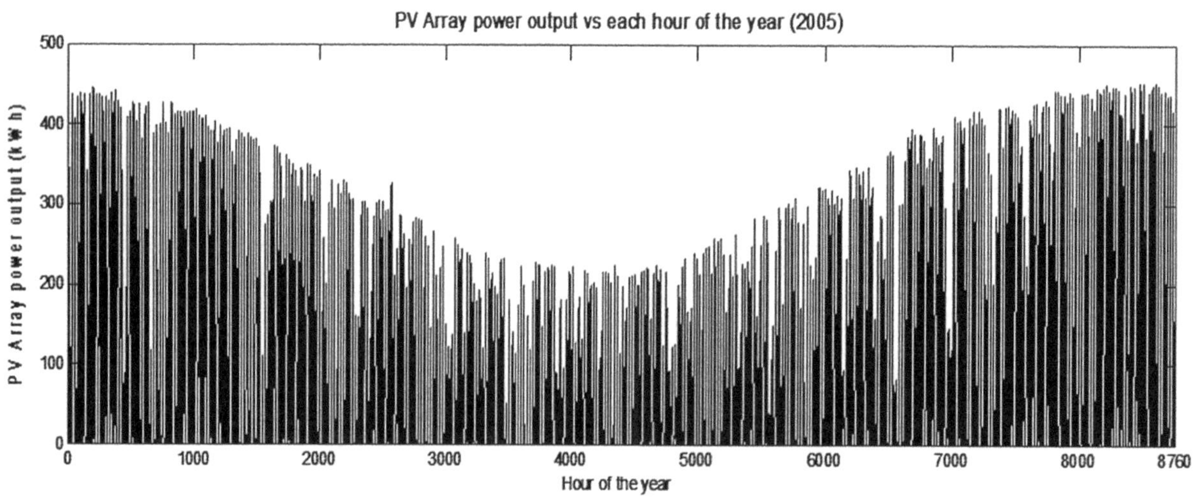

Test System and Case Study

The reliability test system network chosen is the IEEE RBTS Bus 2. In order to analyse the influence of solar PV distributed generation on the distribution system network correctly, it is necessary to place the test system network on a location where historical data is available. The advantage of this system is the diversity of load on each feeder, which is common on distribution system networks, and it is has a radial

Figure 7. One line diagram of the RBTS - Bus 2

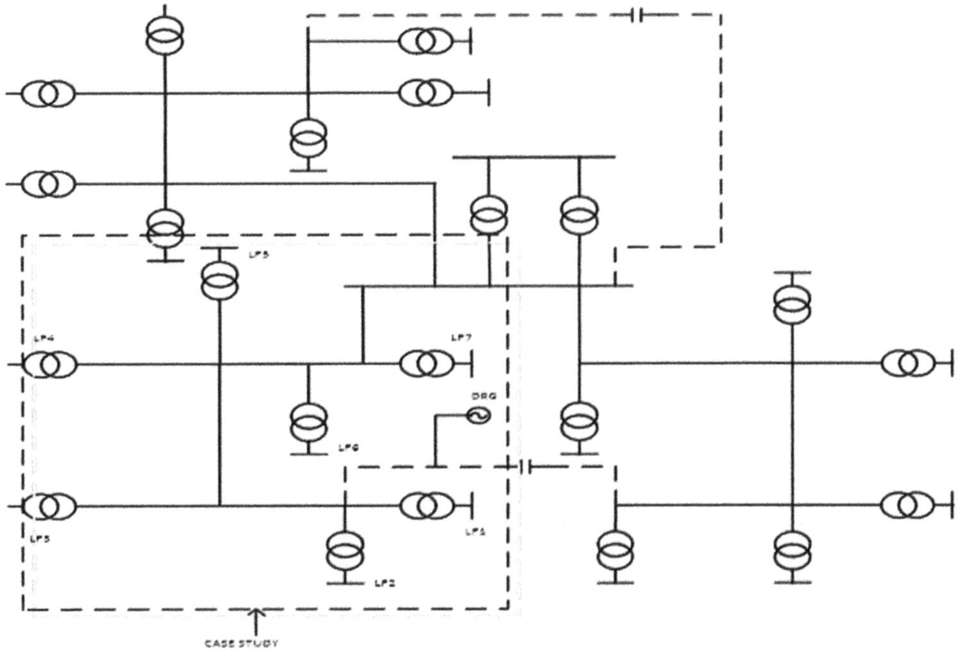

topology. It also contains substation components as well as distribution transformers. Thus, the selection is convenient and provides an excellent base case from which further investigation should be done.

The time-dependent solar PV distributed generation unit is installed in the test distribution system network in backup mode, supplying total or partial load during failures. In the base case scenario, with no solar PV distributed generation unit, some load points lose supply completely depending on the topology and reconfiguration procedures. The solar PV distributed generation is introduced to alleviate and shorten as much as possible the interruption period, hence improves the network supply reliability. In the time spanned from the breaker trip to the repaired state (initial condition), the solar PV distributed generation has the task to supply as much load as it can in order to shorten the duration of blackouts. Once the repair has been performed and the network returns to its initial condition, some loads might not have experienced any switching, thus reliability could improve avoiding unnecessary interruptions for them.

Components Reliability Model

The component reliability data used in the analysis is presented in Table 1 and is taken from Dzobo, Gaunt, Herman (2011). The probability distribution function (PDF) for the time to failure was assumed to be exponential with the failure rate given in Table 1. The failure rate of overhead lines is calculated at 0.065 failures per year per kilometre. The PDF for load point interruption durations (repair/replacement time, switching time) is assumed to be lognormal (Billinton & Allan, 1996; Dzobo, Gaunt, & Herman, 2011). The disconnection assets (circuit breakers, switches etc) are considered highly reliable. This assumption is realistic with exceptional cases of mis-operation or severe faulty conditions. The

Table 2. Reliability parameters

Component	Description	Failure Rate per Year (f/yr)	Replacement/Repair Time (hr)	Switching Time (hr)	Probability Distribution
Solar PV array	-	0.010	2.0 (0)	1.00 (0.40	Lognormal
Transformer	11/0.4kV	0.015	10.0 (1)	1.00 (0.4)	Lognormal
Breaker	11kV	0.006	4.0 (0.4)	1.00 (0.4)	Lognormal
Transmission lines	L1 and L8	0.065	5.0 (1)	1.00 (0.4)	Lognormal
	L2, L3 and L10	0.182	5.0 (1)	1.00 (0.4)	Lognormal
	L4 and L7	0.117	5.0 (1)	1.00 (0.4)	Lognormal
	L5, L6 and L9	0.0975	5.0 (1)	1.00 (0.4)	Lognormal

power supply at 33 kV is available at all times and provides the demand as required. It is assumed that the protection systems have been coordinated adequately and the transfer switches and power electronic inverters of the solar PV distributed generation unit are 100% reliable. No energy storage system is considered in the analysis.

Customer Interruption Cost Model

Customer interruption cost (CIC) models are utilized to determine the supply reliability worth of power system networks. The CIC model used in the analysis is obtained from the Cape Town customer survey presented in Dzobo, Gaunt, & Herman (2012a) and Dzobo, Gaunt, & Herman (2012b). The direct cost estimate method was used in collecting interruption costs. The approach uses direct cost estimate questions to ask respondents to estimate their financial losses for a given power interruption duration that varies with time of day, day of week and season. Table 3 indicates the interruption cost estimates for each customer type and interruption duration. The cost for any given duration is estimated from the measured durations using linear interpolation. These cost estimates are normalized using average monthly energy costs.

Table 3. Customer interruption cost data for different customer segments

Customer Segment	Duration (hr)			
	1	2	4	8
Garage	0.74 (0.64)	1.56 (1.23)	3.06 (2.76)	4.82 (3.66)
Retail	0.82 (0.53)	1.02 (0.50)	1.72 (1.12)	5.52 (3.67)
Metal	4.65 (6.37)	8.77 (12.66)	14.19 (13.08)	22.01 (33.62)
Clothing	2.20 (1.44)	5.48 (3.54)	10.08 (6.82)	20.30 (14.88)
Hotel	0.45 (0.18)	0.92 (0.32)	1.93 (0.95)	3.70 (1.93)

Load Model

The load model used in this study is given in terms of the variation of average monthly energy cost for the different customer segments investigated and was adopted from similar work (Dzobo, Gaunt, & Herman, 2011; 2012a; 2012b). A working month of 30 days was assumed for all months. Table 4 presents the load point data used for the supply reliability worth analysis in this study.

Time Sequential Monte Carlo Simulation

The simulation algorithm used to analyse the RBTS was developed using MATLAB software and is shown in the flowchart in Figure 8.

The simulation is performed on an annual basis and the total annual hours are taken as 8 760 hours. The analysis is done considering events consisting of failure in one component, no second order failures are considered. The energy not served (ENS) value due to a component failure is superimposed with the respective solar PV distributed generation output power profile value and the difference between these two profile values will give the new ENS value. There will be cases with surplus renewable energy not used which can be stored if the technology is available but for the present simulation storage is not considered. The new ENS value is then multiplied by the customer interruption costs of the respective customer segments at the affected load point to get the ECOST value. More than 400 years were simulated and the resultant ECOST index values were fitted to a Beta PDF. The Beta PDF was chosen because of its versatility. The approach was adopted from similar work as in Cross, Herman, & Gaunt (2006), Dzobo, Gaunt, & Herman (2011), and Edimu, Gaunt, & Herman (2011).

SOLUTIONS AND RECOMMENDATIONS

In all the analyses, a beta PDF was fitted to ECOST index values computed. This aimed to capture any skewness in the distribution of the supply reliability worth index. Figure 8 – 13 show the shapes of the ECOST index PDFs derived in the analyses. It is clear from the shapes of the PDFs that there exists inherent skewness in the performance index. Using average values to represent the resultant index therefore neglects the information on its variance and skewness. Considering the base case values, a significant

Table 4. Average monthly energy cost for each customer segment

Load Point	Number of Customers	Customer Segment	Average Monthly Energy Cost (Rand)
LP1	50	Retail	2200
LP2	100	Garage	2000
LP3	40	Clothing	1150
LP4	150	Retail	2200
LP5	60	Metal	4500
LP6	90 50	Retail Garage	2200 2000

Figure 8. Flowchart for the time sequential Monte Carlo simulation

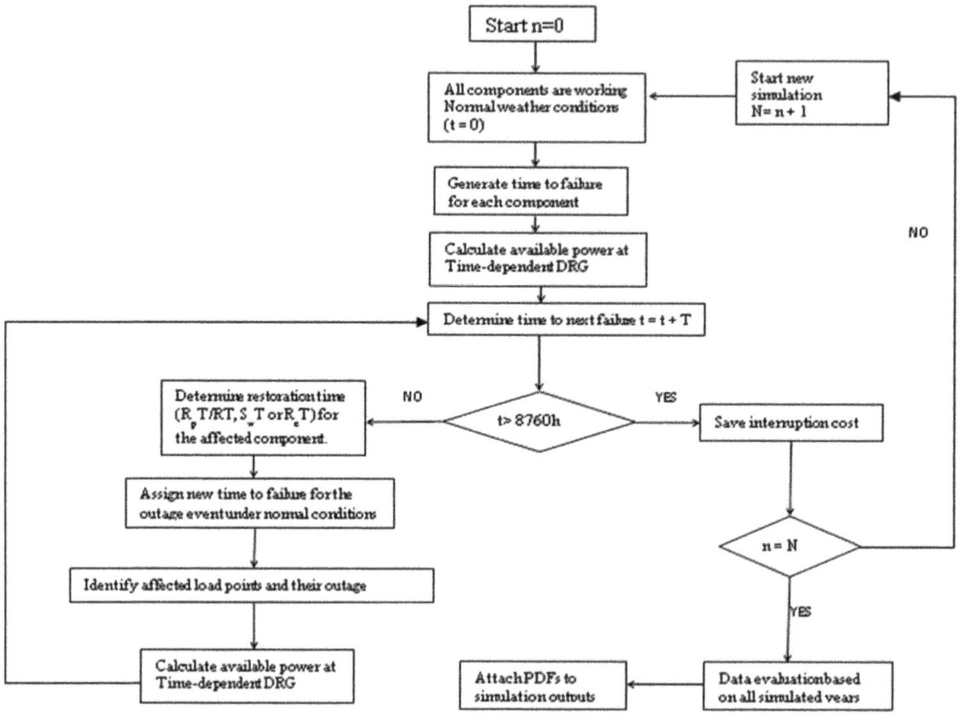

difference in the index values is noted in comparison with the results from the other scenarios. The value of the index reduces after inclusion of time variation in the analysis.

Figures 8 – 10 show a significant difference in ECOST index values between the base case and corresponding data sets. There is a 20% – 79% reduction in ECOST index values after the inclusion of solar PV distributed generation unit in the power system network. The reduction is a result of the power from the DRG which is fed to the different loads in case of a fault. However, in some cases the power output of the DRG may not be able to feed all the load points and hence some load points may still experience load curtailment for the entire period of failure of the transmission line. The solar PV distributed generation improves the supply reliability of load points closer to its power system network connection location or supply point. This is shown by the high percentage reduction of ECOST values for LP1 and LP2 which is 78.8% and 60.3% as compared to a reduction of 2.7% and 20.7% for LP6 and LP7 respectively. This implies that the load points with the highest priority level or in this case closer to the supply point will have lower ECOST index values as compared to low priority load points further away from the supply point. The load points closer to the DRG must always be those with highest priority as this would reduce economic losses due to the interruption of electricity to electricity customers. The reduction in ECOST index values therefore depends on the priority level set to the load points during failure conditions. Power system planners and operators can thus use the results to measure the level of supply reliability for the different electricity customers connected to the power system network. Highest priority level will be given to load points or electricity customers who are able to pay a premium for

Figure 9. Probability distribution of ECOST for LP2

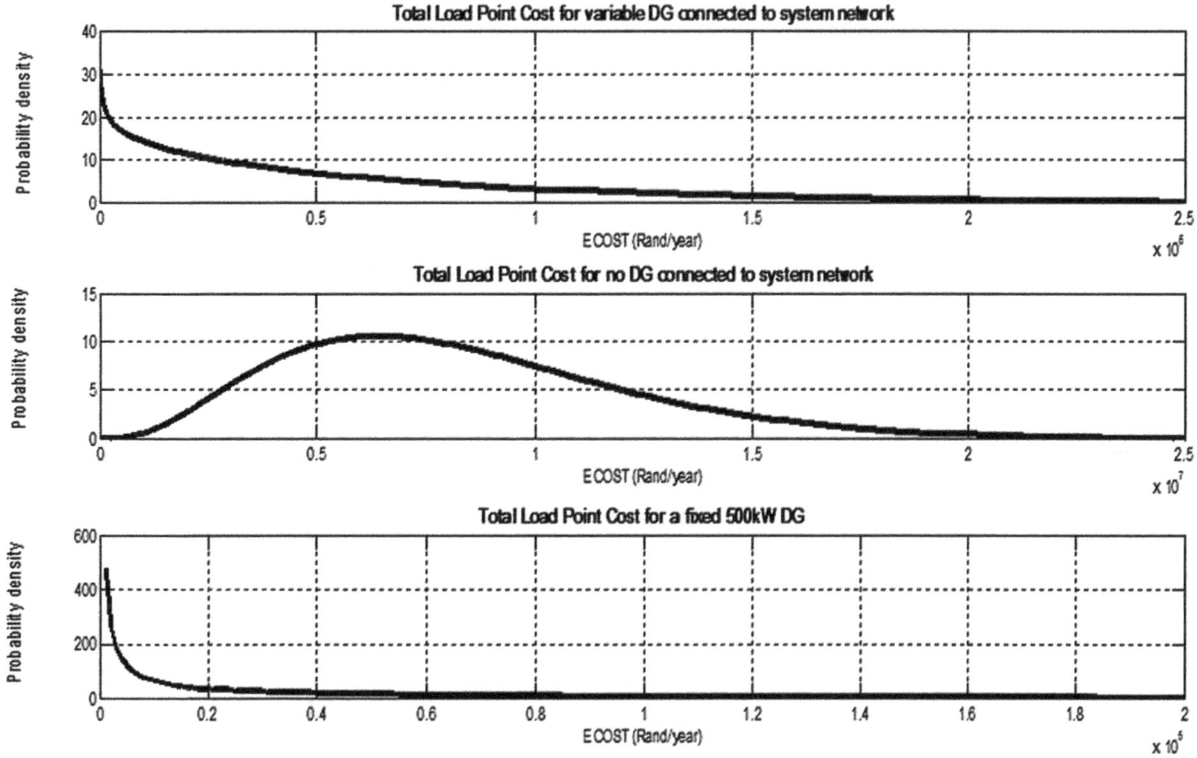

the increase in the supply reliability. For example, large manufacturing or mining companies can have distributed generation unit stationed closer to their premises with a dedicated feeder so that in the event of a power system network fault or when demand surpasses supply from conventional generation units, the power utility can supply the contract electricity customers from the distributed generation unit. In this way, the power system management will balance the reliability cost (investment cost of the distributed generation unit) and electricity tariff.

The results were also analyzed from the solar PV distributed generation penetration level perspective. This feature allows understanding the trend in reliability worth index when increasing solar PV panel power output. Four fixed penetration levels were analysed. Increasing the penetration level reduced the ECOST index values on all the load points. Figure 12 and 13 show the impact of the different penetration levels on ECOST index values for load points LP5 and LP7. The reason for the reduction in ECOST index values is that more electricity customers were able to be supplied with electricity as the capacity of the distributed generation increases. In addition, the assumption that the solar PV distributed generation unit is always available at all times and is operating at full capacity also reduced the ECOST index values. Power system planners and operators can use the results to calculate the optimal size of storage facility that may be required to reach a desired level of supply reliability. The ECOST index values of different penetration levels can be extracted from the PDFs and the reduction in this index value can be used as a rate of return value that can be equated to the cost of investing in storage facility. The smaller the percentage in cost difference the easier it is to justify the investment in the storage facility.

Figure 10. Probability distribution of ECOST for LP4

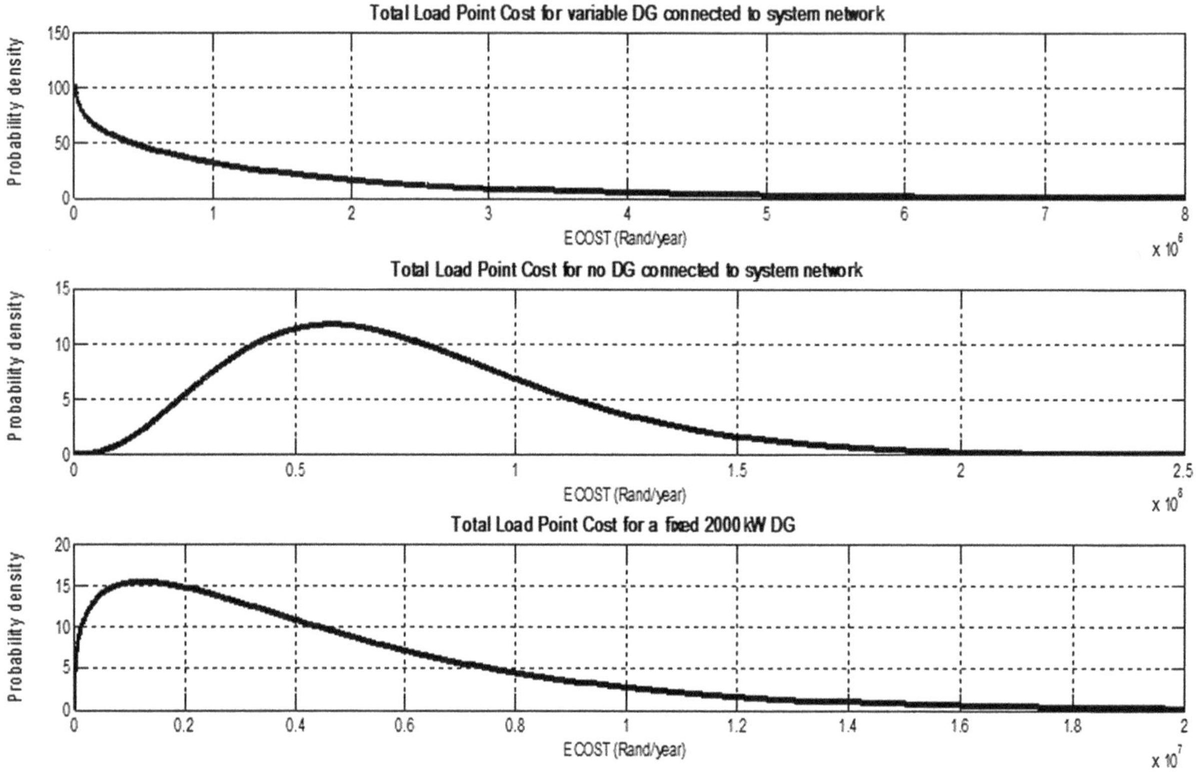

Figure 11. Probability distribution of ECOST for LP6

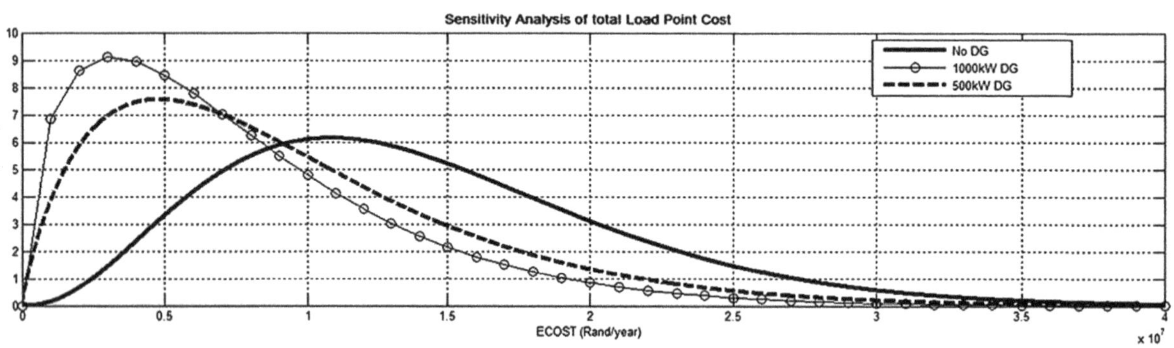

The supply reliability worth index, ECOST, at each load point is expressed in a probabilistic manner. This has the advantage that singular values can be extracted from the distributions based on the level of confidence and/or conversely level of risk. Power system planners and utility owners can have their decisions of investment in distributed generation for power system network improvement based on index values that have risk levels attached. The use of average values means that power system planners and utility owners allows for a 50% risk level on the index values used and thus exclude the likelihood of occurrence of extreme events. PDFs provide more information and would allow the comparison of

Figure 12. Probability distribution of ECOST when no DRG is connected to the power system network

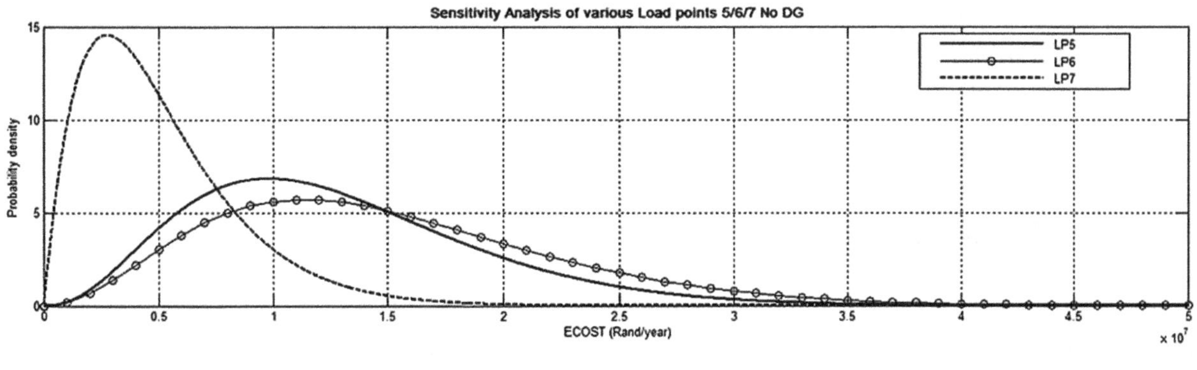

Figure 13. Probability distribution of ECOST for LP5

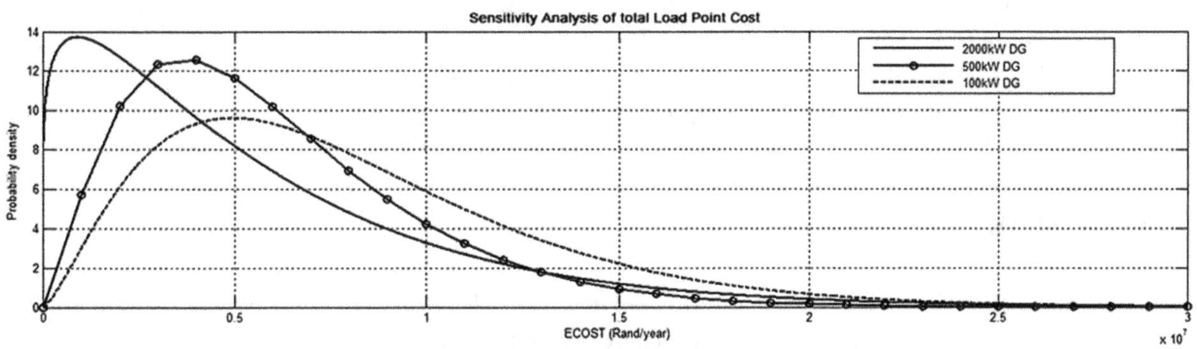

Figure 14. Probability distribution of ECOST for LP7

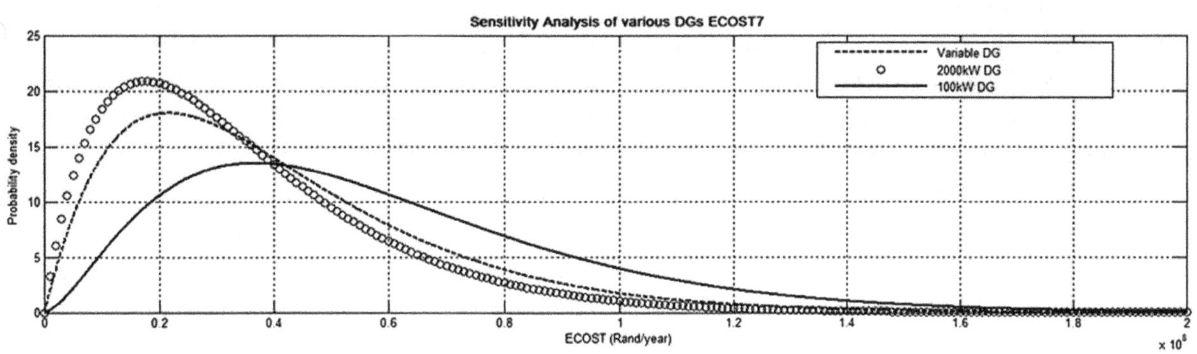

Table 5. Average ECOST results of the seven load points (kRand)

Load Point	LP1	LP2	LP3	LP4	LP5	LP6	LP7
Base case	5 506.60	7 331.30	56 852.00	10 274.00	8 670.90	1 008.50	5 074.90
Variable DRG connected	1 164.30	2 903.40	33 807.00	8 216.10	8 049.00	981.01	4 023.80

low risk level (high confidence) index values with the high risk (low confidence) index values to justify distributed generation investments in the power system network. These results could be taken as inputs for optimal placement of distributed generation study. In this case, every possible result is analyzed to achieve the best outcome on every section of the distribution system network and thus, maximizing the benefit of distributed generation in the power system network.

FUTURE RESEARCH DIRECTIONS

As more distributed generation units are connected to the conventional grid coupled with advancement in communication technology, storage of energy and its redistribution between different distributed generation has become more important. The use of virtual power plants to reconfigure the power system network during time of supply shortage in certain parts of the power system network needs to be explored. Electrical vehicles and their storage capabilities are also able to shift the demand of the electricity consumers as they can be used as backup power supply during peak periods when electricity prices are high. Different demand response strategies can also be included in the analysis and its effect on the power system reliability index determined.

CONCLUSION

The penetration of DRGs will increase in future smart grids. DRG inherent characteristics should therefore be considered when practical power system network analyses are carried out. A preliminary analysis was carried out on how time dependent DRG affect the configuration of power system network using time sequential MCS technique. The results showed that including DRG in the power system network improves supply reliability to electricity customers. However, the improvement is dependent on the priority order adopted in the power system network reconfiguration during failure conditions. It also revealed that ignoring time variation of the output from the DRG units can severely underestimate the ECOST index values. The technique and results obtained from the analysis demonstrate that reliability worth index predicted with a defined level of confidence and or conversely level of risk can provide more useful information to system planners and operators. Each fixed penetration level of solar PV distributed generation considered has different impact on the supply reliability and contributes differently to each type of load. It has been shown that on average, the availability of the network will improve positively as the penetration level increases. The information provided, and the examples illustrated in this paper should prove useful to power system planners and distributed renewable generation developers to assess the reliability benefit of distributed renewable generation in electric power system networks.

REFERENCES

Bae, I. S., Kim, J. O., & Kim, J. C. (2007). Reliability evaluation of distributed generation based on operation mode. *IEEE Transactions on Power Systems*, *22*(2), 785–790. doi:10.1109/TPWRS.2007.894842

Bae, I. S., Kim, J. O., Kim, J. C., & Singh, C. (2004). Optimal operating strategy for distributed generation considering hourly reliability worth. *IEEE Transactions on Power Systems*, *19*(1), 287–292. doi:10.1109/TPWRS.2003.818738

Barker, P. P., & De Mello, R. W. (2000). Determining the impact of distributed generation on power systems: Part 1 - Radial distribution systems. *Proceedings of the IEEE Power Engineering Society Transmission and Distribution Conference*, 3, 1645-1656. 10.1109/PESS.2000.868775

Billinton, R., & Allan, R. N. (1996). *Reliability Evaluation of Power Systems* (2nd ed.). Plenum Press. doi:10.1007/978-1-4899-1860-4

Bizon, N. (2013). Energy harvesting from the PV Hybrid Power Source. *Energy*, *52*, 297–307. doi:10.1016/j.energy.2013.02.006

Caballero, F., Sauma, E., & Yanine, F. (2013). Business optimal design of a grid-connected hybrid PV (photovoltaic)-wind energy system without energy storage for an Easter Island's block. *Energy*, *61*, 248–261. doi:10.1016/j.energy.2013.08.030

Cabral, C. V. T., Filho, D. O., Diniz, A. S. A. C., Martins, J. H., Toledo, O. M., & Neto, L. V. B. M. (2010). A stochastic method for stand-alone photovoltaic system sizing. *Solar Energy*, *84*(9), 1628–1636. doi:10.1016/j.solener.2010.06.006

Collares-Pereira, M., & Rabl, A. (1979). Derivation of method for predicting long term average energy delivery of non-concentrating and concentrating solar collectors. *Solar Energy*, *22*, 155–170. doi:10.1016/0038-092X(79)90100-2

Cross, N., Herman, R., & Gaunt, C. T. (2006). *Investigating the usefulness of beta pdf to describe parameters in reliability analyses. In Proceedings of Probabilistic Methods Applied to Power Systems*. PMAPS.

Department of Minerals and Energy - Republic of South Africa. (2003, November). *White paper on renewable energy.*

Dursun, E., & Kilic, O. (2012). Comparative evaluation of different power management strategies of a stand-alone PV/Wind/PEMFC hybrid power system. *International Journal of Electrical Power & Energy Systems*, *34*(1), 81–89. doi:10.1016/j.ijepes.2011.08.025

Dzobo, O., Gaunt, C. T., & Herman, R. (2011). Investigating the use of probability distribution functions in reliability-worth analysis of electric power systems. *International Journal of Electrical Power & Energy Systems*, *37*(1), 110–116. doi:10.1016/j.ijepes.2011.12.013

Dzobo, O., Gaunt, C. T., & Herman, R. (2012a). *Customer interruption cost for composite reliability analysis, Proceedings of Probabilistic Methods Applied to Power Systems*. PMAPS.

Dzobo, O., Gaunt, C. T., & Herman, R. (2012b). Reliability worth assessment of electricity consumers: A South African case study. *Journal of Energy in Southern Africa*, *23*(3), 31–39. doi:10.17159/2413-3051/2012/v23i3a3170

Edimu, M., Gaunt, C. T., & Herman, R. (2011). Applying Probability Distribution Functions to Model System Failures Due to Adverse Weather. *17th Power Systems Computation Conference.*

European Commission. (2003). EUR 20871 – Renewable energy technologies and Kyoto protocol mechanisms. Author.

Hegazy, Y. G., Salama, M. M. A., & Chikhani, A. Y. (2003). Adequacy assessment of distributed generation systems using Monte Carlo simulations. *IEEE Transactions on Power Systems*, *18*(1), 48–52. doi:10.1109/TPWRS.2002.807044

Hove, T. (2000). A method for predicting long-term average performance of photovoltaic systems. *Renewable Energy*, *21*(2), 207–229. doi:10.1016/S0960-1481(99)00131-7

Mohamed, F. A., & Koivo, H. N. (2010). System modelling and online optimal management of MicroGrid using Mesh Adaptive Direct Search. *International Journal of Electrical Power & Energy Systems*, *32*(5), 398–407. doi:10.1016/j.ijepes.2009.11.003

Moradi, M. H., & Abedini, M. (2012). A combination of genetic algorithm and particle swarm optimization for optimal DG location and sizing in distribution systems. *International Journal of Electrical Power & Energy Systems*, *34*(1), 66–74. doi:10.1016/j.ijepes.2011.08.023

Niknam, T., Golestaneh, F., & Malekpour, A. (2012). Probabilistic energy and operation management of a microgrid containing wind/photovoltaic/fuel cell generation and energy storage devices based on point estimate method and self-adaptive gravitational search algorithm. *Energy*, *43*(1), 427–437. doi:10.1016/j.energy.2012.03.064

Papaefthymiou, G., Schavemaker, P. H., van der Sluis, L., Kling, W. L., Kurowicka, D., & Cooke, R. M. (2006). Integration of stochastic generation in power systems. *International Journal of Electrical Power & Energy Systems*, *28*(9), 655–667. doi:10.1016/j.ijepes.2006.03.004

Singh, G. K. (2013). Solar power generation by PV (photovoltaic) technology: A review. *Energy*, *53*, 1–13. doi:10.1016/j.energy.2013.02.057

Solar Energy Services for Professionals. (n.d.). Retrieved from http://www.soda-is.com/eng/services/service_invoke/gui.php?xml_descript=hc3v5_invoke_hour_demo.xml#parameters

Sun, Y., Bollen, M. H. J., & Ault, G. W. (2006). Probabilistic reliability evaluation of distribution systems with DER and Microgrids. *Proceeding of International Conference on Probabilistic Methods Applied to Power Systems*, 1-8. 10.1109/PMAPS.2006.360214

Tan, X. A., Li, Q. B., & Wang, H. A. (2013). Advances and trends of energy storage technology in Microgrid. *International Journal of Electrical Power & Energy Systems*, *44*(1), 179–191. doi:10.1016/j.ijepes.2012.07.015

Tazvinga, H., & Dzobo, O. (2018, April). Hybrid energy systems for rural communities in Zimbabwe. *EAI International Conference for Research, Innovation and Development for Africa*, 480-492. 10.4108/eai.20-6-2017.2270759

Tazvinga, H., Xia, X., & Zhang, J. (2013). Minimum cost solution of photovoltaic–diesel–battery hybrid power systems for remote consumers. *Solar Energy*, *96*, 292–299. doi:10.1016/j.solener.2013.07.030

Tina, G., & Gagliano, S. (2011). Probabilistic modelling of hybrid solar/wind power system with solar tracking system. *Renewable Energy*, *36*(6), 1719–1727. doi:10.1016/j.renene.2010.12.001

Winkler, H. (2005). Renewable energy policy in South Africa: Policy options for renewable electricity. *Energy Policy*, *33*(1), 27–38. doi:10.1016/S0301-4215(03)00195-2

Yang, H. X., Lu, L., & Burnett, J. (2003). Weather data and probability analysis of hybrid photovoltaic-wind power generation systems in Hong Kong. *Renewable Energy*, *28*(11), 1813–1824. doi:10.1016/S0960-1481(03)00015-6

Zeinalzadeh, A., Mohammadi, Y., & Moradi, M. H. (2015). Optimal multi objective placement and sizing of multiple DGs and shunt capacitor banks simultaneously considering load uncertainty via MOPSO approach. *International Journal of Electrical Power & Energy Systems*, *67*, 336–349. doi:10.1016/j.ijepes.2014.12.010

KEY TERMS AND DEFINITIONS

Customer Interruption Cost: The financial value incurred by an electricity consumer as a result of power supply interruption to its daily activities.

Distributed Generation: Production of energy from renewable energy sources at different geographical places. The production of energy is not as large as that of conventional power generation that makes up the conventional grid network.

Intermittent Power Supply: The random generation of energy from renewable energy sources which is dependent on the availability of the energy source (e.g., wind, solar).

Monte Carlo Simulation Technique: Is a random sampling process that is repeated for many iterations.

Power System Reliability: The ability of a power system network to perform its function for a given time period under certain conditions.

Solar Irradiance: The power per unit received from the sun. The units of solar irradiance is W/m^2.

Time-to-Failure: Is the time taken by a system component in the UP state to reach its failure state.

Time-to-Repair: Time take by a system component in the DOWN state to restore to normal condition (UP state).

Chapter 5
Electric Vehicle Infrastructure Planning:
A Distribution Side Perspective

Tushar Kumar
MANIT Bhopal, India

Tripta Thakur
MANIT Bhopal, India

ABSTRACT

Widespread adoption of electric vehicles would bring a paradigm shift in the way distribution infrastructure is planned and electricity markets operate. Electric vehicle adoption could help in meeting the worldwide targets for greenhouse gas emissions. Moreover, the health benefits for the public would be immense as the source of emissions would be far away from the massively populated areas. For electricity markets, electric vehicles can serve as a distributed plug in facility of energy storage at low cost requiring minimal capital investment from grid utilities. However, widespread electric vehicle adoption faces a number of hurdles such as limited range in comparison to Internal combustion engines, but from the grid perspective, it faces issues such as limitations of available charging infrastructure to charge large number of electric vehicles and longer charging time currently as compared to refueling fuel driven vehicles. This chapter explores such issues and their remedies in the current literature.

INTRODUCTION

Transportation of goods and services effectively is the rock bed of any thriving economy. Increased population and especially population density in urban areas has driven the demand for vehicles. Increased population density in urban areas has driven the demand for vehicles but the bulk share of this demand is taken up by conventional vehicles driven by fossil fuels and the upswing in the demand is matched by the upswing in CO2 emissions and pollution levels in the densely populated areas. Thus this population is directly exposed to the long-term risk of health deterioration. Electric vehicle adoption could help in

DOI: 10.4018/978-1-5225-8551-0.ch005

meeting the worldwide targets for greenhouse gas emission reduction. Moreover, the health benefits for the public would be immense as the source of emissions would be far away from the massively populated areas. The International Energy Agency estimates that the sales of passenger light-duty EVs/plug-in hybrid EVs (PHEVs) will increase significantly from 2020 onwards and might reach more than 100 million EVs/PHEVs sold per year worldwide by 2050 (International Energy Agency, 2011). Widespread adoption of Electric vehicles would bring a paradigm shift in the way distribution infrastructure is planned and electricity markets operate. For electricity markets Electric vehicles can serve as a distributed plug in facility of energy storage at low cost requiring minimal capital investment from grid utilities.

However widespread Electric vehicle (EV) adoption faces a number of hurdles (Saxena, MacDonald & Moura, 2015), such as limited range in comparison to IC engines but from the grid perspective it faces following issues:

- Limitations of available charging infrastructure to charge large number of Electric vehicles.
- Charging time for Electric vehicles is much longer currently as compared to refueling fuel driven vehicles.

Improper management of charging loads can negatively impact grid operations and reliability, increase emissions and electricity prices (He, Wu, Yin & Guan, 2013; Parks, Denholm & Markel, 2007). Uncontrolled charging refers to charging when little or no information is available about the electricity prices and the owner decides the time to plug in, and the charger draws maximum power from point of plug-in till the battery is charged fully. Controlled charging is the practice of directing EV charging behavior by providing price signals or other incentives to the consumer, in order to limit negative grid impacts (M.J Bradley and Associates LLC, 2013).With controlled charging, EVs can levelize the overall load, make better use of baseload generating units, and require no extra installed capacity. On the other hand, large numbers of EVs charging uncontrolled can result in voltage limit violations, increased line losses and transformer overloads (Richardson, 2013). Huge penetration of the RES into the existing power infrastructure would require large scale energy storage to support the electrical power demand and to meet the operating standards in place currently. Electric vehicle fleets could help in meeting this challenge head on. Plug in Electric vehicle charging patterns have also been explored in recent studies to tackle the negative effects of such patterns on the distribution networks (Zhou, Wang & Gao, 2018). Moreover large scale data management for Electric vehicles in itself poses a challenge of protecting privacy of customers as their usage patterns could be monitored.

Electric vehicle owners would like to know how to minimize their cost of charging, so selling cost of power should be known to the electric vehicle owners for providing V2G services.

Similarly for responding to frequency regulation and Demand Side Management (DSM) issues the real time information regarding the electric vehicle parameters such as overall capacity, mobility patterns etc. is essential. Electricity cost would determine the interaction of electric vehicle owners with the grid whereas grid operators would consider the load profile. Demand response initiatives and inbuilt vehicle timers could help in easing the stress on the power grid. However such schemes are effective for residential charging points but fast charging stations can draw huge amount of power for shorter time intervals because they may not be frequented by drivers. As a result they could incur high demand charges which depends on the highest capacity used which would in turn lead to increased charging cost for drivers.

TECHNICAL STANDARDS FOR EV

The duration for EV charging depends on the charging characteristics as well as the battery storage capacity. Currently there is no single charging standard that is unanimously accepted throughout the world but various systems emerged for higher-current charging connectors with the standardization in IEC 62196 and IEC 61851 for electric vehicle conductive charging systems, parts of which are currently still under development. SAE combo standard developed by Society of Automotive engineers (SAE) consists of a combination of fast DC charging and AC charging points in the same connector (Society of Automotive Engineers, 2012). Apart from this CHAdeMO, a fast charging standard developed by TEPCO (Tokyo Power Electric Company) was revised to CHAdeMO 2.0 which would allow up to 400 kW charging (ChAdeMO Association, 2019). On the other hand tesla has its own standard tesla supercharger that can currently supply 120 kW for each car. Table 1 shows the performance features of Electric vehicles currently in market.

The impact of EV charging would appear on the existing distribution grid as an additional burden. If this load is uncontrolled then it may lead to premature aging of power system equipment and relay tripping for overload conditions especially where the infrastructure has not been upgraded. However 60% to 70% of the additional investment cost needed for the power distribution infrastructure could be avoided if smart EV charging schemes are adopted (Pieltain Fernandez et al., 2011).

Table 1. Features of Battery Electric vehicles (cars) currently in market or to be introduced into the market

EV Brand	EV Model	Battery Capacity (kWh)	Range (Km)	Consumption (kWh/100 km)
BMW	i3	33	182.4	28.95
Tesla	S 75 D	75	414.4	28.96
Tesla	S 100 D	100	536	29.85
Nissan	leaf	40	241.6	26.49
Chevrolet	Bolt EV	60	380.8	25.21
Ford	Focus EV	33.5	184	29.13
Volkswagen	e-golf	35.8	200	28.64
Volkswagen	I.D. Neo	48	328	23.41
Kia	Soul electric	30.5	177.6	27.48
Kia	Niro Electric	64	382.4	26.78
Hyundai	Kona	64	412.8	24.81
Mitsubishi	I-MIEV	16	99.2	25.81
FIAT	500e	24	134.4	28.57
Hyundai	Ioniq electric	28	198.4	22.58
Honda	Clarity EV	25.5	142.4	28.65
Jaguar	I-PACE	90	374.4	38.46
Renault	Zoe	41	297.6	22.04
Mercedes Benz	B250e	36	139.2	41.38

("Find and Compare Cars", 2018; "List of Electric Vehicles (2018)", 2018)

ELECTRIC VEHICLE INTEGRATION INTO ELECTRIC GRID

One of the approaches proposed in the literature for integrating large electric vehicle fleets involves aggregator coordinating operations like communication between distribution system operators and electric vehicles. Another approach in involves individual electric vehicle assets participating directly in the energy market. Such model of electric vehicle integration is also mentioned in the literature (Ma, Callaway and Hiskens, 2013) in which the charging price is optimized leading to reduction of the stresses in the grid along with minimization of charging cost for EV owners. Reliability of such schemes could be questioned though as the direct interaction with the electric vehicle owners would result in increased complexity of energy planning and control.

Improvement in battery technologies has been driving the participation of electric vehicles in energy market. While research on these technologies would improve driving range and reduce initial costs, effect of V2G (Vehicle to Grid) application on the lifetime and performance of batteries must also be taken into consideration. These effects have been investigated by researchers (Bishop et al., 2013) and a study by Peterson, Apt and Whitacre, (2010) found that lithium-iron-phosphate battery cells withstand frequent charging-discharging cycles with minimal loss of capacity. However a study by Guenther et al., (2013) examined the battery deterioration characteristics of lithium ion battery taking into account combinations of charging schemes, V2G transactions and driving conditions. Their results reveal that prolonged discharging cycles along with greater cycle depths can reduce the battery lifetime up to 3 years. This issue could be resolved by adopting intelligent charging schemes. For any battery model to be realistic it should consider aging cycles, calendar aging and self-discharging combinedly.

Operating costs of electric vehicles could vary a lot depending on the charging location and the policies that are in effect. For example under a tiered electricity rate mechanism it is possible that if the users charge their electric vehicles at home then their electricity consumption may cross over into a higher priced tier. This could make the pure electric vehicle more expensive to drive than a conventional vehicle as shown by Huang et al., (2011) in case of California electricity markets.

Individual EVs on their own have power capacities that are quite low from the grid perspective for realizing vehicle to grid (V2G) services. But aggregating a fleet of EVs through various objectives and control schemes can help in implementing the V2G services (Sortomme and El-Sharkawi, 2012). A study made by Fasugba and Krein (2011) compared the maximum hourly regulation capacity for EV system considering bidirectional and unidirectional chargers. They found that for a unidirectional charger the battery capacity required would be twice that of the bidirectional charger to supply for the same demand for the grid. However due to extra cost burden of additional metering and protection measures this advantage is not significant.

The concept of virtual power plants could facilitate the control and information exchange between the EVs and utilities for V2G services. Many control schemes involving virtual power plants (VPP) have been proposed to tackle the EV – grid integration issues. Broadly three approaches for VPP control are available in literature namely – distributed, centralized and hierarchical. The EV aggregator operates as a VPP in the energy market. Vehicle status like available power and SOC (State Of Charge) is provided by the clusters of EV to the charging management system (CMS) which in turn communicates with local VPP center. The aggregated power can then be dispatched by the VPP control center to deliver the ancillary services to the grid. Musio, Lombardi and Damiano (2010) have envisaged the optimization of VPP operation for minimizing the operating cost considering both charging and discharging cycles of the battery along with purchasing cost for EV. A reduction of 26.5% in cost was observed when the fleet is operated as a DSM unit as well as an energy storage system.

EV CHARGING INFRASTRUCTURE OPERATION

Various charging levels for EVs exist which indicate the duration of charging and power capacity of charging infrastructure (Yilmaz & Krein, 2013). AC level 1 charging requires a voltage of 120 VAC an can supply up to 1.9 kW which means the full charging duration could range between 8 to 18 hrs depending on the battery capacity. However extra costs for setting up an infrastructure are saved as the domestic outlets could handle these power levels. AC level 2 charging on the other hand requires a voltage of 240 VAC and supplies up to 19.2 kW so the full charging duration varies between 3 to 7 hrs. Table 2 shows the current AC/DC charging level characteristics based on the duration of charging.

Public charging stations could provide this facility or the owners could upgrade their outlets at home or workplace. All AC level charging schemes have a general configuration in which the charging outlet provides 1Φ or 3Φ supply point with protection units and the charging system (including rectifier, filters, DC/DC converter) is in the EV itself.

But the slow speed of charging coupled with the fact that EV charging stations are still not as commonly available as gas stations results in range anxiety among the EV owners. For this reason research has intensified towards developing faster DC charging outlets in which along with the supply point and protection units, the charging system (including rectifier, filters, DC/DC converter) is set up. DC level 1

Table 2. AC/DC charging levels characteristics

Power Level Type	Voltage Level [V]	Current Capacity [A]	Power Capacity [kW]	Remark(s)
AC level 1	120V AC	12-16	1.4-1.9	1-phase supply (EV contains an on-board charger) Charging time: 9-12 hrs
AC level 2	240V AC	Up to 80	19.2	1 or 3-phase supply (EV contains an on-board charger) Charging time: 4-6 hrs
DC level 1	200-500V DC	up to 80	up to 40	3-phase supply (EVSE contains an off-board charger) Charging time: 1-2 hrs
DC level 2	200-500V DC	up to 200	up to 100	3-phase supply (EVSE contains an off-board charger) Charging time: 10-20 mins
DC level 3	> 200V DC	> 200	> 240	3-phase supply (EVSE contains an off-board charger) Under development Expected Charging time: 3-6 mins

(Mwasilu et al., 2014)

charging with a voltage of 200-500VDC can supply up to 40kW and charge compatible EVs fully in 1-2 hrs. DC level 2 charging with a voltage of 200-500VDC can supply up to 100kW and charge compatible EVs fully in 30 minutes to 1 hr.

Currently DC level 3 chargers are being developed that are expected to compete with the refueling time of a typical gasoline based vehicles. Figure 1 and 2 show respectively the on board and off board configurations of charging schemes.

Recently a Porsche-BMW charger prototype having a capacity of 450 kW at a voltage of 900V and 500A has been unveiled that can charge 62 miles in 3 minutes (Gilboy, 2019). However as of now no EV on road can handle such high currents. Since a high charging rate can permanently affect the battery capacity of lithium – ion type, the charging rate is limited by manufacturers to prevent thermal issues and lifetime reduction of batteries. Such a high charging rate also puts tough demands on the charging hardware as well. The fast charging prototype was fitted with liquid cooled charging cable and the EV battery cooling system was also modified to handle the heavy load.

Charging infrastructure can have 'modes' of operation based on the type of electric and communications connection between the charging point and the vehicle (Bräunl, 2012). Table 3 shows the modes of operation and among these mode 3 and 4 enables smart charging, the coordination of charging according to renewable energy availability, fleet schedules or utility needs.

As the charging time for EVs reduce, a need to understand the impact of fast charging on the electrical grid arises.

Figure 1. AC level 1 and 2 configuration for EV charging
(Mwasilu et al., 2014)

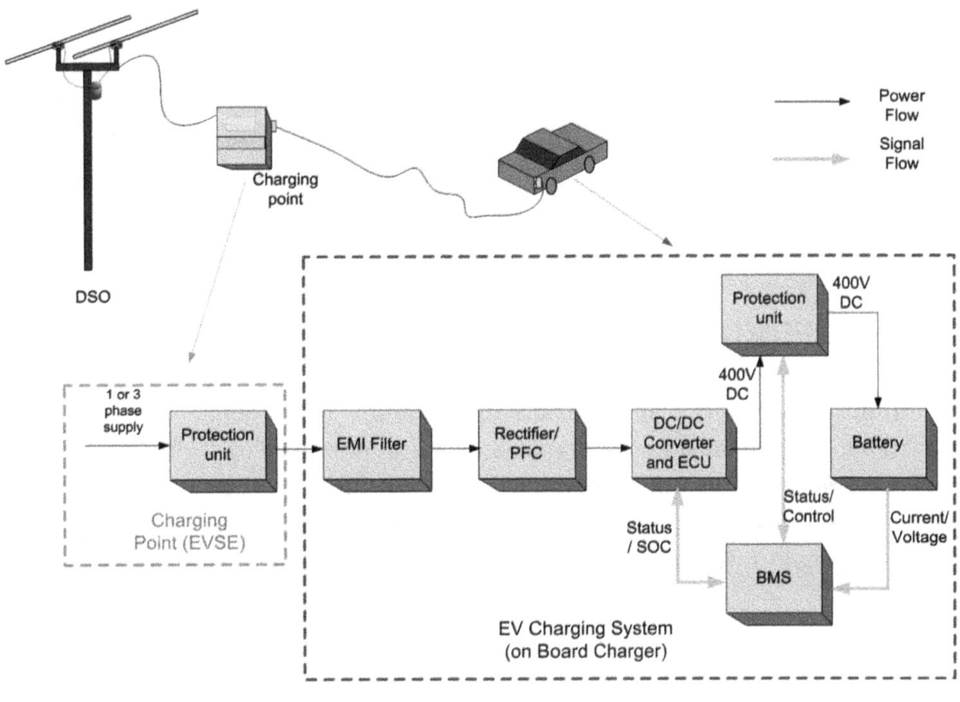

Figure 2. DC level configuration for EV charging
(Mwasilu et al., 2014)

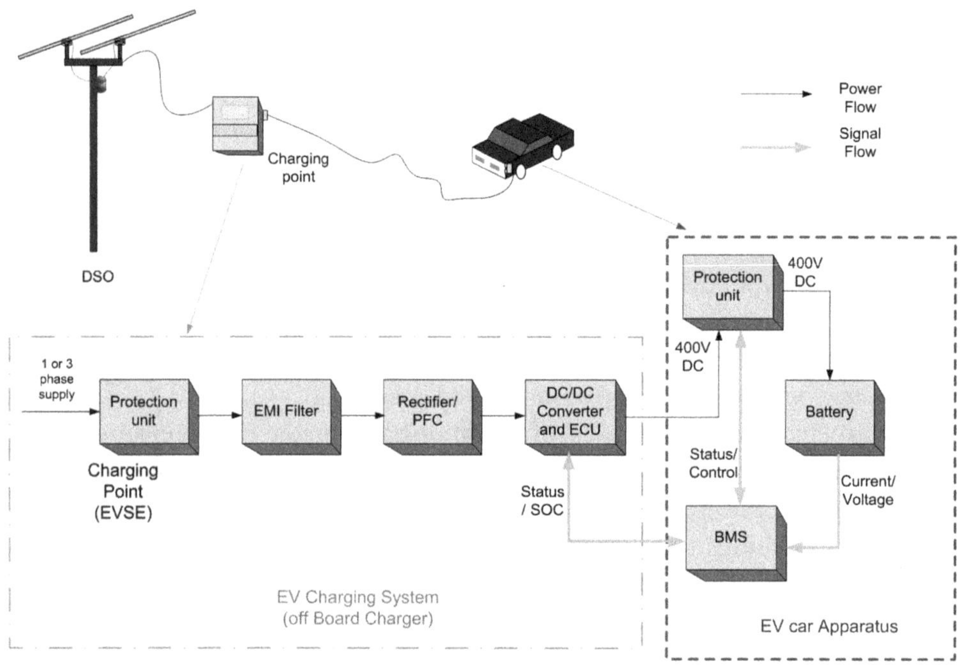

Table 3. Modes of Operation for Charging Infrastructure

Type of Mode	Voltage Level (V)	Current Capacity (A)	Setting
Mode 1	120/240V	Up to 16A	a common circuit without safety protocols
Mode 2	120/240V	Up to 32A	from a standard outlet, on a common or customized circuit, with safety protocols such as thermal limits, overcurrent protection, grounding detection
Mode 3	240V	Any amperage	on a wired-in charging station on a customized circuit, with the same safety protocols as Mode 2 and an active communication line with the vehicle
Mode 4	400V	Any amperage	Basically for DC fast charging and requires wired-in connection, and more advanced safety and communications protocols

(Bräunl, 2012)

MARKET CONSIDERATIONS FOR CHARGING INFRASTRUCTURE

Most electric utilities for generation and transmission up to now have operated under the assumption of perpetual growth but as the demand gets stagnated owing to improved efficiency of the systems along with increased penetration of renewable energy resources, the utilities need to adapt to a new paradigm of operation. Earlier utilities were concerned with this trend but now they have in fact begun to invest in electric vehicle charging infrastructure as they could offset the decline of electricity demand in coming years.

In Germany, the EV adoption program provides €4,000 off to the buyers on purchasing fully electric vehicles. Apart from the initial purchase bonus, EV owners also get 10 years exemption on ownership tax. Although incentives for purchase of EV is in place in many countries, such schemes would not affect the long term adoption unless proper EV usage incentive schemes are in implemented with a focus on the local market conditions. Programs that engage various stakeholders by utilizing smart charging schemes, integrating EV owner's feedback on charger deployment and devising partnerships between public and private utilities are more likely to be effective (Hall & Lutsey, 2017). In addition governments need to adopt strategies that suit the local requirements such as intercity fast charging points to cater to the heavy traffic between nearby big cities.

Inspite of the recent growth in EV technology and market, EV charging infrastructure is mostly fragmented due to inconsistent data availability and absence of consistent standards for operation. Therefore governments could push for open standards implementation for EV–charging point communication along with data collection. Such initiatives could enable interoperability between charging systems and have been implemented in Netherlands (Hall & Lutsey, 2017). It is important to identify and tackle specific charging requirements of a geographical region because with further evolution of the EV industry, assumptions about mobility patterns and state of the art technology may change. Hence the focus must be preferably on establishing infrastructure for long term needs like residential charging or DC fast charging stations between cities (Hall & Lutsey, 2017).

CONCLUSION

The battery capacities of EVs have increased in recent years to improve the range of travel and many models now have capacities in the range of 60-100 kWh. As the number of such EVs increases in the market, the charging demand could surpass the demand of a typical household. Such high demand levels could cause overloading of existing distribution equipment and voltage sags. Mitigation of such issues requires impact analysis of the widespread adoption of EVs on the infrastructure which helps in optimal scheduling of charging operations and planning of reinforcements required for handling such a demand. Factors such as EV owner preferences, battery characteristics and travel patterns influence the impact of EV charging on the grid. A systematic assessment of the impact including data collection, statistical cluster analysis, modeling of feeder, study of charging scenarios and feeder analysis is necessary for planning and designing charging infrastructure that would power the future transportation.

REFERENCES

Bishop, J., Axon, C., Bonilla, D., Tran, M., Banister, D., & McCulloch, M. (2013). Evaluating the impact of V2G services on the degradation of batteries in PHEV and EV. *Applied Energy*, *111*, 206–218. doi:10.1016/j.apenergy.2013.04.094

Chademo Association. (2019). *CHAdeMO releases the latest version of the protocol enabling up to 400KW*. Available at: https://www.chademo.com/chademo-releases-the-latest-version-of-the-protocol-enabling-up-to-400kw

Fasugba, M., & Krein, P. (2011). Cost benefits and vehicle-to-grid regulation services of unidirectional charging of electric vehicles. *2011 IEEE Energy Conversion Congress And Exposition*. doi:10.1109/ecce.2011.6063856

Find and Compare Cars. (2018). Retrieved from https://www.fueleconomy.gov/feg/findacar.shtml

Gilboy, J. (2019). *BMW, Porsche Reveal Prototype EV Fast Charger That Gives 62 Miles of Range in 3 Minutes*. Retrieved from http://www.thedrive.com/tech/25468/bmw-porsche-reveal-prototype-ev-fast-charger-that-gives-62-miles-of-range-in-3-minutes

Guenther, C., Schott, B., Hennings, W., Waldowski, P., & Danzer, M. (2013). Model-based investigation of electric vehicle battery aging by means of vehicle-to-grid scenario simulations. *Journal of Power Sources*, *239*, 604–610. doi:10.1016/j.jpowsour.2013.02.041

He, F., Wu, D., Yin, Y., & Guan, Y. (2013). Optimal deployment of public charging stations for plug-in hybrid electric vehicles. *Transportation Research Part B: Methodological*, *47*, 87–101. doi:10.1016/j.trb.2012.09.007

Huang, S., Hodge, B., Taheripour, F., Pekny, J., Reklaitis, G., & Tyner, W. (2011). The effects of electricity pricing on PHEV competitiveness. *Energy Policy*, *39*(3), 1552–1561. doi:10.1016/j.enpol.2010.12.029

International Energy Agency. (2011). *Technology Roadmaps - Electric and plug-in hybrid electric vehicles*. Retrieved from http://www.ieahev.org/assets/1/7/EV_PHEV_Roadmap.pdf

List of Electric Vehicles. (2018). (2018). Retrieved from https://evrater.com/evs

Ma, Z., Callaway, D., & Hiskens, I. (2013). Decentralized Charging Control of Large Populations of Plug-in Electric Vehicles. *IEEE Transactions on Control Systems Technology*, *21*(1), 67–78. doi:10.1109/TCST.2011.2174059

M.J Bradley and Associates LLC. (2013). Electric Vehicle Grid Integration in the U.S., Europe, and China. *International Council on Clean Transportation*. Retrieved from https://www.theicct.org/sites/default/files/publications/EVpolicies_final_July11.pdf

Musio, M., Lombardi, P., & Damiano, A. (2010). Vehicles to grid (V2G) concept applied to a Virtual Power Plant structure. *The XIX International Conference on Electrical Machines - ICEM 2010*. 10.1109/ICELMACH.2010.5608261

Mwasilu, F., Justo, J., Kim, E., Do, T., & Jung, J. (2014). Electric vehicles and smart grid interaction: A review on vehicle to grid and renewable energy sources integration. *Renewable & Sustainable Energy Reviews*, *34*, 501–516. doi:10.1016/j.rser.2014.03.031

Parks, K., Denholm, P., & Markel, T. (2007). Costs and Emissions Associated with Plug-In Hybrid Electric Vehicle Charging in the Xcel Energy Colorado Service Territory. *National Renewable Energy Laboratory*. Retrieved from https://www.nrel.gov/docs/fy07osti/41410.pdf

Peterson, S., Apt, J., & Whitacre, J. (2010). Lithium-ion battery cell degradation resulting from realistic vehicle and vehicle-to-grid utilization. *Journal of Power Sources*, *195*(8), 2385–2392. doi:10.1016/j.jpowsour.2009.10.010

Pieltain Fernandez, L., Gomez San Roman, T., Cossent, R., Mateo Domingo, C., & Frias, P. (2011). Assessment of the Impact of Plug-in Electric Vehicles on Distribution Networks. *IEEE Transactions on Power Systems*, *26*(1), 206–213. doi:10.1109/TPWRS.2010.2049133

Richardson, D. (2013). Electric vehicles and the electric grid: A review of modeling approaches, Impacts, and renewable energy integration. *Renewable & Sustainable Energy Reviews*, *19*, 247–254. doi:10.1016/j.rser.2012.11.042

Saxena, S., MacDonald, J., & Moura, S. (2015). Charging ahead on the transition to electric vehicles with standard 120 V wall outlets. *Applied Energy*, *157*, 720–728. doi:10.1016/j.apenergy.2015.05.005

Society of Automotive Engineers (2012), Electric vehicle and plugin hybrid electric vehicle conductive charge coupler. *SAE Standard J, 1772*.

Sortomme, E., & El-Sharkawi, M. (2012). Optimal Scheduling of Vehicle-to-Grid Energy and Ancillary Services. *IEEE Transactions on Smart Grid*, *3*(1), 351–359. doi:10.1109/TSG.2011.2164099

Yilmaz, M., & Krein, P. (2013). Review of Battery Charger Topologies, Charging Power Levels, and Infrastructure for Plug-In Electric and Hybrid Vehicles. *IEEE Transactions on Power Electronics*, *28*(5), 2151–2169. doi:10.1109/TPEL.2012.2212917

Zhou, X., Wang, P., & Gao, Z. (2018). ADMM-Based Decentralized Charging Control of Plug-In Electric Vehicles with Coupling Constraints in Distribution Networks. In *37th Chinese Control Conference (CCC)* (pp. 2512-2517). Wuhan, China: IEEE. 10.23919/ChiCC.2018.8483082

Chapter 6
Voltage Stability Assessment Techniques for Modern Power Systems

Mahiraj Singh Rawat
National Institute of Technology Uttarakhand, India

Shelly Vadhera
National Institute of Technology Kurukshetra, India

ABSTRACT

In recent times, most of the power systems are made to operate close to their operating limits owing to various reasons like slow pace of transmission line expansion, environmental constraints, deregulated electricity market, etc. Therefore, the issue of maintaining the system stability has become the primary objective of the utility companies. The recent development and integration of renewable energy sources have further pushed the modern power systems to system security risks. The voltage instability had been the major cause of recent blackouts around the world. The timely assessment of voltage stability can prevent the blackouts in the power systems. This chapter explores the classical as well as newly developed static voltage stability assessment techniques proposed by various researchers in recent years. Also, the chapter cater to the needs of undergraduate as well as graduate students, professional engineers, and researchers who all are working in the domain of power system voltage stability.

INTRODUCTION

The phenomena of power system voltage stability is being continuously studied through various offline and online computer simulation programs at energy control centers. In literature, the authors have proposed and implemented various static and dynamic techniques for the assessment of voltage stability. The static voltage assessment techniques utilize the Newton-Raphson (N-R) based power flow programs, whereas the dynamic techniques requires the time domain simulation with mathematical modeling of various power system components such as automatic voltage regulators (AVRs), generators, governors,

DOI: 10.4018/978-1-5225-8551-0.ch006

and under load tap changing transformers (ULTC) etc. (Kundur et al., 2004). Plotting the P-V/ Q-V curves for the particular load buses selected are the most widely used methods for assessment of voltage stability (Ajjarapu, 2009). Traditionally, when P-V/Q-V curves are to be plotted, the power flow solutions are executed repeatedly with increasing the loads in steps until solution diverges. The system load at which the Jacobian of N-R method becomes singular is considered as maximum loading/critical point. The problem of divergence in the power flow results are mostly due to mathematical error or if the system states reach to a critical point. Therefore, the main drawback of this method is the recognition of the critical point. The problem of divergence in power flow solutions is resolved by the continuation power flow (CPF) technique. In CPF, the power flow equations are reformulated and locally parameterized continuation technique is applied (Ajjarapu & Cristy 1992).

The voltage stability is not only about the identification of critical point, but also it is necessary to find the factors that influence it. The bus voltage sensitivity with respect to change in reactive load demand (dV/dQ) is used as a tool for assessment of voltage stability by Faltabo, Ognedal, & Carlson (1990). The modal analysis technique which is an indirect method for calculating dV/dQ sensitivities is proposed by Gao, Morison, & Kundur (1992). In order to determine the stability margin between current operating point to the critical point, several voltage stability indices (VSIs) are proposed in the literature. Recently, a comprehensive review on VSIs had been presented by Modarresi et al. (2016). Other voltage stability methods such as bifurcation analysis (Ajjarapu 1992; Dobson 1993; Alvarado, 1994), direct methods and energy function (Overbye & DeMarco, 1991) methods can be found in literature. In recent years, wide deployment of phasor measurement units (PMUs) in power systems had opened a new perspective for developing voltage stability assessment techniques. The PMUs based voltage assessment techniques can be classified into local (Corsi & Tarantu, 2008) and wide area monitoring (Glavic 2009; Beiraghi 2013). Since, PMUs capture large amount of data, therefore for large power systems, Mohammadi & Behghani (2015), Diao et al., (2009) have proposed computational intelligence based techniques such as decision trees for online voltage stability assessment. The chapter reflects the major assessment techniques that are useful for the voltage stability of power system.

BACKGROUND

According to IEEE/CIGRE joint task force report, power system stability is defined as "*Power system stability is the ability of an electric power system, for a given initial operating condition, to regain a state of operating equilibrium after being subjected to a physical disturbance, with most system variables bounded so that practically the entire system remains intact*" (Kundur et al., 2004, p. 1388). Power system instability may be evident in various ways depending upon the power system configuration and operating mode. Different phenomenon that lead to instability are presented in Kundur, (1994), Kundur (2006). These phenomenon are referred to as: 1) rotor angle stability, 2) voltage stability and 3) frequency stability. The broad classification of power system stability is shown in Figure1. Rotor angle stability is related to the ability of the synchronous generators in an interconnected power network to maintain synchronism after large disturbances such as sudden loss of large loads, line tripping due to faults etc. The frequency instability arises due to imbalance between total generation and load demand. In recent years, the voltage instability was identified as the primary cause of many blackouts around the world.

Figure 1. Classification of power system stability

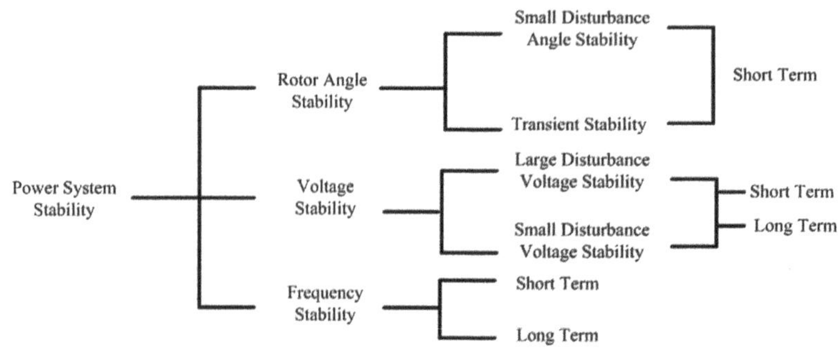

The primary causes of voltage instability are load characteristics, reactive power limits on generators, reactive power compensation devices and operation of voltage control devices such as under load tap changer (ULTC) transformer, etc.

Voltage stability is considered as a local phenomena which gets driven by load characteristics. In case of disturbance, the power consumed by loads try to restore the power by the action of slip adjustment in induction motor loads, voltage regulators and tap changing transformers at distribution and thermostats installed at air conditioner units. Restored loads further increase the reactive load demand on transmission network and cause a further voltage reduction. Voltage stability depends upon the ability of the system to maintain equilibrium between load demand and load supply from the power system.

Voltage stability as per the IEEE/CIGRE joint task force report is, "*The ability of a power system to maintain steady voltages at all buses in the system after being subjected to a disturbance from a given initial operating condition*" (Kundur et al., 2004, p. 1390). Voltage instability may appear in the form of rise of voltage or fall in voltage on some buses. The voltage instability leads into blackouts or low voltages in a large area because of cascaded events such as tripping of transmission line, loss of large load and generator tripping by their protective systems. According to disturbance type, the voltage stability can be classified into large and small disturbance voltage stability, whereas if time frame is important, the voltage stability may be either long term or short term. In the literature, various static voltage stability assessment methods have been proposed and implemented on the real power systems.

The long term static voltage stability assessment techniques employ the non-linear power flow solutions. However, the dynamic studies, where both the short and long term voltage stability can be detected usually make use of time domain simulation with dynamic modeling of generators, governors, loads, automatic voltage regulators (AVRs) and online tap changers (OLTCs) etc.

Large Disturbance Voltage Stability

In this type of voltage stability, it is must for a system to maintain the voltage stability even after large disturbances such as loss of large generators, tripping of transmission lines, system faults etc. For identification of this type of voltage stability, the time domain analysis of the nonlinear power system is required to capture the interaction of devices e.g., induction motor load, under load tap changer (ULTC), and over excitation limiters (OXLs) at generation unit. The time period required for analysis of long term voltage stability varies widely from a few seconds to those in several minutes.

Small Disturbance Voltage Stability

The small disturbance voltage stability is the ability of the power system to maintain the voltage stable at all its buses during small disturbances such as a gradual change in the system load demand. The small signal voltage stability depends upon the load characteristics, discrete and continuous control operations at a specific instant of time. To investigate this type of voltage stability problem, the nonlinear system equations can be linearized with appropriate assumptions. However, this technique cannot account the effect of tap changer control (discrete tap step, time delays and dead time). The time interval for this type of voltage stability problem varies between few seconds to tens of minutes.

Long Term Voltage Stability

It involves the study of power system with equipments such as tap changing transformer, over excitation limiter at the generator and thermostatically controlled load such as air conditioners. This type of stability problem evolves due to an outage of the equipments instead of any severity of the initial disturbances. The cause of such type of voltage instability is due to the loss of equilibrium for long term (e.g., those loads which try to absorb more power than that of combined capability of generation and transmission systems), the post disturbance steady state operating point being unstable due to small disturbances or lack of attraction towards the stable post disturbances equilibrium point (e.g., when the action taken as remedy is delayed). In such cases, the static analysis can be implemented to determine stability margins, to find the various factors that can influence stability and to rank the broad range of system conditions and a variety of scenarios.

Short Term Voltage Stability

This type of voltage stability problem is associated with fast dynamics of load components such as induction motors, high voltage DC converters and electronically controlled loads. The time interval for the short term voltage stability problem is in the order of several seconds. To investigate the stability problem, the solution of detailed system, differential equations is required. The dynamic load model is often essential.

CLASSIFICATION OF STATIC VOLTAGE STABILITY ASSESSMENT TECHNIQUES

The classification of static voltage stability assessment techniques which are widely used in literature are represented in Figure 2. The voltage stability assessment techniques are classified in broadly three categories (1) the methods based on the reformulation of the Jacobian matrix of power flow (2) voltage stability indices (3) measurements based techniques.

Figure 2. Classifications of voltage stability assessment techniques

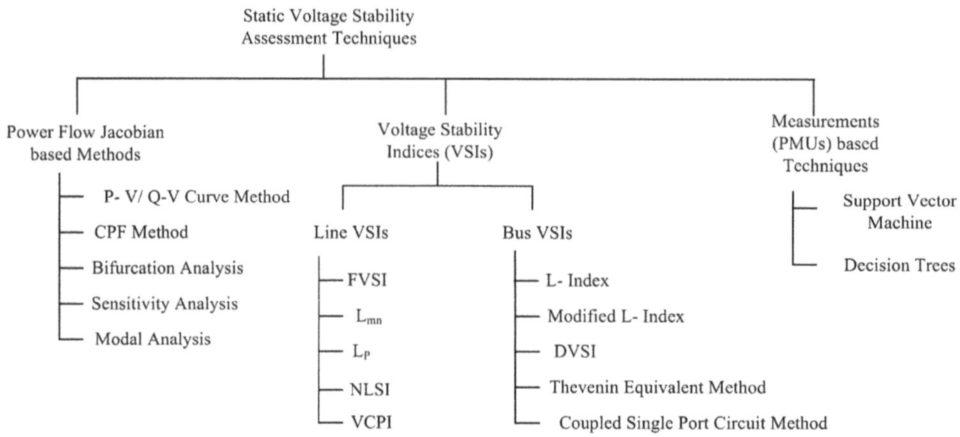

POWER FLOW BASED TECHNIQUES

The electric utility companies largely depend upon the power flow programs for static assessment of voltage stability. The voltage stability gets evaluated by computation of the P-V/ Q-V curves at selected load buses (Ajjarapu, 2009, Cutsem, 2000). Generally, P-V/Q-V curves are obtained after execution of a large number of power flow solutions by making use of conventional models. For determining the voltage stability margins between current operating point to the voltage collapse point, repetitively the power flow solutions have to be carried out by increasing the load demand in steps until the power flow solution diverges. At the point near to voltage collapse, the Jacobian matrix of Newton Raphson based power flow becomes singular. Therefore, the exact location of a voltage collapse point cannot be determined. The problem of singularity at voltage collapse point is solved with the continuous power flow (CPF) method by restructuring the power flow equations and by applying the local parameterization technique (Ajjarapu & Christy, 1992). The major drawback of the P-V/ Q-V curve technique is that to check the system voltage stability margin, the curves have to be generated for each load bus, which is a laborious task. Also, these methods do not provide any sensitive information useful in making design decisions.

The sensitive information refers to the identification of those factors which can influence the voltage stability margins. The sensitivity information obtained from the power flow Jacobian matrix was evaluated using dV/dQ method for assessment of voltage stability margins (Flatabo, Ognedal, & Carlson, 1990). A static voltage stability assessment method based on modal analysis technique, which utilize the eigenvalues and associated eigenvectors of the reduced Jacobian matrix was proposed (Gao, Morison, & Kundur, 1992). The modal analysis technique is an indirect way to calculate the dV/dQ sensitivities. In order to get the stability margin between the current operating point and critical point where the system becomes unstable, several voltage collapse indices are proposed in the literature for assessment of voltage stability (Modarresi, Gholipour, & Khodabakhshian 2016).

P-V and Q-V Curve Based Techniques

P-V Curve Tracing

To draw the P-V curve, a series of power flow solutions are executed for increasing power transfer of MW and the voltages at a certain bus are noted for each increment of MW load. The typical P-V curve is shown in Figure 4. for various power factor loads. On tracing the P-V curve, it is observed that voltage magnitude decreases from its nominal value with each increment of active power (in MW) at load bus. After reaching the nose point of the curve, no more active power can be transferred to the load. This point is referred to as critical voltage point ($V_{critical}$) or maximum power transfer point beyond that system voltage will collapse. The mathematical representation of critical voltage or maximum power transfer can be clearly understood from an example of a two bus system as framed in Figure 3. It is considered that a generator is supplying power to the load through lossless transmission line. The receiving end voltage (E_2) can be represented by Equation (1).

$$\overline{E_2} = \overline{E_1} - jX\overline{I_{12}} \tag{1}$$

Where, $\overline{E_1}$ and $\overline{E_2}$ are voltages at generator and load terminal respectively; X is transmission line reactance; I_{12} is current flowing through transmission line. From the Figure 3, the load current I_L is equal to I_{12} i.e., $\overline{I_L} = \overline{I_{12}}$. The complex power transferred to load is represented by Equation (2).

$$S_L = P_L + jQ_L = \overline{E_2}\,\overline{I_L^*} = \overline{E_2}\left(\frac{\overline{E_1} - \overline{E_2}}{jX}\right)^* \tag{2}$$

Substituting the value of I_L from Equation (1).

Figure 3. A simplified two bus system for lossless transmission line

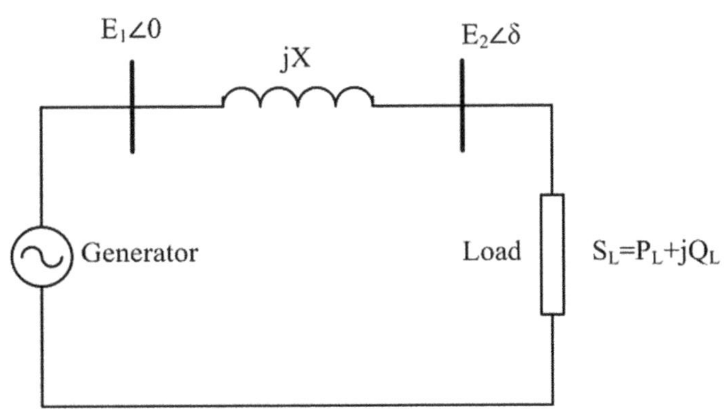

Figure 4. The P-V curve generated at load bus at different power factors

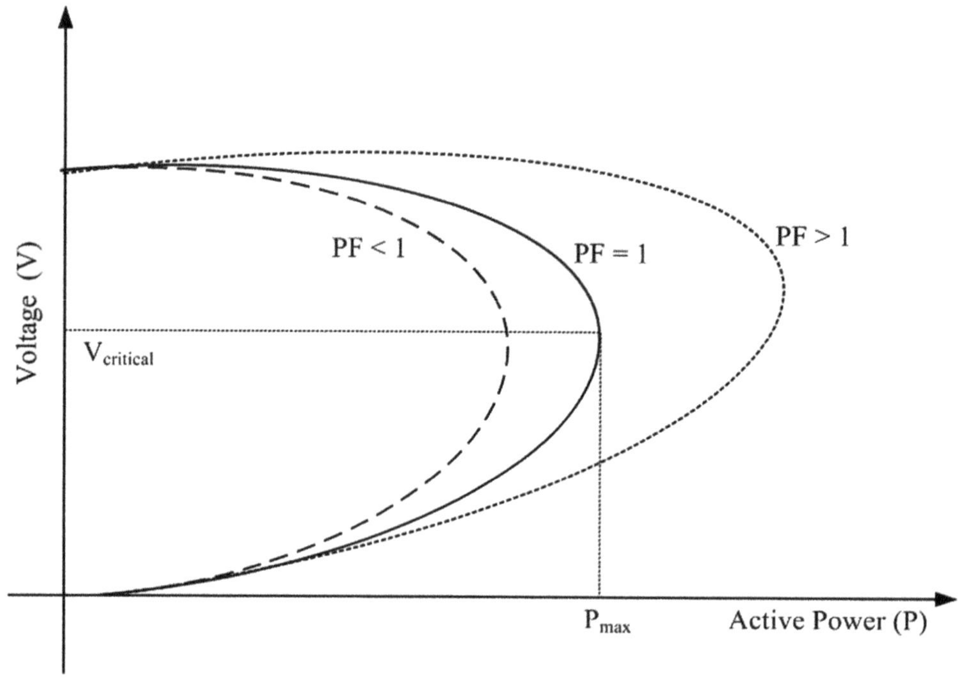

$$S_L = \overline{E_2}\left(\frac{\overline{E_1^*} - \overline{E_2^*}}{-jX}\right) = \left|E_2\right|\angle\delta\left(\frac{\left|E_1\right|\angle 0 - \left|E_2\right|\angle - \delta}{-jX}\right)$$

$$S_L = \frac{j}{X}\left(\left|E_1\right|\left|E_2\right|\cos\delta + j\left|E_1\right|\left|E_2\right|\sin\delta - \left|E_2\right|^2\right)$$

$$P_L + jQ_L = -\frac{\left|E_1\right|\left|E_2\right|\sin\delta}{X} + \frac{j}{X}\left(\left|E_1\right|\left|E_2\right|\cos\delta - \left|E_2\right|^2\right) \tag{3}$$

Comparing the real and imaginary part of left and right side of Equation (3).

$$P_L = -\frac{\left|E_1\right|\left|E_2\right|\sin\delta}{X} \tag{4}$$

$$Q_L = -\frac{\left|E_2\right|^2}{X} + \frac{\left|E_1\right|\left|E_2\right|\cos\delta}{X} \tag{5}$$

Eliminating sin δ and cos δ term from Equation (4) and (5) using trigonometry identity $\sin^2\delta + \cos^2\delta = 1$.

$$\sin\delta = -\frac{P_L X}{|E_1||E_2|} \tag{6}$$

$$\cos\delta = \frac{Q_L X + |E_2|^2}{|E_1||E_2|} \tag{7}$$

$$\left(-\frac{P_L X}{|E_1||E_2|}\right)^2 + \left(\frac{Q_L X + |E_2|^2}{|E_1||E_2|}\right)^2 = 1 \tag{8}$$

$$\frac{P_L^2 X^2}{E_1^2 E_2^2} + \left(\frac{Q_L^2 X^2 + E_2^4 + 2E_2^2 Q_L X}{E_1^2 E_2^2}\right) = 1 \tag{9}$$

$$P_L^2 X^2 + Q_L^2 X^2 + E_2^4 + 2E_2^2 Q_L X = E_1^2 E_2^2 \tag{10}$$

$$\left(E_2^2\right)^2 + \left(2Q_L X - E_1^2\right) E_2^2 + X^2\left(P_L^2 + Q_L^2\right) = 0 \tag{11}$$

For obtaining the feasible solution of E_2, $b^2 - 4ac \geq 0$.

$$\left(2Q_L X - E_1^2\right)^2 - 4X^2\left(P_L^2 + Q_L^2\right) \geq 0 \tag{12}$$

The above equation can be written in the simplified form given by Equation (13)

$$-P^2 - \frac{E_1^2}{X} Q + \left(\frac{E_1^2}{2X}\right)^2 \geq 0 \tag{13}$$

The solution of E_2 is given by Equation (14).

$$E_2 = \sqrt{\frac{E_1^2}{2} - QX \pm \sqrt{\frac{E_1^4}{4} - X^2P^2 - X\,E_1^2\,Q}} \tag{14}$$

$$E_2 = \sqrt{\frac{E_1^2}{2} - \left(P\tan\theta\right)X \pm \sqrt{\frac{E_1^4}{4} - X^2P^2 - X\,E_1^2\left(P\tan\theta\right)}} \tag{15}$$

Where, θ is the power factor angle. The relationship between E_2 and P for constant power factor are represented by P-V curves.

Q-V Curve Tracing

The Q-V curve at any bus show the variations between bus voltage and the reactive power injected (Q) at the same bus. To generate the Q-V curve at any bus, a fictitious generator is placed on that bus. The fictitious generator is a synchronous condenser that injects a variable reactive power into the bus. With the variation of bus voltage, the reactive power output from fictitious generator is calculated in order to maintain the bus voltage at set point. The typical Q-V curve characteristics are shown in Figure 5. The vertical axis (y axis) represents the MVAr output of synchronous condenser (fictitious generator) and the horizontal axis (x axis) reflects the bus voltage. When, the system is at nominal load, the Q-V curve is as shown by curve 1 (Figure 5). Tracing the curve, it is observed that with voltage set point being varied from higher to lower voltages, the fictitious generator's reactive power output gets decreased, which is represented in the form of increment in MVAr load. The bottom of the curve will be reached when the reactive power output from fictitious generator stop decreasing. This minimum point represents the maximum MVAr value that a load can possess before voltage collapse. The curve 2 represents the system with no reactive power reserve and with small increment in system load could result into voltage instability. Similarly, curve 3 represents the system is under shortage of reactive power reserve and the situation of voltage collapse could be avoided if additional reactive power reserve i.e., capacitor banks, synchronous condensers, flexible AC transmission system (FACTS) devices are switched on within a certain period of time. For tracing the Q-V curve for particular load bus, P is assumed constant in Equation (14).

Continuous Power Flow Technique

The difficulty faced in Newton Raphson (N-R) based power flow method while generating the P-V curve at any load bus is that the Jacobian in N-R power flow becomes singular at a critical point. To avoid the singularity problem in the N-R Jacobian, Ajjarapu and Christy have proposed continuous power flow (CPF) method. In CPF method, the power flow nonlinear equations are reformulated and also a locally parameterized continuation technique is applied so that singularity problem in Jacobian matrix should not encountered. The formulation of CPF method is given as follows.

In continuation power flow method, 'λ' the load factor is inserted in power flow equations in order to apply local parameterization technique. The active and reactive power balance equations at any bus i are given by Equation (16) and (17).

Figure 5. The Q-V curves generated for different cases

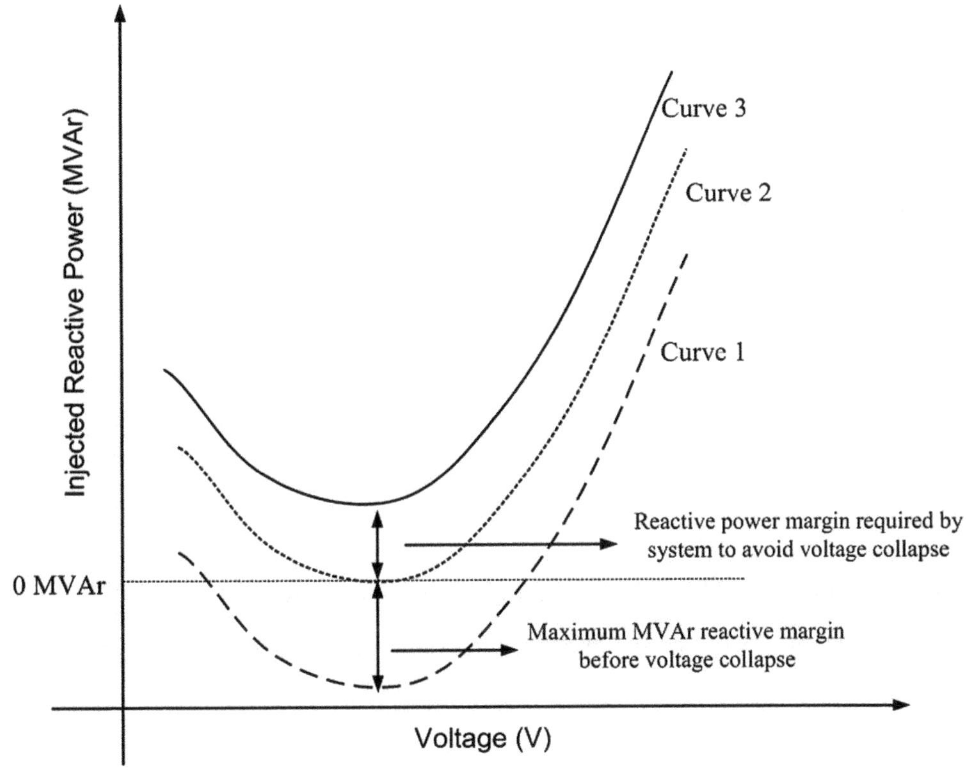

Figure 6. Predictor corrector used in CPF method

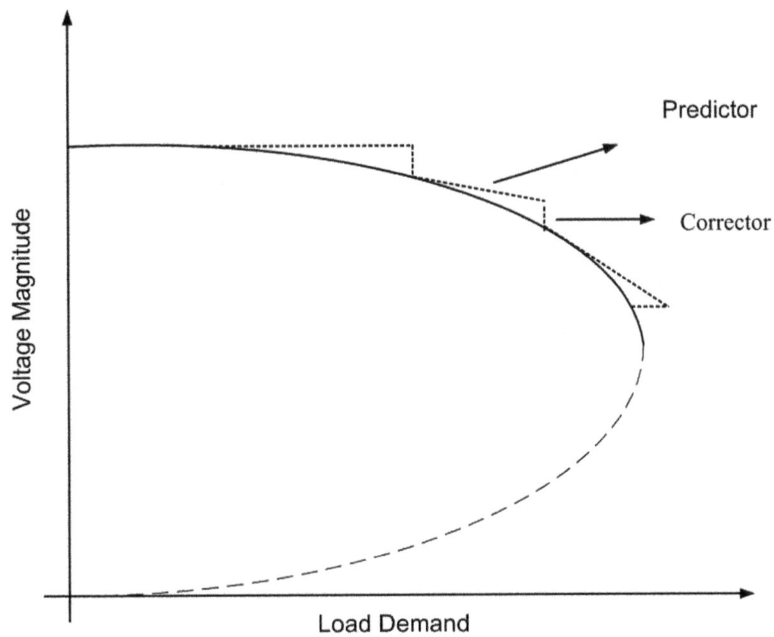

$$0 = P_{G_i} - P_{L_i} - P_{cal_i} \tag{16}$$

$$0 = Q_{G_i} - Q_{L_i} - Q_{cal_i} \tag{17}$$

Where, P_{cal_i} and Q_{cal_i} are given by Equation (18) and (19).

$$P_{cal_i} = \sum_{j=1}^{m} |V_i||V_j||Y_{ij}|\cos\left(\delta_i - \delta_j - \theta_{ij}\right) \tag{18}$$

$$Q_{cal_i} = \sum_{j=1}^{m} |V_i||V_j||Y_{ij}|\sin\left(\delta_i - \delta_j - \theta_{ij}\right) \tag{19}$$

Where, the voltage at bus i and j for m bus systems are represented by $|V_i|\angle\delta_i$ and $|V_j|\angle\delta_j$ respectively. $|Y_{ij}|\angle\theta_{ij}$ is the (i, j)[th] element of system admittance matrix Y_{bus}. The load change in the load flow simulation program can be modified by using Equation (20) and (21).

$$P_{L_i} = P_{L_{i0}} + \lambda\left(K_{L_i} S_{\Delta base} \cos\varphi_i\right) \tag{20}$$

$$Q_{L_i} = Q_{L_{i0}} + \lambda\left(K_{L_i} S_{\Delta base} \sin\varphi_i\right) \tag{21}$$

First term on right side of Equation (20) and (21) represents the base case load at i[th] bus and the second term stands for load change brought by change in load factor (λ). $P_{L_{i0}}$ and $Q_{L_{i0}}$ are representing the active and reactive loads at i[th] bus under base load condition. K_{L_i} is the multiplying factor representing the rate of change in load power as λ changes. $S_{\Delta base}$ is the apparent power selected for scaling of λ. φ_i is the power factor angle of load change at i[th] bus. The active power generated can now be obtained as per Equation (22).

$$P_{G_i} = P_{Gi0}\left(1 + \lambda K_{Gi}\right) \tag{22}$$

Where, P_{Gi0} is the active power generated at i[th] bus under base case; K_{Gi} is multiplying factor representing the change in the generation rate with respect to λ. The whole set of equations in a reformulated power flow can be expressed by Equation (23).

$$\overline{F}\left(\bar{\delta},\overline{V},\lambda\right)=0\ 0\leq\lambda\leq\lambda_{Critical} \tag{23}$$

Where, $\bar{\delta}$ and $\overline{V}$ represents the bus voltage angle and magnitude respectively. The base case solution (when λ=0) can be obtained with traditional N-R power flow solutions and the solution for range of λ is being sought. The CPF method starts with a known solution and make use of predictor and corrector steps in order to achieve solutions for different values of load demand. The predictor and corrector steps are given as follows.

Predictor Step

After obtaining the power flow solution in the base case (λ=0), the prediction for the next solution is obtained by considering an appropriate sized step in a direction tangent to solution path. The tangent vector is obtained by first taking the derivative of Equation (23) on both sides.

$$d\,\overline{F}\left(\bar{\delta},\overline{V},\lambda\right)=\overline{F}_{\delta}\overline{d\delta}+\overline{F}_{V}\,\overline{dV}+\overline{F}_{\lambda}\,d\lambda=0 \tag{24}$$

$$\begin{vmatrix}\overline{F}_{\delta} & \overline{F}_{V} & \overline{F}_{\lambda}\end{vmatrix}\begin{vmatrix}\overline{d\delta}\\ \overline{dV}\\ d\lambda\end{vmatrix}=0 \tag{25}$$

The left side of Equation (25) is the power flow Jacobian matrix augmented by one column multiplied by a tangent vector being sought. One problem arises while solving the Equation (25) that the additional unknown are added to the power flow equation when load factor λ is added to power flow equations. But, the number of equations remain unchanged. To overcome this problem, one component of tangent vector is chosen to a value of that of a non-zero magnitude (normally one). In other words, tangent vector $\bar{t}$ is represented by Equation (26).

$$\bar{t}=\begin{vmatrix}\overline{d\delta}\\ \overline{dV}\\ d\lambda\end{vmatrix};\ t_{k}=\pm1 \tag{26}$$

Now, Equation (25) can be modified by Equation (27).

$$\begin{vmatrix}\overline{F}_{\delta} & \overline{F}_{V} & \overline{F}_{\lambda}\\ & \overline{e}_{k} & \end{vmatrix}\begin{vmatrix}\bar{t}\end{vmatrix}=\begin{vmatrix}0\\ \pm1\end{vmatrix} \tag{27}$$

Where, e_k represents the row vector which has all the elements zero except kth which is equal to 1. Choosing the suitable value of $t_k = \pm 1$ imposes a nonzero norm on the tangent vector which guarantees that the augmented Jacobian matrix will be non-singular at the critical point. The selection of t_k value depends upon the variation of kth state variable with respect to change in solution path being traced. +1, is used in case it is increasing otherwise -1 to be used. After obtaining the tangent vector, the predictions are made by using Equation (28).

$$\begin{vmatrix} \bar{\delta}^* \\ \bar{V}^* \\ \lambda^* \end{vmatrix} = \begin{vmatrix} \bar{\delta} \\ \bar{V} \\ \lambda \end{vmatrix} + \sigma \begin{vmatrix} d\bar{\delta} \\ d\bar{V} \\ d\lambda \end{vmatrix} \tag{28}$$

Correction Step

Now, a need arises for a method to correct the solution that has been predicted from the last step.

$$\text{Let, } \bar{x} = \begin{vmatrix} \bar{\delta} \\ \bar{V} \\ \lambda \end{vmatrix} \text{ and } x_k = \eta \tag{29}$$

Where, η is the approximate value for the kth element in $\bar{x}$. Equation (30) gives the new set of equations.

$$\begin{vmatrix} \bar{F}\left(\bar{x}\right) \\ x_k - \eta \end{vmatrix} = |0| \tag{30}$$

Now, once a suitable value of η and index k are selected, the modified power flow N-R method can be solved to find a correct solution.

Sensitivity Analysis

In this section, the sensitivity analysis proposed by Flatabo et al. (1990) has been discussed. The assessment of voltage stability is not limited in identifying the critical point itself, but the onus also lies in identifying all the system conditions that affects the critical point. The influence of variations in control parameters are studied through sensitivity analysis. The common methodology adopted under any sensitivity analysis is firstly to define the voltage stability index and secondly to study the effect of different system parameters and control through this index. The sensitivity based voltage index defined by Flatabo et al., are categoried as follows.

1. Reactive power sensitivity
2. Voltage sensitivity
3. Determination of MVAr margin to voltage collapse
4. Determination of MW margin to voltage collapse

The power flow equation is given by Equation (31).

$$f\left(x, u, p\right) = 0 \tag{31}$$

Where, x is the vector of state variables, i.e., voltage magnitude and angles at load bus; u is the vector of control variables, i.e., voltages at generator bus, MW generation and MVAr compensation; p represents parameters, i.e., MW and MVAr load. Applying Taylor series expansion to Equation (31).

$$\Delta f = f_x \Delta x + f_u \Delta u + f_p \Delta p = 0 \tag{32}$$

The sensitivity of the state variable is represented by Equation (33).

$$\Delta x = -f_x^{-1} f_u \Delta u - f_x^{-1} f_p \Delta p \tag{33}$$

$$\Delta x = S_{xu} \Delta u + S_{xp} \Delta p \tag{34}$$

Where, S_{xu} and S_{xp} are sensitivities of state variable x with respect to control variable u and input parameter p respectively. The dependent variable *w* is considered to be representing the generator reactive power production and reference bus MW production. The dependent variable *w* is also a function of state variable x, control variable u and parameter p and can be represented by Equation (35).

$$w = w\left(x, u, p\right) \tag{35}$$

Applying Taylor series expansion to Equation (35).

$$\Delta w = w_x \Delta x + w_u \Delta u + w_p \Delta p \tag{36}$$

Substituting, Δx value from Equation (33) to Equation (36).

$$\Delta w = \left(-w_x f_x^{-1} f_u + w_u\right) \Delta u - w_x f_x^{-1} f_p \Delta p \tag{37}$$

$$\Delta w = S_{wu}\,\Delta u + S_{wp}\,\Delta p \tag{38}$$

Where, S_{wu} and S_{wp} are sensitivities of dependent variable w with respect to control variable u and input parameter p.

Reactive Power Sensitivity

The sensitivity of reactive power generated with respect to active and reactive load demand variations in a specified bus is represented by the sensitivity matrix (S_{wp}). The variations in the reactive power generation at j[th] bus due to variations in reactive power load demand at i[th] bus can be determined from Equation (39).

$$\Delta Q_{ji_q} = \frac{dQ_{ji}}{dQ_i}\Delta Q_i \tag{39}$$

$$\frac{dQ_{ji}}{dQ_i} = S_{wp}\left(j, i_q\right) \tag{40}$$

Where, $S_{wp}\left(j, i_q\right)$ are the matrix elements which relates the reactive power generated at j[th] bus and reactive power load demand at i[th] bus. The reactive power generated (ΔQ_{gen}) at all the generators in the system when a reactive load at a specified bus is incremented by ΔQ_i is given by Equation (41).

$$\Delta Q_{gen} = \sum_{i=1}^{n_g} \frac{dQ_{ji}}{dQ_i}\Delta Q_i = \sum_{i=1}^{n_g} S_{wp}\left(j, i_q\right)\Delta Q_i = Q_{S_i}\Delta Q_i \tag{41}$$

Where, n_g is the number of generators; Q_{S_i} is a sensitivity factor whose value will become infinite as when the system approaches near to a critical point.

Voltage Sensitivity

The voltage sensitivity at i[th] bus, the voltage sensitivity (VS_i) is represented by Equation (42). The voltage sensitivity factor is corresponding to an element related to ΔV_i and ΔQ_i in the sensitivity matrix S_{xp}. The value of slope approaches to infinity when the voltage reaches near to collapse point.

$$VS_i = \left.\frac{dV}{dQ}\right|_i \tag{42}$$

Determination of MVAr Margin to Voltage Collapse

The sensitivity factor, i.e., Q_{si} and VS_i are linearized around an operating point which varies with change in the network conditions. These factors are only the snapshots of the system conditions. To obtain the MVAr margin to voltage collapse, the multiple power flow solutions have to be carried out. As the system reaches near to voltage collapse, N-R Jacobian becomes singular. This singularity may occur either due to the mathematical problem or if voltage has reached to a critical point. Hence, divergence in power flow cannot be considered as an indicator for voltage instability. The MVAr margin is defined as the maximum reactive power load (in MVAr) that can be loaded on a certain bus before the system reaches voltage instability. The MVAr margin to voltage collapse is determined by following steps.

Step 1. Determine the sensitivity matrices S_{wp} and S_{xp} at initial condition.
Step 2. Search for the minimum value of load ΔQ_i at a specific bus i that can be incremented before reaching maximum MVAr generation capacity of a MVAr generation plant (i.e., generators, synchronous compensator, SVC)
Step 3. Update the dependent variables w and x using Equation (43) and (44).

$$\Delta w = S_{wp} \Delta Q_i^n \tag{43}$$

$$\Delta x = S_{xp} \Delta Q_i^n \tag{44}$$

Step 4. Update vector x, u, w.
Step 5. Recalculate the sensitivity matrices S_{xp} and S_{wp}.
Step 6. Observe the change in sign of any VS_i. If the system is stable return to step 2, otherwise go to step 7.
Step 7. Represent the MVAr margin to voltage collapse as per Equation (45).

$$\Delta Q_{ic} = \sum_{n=1}^{n_{\max i_q}} \Delta Q_i^n \tag{45}$$

The available reactive power reserve at i[th] bus can be determined by Equation (46).

$$\Delta Q_{res(i_q)} = \sum_{n=1}^{n_{\max i_q}} \left[\sum_{j=1}^{n_q} S_{wp}\left(j, i_q\right) \Delta Q_i^n \right] \tag{46}$$

Where, n_q is the number of MVAr generation plant; n_{maxiq} is the number of steps used in MVAr distance calculations, when the reactive load is increased at i[th] bus.

Determination of MW Margin to Voltage Collapse

The change in the reactive power generation at j[th] bus is given by Equation (47) when MW load is increased at i[th] bus.

$$\Delta Q_{ji_p} = \frac{dQ_{ji}}{dP_i}\Delta P_i = S_{wp}\left(j, i_p\right)\Delta P_i \tag{47}$$

Where, $S_{wp}(j, i_p)$ is the matrix element which represents the generated reactive power at bus j to active power load demand at bus i. The available reactive power reserve at the bus 'i' with respect to increasing active power demand up to the voltage collapse point can be determined by Equation (48).

$$\Delta Q_{res\left(i_p\right)} = \sum_{n=1}^{n_{\max i_p}}\left[\sum_{j=1}^{n_q} S_{wp}\left(j, i_p\right)\Delta P_i^n\right] \tag{48}$$

The MW distance is evaluated by Equation (49).

$$\Delta P_{ic} = \sum_{n=1}^{n_{\max_{i_p}}} \Delta P_i^n \tag{49}$$

Where, n_{max_ip} is the maximum number of steps required to determine the MW distance at bus i from current operating point to voltage collapse point when active power load is increased in steps. The step length is given by Equation (50).

$$\Delta P_i^n = \min\left[\frac{\left(Q_{j\max} - Q_j\right)}{S_{wp}\left(j, i_p\right)}\right] \tag{50}$$

Modal Analysis

Gao, Morison, & Kundur (1992) have proposed voltage stability assessment of the system using modal analysis. The system is considered to be voltage stable if Δ|V|/ ΔQ is positive whereas it is considered unstable if this ratio is negative for at least one bus. However, in modal analysis, active power is taken as constant at each operating point. Considering this approach, the linearized power mismatch equation is given as per Equation (51).

$$\begin{bmatrix}\Delta P\\ \Delta Q\end{bmatrix} = J\begin{bmatrix}\Delta\delta\\ \Delta|V|\end{bmatrix} = \begin{bmatrix}J_1 & J_2\\ J_3 & J_4\end{bmatrix}\begin{bmatrix}\Delta\delta\\ \Delta|V|\end{bmatrix} \tag{51}$$

Where, ΔP, Δ Q represents the incremental change in active and reactive power respectively, and Δ |V|, Δδ represents the incremental change in voltage magnitude and angle respectively. J is the Newton Raphson Jacobian matrix which consists of partial derivative terms. Let ΔP =0, then Equation (51) can be solved as follows.

$$\begin{bmatrix} 0 \\ \Delta Q \end{bmatrix} = \begin{bmatrix} J_1 & J_2 \\ J_3 & J_4 \end{bmatrix} \begin{bmatrix} \Delta\delta \\ \Delta|V| \end{bmatrix} \tag{52}$$

$$0 = J_1 \Delta\delta + J_2 \Delta|V|$$

$$\Delta\delta = -J_1^{-1} J_2 \Delta|V| \tag{53}$$

$$\Delta Q = J_3 \Delta\delta + J_4 \Delta|V| \tag{54}$$

Substituting Δδ from Equation (53) into Equation (54),

$$\Delta Q = \left[J_4 - J_3 J_1^{-1} J_2\right] \Delta|V| \tag{55}$$

$$\Delta Q = J_{red} \Delta|V| \tag{56}$$

Where $J_{red} = [J_4 - J_3 J_1^{-1} J_2]$ and the modes of instability can be defined by the eigenvalues and eigenvector of J_{red}, given by Equation (57).

$$J_{red} = \xi \wedge \eta \tag{57}$$

Where, ξ = right eigenvector matrix of J_{red}; Λ = diagonal eigenvalue matrix of J_{red}; η = left eigenvector matrix of J_{red}.

$$\Delta|V| = J_{red}^{-1} \Delta Q \,;\; J_{red} = \xi \wedge^{-1} \eta \tag{58}$$

$$\Delta|V| = \sum_i \frac{\xi_i \eta_i}{\lambda_i} \Delta Q \tag{59}$$

Where, ξ_i = i[th] column right eigenvector of J_{red}; η_i= i[th] row left eigenvector of J_{red}

The eigenvalue (λ_i), the right eigenvectors (ξ_i) and left eigenvector (η_i) defines the ith mode of the system. The small eigenvalues represent weak modal voltage. For a voltage stable system, all eigenvalues of the reduced Jacobian matrix must be positive.

Bus Participation Factor

The participation factor of bus k in mode i is given by Equation (60).

$$P_{ki} = \xi_{ki}\eta_{ik} \tag{60}$$

Where, P_{ki} indicates the contribution of the each eigenvalues to the V-Q sensitivity at bus k. Larger the value of P_{ki}, the more λ_i contributes in determining the V-Q sensitivity at bus k. The high value of bus participation factors associated with small eigenvalues determines the weak areas in the power systems.

Generation and Branch Participation Factor

Due to variations in reactive power injection ΔQ_{mi}, the system voltage variation is ΔV_{mi} and i[th] modal angle variation is $\Delta\theta_{mi} = -J_{P\theta}^{-1}J_{PV}\Delta V_{mi}$ with ΔV and $\Delta\theta$ known. Let, ΔQ_{lmn} be the reactive power loss variation in branch (transmission line) m-n and ΔQ_{gki} be the reactive power output variations of generator gk. The branch participation factor at mode i is given by Equation (61).

$$P_{lmn} = \frac{\Delta Q_{lmn}}{\Delta Q_{\mathrm{lmax}_i}} \tag{61}$$

Where, $\Delta Q_{\mathrm{lmax}_i} = \max\left(\Delta \mathrm{Q}_{\mathrm{lmn}}\right)$.

For each mode, the branch participation factor indicates the branch which has the maximum reactive power loss for a fixed incremental change in Q at load buses. The branches with high P_{lmn} are the ones cause the mode i to be weak.

In mode i, the participation factor for generator is given by Equation (62).

$$P_{gki} = \frac{\Delta Q_{gki}}{\Delta Q_{\mathrm{gmax}_i}} \tag{62}$$

For each mode i, the generator participation factor indicates that generator which is supplying the highest reactive power output with respect to incremental change in Q at load buses. The generators with higher P_{gki} value are most important from voltage stability point of view.

VOLTAGE STABILITY INDICES

In literature, various voltage stability indices are proposed. These indices can be broadly classified into line and bus voltage stability indices.

Line Voltage Stability Indices

Voltage Collapse Proximity Index (VCPI)

Moghavvemi & Faruque (1998) have proposed VCPI index based on the concept of maximum power transfer to the load. The maximum power that can be transferred to the load is calculated using $Z_r/Z_s = 1$ and moreover this ratio is also applied as a voltage collapse indicator. In Figure 7, the loading of the line is considered as the power transferred to the receiving end through transmission lines. The load is assumed to have a constant power factor. With increasing load demand, the Z_r decreases and current I increases, which further causes a voltage drop at the receiving end.

The line current is represented by Equation (63).

$$I = \frac{V_s}{\sqrt{\left[\left(Z_s \cos\alpha + Z_r \cos\beta\right)^2 + \left(Z_s \sin\alpha + Z_r \sin\beta\right)^2\right]}} \tag{63}$$

The receiving end voltage is given by Equation (64).

$$V_r = Z_r I = \frac{Z_r}{Z_s} \frac{V_s}{\sqrt{\left[1 + \left(\frac{Z_r}{Z_s}\right)^2 + 2\left(\frac{Z_r}{Z_s}\right)\cos\left(\alpha - \beta\right)\right]}} \tag{64}$$

The active and reactive powers at receiving end are represented by Equation (65) and (66) respectively.

Figure 7. The line model with loading

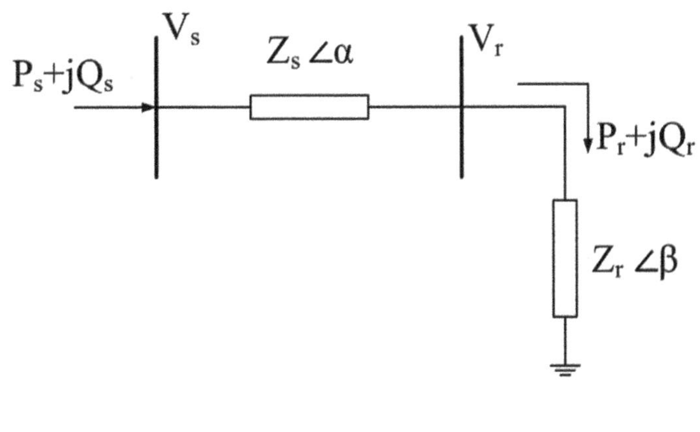

$$\mathrm{P}_r = V_r I \cos \beta \tag{65}$$

$$Q_r = V_r I \sin \beta \tag{66}$$

Equation (65) and (66) can be written as follows.

$$P_r = \frac{\left(V_s\right)^2 / Z_s}{1 + \left(\frac{Z_r}{Z_s}\right)^2 + 2\left(\frac{Z_r}{Z_s}\right)\cos\left(\alpha - \beta\right)} \frac{Z_r}{Z_s} \cos \beta \tag{67}$$

$$Q_r = \frac{\left(V_s\right)^2 / Z_s}{1 + \left(\frac{Z_r}{Z_s}\right)^2 + 2\left(\frac{Z_r}{Z_s}\right)\cos\left(\alpha - \beta\right)} \frac{Z_r}{Z_s} \sin \beta \tag{68}$$

The active and reactive power losses in the transmission line are represented by Equation (69) and (70) respectively.

$$P_{TL} = I^2 Z_s \cos \alpha \tag{69}$$

$$Q_{TL} = I^2 Z_s \sin \alpha \tag{70}$$

Equation (69) and (70) can be written as follows.

$$P_{TL} = \frac{\left(V_s\right)^2 / Z_s}{1 + \left(\frac{Z_r}{Z_s}\right)^2 + 2\left(\frac{Z_r}{Z_s}\right)\cos\left(\alpha - \beta\right)} \frac{Z_r}{Z_s} \cos \alpha \tag{71}$$

$$Q_{TL=} \frac{\left(V_s\right)^2 / Z_s}{1 + \left(\frac{Z_r}{Z_s}\right)^2 + 2\left(\frac{Z_r}{Z_s}\right)\cos\left(\alpha - \beta\right)} \frac{Z_r}{Z_s} \sin \alpha \tag{72}$$

Using boundary conditions, the maximum real power transferred to receiving end can be obtained as $\partial P_r / \partial Z_r = 0$ which leads into $Z_r / Z_s = 1$. Now by substituting $Z_r / Z_s = 1$ into Equation (67) the maximum real power that can be transferred is as per Equation (73) whereas the maximum reactive power transferred is given by Equation (74).

$$P_{r(\max)} = \frac{V_s^2}{Z_s} \frac{\cos\beta}{4\cos^2\left(\frac{\alpha-\beta}{2}\right)} \tag{73}$$

$$Q_{r(\max)} = \frac{V_s^2}{Z_s} \frac{\sin\beta}{4\cos^2\left(\frac{\alpha-\beta}{2}\right)} \tag{74}$$

The real and reactive power losses in the transmission line can be given by Equation (75) and Equation (76) respectively.

$$P_{TL(\max)} = \frac{V_s^2}{Z_s} \frac{\cos\alpha}{4\cos^2\left(\frac{\alpha-\beta}{2}\right)} \tag{75}$$

$$Q_{TL(\max)} = \frac{V_s^2}{Z_s} \frac{\sin\alpha}{4\cos^2\left(\frac{\alpha-\beta}{2}\right)} \tag{76}$$

After obtaining the maximum permissible quantities for transmission line, the voltage collapse proximity indicator (VCPI) is given as follows.

$$VCPI(1) = \frac{P_r}{P_{r(\max)}} \tag{77}$$

$$VCPI(2) = \frac{Q_r}{Q_{r(\max)}} \tag{78}$$

$$VCPI(3) = \frac{P_{TL}}{P_{TL(\max)}} \tag{79}$$

$$VCPI(4) = \frac{Q_{TL}}{Q_{TL(\max)}} \tag{80}$$

Where, P_r, Q_r, P_{TL} and Q_{TL} are obtained from the power flow program.

Fast Voltage Stability Index (FVSI)

FVSI is a line voltage stability index proposed by Musirin & Rahman (2002) in order to determine the weak transmission line in the power network. The sending end bus is considered as reference bus and hence $\theta_s = 0$ and $\theta_r = \theta$. The current flowing through transmission line in Figure 8 can be defined as per Equation (81).

$$I_{sr} = \frac{V_s\angle 0 - V_r\angle -\theta}{R + jX} \tag{81}$$

$$I_{sr} = \left(\frac{S_r}{V_r}\right)^* = \frac{P_r - jQ_r}{V_r\angle\theta} \tag{82}$$

Where, V_s and V_r represents the sending and receiving bus voltages respectively; P_s and P_r are the active powers at sending and receiving bus respectively; Q_s and Q_r are the reactive powers injected at sending and receiving bus respectively; S_s and S_r are the apparent power injected at sending and receiving bus respectively; θs, θr represents the voltage angle at sending and receiving end respectively; $z\angle\alpha$ is the transmission line impedance; R and X are the resistance and reactance of transmission line respectively.

Using Equation (81) and (82).

Figure 8. Equivalent circuit of two bus power system

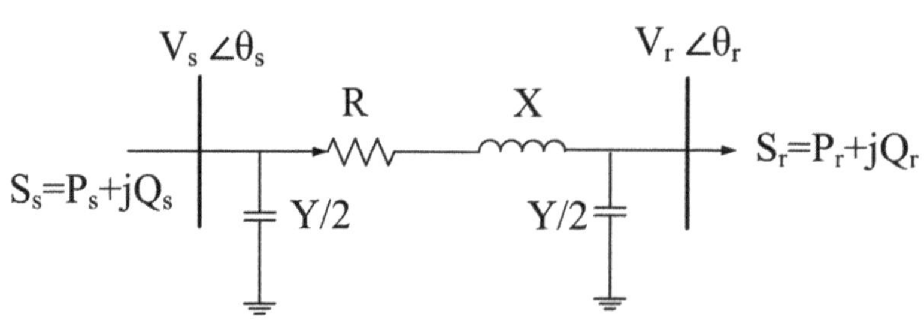

$$\frac{V_s\angle 0 - V_r\angle -\theta}{R + jX} = \frac{P_r - jQ_r}{V_r\angle\theta} \tag{83}$$

$$V_s V_r\angle\theta - V_r^2 = \left(P_r - jQ_r\right)\left(R + jX\right) \tag{84}$$

Equation (84) can be modulated into following quadratic form.

$$V_r^2 + V_s\left(\frac{R\sin\theta}{X} - \cos\theta\right)V_r + Q_r\left(\frac{R^2 + X^2}{X}\right) = 0 \tag{85}$$

Comparing Equation (85) to the standard quadratic equation, we get

$$A = 1;\ B = V_s\left(\frac{R\sin\theta}{X} - \cos\theta\right);\ C = Q_r\left(\frac{R^2 + X^2}{X}\right)$$

For feasible solutions Equation (85), the inequality must be true. $B^2 - 4AC \geq 0$

$$V_s^2\left(\frac{R\sin\theta}{X} - \cos\theta\right)^2 - 4Q_r\left(\frac{R^2 + X^2}{X}\right) \geq 0 \tag{86}$$

The voltage stability index is given by Equation (87). FVSI value must be ≤ 1 for the voltage stable system.

$$FVSI_{sr} = \frac{4Z^2 Q_r X}{\left(V_s\right)^2\left(R\sin\theta - X\cos\theta\right)^2} \leq 1 \tag{87}$$

Assume, $R\sin\theta \approx 0$ and $X\cos\theta \approx X$.

$$FVSI_{sr} = \frac{4Z^2 Q_r}{V_s^2 X} \tag{88}$$

Line Stability Index (LQP)

The formulation of line voltage stability index (LQP), is explained by Mohamed, Jasmon, & Yusoff, 1989. From the Equation (81) and (82), the active and reactive powers at receiving bus are given as follows.

$$P_r = \frac{V_s V_r \cos\theta - V_r^2 - Q_r X}{R} \tag{89}$$

$$Q_r = \frac{P_r X - V_s V_r \sin\theta}{R} \tag{90}$$

Substituting Equation (89) into Equation (90) and vice-versa. Line is considered to be lossless i.e., $\frac{R}{X} \ll 1$. The Equation (89) and (90) can be written in the simplified form.

$$P_r = \frac{V_s V_r \sin\theta}{X}, \text{ hence } \sin\theta = \frac{X P_r}{V_s V_r} \tag{91}$$

$$Q_r = \frac{V_s V_r \cos\theta - V_r^2}{X}, \text{ hence } \cos\theta = \frac{X Q_r + V_r^2}{V_s V_r} \tag{92}$$

Applying trigonometric identity, i.e., $Sin^2\theta + Cos^2\theta = 1$.

$$\left(\frac{X P_r}{V_s V_r}\right)^2 + \left(\frac{X Q_r + V_r^2}{V_s V_r}\right)^2 = 1 \tag{93}$$

$$V_r^4 + \left(2XQ_r - V_s^2\right)V_r^2 + X^2Q_r^2 + P_r^2X^2 = 0 \tag{94}$$

For feasible solutions of Equation (94), the inequality must be true i.e., $B^2 - 4AC \geq 0$.

$$\left(2XQ_r - V_s^2\right)^2 - 4\left(X^2Q_r^2 + P_r^2X^2\right) \geq 0 \tag{95}$$

$$4\left(\frac{X}{V_s^2}\right)\left(\frac{P_s^2 X}{V_s^2} + Q_r\right) \leq 1 \tag{96}$$

The static voltage stability index LQP is given by Equation (97).

$$LQP_{sr} = 4\left(\frac{X}{V_s^2}\right)\left(Q_r + \frac{P_s^2 X}{V_s^2}\right) \tag{97}$$

For voltage stability of transmission lines, the LQP should be ≤ 1.

Line Stability Index (Lp)

Moghavvemi & Omar (1998) had also presented line stability index (Lp) from two bus transmission line model shown in Figure 3. Substituting Equation (90) into Equation (89), then taking Vr as quadratic equation and solving for the real solution yields into Equation (98).

$$-V_s R\cos\theta - V_s X\sin\theta - 4RP_r\left(R^2 + X^2\right) \geq 0 \tag{98}$$

$$\frac{4RP_r\left(R^2 + X^2\right)}{V_s^2\left(R\cos\theta + X\sin\theta\right)} \leq 1 \tag{99}$$

The line index L_p is defined as follows.

$$L_{P_{sr}} = \frac{4RP_r}{\left[V_s\cos\left(\alpha - \theta\right)\right]^2} \tag{100}$$

Line Stability Index (L_{mn})

Moghavvemi & Faraque (2001) have proposed line voltage stability index as L_{mn}. The mathematical formulation of L_{mn} is given as follows.

The power flowing in transmission line at the receiving end is given by Equation (101).

$$S_r = P_r + jQ_r = V_r I_{sr}^* \tag{101}$$

$$I_{sr} = \frac{V_s\angle\theta_s - V_r\angle\theta_r}{Z\angle\alpha} \tag{102}$$

Where, θ_s and θ_r are voltage angle at sending and receiving end of the line; $Z\angle\alpha$ is the impedance of the transmission line. Substituting line current (I_{sr}) value from Equation (102) in Equation (101).

$$S_r = \frac{V_s V_r \angle\left(-\theta_s + \theta_r\right) - V_r^2 \angle 0}{Z\angle - \alpha} \tag{103}$$

$$S_r = \frac{V_s V_r}{Z} \angle\left(\alpha - \theta_s + \theta_r\right) - \frac{V_r^2}{Z} \angle \alpha \tag{104}$$

Let, $\theta = \theta_s - \theta_r$, then the active ($P_r$) and reactive ($Q_r$) powers at receiving end can be given by Equation (105) and (106) respectively.

$$P_r = \frac{V_s V_r}{Z} \cos\left(\alpha - \theta\right) - \frac{V_r^2}{Z} \cos\alpha \tag{105}$$

$$Q_s = \frac{V_s V_r}{Z} \sin\left(\alpha - \theta\right) - \frac{V_r^2}{Z} \cos\alpha \tag{106}$$

Solving Equation (105) and (106) for V_r.

$$V_r = \frac{V_s \sin\left(\alpha - \theta\right) \pm \sqrt{\left(-V_1 \sin\left(\alpha - \theta\right)\right)^2 - 4Z \sin\theta Q_r}}{2\sin\alpha} \tag{107}$$

To obtain the real value of V_r, Equation (107) must have real roots and also it is assumed that $Z\sin\alpha = X$.

$$\left(V_s \sin\left(\alpha - \theta\right)\right)^2 - 4XQ_r \geq 0 \tag{108}$$

The line index L_{mn} can be represented by Equation (109).

$$L_{mn_{sr}} = \frac{4XQ_r}{\left(V_s \sin\left(\alpha - \theta\right)\right)^2} \tag{109}$$

For the voltage stable system, the value of line index L_{mn} should be less than 1.

Novel Line Stability Index (NLSI)

The mathematical formulation of line stability index NLSI proposed by Goharrizi & Asghari (2007) is given as follows.

From Figure 8, the complex power at receiving end is given by Equation (110).

$$S_r = V_r I_{sr}^* \tag{110}$$

$$I_{sr} = \frac{V_r \angle 0 - V_s \angle \delta}{R + jX} \tag{111}$$

Substituting the value of I_{sr} in Equation (110) which yields into Equation (112) and (113).

$$V_s V_r \sin\theta - RQ_r + XP_r = 0 \tag{112}$$

$$V_r^2 - V_s V_r \cos\theta + P_r R + Q_r X = 0 \tag{113}$$

The receiving end voltage is given by Equation (111).

$$V_r = \frac{V_s \cos\theta \pm \sqrt{V_s^2 \cos^2\theta - 4\left(P_r R + Q_r X\right)}}{2} \tag{114}$$

For real value of V_r, the terms under square root should be greater than or equal to zero. Equation (115) represents the line stability index NLSI.

$$NLSI_{sr} = \frac{4\left(P_r R + Q_r X\right)}{V_s^2 \cos^2\theta} \leq 1 \tag{115}$$

Since, θ is very small and hence $\cos^2\theta = 1$.

$$NLSI_{sr} = \frac{4\left(P_r R + Q_r X\right)}{V_s^2} \tag{116}$$

Bus Voltage Stability Index

Bus Voltage Collapse Prediction Index ($VCPI_{bus}$)

Balamourougan, Sidhu, & Sachdev (2004) have come up with the voltage collapse prediction index ($VCPI_{bus}$) which is calculated from the power flow equations. The $VCPI_{bus}$ value approaching to 1, indicates that the system has collapsed. By using the information of voltage phasors at each bus and the system admittance matrix (Y_{bus}), the $VCPI_{bus}$ can be calculated at each bus.

For a N bus test system, the current injected into the kth bus is given by Equation (117).

$$I_k = V_k \sum_{\substack{m=1 \\ m \neq k}}^{N} Y_{km} - \sum_{\substack{m=1 \\ m \neq k}}^{N} V_m Y_{km} \tag{117}$$

Where, Y_{km} is the mutual admittance between k[th] and m[th] bus; Y_{kk} is the self admittance at the kth bus; V_k and V_m are the voltage at k[th] and m[th] bus respectively. The complex power (S_k) injected at k[th] bus is given as follows.

$$S_k = V_k I_k^* \tag{118}$$

Substituting, I_k value from Equation (117) in Equation (118), we get.

$$S_k^* = \left|V_k\right|^2 \sum_{\substack{m=1 \\ m \neq k}}^{N} Y_{km} - V_k^* \sum_{\substack{m=1 \\ m \neq k}}^{N} V_m Y_{km} \tag{119}$$

Let,

$$Y_{kk} = \sum_{\substack{m=1 \\ m \neq k}}^{N} Y_{km} \tag{120}$$

Substituting Equation (120) into Equation (119), it yields into Equation (121).

$$S_k^* = \left|V_k\right|^2 Y_{kk} - V_k^* \sum_{\substack{m=1 \\ m \neq k}}^{N} V_m' Y_{kk} \tag{121}$$

Where,

$$V_m' = \frac{Y_{km}}{\sum_{\substack{i=1 \\ i \neq k}}^{N} Y_{ki}} V_m = \left|V_m'\right| \angle \delta_m' \tag{122}$$

Rearranging Equation (122),

$$\frac{S_k^*}{Y_{kk}} = \left|V_k\right|^2 - V_k^* \sum_{\substack{m=1 \\ m \neq k}}^{N} V_m^{'} \tag{123}$$

Equation (123) can be written as follows.

$$\frac{S_k^*}{Y_{kk}} = \left|V_k\right|^2 - \left(\left|V_k\right| \cos\delta_k - j\left|V_k\right| \sin\delta_k\right) \left[\sum_{\substack{m=1 \\ m \neq k}}^{N} \left(\left|V_m^{'}\right| \cos\delta_m^{'} + j\left|V_m^{'}\right| \sin\delta_m^{'}\right)\right] \tag{124}$$

$$\frac{S_k^*}{Y_{kk}} = \left[\left|V_k\right|^2 - \left(\sum_{\substack{m=1 \\ m \neq k}}^{N} \left|V_m^{'}\right|\left|V_k\right| \cos\left(\delta_k - \delta_m^{'}\right)\right) + j\left(\sum_{\substack{m=1 \\ m \neq k}}^{N} \left|V_m^{'}\right|\left|V_k\right| \sin\left(\delta_k - \delta_m^{'}\right)\right)\right] \tag{125}$$

The RHS of Equation (125) is a complex number which can be represented as Equation (126). Where, a and b are two real numbers.

$$\frac{S_k^*}{Y_{kk}} = a - jb \tag{126}$$

By comparing Equation (125) and (126), it results into Equation (127).

$$a - jb = \left(\left|V_k\right|^2 - \sum_{\substack{m=1 \\ m \neq k}}^{N} \left|V_m^{'}\right|\left|V_k\right| \cos\left(\delta_k - \delta_m^{'}\right)\right) + j\left(\sum_{\substack{m=1 \\ m \neq k}}^{N} \left|V_m^{'}\right|\left|V_k\right| \sin\left(\delta_k - \delta_m^{'}\right)\right) \tag{127}$$

Where,

$$a = \left|V_k\right|^2 - \sum_{\substack{m=1 \\ m \neq k}}^{N} \left|V_m^{'}\right|\left|V_k\right| \cos\left(\delta_k - \delta_m^{'}\right) \tag{128}$$

$$b = \sum_{\substack{m=1 \\ m \neq k}}^{N} \left|V_m^{'}\right|\left|V_k\right| \sin\left(\delta_k - \delta_m^{'}\right) \tag{129}$$

Let, $\delta_k - \delta_m = \delta$, then the Equation (128) and (129) can be represented with two unknown variables (V_k, δ) and given as follows.

$$f_1\left(\left|V_k\right|,\delta\right) = \left|V_k\right|^2 - \sum_{\substack{m=1 \\ m\neq k}}^{N} \left|V'_m\right|\left|V_k\right|\cos\delta \tag{130}$$

$$f_2\left(\left|V_k\right|,\delta\right) = \sum_{\substack{m=1 \\ m\neq k}}^{N} \left|V'_m\right|\left|V_k\right|\sin\delta \tag{131}$$

The Jacobian matrix of partial derivative for the solutions of unknown variables (V_k, δ) is represented by Equation (132).

$$J = \begin{vmatrix} 2\left|V_k\right| - \sum\limits_{\substack{m=1 \\ m\neq k}}^{N} \left|V'_m\right|\cos\delta & \left|V_k\right|\sum\limits_{\substack{m=1 \\ m\neq k}}^{N} \left|V'_m\right|\sin\delta \\ \sum\limits_{\substack{m=1 \\ m\neq k}}^{N} \left|V'_m\right|\sin\delta & \sum\limits_{\substack{m=1 \\ m\neq k}}^{N} \left|V'_m\right|\cos\delta \end{vmatrix} \tag{132}$$

At the voltage collapse point, the Jacobian matrix becomes singular or in other words the determinant of Jacobian matrix should be zero. Solving and equating the Equation (132) to zero results into Equation (133).

$$\frac{\left|V_k\right|\cos\delta}{\sum\limits_{\substack{m=1 \\ m\neq k}}^{N} \left|V'_m\right|} = \frac{1}{2} \tag{133}$$

Equation (133) can be written as follows.

$$\frac{V_k}{\sum\limits_{\substack{m=1 \\ m\neq k}}^{N} V'_m} = \frac{1}{2} + jk \tag{134}$$

The $VCPI_{bus}$ index is formulated as follows.

$$\left| 1 - \frac{\sum_{\substack{m=1 \\ m \neq k}}^{N} V'_m}{V_k} \right| = 1 \tag{135}$$

L-Index

Kessel & Glavitsch (1986) had proposed L-index based on power flow solutions. A 'n' bus power system is considered whereby the generators and loads are connected at 1, 2, 3,..., m and m+1, m+2,......, n respectively. Based on power flow results, the L-index is given by Equation (136).

$$L_j = \left| 1 - \sum_{i=1}^{g} F_{ji} \left(\frac{V_i}{V_j} \right) \right| \tag{136}$$

Where, $j = m+1,\ m+2,\ ....,\ n$; Note that V_i, V_j and F_{ji} are the complex quantities. The value of F_{ji} can be evaluated by solving Equation (137).

$$\begin{bmatrix} I_G \\ I_L \end{bmatrix} = \begin{bmatrix} Y_{GG} & Y_{GL} \\ Y_{LG} & Y_{LL} \end{bmatrix} \begin{bmatrix} V_G \\ V_L \end{bmatrix} \tag{137}$$

Where, V_G, V_L, I_G, I_L represents the voltage and current vectors at the generator and load buses respectively. Solving the Equation (137), we get

$$\begin{bmatrix} V_L \\ I_G \end{bmatrix} = \begin{bmatrix} Z_{LL} & F_{LG} \\ K_{GL} & Y_{GG} \end{bmatrix} \begin{bmatrix} I_L \\ V_G \end{bmatrix} \tag{138}$$

Where, $F_{LG} = -\left[Y_{LL}\right]^{-1}\left[Y_{LG}\right]$ are the required values. The F_{LG} matrix is obtained by the LU decomposition technique described. Hence, for all the load buses, the L-index is computed at a given load condition.

Modified L-Index

The mathematical formulation of modified L-index (Wang et al., 2013) as proposed by Wang et al. is given as follows.

Consider a power system connected with 'm' generators and 'k' connected loads. The generator is represented at kth bus by a voltage source (E_{gk}) behind synchronous impedance (Z_{gk}). The voltage and current relationship at buses can be described by Equation (146).

$$\begin{bmatrix} I_L \\ 0 \\ I_G \end{bmatrix} = \begin{bmatrix} Y_{LL} & Y_{LG} & 0 \\ Y_{GL} & Y_{GG} + Y_{gg} & -Y_{gg} \\ 0 & -Y_{gg} & Y_{gg} \end{bmatrix} \begin{bmatrix} V_L \\ V_G \\ E_G \end{bmatrix} \tag{139}$$

Where, the subscripts L and G stand for load and generator buses respectively. As shown in Figure 9, Y_{LL}, Y_{LG}, Y_{GL} and Y_{GG} are the sub matrices of system admittance matrix Y_N without considering the generator equivalent model. Thus, Y_{gg} can be represented by Equation (140).

$$Y_{gg} = diag\left(\frac{1}{Z_{g_1}}, \frac{1}{Z_{g_2}} \ldots\ldots\ldots, \frac{1}{Z_{g_m}}\right) \tag{140}$$

Eliminating V_G from Equation (139).

$$I_L = \left[Y_{LL} - Y_{LG}\left(Y_{GG} + Y_{gg}\right)^{-1} Y_{GL}\right] V_L + Y_{LG}\left(Y_{GG} + Y_{gg}\right)^{-1} Y_{gg} E_G \tag{141}$$

$$I_L = Y_{LL}^{'} V_L + Y_{LG}^{'} E_G \tag{142}$$

Where,

$$Y_{LL}^{'} = Y_{LL} - Y_{LG}\left(Y_{GG} + Y_{gg}\right)^{-1} Y_{GL}$$

and

Figure 9. Typical power system with m generators and l load demand.

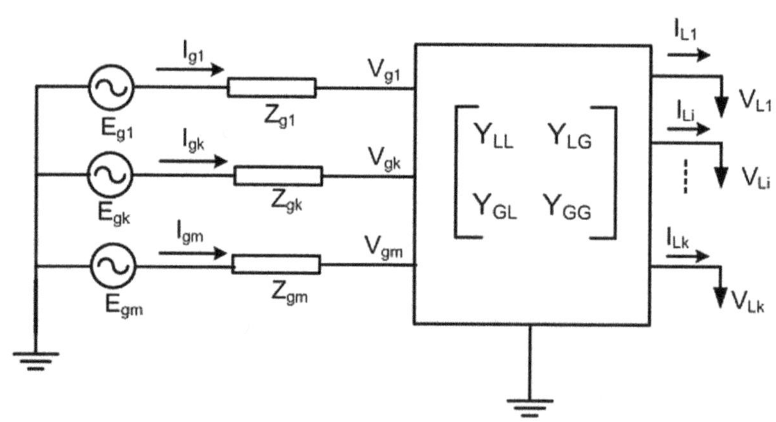

$$Y'_{LG} = Y_{LG}\left(Y_{GG} + Y_{gg}\right)^{-1} Y_{gg}.$$

The voltage at load buses can be represented by Equation (143).

$$V_L = Z'_{LL} I_L - Z'_{LL} Y'_{LG} E_G = Z'_{LL} I_L + F'_{LG} E_G \tag{143}$$

Where,

$$Z'_{LL} = \left(Y'_{LL}\right)^{-1} \text{ and } F'_{LG} = -Z'_{LL} Y'_{LG}.$$

The voltage at a specific load bus j can be represented as follows.

$$V_j = \sum_{k\in\alpha_L} Z'_{jk} I_{Lk} + \sum_{i\in\alpha_G} F'_{ji} E_{Gi} \tag{144}$$

Where, Z'_{jk} and F'_{ji} are the elements of the matrix Z'_{LL} and F'_{LG} respectively; α_L and α_G are the sets of loads and generators buses respectively; E_{Gi} is the internal voltage of generator i; I_{LK} represents the current injection at load bus k. The modified L index can be defined as follows.

$$L_j = \left| 1 - \frac{\sum_{i\in\alpha_G} F'_{ji} E_{Gi}}{V_j} \right| \tag{145}$$

The modified L-index at the jth bus will approach to 1, when j[th] bus voltage arrives near to the critical point. The voltage instability of the whole system is decided by the maximum value of L_j at the jth bus. Thus, the system voltage stability is described by Equation (146).

$$L = \max_{j\in\alpha_L}\left\{L_j\right\} \tag{146}$$

Distribution Network VSI

For the voltage stability assessment of distribution network, the VSI proposed by Chakrovorty & Das (2001) is described in this section. From the equivalent circuit of the radial distribution network as shown in Figure 10, the following expressions can be deduced as per Equation (147) and Equation (148).

Figure 10. Equivalent circuit of radial network.

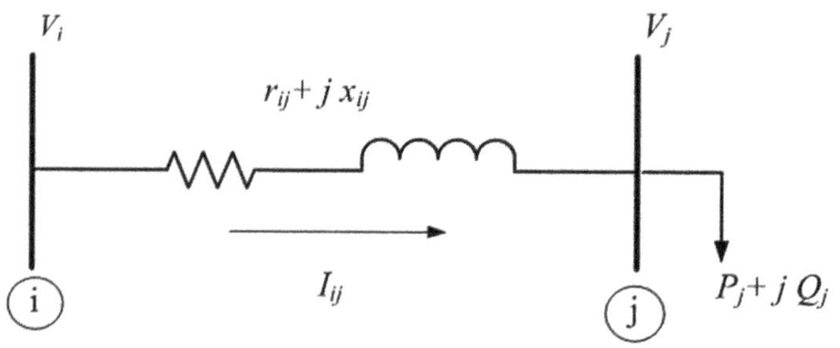

$$I_{ij} = \frac{V_i - V_j}{r_{ij} + jx_{ij}} \tag{147}$$

$$P_j - jQ_j = V_j^* I_{ij} \tag{148}$$

Where, i, j are the sending and receiving end respectively; I_{ij} represents the branch current; V_i and V_j are the voltages of node i and j, respectively; P_j and Q_j are the total real and reactive power loads fed through node j. Further, Equation (147) and (148), the following expression can be deduced.

$$\left|V_j\right|^4 - \left\{\left|V_i\right|^2 - 2P_j r_{ij} - 2Q_j x_{ij}\right\}\left|V_j\right|^2 + \left\{P_j^2 + Q_j^2\right\}\left\{r_{ij}^{\ 2} + x_{ij}^{\ 2}\right\} \tag{149}$$

Let,

$$b = \left|V_i\right|^2 - 2P_j r_{ij} - 2Q_j x_{ij} \tag{150}$$

$$c = \left\{P_j^2 + Q_j^2\right\}\left\{r_{ij}^{\ 2} + x_{ij}^{\ 2}\right\} \tag{151}$$

$$\left|V_j\right|^4 - b\left|V_j\right|^2 + c = 0 \tag{152}$$

The feasible solution of Equation (149) is unique and can be obtained as follows.

$$\left|V_j\right| = 0.707\sqrt{b + \sqrt{b^2 - 4c}} \tag{153}$$

$$b^2 - 4c \geq 0 \tag{154}$$

From Equation (150), (151) and (154).

$$\left(\left|V_i\right|^2 - 2P_j r_{ij} - 2Q_j x_{ij}\right)^2 - 4\left(P_j^2 + Q_j^2\right)\left(r_{ij}^2 + x_{ij}^2\right) \geq 0 \tag{155}$$

Rearranging Equation (155).

$$\left|V_i\right|^4 - 4\left(P_j x_{ij} - Q_j r_{ij}\right)^2 - 4\left(P_j r_{ij} + Q_j x_{ij}\right)\left|V_i\right|^2 \geq 0 \tag{156}$$

Voltage stability index of node j can be expressed as follows.

$$VSI = \left|V_i\right|^4 - 4\left(P_j x_{ij} - Q_j r_{ij}\right)^2 - 4\left(P_j r_{ij} + Q_j x_{ij}\right)\left|V_i\right|^2 \tag{157}$$

The minimum value of the stability index at any node represents that the node is more sensitive to voltage collapse. The stable operation of the radial distribution network can be obtained with VSI ≥ 0.

Thevenin Equivalent based Technique

Voltage instability occurs in the system when the load impedance matches with the Thevenin impedance seen from the measured load point. The Thevenin equivalent technique can be summarized as follows. For a typical power system with multiple generators and loads, the current and voltage phasors are measured at load point and then the Thevenin voltage and impedance of entire system seen from this bus are derived (Figure 11). Based on network theory, the maximum power can be delivered to a load under the condition of impedance match. This voltage stability assessment technique requires the Thevenin equivalent parameters Z_{Th} and E_{Th} that can be defined as follows.

$$Z_{Th} = \frac{V_m^t - V_m^{t+1}}{I_m^{t+1} - I_m^t} \tag{158}$$

Where, V_m^t, and I_m^t represents the voltage and current phasor at load bus at time t; V_m^{t+1} and I_m^{t+1} represents the load voltage and current phasor at time t+1. The Thevenin voltage can be given by Equation (159). It is assumed that the Thevenin equivalent parameters remains constant during t and t+1.

$$E_{Th} = Z_{Th} I_m + V_m \tag{159}$$

The maximum power delivered to load can be evaluated using maximum power transfer theorem.

Figure 11. Thevenin equivalent representation in a power system

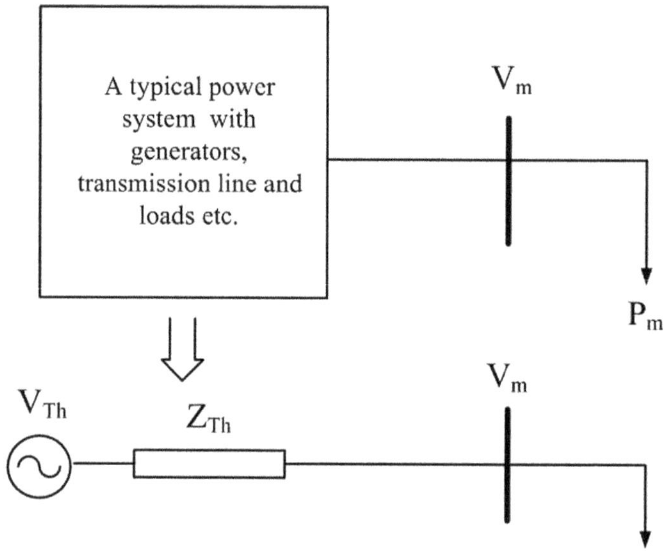

$$|Z_L| = |Z_{Th}| \tag{160}$$

$$S_{\max} = |I_L^2||Z_L| = |I_L^2||Z_{Th}| \tag{161}$$

Equation (161) can be solved as follows.

$$S_{\max} = \left| \frac{E_{Th}}{Z_{Th} + |Z_{Th}| \angle \delta} \right|^2 |Z_{Th}|$$

$$S_{\max} = |E_{Th}|^2 \frac{|Z_{Th}|}{\left(R + |Z_{Th}| \cos \delta\right)^2 + \left(X + |Z_{Th}| \cos \delta\right)^2}$$

$$S_{\max} = |E_{Th}|^2 \frac{|Z_{Th}|}{2|Z_{Th}|^2 + 2R|Z_{Th}| \cos \delta + 2X|Z_{Th}| \sin \delta}$$

$$S_{\max} = |E_{Th}|^2 \frac{1}{2|Z_{Th}| + 2R \cos \delta + 2X \sin \delta}$$

$$S_{\max} = \frac{\left|E_{Th}\right|^2}{2} \frac{\left(\left|Z_{Th}\right| - \left(R\cos\delta + X\sin\delta\right)\right)}{\left|Z_{Th}\right|^2 - \left(R\cos\delta + X\sin\delta\right)^2}$$

$$S_{\max} = \frac{\left|E_{Th}\right|^2}{2} \frac{\left(\left|Z_{Th}\right| - \left(R\cos\delta + X\sin\delta\right)\right)}{R^2\left(\cos^2\delta + \sin^2\delta\right) + X^2\left(\cos^2\delta + \sin^2\delta\right) - \left(R\cos\delta + X\sin\delta\right)^2}$$

$$S_{\max} = \frac{\left|E_{Th}\right|^2}{2} \frac{\left(\left|Z_{Th}\right| - \left(R\cos\delta + X\sin\delta\right)\right)}{\left(X\cos\delta - R\sin\delta\right)^2} \tag{162}$$

$$\text{margin}_{\text{n}} = \frac{S_{\max_n} - S_{L_n}}{S_{L_n}} \times 100 \tag{163}$$

Where, S_{max} represents the maximum apparent power at n[th] bus; δ represents the power factor angle at load bus; S_{Ln} represents the apparent power at the load bus in the base case.

Demerits of Thevenin Equivalent Technique

1. In the Thevenin equivalent technique, the Thevenin parameters are considered constant during evaluation. But this assumption cannot be satisfied while system parameters are changing such as tripping of transmission line, switching of capacitor banks, generators at their reactive power limits, etc.
2. As system load in nonlinear and dynamic in nature and it is inconvenient to represent them with single Thevenin impedance (Z_{Th}).

Coupled Single Port Circuit Technique

The Thevenin based voltage stability assessment method has a major drawback that while calculating the equivalent voltage and impedance as seen from the concerned load bus, the method involves the remaining loads into equivalent model. The loads have nonlinear characteristics and also varies with respect to time. It is inconvenient to represent them as a single Z_{Th}. To overcome this difficulty, the concept of multiport network was proposed by Liu & Chu (2014). In a multiport network, the loads and generators are considered to be outside of equivalent network and moreover the transmission line is represented as equivalent impedance matrix Z_{LL} (Figure 12).

The mathematical model of multiport network can be given by Equation (164).

Figure 12. Multiport network equivalent system model

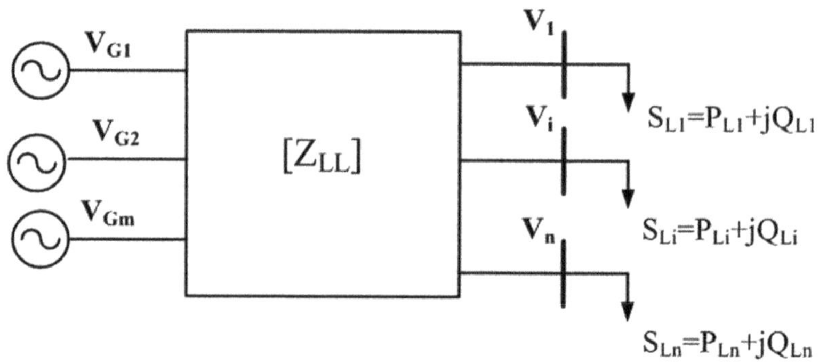

$$\begin{bmatrix} -I_L \\ 0 \\ I_G \end{bmatrix} = \begin{bmatrix} Y \end{bmatrix} \begin{bmatrix} V_L \\ V_T \\ V_G \end{bmatrix} = \begin{bmatrix} Y_{LL} & Y_{LT} & Y_{LG} \\ Y_{TL} & Y_{TT} & Y_{TG} \\ Y_{GL} & Y_{GT} & Y_{GG} \end{bmatrix} \begin{bmatrix} V_L \\ V_T \\ V_G \end{bmatrix} \tag{164}$$

Where, system admittance matrix is represented by Y matrix that is constructed from network topology and parameters. The vectors I and V represent the current and voltage vectors and the subscript G, L, T represents the generator, load and tie bus respectively. Since, the injection current to tie bus is zero thus Equation (164) can be expressed as follows.

$$V_L = KV_G - Z_{LL} I_L \tag{165}$$

Where,

$$Z_{LL} = \left(Y_{LL} - Y_{LT} Y_{TT}^{-1} Y_{TL} \right)^{-1}$$

and

$$K = Z_{LL} \left(Y_{LL} - Y_{LT} Y_{TT}^{-1} Y_{TL} \right)$$

The voltage on a jth load bus is represented as follows.

$$V_{L_j} = E_{Th_j} - Z_{Th_j} I_{L_j} - E_{coupled_j} \tag{166}$$

Where,

$$Z_{Th_j} = Z_{LL_{jj}};$$

Figure 13. The coupled term modeled as an extra impedance

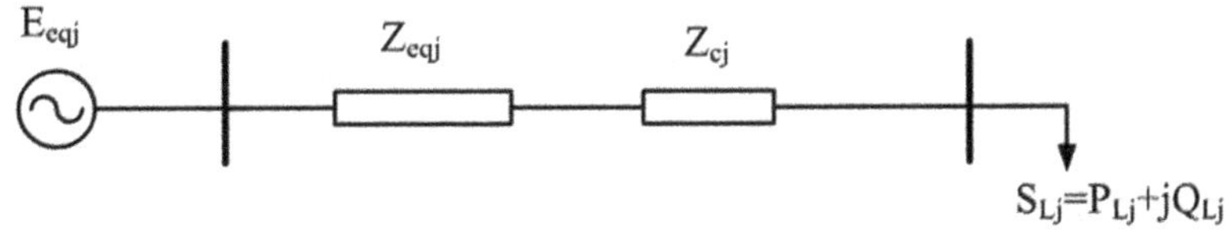

$$E_{Th_j} = \left[KV_G \right]_j$$

and

$$E_{coupled_j} = \sum_{i=1, i \neq j}^{n} Z_{LL_{ji}} I_{Li}$$

Where, Z_{Thj} represents the Thevenin impedance at bus j without considering the remaining load. The Z_{Thj} represents the diagonal element of equivalent impedance matrix Z_{LL}. The term $E_{coupledj}$ represents the influence of remaining loads on bus j also called the coupled effect. Compared to the traditional Thevenin method which is a single port equivalent network method, the difference is the additional coupling term $E_{coupledj}$. This new equivalent circuit model is called the coupled single port circuit in which the power network can be separated into a set of single port circuits that represents the impact of other loads explicitly. The three approaches had been studied by Liu et al., in their research work to model coupling term while preserving the single port structure. These approaches are given as follows.

1. Taken as an additional power demand
2. Taken as an additional voltage source
3. Taken as an additional impedance

Since, the only additional impedance approach was found suitable to represent coupling term and hence taken into consideration. In this approach, the coupling term $E_{coupledj}$ has been represented by additional impedance called virtual impedance (Z_{cj}). The virtual impedance is given by Equation (167) in which the ratio α_{ij} of the load is kept constant. Moreover the bus voltages are also changing proportionally when the power system is scaled up, which represents that the ratio of the bus voltages remains constant.

$$Z_{cj} = \frac{E_{coupled-j}}{I_{L_j}} = \sum_{i=1, i \neq j}^{n} Z_{LL_{ji}} \frac{I_{Li}}{I_{Lj}} \tag{167}$$

$$Z_{cj} = \sum_{i=1, i \neq j}^{n} \left(Z_{LL_{ji}} \frac{S_{L_i}}{S_{L_j}} \times \frac{V_{L_j}^*}{V_{L_i}^*} \right) \tag{168}$$

$$Z_{cj} = \sum_{i=1, i \neq j}^{n} \left(Z_{LL_{ji}} \alpha_{ij} \times \frac{V_{L_j}^*}{V_{L_i}^*} \right) \tag{169}$$

The voltage stability margin at any bus is calculated based on impedance matching criteria. The voltage stability margin will be different for individual coupled port circuits. The smallest voltage stability margins among all the buses represent the system margin.

$$\text{Margin}_{\text{sys}} = \min\left(\text{margin}_1, \text{margin}_2, \ldots.., \text{margin}_n\right) \tag{170}$$

The weakest bus of the entire network is the one which has smallest margin and the other buses that have margins near to the weakest bus shall also be identified as weak buses.

WIDE AREA MEASUREMENTS BASED VOLTAGE ASSESSMENT TECHNINQUES

Online Voltage Stability Assessment Techniques Using Decision Trees

The state estimation plays an important role during online monitoring and reliable operations of the power system. Traditionally, the supervisory control and data acquisition (SCADA) are being used in the energy management systems (EMS) to determine the system status. From, the last two decades, although the SCADA is serving the power sector in a useful manner, but its incapability to take the time synchronized measurement data across the whole power system is a bottleneck. The evolution of phasor measurement units (PMUs) has revolutionized the field of state monitoring in power system. The PMUs have several advantages over SCADA system regarding both accuracy and speed of measured data. The PMUs generate the time stamped phasor data which is synchronized with global positioning system (GPS) and converted into a series of data stream for communications to energy control centers. In literature, the various authors have utilized the PMUs measurement data for the online voltage instability assessment using the methods such as L-index, Thevein equivalent method, coupled single port circuits, decision trees and support vector machine. Here, the decision tree and support vector machine are explained in details as these methods hold the futuristic appeal.

Decision Tree (DT)

The DT technique is a data mining tool used for classifications. The decision tree is a binary tree with two types of nodes, the "internal node" with two successors and the "terminal node" without successors. A critical splitting rule (CSR) is set for each internal node to decide which successor to follow. The CSR can be represented as a numerical value with a threshold or the checking can be categorically made that whether the current numerical value belong to a particular set of data. At each terminal node, the classification results are labeled in the form of objectives, e.g., "stable state" and "unstable state". In order to train the decision tree, a learner set and test set are required. For all the known cases, the decision tree processes the predominant system parameters that contribute towards the final objectives and also optimizes the forecast ability of the new unknown cases. In the first step, the maximum numbers of decision trees are trained from the learner set by recursively splitting a parent node in pure child node with impurity reduction. The nodes start to split continuously from parent node until further splitting cannot enhance the overall performance of decision trees or predefined threshold has reached. In the second step, using the test sets, small decision trees are generated in the form of the misclassification cost of the test set. The decision tree with lowest misclassification cost is considered as an optimal decision tree. Once decisions trees are trained properly then an optimal decision tree becomes suitable for identifying the critical system states from various system states that are related to the problem of power system security. The decision trees are used in several power system applications such as transient stability detection, voltage stability monitoring and estimations, frequency control etc.

Diao et al. (2009) have applied the decision tree methodology for the online voltage stability assessment in the power system considering data from PMU. The proposed scheme had been divided into three steps.

1. Offline decision tree training
2. Periodic decision tree updates
3. Online decision tree applications

The two assumptions are made 1) All the bus voltages are to be within limits at the base cases 2) For voltage instability cases the critical contingency considerations are to be initiated.

Offline Decision Tree Training

All the system operating states (N_{os}) from the past recorded data and those which have been forecasted for the next day are collected. N is considered as the number of critical contingency cases from voltage instability viewpoint, then, Nc contingency cases are run for each operating state. For each operating case, a voltage security level ('Stable' or 'Unstable') is assigned. Total N_{os}*N_c number of cases are created in the dataset. The collected data from PMUs are used to predict the voltage stability from the pre disturbance state, whereas for stability application for next day, the offline training of the decision trees are done.

Periodic Decision Tree Updates

After offline training of decision trees, the system operating states are checked periodically and are also updated on hour to hour basis so that the offline trained decision trees gives high performance on the new system states. If the system states have changed significantly due to changes in network topologies, generator outages or load levels, then the simulation of voltage instability scenarios is run on these changes and new cases are generated for decision tree tests. If decision tree performance is found to be satisfactory, then no modifications are required. Otherwise, to achieve better accuracy the newly developed cases are mixed with the original cases in order to build new DTs.

Online Decision Trees Applications

In the final decision trees, the real time measurement data collected from PMUs are then compared with CSRs. If the particular CSR violates the threshold numeric value, then preventive and corrective action should be initiated. Instead of thousand of system states, only few critical predisturbance system parameters can effectively characterize the severity of the current operating states. The method takes directly the inputs from measurement of PMUs and compares it with CSRs to obtain the stability status.

Online Voltage Stability Assessment Techniques Using Support Vector Machine Model

In literature, the support vector machine based algorithms have also been utilized by various authors for the assessment of online voltage stability. Sajan et al. (2015) had proposed the genetic algorithm based support vector machine (GA-SVM) model for the online assessment of voltage stability. The SVM is a powerful machine learning technique. In this method, the voltage phasor obtained from PMUs are taken as input to SVM and the voltage stability margin index (VSMI) is taken as an output vector.

SVM Approach

The support vector machine (SVM) is a supervised machine learning algorithm which is based on statistical learning theory (SLT) and structural risk minimization principle. It can be applied to classification as well as regression problem.

SVM Regression Theory

Lets, consider a training data set which consists of input/output parameters $(x_1, y_1),\ldots,(x_p, y_p),\ldots,(x_n, y_n)$. The input x_p represents the voltage phasor for the current operating point and y_p represents the output parameter, i.e., VSMI. Equation (171) represents the VSMI for the system.

$$VSMI = \min\left(\frac{VSM_p}{P_{\max_p}}\right) \tag{171}$$

Where, VSM_p is the voltage stability margin obtained from the P-V curve method at the pth operating state and P_{max} represents the maximum active power that has been delivered to the load bus. The VSMI varies between 1 (no load) to 0 (critical loading). The continuous power flow (CPF) is made use of in generating the P-V curves. The idea behind the regression problem is to find the linear relationship between the operating states and to calculate the VSMI with reference to the new operating state. The linear relationship can be defined as follows.

$$f\left(x\right) = w\phi\left(x\right) + b \tag{172}$$

Where, $\phi(x)$ represents the high dimensional feature space mapped into input space x; b and w represents the bias and weight factor respectively. The objective is to determine the value of w and b in such a way that the regularized risk function (Equation 173) should be minimum.

$$R_{reg}(f) = C\frac{1}{n}\sum_{l=0}^{n}\xi\left(f(x_l) - y_l\right) + 0.5 * \|w\|^2 \quad (173)$$

Where, $\|w\|$ represents the Euclidean norm; ξ(.) represents the loss function and the parameter C represents the cost function measuring the empirical risk. The most widely used ε-insensitive loss function is given by Eq (174). This insensitive loss function is applied to reduce noise.

$$\xi\left(f(x_l), y_l\right) = \begin{cases} \left|f(x) - y\right| - \varepsilon & \left|f(x) - y\right| \geq \varepsilon \\ 0 & \text{Otherwise} \end{cases} \quad (174)$$

For determination of the parameter values b and w, Equation (174) is converted to primal function given by Equation (175).

$$\begin{aligned} &\text{minimize } R_{reg}(w, \xi^*) = \frac{1}{2}\|w\|^2 + C\sum_{l=1}^{n}(\xi_l + \xi_l^*) \\ &\text{subjected to } \begin{Bmatrix} f(X_l) - w\phi(x_l) - b_l\varepsilon + \xi_l \\ w\phi(x_l) + b_l - f(x_l)\varepsilon + \xi_l^* \end{Bmatrix} \xi_l^* \geq 0 \end{aligned} \quad (175)$$

Where, ξ_l, ξ_l^* are the slack variables that find the upper and lower side errors. Equation (175) indicates that the increasing value of epsilon (ε) decreases the corresponding ξ_l and ξ_l^*. The formulation of Equation (175) is represented as a maximization optimization problem and given by Equation (176).

$$w(\alpha_i, \alpha_i^*) = \max\left(-\frac{1}{2}\sum_{l,m=1}^{n}(\alpha_l - \alpha_m^*)(\alpha_m - \alpha_l^*)K(x_l \cdot x_m) - \varepsilon\sum_{l=1}^{m}(\alpha_l + \alpha_l^*) + \sum_{l=1}^{m} y_l(\alpha_l - \alpha_l^*)\right) \quad (176)$$

Subjected to

$$\sum_{l=1}^{i}\alpha_l - \alpha_l^* = 0, \quad \alpha_l, \alpha_l^* \in [0, C] \quad (177)$$

The Lagrange multiplier α_i and α_i^* represents the solutions to the above problem that pushes the predictors towards the targeted output value y_i. To forecast the regression lines (support vectors), the only non-zero values of α_i and α_i^* are useful. The data points inside the epsilon (ε) tube, do not contrib-

Figure 14. SVM for regression

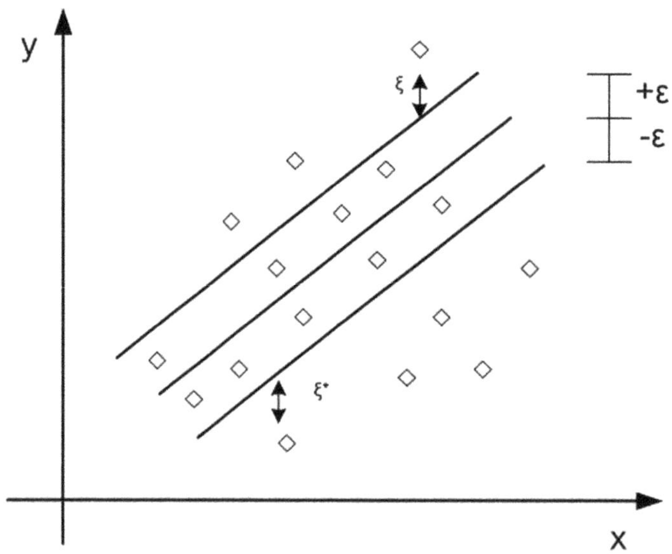

ute to the regression function. Kernel function is denoted by $K(x_i, x_m)$ in Equation (176). The various types of kernel functions used in machine learning algorithms are given as follows.

1. Linear kernel: $K\left(x_i, x_j\right) = x_i^T x_j$
2. Polynomial kernel: $K\left(x_i, x_j\right) = \left(1 + x_i x_j\right)^d$
3. Gaussian kernel: $K\left(x_i, x_j\right) = \exp\left(-\frac{\left\|x_i, x_j\right\|^2}{2\sigma^2}\right)$

Where, x_i & x_j represents the input space vectors, d represents the degree of the polynomial and σ^2 is the bandwidth of radial basis function (RBF) kernel. The accurate selection of these parameters are vital as final solutions depends on it.

SVM Parameter Selection

The parameters used in SVM for regression are defined as follows.

1. The kernel function is applied to generate a non-linear hyper-surface on the input data space in support vector regression (SVR) model. The various types of kernel functions are used in SVR but only Gaussian radial basis function (RBF) yields the superior estimate performance.
2. Regularization parameter C controls the tradeoff between minimizing the model's complexity and training error.
3. Bandwidth (σ^2) of the RBF kernel function represents the variance in the Gaussian kernel function.
4. The epsilon (ε) is the radius of a tube. Where, ε is the insensitive loss function within which the regression function lies.

GA-SVM Model

To find the optimal solutions of the SVM parameters, i.e., C, σ^2 and ε, the genetic algorithm (GA) has been utilized. In order to achieve the optimal solutions to SVM parameters, the initial population of chromosomes is generated randomly with GA. The parameters like C, σ^2 and ε are directly coded with their real value data in the chromosomes. The steps adopted in GA-SVM model are given as follows.

Step 1: Initialization: The three SVM parameters, i.e. C, σ^2 and ε are coded to generate the chromosomes. Let chromosome X is presented by $X=\{x_1, x_2, x_3\}$, where x1, x2 and x3 denotes the parameter C, σ^2 and ε respectively.

Step 2: The k fold cross validation technique is applied to obtain the optimum SVM parameters in order to avoid the over an under fitting of GA-SVM model. In this technique, the training set data are randomly divided into k subsets of equal size. The performance of SVM parameter is checked into k^{th} subset. Repeat the procedure in order to check the performance of SVM parameters in all subset.

Step 3: Population initialization: The population size can be selected based on the convergence time and population diversity.

Step 4: The fitness function, i.e., mean absolute percentage error (MAPE) for each randomly generated chromosome in step 3 is obtained using the following equations.

$$\min f = \mathrm{MAPE}_{\mathrm{cross\_val}} \tag{178}$$

$$\mathrm{MAPE}_{\mathrm{cross\_val}} = \frac{\sum_{l=1}^{n}\left|\frac{A_l - P_l}{A_l}\right|}{n} * 100\% \tag{179}$$

Where, n represents the number of training set data; A_l and P_l represents the actual and predicted value.

Step 5: New population replaces the current population by undergoing selection, crossover and mutation processes. In the recombination pool only those chromosomes are selected which have better fitness value. The selection is done with the help of the roulette wheel. In order to achieve better solution, the new offsprings are obtained by exchanging the genes between two parent chromosomes. The probability of creating new chromosomes in each pair is set to be 0.8. Alternation of binary code with probability of 0.05 is attained through mutation process.

Step 6: Termination criteria: The procedure is repeated until the generation count reaches its limits.

Step 7: For complete evaluation of the performance of the overall GA-SVM, the different performance indices have been considered.

1. Wilmott's index of agreement (WIA)
2. Mean absolute percentage error (MAPE)
3. Normalized mean square error (NMSE)

FUTURE RESEARCH DIRECTION

Over the last few years, the installed capacity of wind energy and solar photovoltaic had increased drastically. Since, the energy output from these energy sources is random in nature. Therefore, the high penetration of these resources into existing power systems is expected to have a significant impact on power system stability. The evolution of phasor measurement units (PMUs) technology can play an important role in real time assessment of voltage instability. Hence, there is a need to develop online computer simulation programs based on wide area measurements of PMUs data in the modern power systems. The handling of the large amount of data collected from various PMUs requires the efficient and fast computer based algorithms (e.g., machine learning algorithms such as decision trees, support vector machine etc.,) which can further identify different phenomena in the real time power system.

CONCLUSION

Worldwide, the actual blackouts are manifestation of primarily due to the voltage instability. Hence it becomes vital to assess the voltage stability of power system from time to time. Various techniques have been established in literature for voltage stability assessment. The chapter highlights the classical methods adopted for voltage stability assessment along with voltage stability indices based approaches. Among the wide spectrum of techniques, the online assessment techniques using decision trees and support vector machines have promising future.

REFERENCES

Ajjarapu, V. (2009). *Computational Techniques for Voltage Stability Assessment and Control*. Springer Science and Business Media.

Ajjarapu, V., & Christy, C. (1992). The continuation power flow: A tool for steady state voltage stability analysis. *IEEE Transactions on Power Systems*, *7*(1), 416–423. doi:10.1109/59.141737

Ajjarapu, V., & Lee, B. (1992). Bifurcation theory and its application theory and its application to non-linear dynamical phenomena in an electrical power system. *IEEE Transactions on Power Systems*, *7*, 424–431. doi:10.1109/59.141738

Alvarado, F., Dobson, I., & Hu, Y. (1994). Computation of closest bifurcations in power systems. *IEEE Transactions on Power Systems*, *9*(2), 918–928. doi:10.1109/59.317655

Balamourougan, V., Sidhu, T. S., & Sachdev, M. S. (2004). Technique for online prediction of voltage collapse. *IEE Proceedings. Generation, Transmission and Distribution*, *151*(4), 453–460. doi:10.1049/ip-gtd:20040612

Beiraghi, M., & Ranjbar, A. M. (2013). Online voltage security assessment based on wide area measurements. *IEEE Transactions on Power Delivery*, *28*(2), 989–997. doi:10.1109/TPWRD.2013.2247426

Chakravorty, M., & Das, D. (2001). Voltage stability analysis of radial distribution networks. *International Journal of Electrical Power & Energy Systems*, *23*(2), 129–135. doi:10.1016/S0142-0615(00)00040-5

Corsi, S., & Taranto, G. N. (2008). A real-time voltage instability identification algorithm based on local phasor measurements. *IEEE Transactions on Power Systems*, *23*(3), 1271–1279. doi:10.1109/TPWRS.2008.922586

Cutsem, T. V. (2000). Voltage instability: Phenomena, countermeasures, and analysis methods. *Proceedings of the IEEE*, *88*(2), 208–227. doi:10.1109/5.823999

Diao, R., Sun, K., Vittal, V., Keefe, R. J., Richardson, M. R., Bhatt, N., ... Sarawgi, S. K. (2009). Decision tree based online voltage security assessment using PMU measurements. *IEEE Transactions on Power Systems*, *24*(2), 832–839. doi:10.1109/TPWRS.2009.2016528

Dobson, I., & Lu, L. (1993). New methods for computing a closest saddle node bifurcation and worst case load power margin for voltage collapse. *IEEE Transactions on Power Systems*, *8*(3), 905–913. doi:10.1109/59.260912

Flatabo, N., Ognedal, R., & Carlson, T. (1990). Voltage stability condition in a power transmission system calculated by sensitivity methods. *IEEE Transactions on Power Systems*, *5*(4), 1286–1293. doi:10.1109/59.99379

Gao, B., Morison, G. K., & Kundur, P. (1992). Voltage stability evaluation using modal analysis. *IEEE Transactions on Power Systems*, *7*(4), 1529–1542. doi:10.1109/59.207377

Glavic, M., & Cutsem, T. V. (2009). Wide area detection of voltage instability from synchronized phasor measurements. Part II: Simulation results. *IEEE Transactions on Power Systems*, *24*(3), 1417–1425. doi:10.1109/TPWRS.2009.2023272

Glavic, M., Novosel, D., Heredia, E., Kosterev, D., Salazar, A., Ashrafi, F. H., & Donnelly, M. (2012). See it fast to keep calm: Real time voltage control under stressed conditions. *IEEE Power & Energy Magazine*, *10*(4), 43–55. doi:10.1109/MPE.2012.2196332

Goharrizi, A. Y., & Asghari, R. (2007). A novel line stability index (NLSI) for voltage stability assessment of power systems. *Proceedings of the International Conference on Power Systems*.

Kessel, P., & Glavitsch, H. (1986). Estimating the voltage stability of a power system. *IEEE Transactions on Power Delivery*, *PWRD-1*(3), 346–354. doi:10.1109/TPWRD.1986.4308013

Kundur, P. (1994). *Power System Stability and Control*. New York: McGraw-Hill.

Kundur, P., Paserba, J., Ajjarapu, V., Anderson, G., Bose, A., Canizares, C., ... Vittal, V. (2004). Definition and classification of power system stability- IEEE joint task force on stability terms and definitions. *IEEE Transactions on Power Systems*, *19*(3), 1387–1401. doi:10.1109/TPWRS.2004.825981

Kundur, P., Paserba, J., Vittal, V., & Anderson, G. (2006). Closure of definition and classification of power system stability. *IEEE Transactions on Power Systems*, *21*(1), 446. doi:10.1109/TPWRS.2005.861952

Liu, J. H., & Chu, C. C. (2014). Wide area measurement based voltage stability indicators by modified coupled single port models. *IEEE Transactions on Power Systems*, *29*(2), 756–764. doi:10.1109/TPWRS.2013.2284475

Modarresi, J., Gholipour, E., & Khodabakhshian, A. (2016). A comprehensive review of the voltage stability indices. *Renewable & Sustainable Energy Reviews*, *63*, 1–12. doi:10.1016/j.rser.2016.05.010

Moghavvemi, M., & Faruque, M. O. (2001). Technique for assessment of voltage stability in illconditioned radial distribution network. *IEEE Power Engineering Review*, *21*(1), 58–60. doi:10.1109/39.893345

Moghavvemi, M., & Faruque, O. (1998). Real-time contingency evaluation and ranking technique. *IEE Proceedings. Generation, Transmission and Distribution*, *145*(5), 517–524. doi:10.1049/ip-gtd:19982179

Moghavvemi, M., & Omar, F. M. (1998). Technique for contingency monitoring and voltage collapse prediction. *IEE Proceedings. Generation, Transmission and Distribution*, *145*(6), 634–640. doi:10.1049/ip-gtd:19982355

Mohamed, A., Jasmon, G. B., & Yusoff, S. (1989). A static voltage collapse indicator using line stability factors. *Journal of Information Technology*, *7*, 73–85.

Mohammadi, H., & Behghani, M. (2015). PMU based voltage securing assessment of power systems exploiting principal component analysis and decision trees. *International Journal of Electrical Power & Energy Systems*, *64*, 655–663. doi:10.1016/j.ijepes.2014.07.077

Musirin, I., & Rahman, T. K. A. (2002). Estimating maximum loadability for weak bus identification using FVSI. *IEEE Power Engineering Review*, *22*(11), 50–52. doi:10.1109/MPER.2002.1045568

Overbye, T. J., & DeMarco, C. L. (1991). Improved techniques for power system voltage stability assessment using energy methods. *IEEE Transactions on Power Systems*, *6*(4), 1446–1452. doi:10.1109/59.116988

Sajan, K. S., Kumar, V., & Tyagi, B. (2015). Genetic algorithm based support vector machine for online voltage stability monitoring. *International Journal of Electrical Power & Energy Systems*, *73*, 200–208. doi:10.1016/j.ijepes.2015.05.002

Wang, Y., Wang, C., Lin, F., Li, W., Wang, L. Y., & Zhao, J. (2013). Incorporating generator equivalent model into voltage stability analysis. *IEEE Transactions on Power Systems*, *28*(4), 4857–4866. doi:10.1109/TPWRS.2013.2273501

Chapter 7
Reactive Power Loss Index for Identification of Weak Nodes and Reactive Compensation Analysis to Improve Steady State Voltage Stability

Tukaram Moger
https://orcid.org/0000-0003-4176-5125
National Institute of Technology Karnataka, India

Thukaram Dhadbanjan
Indian Institute of Science Bangalore, India

ABSTRACT

This chapter presents a new reactive power loss index for identification of weak buses in the system. This index can be used for identification of weak buses in the systems. The new reactive power loss index is illustrated on sample 5-bus system, and tested on sample 10-bus equivalent system and 72-bus equivalent system of Indian southern region power grid. The validation of the weak buses identification from the reactive power loss index with that from other existing methods in the literature is carried out to demonstrate the effectiveness of the index. Simulation results show that the identification of weak buses in the system from the new reactive power loss index is completely non-iterative, and thus requires minimal computational efforts as compared with other existing methods in the literature.

INTRODUCTION

The present day power system is being operated under stressed conditions due to rapidly growing power demand, and lacks of upgradation/augmentation of the existing infrastructure such as generation and transmission capacity in the system because of various operational, economical and environmental con-

DOI: 10.4018/978-1-5225-8551-0.ch007

straints. In addition, the situation becomes more worst if the system is perturbed by critical/severe network contingencies such as tripping of heavily loaded transmission lines or outage of large generating units. Under such situations, power system may result in voltage instability and susceptible to voltage collapse due to insufficient amount of reactive power reserve available in the system to support the voltage. Not only is reactive power necessary to operate the transmission system reliably, but it can also substantially improve the efficiency with which real power is delivered to customers/end users. Hence, the secure and reliable operation of power system has always been concern to the system operator.

In recent years, it has been observed that the voltage instability problem is the root cause for several major network blackouts in different countries such as France, Belgium, Sweden, Germany, Iran, Japan, USA and India (FERC, 2005; Srivastava, Velayutham, Agrawal, & Bakshi, 2012.). A system may be voltage unstable if it includes at least one voltage unstable bus (Ca˜nizares, De Souza, & Quintana, 1996). Therefore, the system operator must make sure that there are enough reactive reserve capacities available for the system to maintain voltage profiles. Properly planned reactive power reserve minimizes the risk of voltage collapse or low voltages as well as reduces transmission loss by keeping appropriate voltage profiles. For the above reason, the identification of critical/weak buses in the system is very much useful for installing additional voltage support devices to prevent possible voltage instability problem.

BACKGROUND

Voltage Stability Analysis

In the literature many voltage stability and voltage collapse prediction methods have been proposed (Ajjarapu & Lee, 1998). Some of these methods are P-V and Q-V curve analysis (Taylor, 1994), determination of how far the system is operating from the point of collapse from continuation power flow (CPF) method (Ajjarapu & Christy, 1992), multiple load flow solutions (Tamura, Mori, & Iwamoto,1983), modal analysis (Gao, Morison, & Kundur, 1992), voltage instability proximity indicator (CIGRE Task Force 38.02.11 Report, 1994), minimum singular value of power flow Jacobian (Lof, Andersson, & Hill, 1993), voltage stability index based on load flow solution (Kessel & Glavitsch, 1986), sensitivity analysis (Begovi´c & Phadke, 1992), energy function (Overbye & DeMarco, 1991), reactive power optimization based methods (J. Singh, S. Singh, & Srivastava, 2007), artificial neural networks (El-Keib & Ma, 1995), and other methods (Moger, 2015). The indices derived from these methods can be used to identify the weakest bus or area in the power system and also provide reliable information about the closeness of the system to voltage collapse.

Traditionally, utilities depended on conventional power flow programs for the static analysis of voltage stability by computing P-V curves (Kundur et al., 2004) and QV curves (Taylor, 1994; Ajjarapu & Christy, 1992) at selected load buses. P-V curves are generated by obtaining power flow solutions for the different loading conditions. The MW load at the buses is increased in small steps while maintaining the power factor of the load and the pattern of generation. The shape of the P-V curve is similar to that of a parabola. The knee point of this parabola gives the critical loading of the bus. The distance between the operating point and the knee point gives the MW and the voltage magnitude stability margins for the given load power factor. Similarly, Q-V curves are used in conjunction with P-V curve. A disadvantage of using the conventional power flow or Newton-Raphson method is that power flow simulation will fail to converge near the nose or maximum power point on the curve. Ajjarapu and Christy (1992) proposed a continuation power flow method to overcome the divergence properties of Newton-Raphson power flow method near maximum loadability of the system. The proposed method overcomes this problem by reformulating the power flow equations so that they remain well-conditioned at all possible loading conditions (Ajjarapu, 2007). The method requires iterative process to identify the weak buses in the system. Hence, it requires more computational efforts. Continuation power flow method for locating the point of voltage collapse of power systems have been presented in (Ajjarapu & Christy, 1992; Canizares & Alvarado, 1993; Souza, 2000; Kundur, Balu, & Lauby, 1994; Van Cutsem, 1991). Similar methods for computing the margin to voltage collapse are demonstrated in (Dobson & LIMING, 1992).

The use of singularity of the power flow Jacobian matrix as an indicator of small disturbance stability is pointed out by Venikov et al. (1975), where the sign of the determinant of the power flow Jacobian matrix is used to determine if the studied operating point is stable or not. Singularity of the power flow Jacobian matrix indicates that the inverse of the Jacobian does not exist. This can be interpreted as an infinite sensitivity of the power flow solution to small perturbations in the parameter values. The point where this will occur is called a static bifurcation point, where several branches of equilibrium may come together, and the studied system will experience a qualitative change in the structure of the solutions due to a small change in parameter values.

The use of minimum singular value of the power flow Jacobian matrix obtained from a full singular value decomposition of the power flow Jacobian matrix is proposed as a measure of static voltage stability by Tiranuchit and Thomas (1988). The reason for this is that at the point of voltage collapse no physically meaningful load flow solution is possible, as the load flow Jacobian will become singular. At this point the distance of the minimum singular value from zero at an operating point is the measure of proximity to voltage collapse. Lof et al. (1992, 1993) presented the theory of singular value decomposition technique and the use of this technique in power systems for identifying voltage weak bus bars. Authors also demonstrated that the singular value analysis of a reduced load flow Jacobian with respect to the reactive power equations provides better results than the analysis of the full load flow Jacobian. The right singular vector corresponding to the minimum singular value of Jacobian matrix, which indicates the sensitivities of the voltages, is used to identify the weak buses. The improved singular value decomposition method based on continuation load flow for identifying voltage weak bus bars in large scale power system is presented in (Li & Song, 2002). Gao et al. (1992) proposed the Q-V sensitivities based modal analysis technique for voltage stability analysis of the power system using eigenvalues of the reduced power flow Jacobian and the bus participation factor associated with the smallest eigenvalue. The method computes the smallest eigenvalue and the associated eigenvectors of a reduced power flow Jacobian matrix, which retains the Q-V relationships in the network. The magnitude of the eigenvalues provides a relative measure of proximity to instability. The eigenvectors, on the other hand, provide information related to the mechanism of loss of voltage stability. The demerit of the method is that the weak buses information can only be obtained near the maximum loading point on the system and it has to go in an iterative manner. Hence, it is time consuming for bigger practical system.

Chen et al. (1995) present two methods for identifying weak buses in electrical power systems. The first method is based on the right singular vector corresponding to a minimum singular value of the power flow Jacobian matrix, which indicates the sensitivities of the voltages. The second method is based on the voltage collapse proximity indicator.

Kessel and Glavitsch (1986) defined a voltage stability index (L-index) to evaluate the critical buses of the system. The indicator uses information of the normal power flow solution. This index value ranges from 0 (no load condition of system) to 1 (voltage collapse). The bus with highest L-index value will be the most vulnerable bus and hence this method helps in identifying the weak areas in the system, which needs critical reactive power support.

Musirin and Rahman (2002) defined a fast voltage stability index (FVSI) to determine the maximum loadability in power systems for identification of weak buses. The indicator is derived from the voltage quadratic equation at the receiving end in a two bus system. The load of a bus (which is to be ranked) is increased till the maximum value of FVSI is reached and this load value is used as an indicator for ranking the bus.

He et al. (2004) proposed a voltage stability margin index (VSMI) based on the relationship between voltage stability and angle difference between sending and receiving end buses. VSMI is used to estimate the voltage stability margin and identify the weak transmission lines and buses at any given operating condition.

Chebbo et al. (1992) proposed a voltage collapse proximity indicator at the load points of the power system based on the optimal impedance solution at maximum power transfer. The maximum power transfer to a bus takes place when the load impedance becomes equal to the driving point impedance. The performance of the indicator is investigated when the load at a particular node and the system load are increased gradually. Some of the limitations of this method are the prediction is acceptable for a single load change and its approximately acceptable for system load change, and the index computed is different when the reference bus (slack)bus chosen is different.

Chen (1996) proposed a weak bus-oriented criterion to determine the candidate buses for installing new VAr sources in the VAr planning problem. First, a voltage collapse proximity indicator is used for identifying weak buses. Then, appropriate VAr planning in those weak buses for enhancing the system security margin is obtained by the goal attainment method based on simulated annealing approach. The weak bus-oriented VAr planning criterion not only enhances the system security margin, but also reduces the solution space.

Begovic and Phadke (1992) proposed the control of voltage stability for a class of voltage instabilities that correspond to saddle node bifurcation of load flow equations. The authors observed from simulation studies that voltage collapse is accompanied by a sharp rise in reactive power generation. It is therefore proposed that the sensitivities of generated reactive power with respect to increase of reactive demand obtained for two different load conditions, be used for assessment of voltage stability margin. An important aspect of power system control presented in their paper is the effect of allocation and the amount of reactive power support on voltage stability margin. It is based on sensitivity analysis of generator reactive power with respect to active and reactive load requirements at various locations of the system.

Kumano et al. (1994) presented a new method of monitoring and control of voltage collapse. The monitoring methodology proposed is based upon the multiple load flow solutions and sensitivity analysis. Voltage instability phenomenon or voltage collapse may occur at the critical load flow point. The structural analysis of the load flow manifold (P-V curve) is used to extract information about the system stability. From the system operator's viewpoint, this critical point must be carefully monitored in a heavily loaded power system. A sub-optimal preventive control is then formulated on the basis of the load flow manifold, to shift the operating point with the aim of increasing the total demand margin between the current operating point and the critical load flow.

Crisan and Liu (1994) proposed a voltage stability index on the basis of an improved sensitivity model. The sensitivity approach can predict voltage collapse and suggest corrective action, thus providing valuable information in both planning and operational environments. The improved sensitivity approach includes the physical constraints on the power system, especially load characteristics, dispatch strategy and reactive power generation limits, which makes the model pertinent to the actual system conditions. The model is flexible for different load increase patterns and can be calculated directly using the Newton-Raphson power flow solution.

Canizares et al. (1999) presented detailed steady state models with controls of two FACTS controllers (SVCs and TCSC) to study their effect on voltage collapse phenomena. Based on results at the point of collapse, design strategies are proposed for these controllers, so that their location, dimension and controls can be optimally defined to increase system loadability.

The weakest bus of the system can be identified by using tangent vector analysis (Canizares et al., 2002; Sode-Yome & Mithulananthan, 2004). Tangent vector is the direction vector of the states on the system voltage profile. It is obtained from the predictor steps in continuation power flow process.

De Souza et al. (1997) used tangent vector index, readily available from continuation power flow solution to find possible locations of critical buses for reactive power support.

Thukaram and Lomi (2000) discuss an approach for selection of static VAr compensator location and its size in EHV network for system voltage stability improvement based on static voltage stability analysis. The proposed approach uses the steady state voltage stability index (L-index), which is a scalar number corresponding to each load bus.

Ajjarapu and Lee (1998) made an exhaustive literature review on various other voltage stability and voltage collapse prediction methods covering all aspects of voltage stability problem reported in the literature.

Reactive Power Dispatch

The objective of the system operator is to ensure secure and economic operation of power system. The challenge is to optimize power system operation, while maintaining system security and quality of supply to customers. The growing demand without matching expansion of generation and transmission facilities and more tightly interconnected power systems contribute to the increased complexity of system operation. The increased environmental concerns has also made transmission, as well as generation systems are to be operated very close to design limits, with smaller safety margins, and hence greater exposure to unsatisfactory operating conditions following a disturbance. Under these disturbed conditions the system operator should ensure quality and reliability of supply to the customers by maintaining the load bus voltages within the permissible limits. Generally, power loss in the transmission of electrical energy causes a loss of revenue. So, even a small percentage of savings in loss will be very much appreciated since the total generated power is in the order of thousands of megawatts. Though real power dispatch with objective of minimizing losses is very well established, in the last few decades more attention was paid to optimal reactive power dispatch (Dommel & Tinney, 1968).

In the past, several methods using sensitivity relationship have emerged to solve the complex problems. Since the problem of reactive power optimization is non-linear in nature, non linear programming (NLP) methods have been used to solve it. These NLP methods work quite well for small power systems but may develop convergence problems as system size increases. The studies performed on some IEEE standard test systems (Pudjianto, Ahmed, & Strbac, 2002) show that NLP based optimal power flow (OPF) is comparatively less robust with respect to convergence under all the random starting points. A bad initial point, for instance near any operating limit of control and/or state variables, as well as a too narrow operating range of control or state variables may limit the permissible step length for those variables and therefore restrict the movement of the other variables. This would then cause non convergence of the NLP based OPF. Linear programming technique with iterative scheme is certainly the most promising tools for solving these types of problems (Thukaram & Yesuratnam, 2006). Effective use of linear programming approximation to this non-linear problem is also reported in the literature (Hcbson, 1980; Deeb & Shahidehpour, 1988, 1990; Tomsovic, 1992; Mangoli, Lee, & Park, 1993; Pudjianto et al., 2002; Thukaram, Bansilal, & Parthasarathy, 1996).

Hobson (1980) developed a method of finding the network constrained reactive power control. The author has used incremental transmission line and transformer models, and linearized networks equations. Then, the problem was solved by a special linear programming (LP) technique by giving priorities to generators in the system. Because the main purpose of this method is to find the values of control variables, which would cause the dependent variables to vary within a certain limit. This method can also be used to study contingency analysis.

Mamandur and Chenoweth (1981) presented a method for optimal control of reactive power flow for improvements in voltage profile and real power loss minimization. The method employs linearized sensitivity relationship of power system to establish both the objective function for minimizing the system losses and the system performance sensitivities relating the dependent and control variables. They have used the dual linear programming technique to determine the optimal adjustments to the control variables, simultaneously satisfying the constraints. These constraints include the reactive power limits of the generators, limit on the load bus voltages and the operating of the control variables viz., the transformer tap positions, generator terminal voltages and adjustable reactive power sources. The authors have used the dual LP algorithm for minimizing the objective function. Zig-zagging of the solution is also reported in the paper. Inspite of that the presented approach has many attractive features in problem formulation and solution techniques. Unlike other optimal power flow approaches, they have used the idea of decomposing the optimal power flow problem into active power optimization and reactive power optimization.

The nonlinear approaches have deficiencies such as unreliability or slowness of convergence, the need for a feasible starting point, gradient step control, using penalty function, difficulties in recognizing in-feasibility and in post optimal sensitivity analysis. However, the linear methods are attractive in view of their reliability and computational fastness. Therefore, the approach has good potential for practical applications of the algorithm. However, the algorithm needs further improvement for implementation to larger power system and real time application.

The Newton-Raphson load flow method has been used by Mamandur and Chenoweth (1981) for the successive power flow solutions. Instead, use of a decoupled load flow method makes the algorithm computationally fast. In the formulation of the sensitivity matrix relating the dependent and control variables for the purpose of optimization problem, they advocated for the inversion of a large Jacobian like matrix, which is obtained from the set of Newton-Raphson load flow equations augmented by equations for power injections at the slack bus and power flow equations for tap regulating transformers. Development of sensitivity matrix relating the dependent variables and control variables by avoiding inversion of large matrices would make the algorithm applicable for large size system. In this connection Thukaram et al. (1984) developed an improved method for reactive power optimization in which the problem is formulated by avoiding the inversion of large matrices. The approach adapted is an iterative scheme with successive power flow analysis using a fast decoupled technique and formulation, and solution of the linear programming problem with only upper-bound limits on the state variables. The voltage dependency characteristics of the loads are also included in this model.

Bijwe et al. (1986) presented a technique for improving system voltage profiles. The control of voltage profile is achieved by minimizing the sum of the weighted voltage deviations at the buses using a first-order gradient technique. A constant symmetric load flow model in rectangular coordinates is employed. The constant Jacobian matrix used in load flow is also used in the calculation of the lagrangian multipliers throughout the optimization procedure. These features result in a computationally deficient algorithm.

Qiu and Shahidehpour (1987) presented a method for minimizing the transmission losses and improving the voltage profile using linear programming. The authors have used restricted step sizes on each control variable to control zig-zagging of the solution within a small range. Large step sizes for the control variables were used in the first optimization cycle and smaller step sizes were used in subsequent optimization algorithm as follows. Transmission loss minimization is not handled efficiently because its objective function is strongly non-separable.

Singh et al. (1996) present a method for minimizing the sum of the square of voltage deviations by a least-square minimization technique, and thus improving the voltage profile in a given system by adjusting control variables, such as tap position of transformers, reactive power injection of VAr sources and generator excitations.

Da Costa (2002) described an approach to the optimal reactive dispatch problem with the objective of minimization of real power losses based on an augmented lagrangian function of the original problem. In this approach, Karush Kuhn Tucker (KKT) optimality conditions are solved by the modified newton method.

Venkatesh et al. (2003) proposed a unified optimal power flow method that optimally schedules real and reactive power controllers. It minimizes the total generation cost and system transmission loss while maximizing the system voltage stability margin (VSM). In this study, the VSM is maximized by maximizing a set of few least singular values of the load flow Jacobian (Lof et al., 1993) by using a successive fuzzy LP technique.

Zhu and Xiong (2003) presented an approach to study the optimal reactive power (VAr) control problem with the objective of real power loss minimization using a modified interior point method. In this study, they have used an analytic hierarchical process and sensitivity analysis methods to select the optimal locations of VAr support service.

Menezes et al. (2004) presented a methodology for the inclusion of an evaluation and improvement of voltage stability margins in the power system pre-dispatch problem by optimizing the reactive power injections of generators and synchronous condensers. The objective is to maximize voltage stability margins maintaining the economical dispatch of active power.

Abido and Bakhashwain (2005) presented a novel multi-objective evolutionary algorithm for optimal reactive power dispatch problem. In this, the problem is formulated as a nonlinear constrained multi-objective optimization problem where the real power loss and the bus voltage deviations are treated as competing objectives.

Zhang and Ren (2005) presented a mathematical model for optimal reactive power dispatch, in which the objective function is to minimize the sum of the active power loss of the whole network and the costs of adjusting controlling devices represented in power.

Zobian and Ilic (1996) proposed a weighted least square minimization of the voltage deviation to put more weight on important load buses. This method does not solve exact load flow equations, and therefore it avoids inversion of full system size matrix.

Hong and Gau (1994) presented the application of newton optimal power flow (OPF) to identify the weakest bus/area, which is most likely to cause voltage collapse. The S-V curve (complex power to voltage) is examined via Newton OPF. Marginal costs (Lagrange multipliers) regarding power flow equations, VAr generations, voltages and taps, etc., to the system

MW loss is obtained from Newton OPF. The weakest bus/area in the system is identified by an indicator achieved with these marginal costs via Kuhn-Tucker theorem. This indicator helps the users to know, which bus/area is most likely to cause voltage collapse.

Thukaram et al. (1996) present an algorithm for monitoring and improving voltage stability in power systems for a base-case and credible contingency conditions. The monitoring methodology proposed is based on the voltage stability L-indices of load buses. The proposed algorithm gives an optimal setting of various control devices like generator excitation, switchable VAr compensator (SVC) and on load tap changing transformer (OLTC).

Thukaram et al. (1998) is mainly concerned with analysis and enhancement of steady state voltage stability based on L-index. They have presented an algorithm for optimization of reactive power control variables using linear programming technique (Thukaram & Yesuratnam, 2006), and some of few other work on steady state voltage stability improvement by the same author have been reported in the literature (Thukaram, Jenkins, & Visakha, 2006; Visakha, Thukaram, & Jenkins, 2004a; Thukaram et al., 1984; Thukaram & Yesuratnam, 2008; Thukaram, Jenkins, Khincha, Yesuratnam, 2004; Yesuratnam & Thukaram, 2007; Lomi & Thukaram, 2012).

OBJECTIVES AND CONTRIBUTION

This chapter presents a new reactive power loss index for identification of weak buses in the system. The new index is computed from the reactive power support and loss allocation algorithm using Y-bus approach for the system under intact condition as well as some critical/severe contingencies cases. The fuzzy logic approach is used to select the important and critical/severe line contingencies from the contingency list. The post-contingent quantities such as bus voltage magnitudes, line loadings and voltage stability indexes (L-index) at load buses are considered as inputs. The severity of each post-contingent quantities is evaluated separately and then network composite overall severity index (NCOSI) is determined. Depending on the system topology and reactive power requirements, a few weak buses in the system are selected (as per the severity order of load buses based on the values of reactive power loss index) for optimal placement of reactive compensation devices to support the voltage. The impact analysis of reactive compensation in the system in terms of various system performance parameters such as voltage profile, losses etc., is carried out. Later, the validation of weak load buses identification from the RPLI index with that from other well known existing methods (Ajjarapu & Christy, 1992; Gao et al., 1992) in the literature is carried out to demonstrate the effectiveness of the proposed index. The proposed index is illustrated on sample 5-bus system for different loading conditions up to the system maximum loadability point to highlight the significance of reactive power loss and its application to the system voltage stability. The sample 10-bus equivalent system and 72-bus equivalent system of Indian southern region power grid are considered to test the proposed index.

Identification of Critical Line Contingency Based on Fuzzy Logic Approach

As systems tend to operate closer to their operational limits, contingency analysis plays an important role as far as power systems security/reliability is concerned. It is crucial to identify the severe network contingencies that would lead the system to abnormal or close to critical operating conditions. Therefore, the identification of weak load buses and system reactive power planning process not only decided based on the system intact condition, but also the severe/critical network contingencies should be taken into consideration. There are many analytical approaches have been proposed in the literature for contingency selection. Among them, the ranking based method with scalar performance index is widely used in the contingency selection (Galiana, 1984). Most of the contingency ranking methods rank the contingencies in an approximate order of severity with respect to a scalar performance index, which quantifies the system stress. The main disadvantage of performance index based contingency selection algorithms is the masking effect. By masking effect, the non severe contingencies case can take the position of a severe one. Further, the researchers attempted the use of fuzzy logic approach in contingency selection to overcome the drawbacks of the analytical approaches (Hsu & Kuo, 1992; Visakha et al., 2004a; Visakha, Thukaram, & Jenkins, 2004b).

In this chapter, fuzzy logic approach is used to rank the network contingencies on a non-discriminatory manner. The input variables are the post-contingent quantities such as line loadings, bus voltage magnitudes and voltage stability index (L-index) at the load buses used separately to evaluate the network contingency ranking using the composite criteria. The output variable is the severity index, which shows the degree of severity of post-contingent quantities. Each post-contingent quantity is divided into different categories and transformed into fuzzy set notations with the help of membership functions.

Line Loading

The post-contingent percentage line loading is divided into four categories using fuzzy set notations; lightly loaded (LL), 0-50%; normal loading (NL), 50-80%; fully loaded (FL), 80-100%; over loaded (OL), above 100%. The input membership function showing the relationship between the line loading (in percentage) and fuzzy linguistic variables is shown in Figure 1. Similarly, the output membership function is used to evaluate the severity of line loading also divided into four categories using fuzzy set notations; less severe (LS), below severe (BS), above severe (AS), and more severe (MS). The output membership function showing the relationship between the severity index and fuzzy linguistic variables is shown in Figure 2. Then, based on the system knowledge, fuzzy rules are developed to evaluate the severity of post-contingent quantities of line loading, which are described below.

IF line loading is LL THEN severity is LS
IF line loading is NL THEN severity is BS
IF line loading is FL THEN severity is AS
IF line loading is OL THEN severity is MS

After obtaining the normalized severity indices of all the lines, the overall severity index (OSI) of line loading for a particular line outage is obtained using the following expression.

Figure 1. Line loading (in percentage) and corresponding linguistic variables

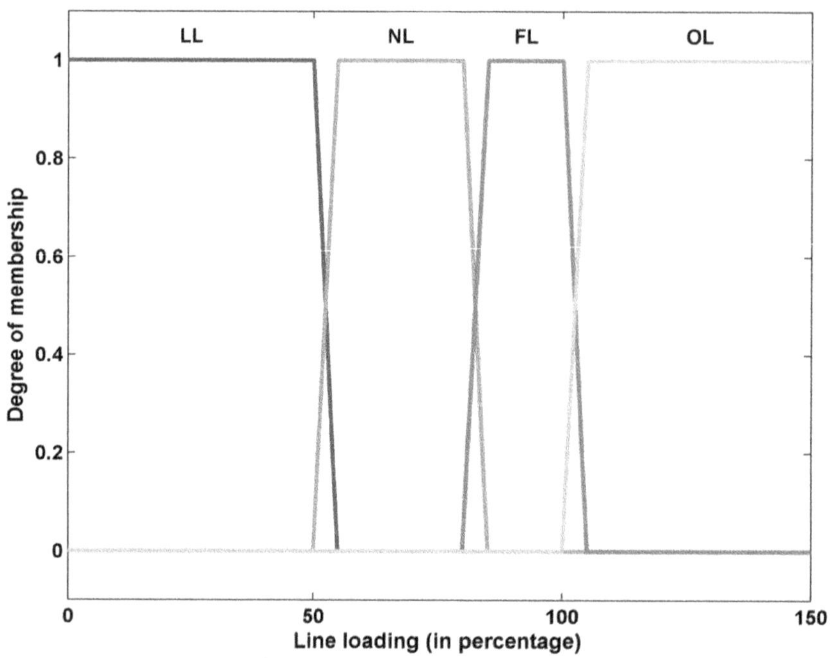

Figure 2. Severity index for line loading and corresponding linguistic variables

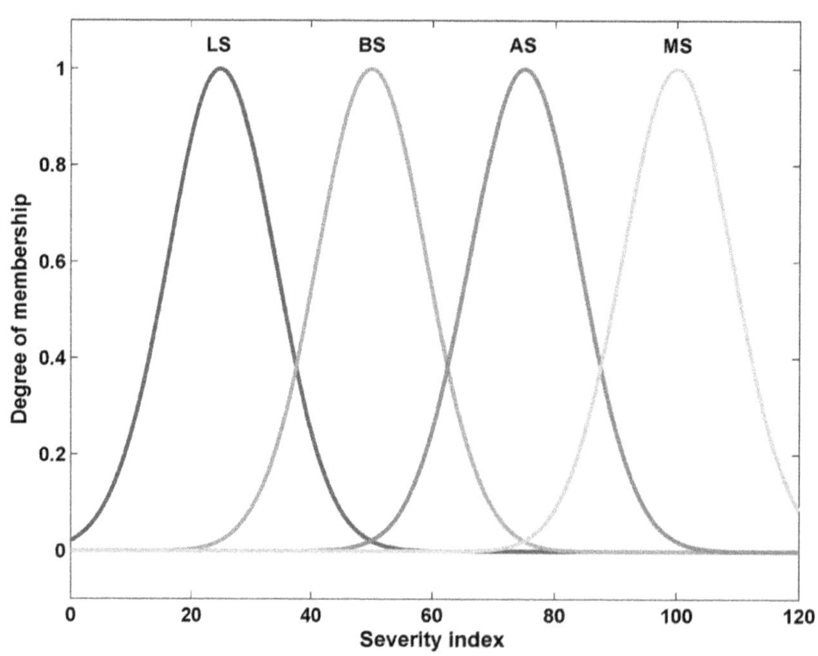

$$OSI_{LL} = \sum_{nl} SI^{n}{}_{LL} \tag{1}$$

Where,

nl is the number of lines,
OSI_{LL} is the overall severity index for line loading,
$SI^{n}{}_{LL}$ is the normalized severity index of post-contingent line loading.

Bus Voltage Profile

The post-contingent bus voltage magnitude is divided into four categories using fuzzy set notations; very low voltage (VLV), below 0.8 p.u.; low voltage (LV), 0.8-0.9 p.u.; normal voltage (NV), 0.9-1.05 p.u.; and over voltage (OV), above 1.05 p.u. The input membership function showing the relationship between the bus voltage profile (p.u.) and fuzzy linguistic variables is shown in Figure 3. The output membership function is used to evaluate the severity of the bus voltage profile also divided into four categories using fuzzy set notation; less severe (LS), below severe (BS), above severe (AS), and more severe (MS). The output membership function showing the relationship between the severity index of bus voltage profile and fuzzy linguistic variables is shown in Figure 4. Then, based on the system knowledge, fuzzy rules are developed to evaluate the severity of post-contingent quantities of bus voltage profile, which are described below.

Figure 3. Voltage profile and corresponding linguistic variables

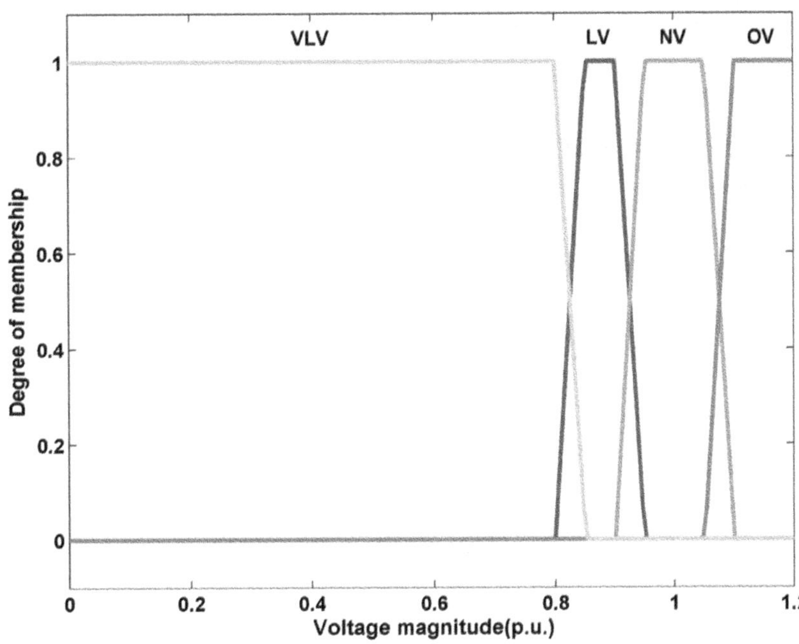

Figure 4. Severity index of voltage profile and the corresponding linguistic variables

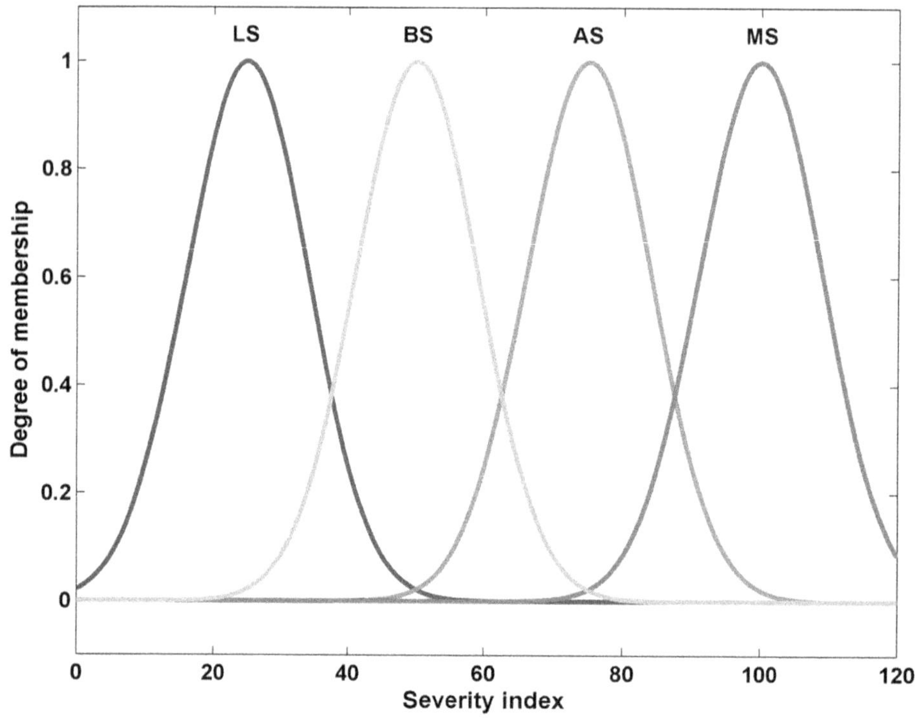

IF bus voltage profile is VLV THEN severity is MS
IF bus voltage profile is LV THEN severity is AS
IF bus voltage profile is NV THEN severity is LS
IF bus voltage profile is OV THEN severity is AS

After obtaining the normalized severity indices of all the load buses, the overall severity index of bus voltage profile for a particular line outage is obtained using the following expression.

$$OSI_{VP} = \sum_{pq} SI^{n}{}_{VP} \quad (2)$$

Where,

pq is the load buses,
OSI_{VP} is the overall severity index for bus voltage profile,
$SI^{n}{}_{VP}$ is the normalized severity index of post-contingent bus voltage profile.

Voltage Stability Index (L-Index)

The post-contingent voltage stability index is divided into five categories using fuzzy set notations; very low index (VLI), 0-0.05; low index (LI), 0.05 -0.2; medium index (MI), 0.2-0.35; high index (HI), 0.35-0.55; and very high index (VHI), above 0.55. The input membership function showing the relationship between the voltage stability index (L-index) and fuzzy linguistic variables is shown in Figure 5. The output membership function used to evaluate the severity of the voltage stability index is also divided into five categories using fuzzy set notation; very less severe (VLS), less severe (LS), below severe (BS), above severe (AS), and more severe (MS). The output membership function showing the relationship between the severity index of the voltage stability index and fuzzy linguistic variables is shown in Figure 6. Then, based on the system knowledge, fuzzy rules are developed to evaluate the severity of post-contingent quantities of voltage stability index, which are described below.

IF voltage stability index (L-index) is VLI THEN severity is VLS
IF voltage stability index (L-index) is LI THEN severity is LS
IF voltage stability index (L-index) is MI THEN severity is BS
IF voltage stability index (L-index) is HI THEN severity is AS
IF voltage stability index (L-index) is VHI THEN severity is MS

After obtaining the normalized severity indices of all the load buses, the overall severity index of voltage stability index (L-index) for a particular line outage is obtained using the following expression.

$$OSL_{VSI} = \sum\nolimits_{pq} SI^{n}{}_{VSI} \tag{3}$$

Where,

Figure 5. Voltage stability index and corresponding linguistic variables

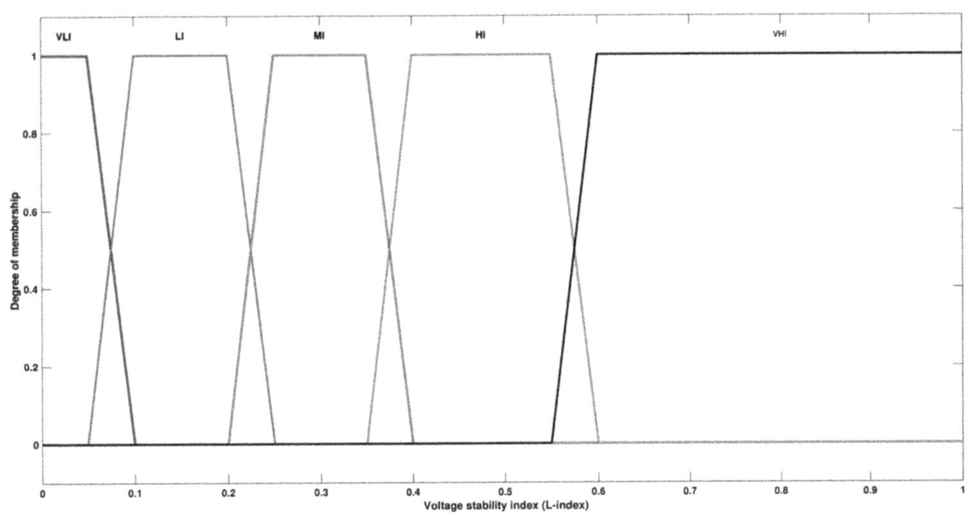

Figure 6. Severity index of voltage stability index and corresponding linguistic variables

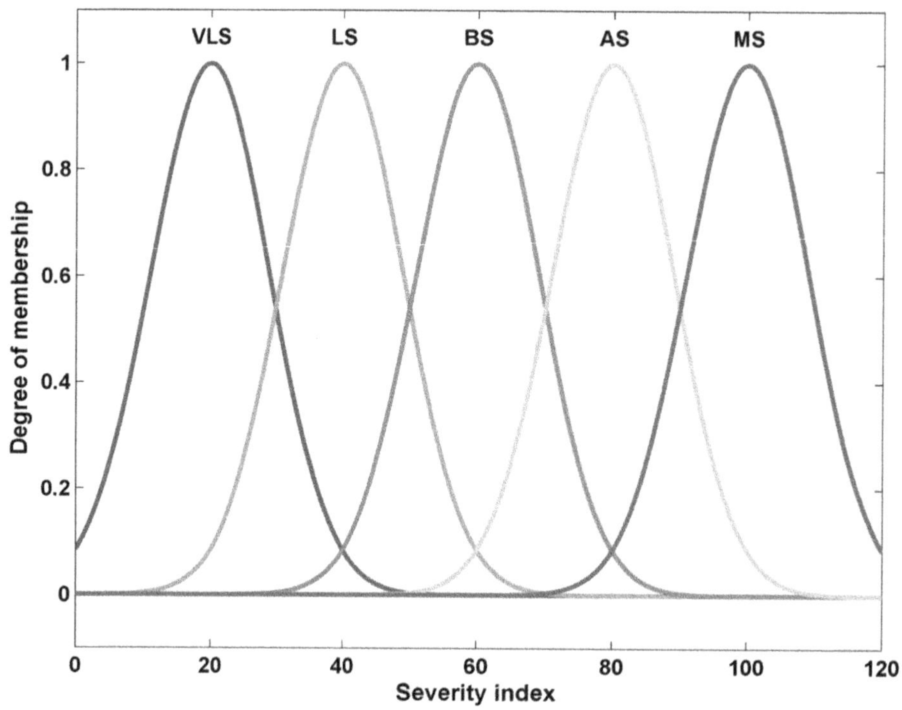

pq is the number of load buses,
OSI_{VSI} is the overall severity index of voltage stability index (L-index),
SI^{n}_{VSI} is the normalized severity index of post-contingent voltage stability index.

Computation of Network Composite Overall Severity Index (NCOSI)

The membership function for each post-contingent quantity of line loading, bus voltage magnitudes, and voltage stability index (L-index) is established, and these membership functions are then used to compute the network composite overall severity index for the particular contingency.

For the given contingency, the post-contingent quantities such as line loadings, voltage profiles, and voltage stability indices are fed to the corresponding fuzzy inference systems as shown in Figure 7. The fuzzy inference systems evaluates the severity of each post-contingent quantity, using the fuzzy rules, and gives the overall severity indices of line loadings (OSI_{LL}), voltage profiles (OSI_{VP}) and voltages stability indices (OSI_{VSI}).

$$NCOSI = OSI_{LL} + OSI_{VP} + OSI_{VSI} \tag{4}$$

Figure 7. Parallel operated fuzzy inference systems (FIS)

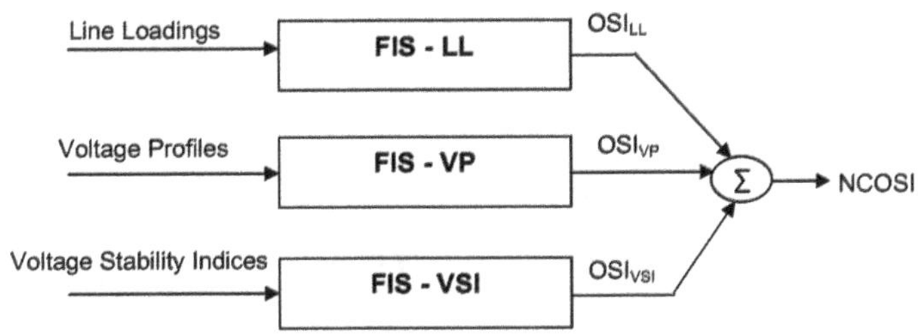

The network composite overall severity index (NCOSI) for each contingency is determined by adding all the individual normalized overall severity indexes. Then, rank the network contingencies according to the normalized values of network composite overall severity index.

METHODOLOGY

Equivalent Model of Transmission Line

The line charging capacitances can be treated as sources of providing reactive power to the system. So, while calculating the contribution of reactive sources towards the reactive sink, the line charging capacitances must be considered as reactive sources. The equivalent model of transmission line is shown in Figure 8. The reactive powers ($Q_{c,m}$ and $Q_{c,n}$) produced by the line shunt admittances ($Ys_{h}/2$) are transferred into the nearby nodes with an assumption that the voltages of the shunt admittances are equal to the nearby nodal voltages. The nodal voltages can be obtained by the power flow calculation.

$$Qc,m = \mathrm{Im}(V_m^{\,2} Ysh / 2) \tag{5}$$

$$Qc,n = \mathrm{Im}(V_n^{\,2} Ysh / 2) \tag{6}$$

The net reactive power support at all generator and load buses are calculated by considering the line charging capacitances and other reactive sources/sinks at the respective buses.

Computation of Reactive Power Loss Allocation to Load Buses

Consider a system comprising of n number of total buses with 1, 2…*g*; *g* be the number of generator buses, and $g + 1, g + 2...n$; $(n - g)$ be the remaining load buses. Under steady state operating condition, for a given system, the network steady state performance equation is given by (Crow, 2002; Stagg, El-Abiad, & El-Abiad, 1968);

$$[I_G] = [Y_{GG}][V_G] + [Y_{GL}][V_L] \tag{7}$$

Figure 8. Equivalent model of transmission line

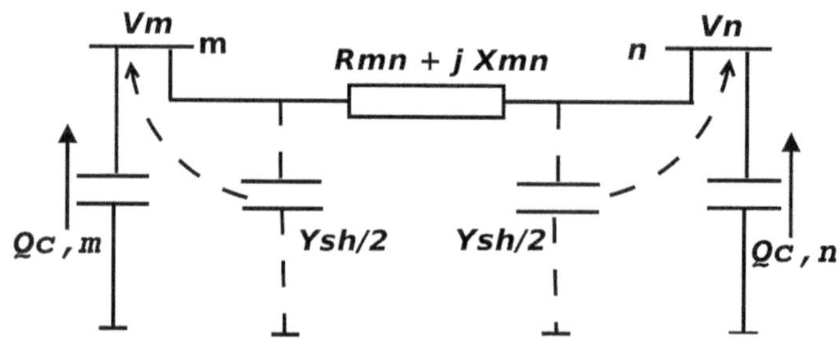

$$[I_L] = [Y_{LG}][V_G] + [Y_{LL}][V_L] \tag{8}$$

where $[I_G]$, $[I_L]$ are the complex bus current injection vectors, $[V_G]$, $[V_L]$ are the complex bus voltage vectors and $[Y_{GG}]$, $[Y_{GL}]$, $[Y_{LG}]$, $[Y_{LL}]$ are the corresponding partitioned matrices of the bus admittance matrix. Equation (7) is rewritten in terms of load bus currents and generator bus voltages is given by,

$$[I_G] = [K_{GL}][I_L] + [Y^e{}_{GG}][V_G] \tag{9}$$

$$[K_{GL}] = [Y_{GL}][Z_{LL}]$$

$$[Y^e{}_{GG}] = [Y_{GG}] - [Y_{GL}][Z_{LL}][Y_{LG}]$$

$$[Z_{LL}] = [Y_{LL}]^{-1}$$

The main aim is to get the generators contribution to meet individual load demand and losses in the system. In order to do so, from circuit theory analysis, the generator voltage V_G in (9) is being replaced as a function of load bus voltages i.e., $V_G = f(V_L)$. A possible way to deduce generator voltage as a function of load bus voltages, is to apply superposition theorem. However, it requires replacing all generators current injection into its equivalent admittances in the circuit. Using readily available load flow results, the equivalent shunt admittance Y_{Gj} of generator node j can be calculated using the following;

$$Y_{Gj} = \frac{1}{V_{Gj}} \left(\frac{-S_{Gj}}{V_{Gj}} \right)^* \tag{10}$$

Where (*) means conjugate, S_{Gj} be the generator apparent power at node j and V_{Gj} be the generator voltage at node j.

Now these equivalences are added to the corresponding diagonal entries of Y-bus matrix. Then, from (7), one can solve for generator voltage V_G as a function of load voltage V_L. This is given as,

$$[V_G] = -[Y'_{GG}]^{-1}[Y_{GL}][V_L] \tag{11}$$

where $[Y'_{GG}]$ is the modified sub-matrices of $[Y_{GG}]$. From (11), it is assumed that

$$[Y_{GL}^B] = -[Y'_{GG}]^{-1}[Y_{GL}] \tag{12}$$

Then, (11) can be written as

$$[V_G] = [Y^B{}_{GL}][V_L] \tag{13}$$

The voltage contribution to generator bus from each load bus voltages is expanded as,

$$V_{Gj} = \sum_{i=1}^{NL} Y^B{}_{GL} * V_{Li} \tag{14}$$

It can be seen from (14) that the original generator voltage at bus j is the sum of individual voltage contribution from all load buses.

By substituting (13) into (9), the generator current can be expressed as,

$$I_G = [K_{GL}][I_L] + [Y^C{}_{GL}][V_L] \tag{15}$$

Where,

$$[Y^C{}_{GL}] = [Y'_{GG}][Y^B{}_{GL}]$$

In order to determine the generator's share/contribution to meet load demand and losses at load bus, the vectors $[I_L]$ and $[V_L]$ should be consider in diagonal matrix. Take a conjugate of (15) and pre-multiplying by $[V_G]$ the diagonal generator voltage matrix, the generators' complex power can be expressed as,

$$[V_G]_{G*G}[I_G^*]_{G*L} = [S_{gen-contrb}]_{G*L}$$

$$= [V_G]_{G*G}[K_{GL}^*]_{G*L}[I_L^*]_{L*L} + [V_G]_{G*G}[Y_{GL}^{C*}]_{G*L}[V_L^*]_{L*L} \tag{16}$$

The reactive power contribution of all generators w.r.t. load buses can be given as,

$$[Q_{gen-contrb}]_{G*L} = \mathrm{Im}\left([S_{gen-contrb}]_{G*L} \right) \tag{17}$$

With further simplification of (17), the reactive power contribution from generator j to load bus I is as follows:

$$Q_{gen-contrb_{(j)}} = \sum_{i=1}^{NL} Q_{gen-contrb_{(ji)}} \tag{18}$$

From (18), the reactive power loss allocated to each load bus i can be expressed as,

$$Qloss_{(i)} = \sum_{j=1}^{NG} Q_{gen-contrb_{(ji)}} - Q_{Li} \tag{19}$$

Where Q_{Li} is the net reactive power load demand at bus i.

Reactive Power Loss Index for Identification of Weak Buses in the System

At each load buses, the reactive power loss (Qloss) is determined as explained in the preceding subsection. It is well known that the reactive power loss in the system not only depend on reactive demand at load buses, but also depend on active power flows/demand and the relative location of load buses with respect to the sources buses in the network (Elgerd, 1982). At high loadings, the reactive power loss can increase significantly with the distance transported. Consequently, the reactive power loss allocated to respective load buses also increases. The detailed evaluation of reactive power support and loss allocation under various system operating conditions, and its impact on system reactive power issues is discussed in (Moger & Dhadbanjan, 2017; Moger & Dhadbanjan, 2013a, 2013b). Thus, the reactive power loss allocation to load buses can act as an indicator to the reactive power deficient/surplus at a bus. Therefore, it can be considered as an index for optimal placement of reactive compensation devices. A weighted sum of normalized values of reactive power loss at each load buses under system intact condition and different severe contingencies is computed, and termed as Reactive Power Loss Index (RPLI). The normalized network composite overall severity index (4) is taken as the relative weightage for different contingencies, as it reflects relative severity of a particular line outage condition. The RPLI at bus i can be expressed as

$$RPLI_i = Qloss_{i,0}^n + \sum_{k=1}^{Nc} (Qloss_{i,k}^n * NCOSI_k) \tag{20}$$

$$Qloss_{i,0}^n = Qloss^n{}_{i,0} \,/\, \max(Qloss_0)$$

and

$$Qloss_{i,k}^n = Qloss^n{}_{i,k} \,/\, \max(Qloss_k)\,,$$

$k \in N_c$ (No. of severe contingencies selected) are the normalized reactive power loss at i^{th} bus under system intact condition and k^{th} most critical line contingency case, respectively.

A severity/merit order of the load buses based on the values of RPLI is determined. The bus having highest RPLI is considered as the weakest bus in the system and the corresponding bus will be considered as the best location for placing the reactive compensation device for additional voltage support. The bus having the lowest RPLI is considered to be more stable, which may not require any further compensation device to support the voltage.

SIGNIFICANCE OF REACTIVE POWER LOSS AND ITS APPLICATION TO THE SYSTEM VOLTAGE STABILITY

Sample 5-Bus System: An Illustrative Example

A sample 5-bus system is considered to illustrate the significance of reactive power loss and its application to the system voltage stability. The sample system showed in Figure 9 has two sources at buses 1, 2 and three loads at buses 3, 4 and 5. It is assumed that the lines L1, L2, L3 and L4 are of 50, 150, 100 and 100 kms length, respectively. The 400 kV line parameters per 100 km are R = 0.002 p.u., X = 0.020 p.u. and b/2 = 0.25 p.u. The initial base-case load is 1100 MW and 555 MVAr. The system data is taken from (Moger, 2016).

For any operating point, the load flow analysis on the system is carried out. Based on the results, calculate the net reactive power support at each generator and load buses considering the other reactive sources/sinks at the respective buses including the line charging capacitances. For discussion purpose, the result of the system under peak load condition is presented. The system has a peak load of 1265 MW and 638.25 MVAr, which is 15% more than the base-case loading condition. The peak load condition is simulated to obtain the stressed condition in the system for the computation of reactive power loss allocation. The load flow results of the system under peak load condition is shown in Table 1 along with the calculation of net reactive power support at each generator and load buses considering the line charging capacitances and other reactive sources/sinks at the respective buses. The system has active and reactive power losses of 20.296MW and 202.964 MVAr, respectively. The allocation of reactive power loss to load buses from the proposed approach for the system under peak load condition is shown in Table 2.

It can be seen from Table 2 that the reactive power loss allocated to bus 5 is more as compared with other two buses. Since bus 5 is situated at equal distance from two generator reactive sources, obviously it contributes more reactive power loss in the lines to meet its load demand. The reactive power support received at bus 5 from generator bus (G1) is 52% and that from other generator bus (G2) is 48% of the total reactive support received at that bus. The reactive power loss allocated to bus 4 is more than that allocated to bus 3 even with more load demand at bus 3. Because, bus 3 is just half the distance from the generator bus (G1) as compared to the distance of bus 4 with respect to the generator bus (G2). Obviously,

Figure 9. Single-line diagram of sample 5-bus system

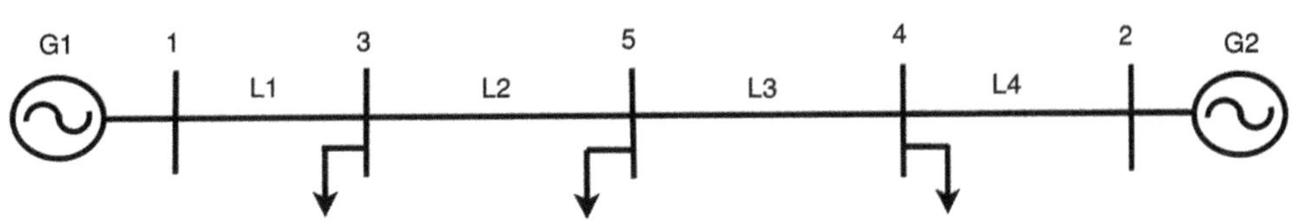

Table 1. Load flow result of sample 5-bus system with the net reactive power support at all buses under peak load condition

Bus No.	Voltage		Generation		Load		Net MVAr
	Mag. (p.u.)	Angle (deg.)	PG (MW)	QG (MVAr)	PD (MW)	QD (MVAr)	
1	1	0	710.2964	366.954	-	-	379.454
2	1	-0.14343	575	296.5375	-	-	321.5375
3	0.95732	-4.02737	-	-	402.5	207	-161.1773
4	0.93055	-6.84352	-	-	345	172.5	-129.204
5	0.90424	-9.71504	-	-	517.5	258.75	-207.6463
Total			1285.2964	663.4914	1265	638.25	
Ploss = 20.29637 MW and Qloss = 202.9637 MVAr							

Table 2. Contribution of generator reactive sources to load buses and reactive power loss allocation for sample 5-bus system under peak load condition

Load Bus	Net Demand (MW, MVAr)	Generator Sources (MVAr)		Total (MVAr)	Qloss (MVAr)	Normalized Qloss/RPLI
		Gen. (G1)	Gen.(G2)			
3	402.5, 161.18	141.53 (72%)	55.406 (28%)	196.93	35.756	0.31173
4	345, 129.2	70.218 (39%)	111.49 (61%)	181.71	52.508	0.45779
5	517.5, 207.65	167.71 (52%)	154.64 (48%)	322.35	114.7	1
Total	1265 498.03	379.45 (54%)	321.54 (46%)	700.99	202.96	–

to meet the load demand at bus 4, the power has to flow double the distance from the nearby generator bus (G2). Hence, it contributes more reactive power loss in the system. Moreover, the load buses 3 and 4 are nearer to generator buses 1 and 2 respectively. Therefore, the reactive power requirements at these buses are met maximum by its nearest generator bus as seen from Table 2. From the results, it can be seen that bus 5 is the weakest in the system.

As loading on the system increases, the power loss taking place in the transmission system must increase. Therefore, the loss allocated to load buses must increase as the system is moving from light load condition to peak load condition. In order to demonstrate the effectiveness of the reactive power loss allocated at load buses and its impact on the system voltage stability, the loading on the system is increased step by step up to the maximum loadability point considering the reactive power limits of the generators. The real and reactive power at load buses are increased in proportion to their initial base-case load levels, step by step up to the maximum loadability point with the help of continuation parameter (loading parameter). The generator output is also increased in proportion to their initial base-case generations in order to meet the increased load. Correspondingly, the reactive power loss allocated to load buses is determined. The Figures 10 and 11 show the voltage profile and reactive power loss allocated to load buses for different loading conditions. It can be observed from Figure 11 that the reactive power loss allocated to load bus 5 is increasing drastically (highly exponential way) as the loading on the system is approaching towards the critical loading point. Similarly, the voltage profile at load bus

5 is also uniformly decreasing as the system is moving towards the critical loading point. However, at one operating point, we noticed that the change in load bus voltages are not uniform as compared with that from other operating conditions because of hitting the reactive power limits of the generator. It can also be seen from the Figures that the large change in reactive power loss or voltage at load bus 5 is not only observed around the critical loading point, there is a significant change around the peak loading condition of the system as well.

A comparative analysis is also carried out from other existing methods (Gao et al., 1992; Ajjarapu & Christy, 1992) in the literature such as V-Q sensitivity based modal analysis (Gao et al., 1992), continuation power flow method (Ajjarapu & Christy, 1992), which have been used for identification of weak buses in the system. The results of the comparison are shown in Table 3. From the other existing methods also, bus 5 is considered to be the weakest bus in the system. The detailed methodology of the existing methods is explained in Appendix.

Table 3. Identification of weak buses for sample 5-bus system: A comparison

Severity Order	RPLI Index		Modal Analysis [Gao et al., 1992]		CPF [Ajjarapu & Christy, 1992]	
	Bus	RPLI	Bus	BPF	Bus	Voltage (p.u.)
1	5	1	5	0.6019	5	0.7342
2	4	0.45779	4	0.3486	4	0.77091
3	3	0.31173	3	0.0495	3	0.9049

Inference

From these discussions, it can be inferred that the reactive power loss allocation at load buses give the clear indication about the system reactive power issues, which in turn give an indication about the system voltage instability/collapse problem. This is clearly observed from Figures 10 and 11. For the system under consideration, load bus 5 is the most critical bus followed by load bus 4.

SYSTEM STUDIES AND DISCUSSIONS

The sample 10-bus 400 kV equivalent system and 72-bus equivalent system, which is a part of Indian southern region power grid (SR 72-bus equivalent system) with different voltage level viz. 400/220 kV are used to verify the effectiveness of the proposed index. The result analysis on the system is carried out in three stages. First, the identification of weak buses in the system based on the proposed index. The proposed index is calculated from the reactive power support and loss allocation algorithm using Y-bus approach for the system under intact condition and considering the impact of few severe network contingencies cases. The fuzzy logic approach is used to select the important and critical/severe network contingencies. Second, depending on the system topology and reactive power needs few weak buses are selected for optimal placement of reactive compensation devices to support the voltage. Then, the impact analysis of reactive compensation in the system in terms of various system performance parameters is

Figure 10. Voltage profile at load buses for different loading conditions up to critical loading point

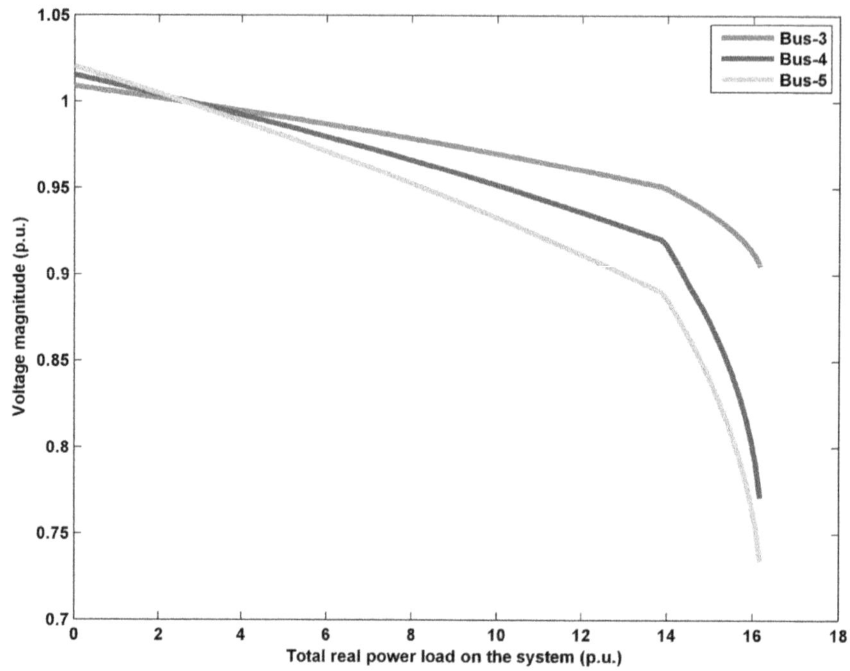

Figure 11. Reactive power loss (Qloss) allocated at load buses for different loading conditions upto critical loading point

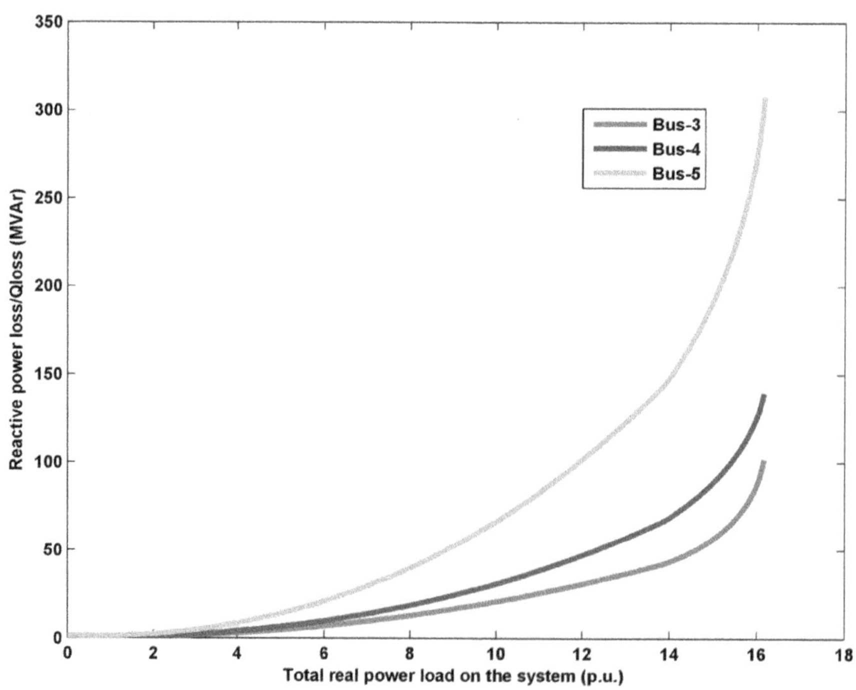

carried out. In this stage, the system operator may execute reactive power optimization algorithm for determining the optimal size of the reactive compensation devices with desired objectives subject to various prevailing system constraints. Lastly, the validation of the proposed index with other well known existing methods (Ajjarapu & Christy, 1992; Gao et al., 1992) in the literature to highlight the robustness/effectiveness of the proposed index.

Sample 10-Bus Equivalent System

The single-line diagram of the system is shown in Figure 12. The network consists of 3 generators, 12 transmission lines and 7 loads. The system data is taken from (Moger, 2016). The system has a base-case load of 1386 MW and 675 MVAr.

Weak Bus Identification and Validation

To identify the weak buses in the system for voltage support, the reactive power loss index (RPLI) is calculated at each load buses, as defined in (20) using the proposed approach. For discussion purpose, the load flow results of the system under intact condition with peak load is shown in Table 4 along with the calculation of net reactive power support at each generator and load buses considering the line charging

Figure 12. Single-line diagram of sample 10-bus equivalent system

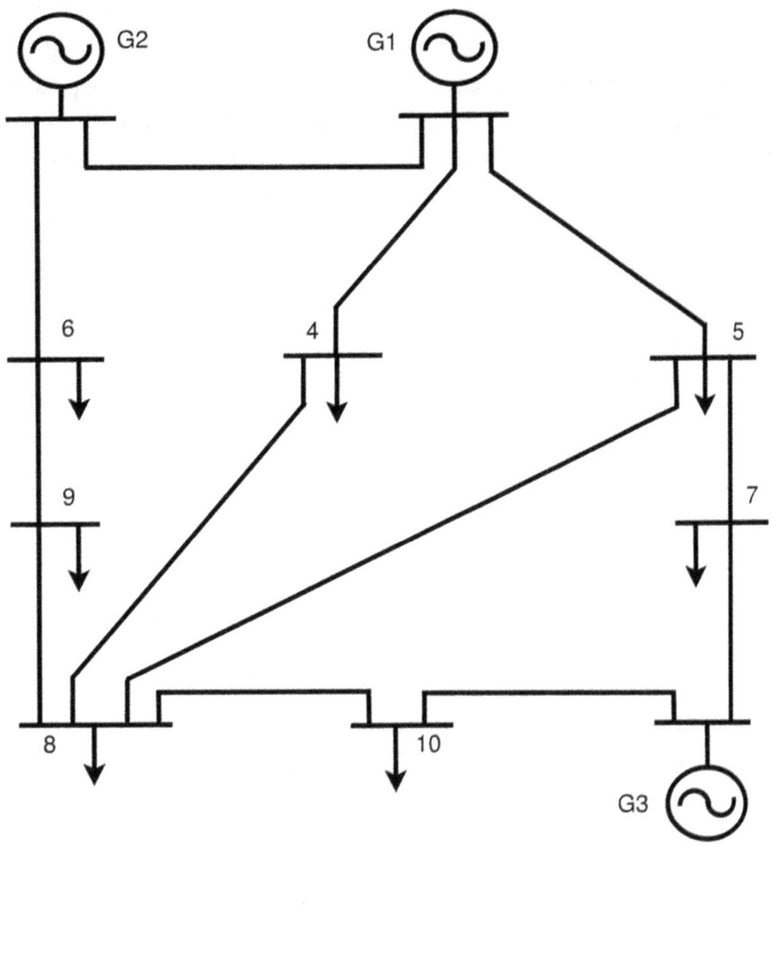

Table 4. Load flow result of sample 10-bus equivalent system under intact condition with peak load

Bus No.	Voltage		Generation		Load		Net MVAr
	Mag. (p.u.)	Angle (deg.)	PG (MW)	QG (MVAr)	PD (MW)	QD (MVAr)	
1	1	0	791.68	305.89	-	-	430.89
2	1	-0.234	472	161.1	-	-	211.1
3	1	-5.992	590	291.74	-	-	366.74
4	0.897	-9.023	-	-	236	118	-198.51
5	0.922	-9.397	-	-	224.2	106.2	-148.72
6	0.953	-5.909	-	-	118	59	-2.23
7	0.948	-10.481	-	-	295	141.6	-62.94
8	0.908	-12.841	-	-	342.2	165.2	-41.55
9	0.901	-13.002	-	-	377.6	177	-126.24
10	0.932	-11.647	-	-	224.2	118	-63.76
Total			1853.68	758.74	1817.2	885	
P-loss = 36.479 MW and Q-loss = 364.79 MVAr							

capacitances and other reactive sources/sinks at the respective buses. Then, the partial reactive power support that each load bus receives from each generator buses and allocation of reactive power loss to load buses under peak load condition without considering the network contingency is shown in Table 5. The sum of partial reactive power contributed by each generator buses to all load buses and the sum of reactive power loss allocated to load buses are in agreement with its net reactive power support calculated at the generator buses and the total reactive power loss calculated by power flow method, respectively.

As it can be seen from Table 5 that the reactive power loss allocated to load bus 9 is maximum i.e., 90.717 MVAr and that for load bus 6 is minimum i.e., 8.1133 MVAr. To meet the power demand at

Table 5. Contribution of generator reactive sources to load buses and reactive power loss allocation for sample 10-bus equivalent system under intact condition with peak load

Load Bus	Net Demand (MW, MVAr)	Generator Sources (MVAr)			Total (MVAr)	Qloss (MVAr)	Normalized Qloss
		Gen. (G1)	Gen. (G2)	Gen. (G3)			
4	236, 198.51	112.49	58.686	79.541	250.72	52.208	0.57551
5	224.2, 148.72	83.971	42.089	65.522	191.58	42.859	0.47245
6	118, 2.2346	7.3908	6.1588	-3.2018	10.348	8.1133	0.089435
7	295, 62.944	38.416	13.23	62.086	113.73	50.788	0.55986
8	342.2, 41.546	50.485	21.395	44.258	116.14	74.592	0.82225
9	377.6, 126.24	96.862	51.441	68.652	216.95	90.717	1.0
10	224.2, 63.758	41.278	18.104	49.883	109.27	45.508	0.50164
Total	1817.2, 643.95	430.89	211.1	366.74	1008.7	364.79	–

load bus 9, the partial reactive power contribution from the generator buses are G1=96.862 MVAr, G2= 51.441 MVAr and G3= 68.652 MVAr. The total partial contribution from all generator reactive sources in a system is 216.95 MVAr. The net reactive power demand at load bus 9 is 126.24 MVAr and rest of the power lost in the transmission corridors while power is being transferred from source to load due to inductive nature of the transmission lines. In case of load bus 6, which is nearer to two generator buses G1 and G2 in comparison with other generator bus G3, and get maximum share from these two nearest generator sources. To meet the power demand at load bus 6, the partial reactive power contribution from the generator buses are G1= 7.3908 MVAr, G2= 6.1588 MVAr and G3= -3.2018 MVAr. As reported in (Kirschen & Strbac, 1999), generators are sources for real power but may be sources or sinks for reactive power. The partial reactive power support received at bus 6 from generator bus G3 is -3.2018 MVAr, so it can be interpreted as generator G3 acts as a sink instead of source for bus 6 for that partial contribution/support. However, the total partial contribution from all the generator reactive source buses to load/sink bus 6 is positive. The proposed approach is based on superposition theorem applied to linearized system model, the partial contribution represents the impact of a particular generator reactive source to meet the load demand in accordance with circuit characteristics.

Since this work mainly focuses on reactive power loss allocation to load buses and its application to system voltage stability problems, the discussion on individual generator partial reactive power contributions to load buses is not in the scope of this Chapter. However, the individual generator partial reactive power contributions to load buses are the basic building block for allocating the reactive power loss to load buses. The detailed discussions on reactive power support and loss allocation under various system operating conditions, and its impact on system reactive power issues is discussed in (Moger, Johnson & Dhadbanjan, 2018).

From these discussions and case studies, it can be inferred that under heavy loading/stressed condition, even though generator may have enough reactive power that cannot be efficiently used if the reactive power requirement in the network is far away from their location due to requirement of reactive power by the transmission lines itself. Thus, the remote bus may be in reactive power deficient area, which may lead to voltage instability if proper reactive power compensation is not provided. Therefore, the reactive power loss allocations at load buses give the clear indication about the system reactive power issues, which in turn give an indication about the system voltage instability/collapse problem. Thus, the amount of reactive power loss allocated to load buses can be considered as an indicator to the reactive power deficit/surplus at the buses and hence, further it can be used for identification of weak buses in the system.

As reported in the literature (Ca˜nizares et al., 1996; Taylor, 1994), system may be voltage unstable if it includes at least one voltage unstable/collapse bus. For this reason, identifying the weak buses in the system is very much important. The identification of weak buses in the system for additional voltage support is not only decided based on the system performance under peak/heavy load condition. Under certain severe network contingencies the system may result in voltage instability. The main factor causing voltage instability is that the reactive power reserve facilities are not sufficient/adequate to meet the reactive power demand. The possibility of voltage instability is more in a system under severe network contingencies than in a system under intact condition. Therefore, the reactive power loss allocation to load buses are also calculated under severe network contingencies for identifying the weakest buses in the system.

The contingency analysis is carried out in the system and line contingency/outage ranking is obtained from the fuzzy logic approach. The overall severity index of each post-contingent quantities and network composite overall severity index (NCOSI) for the particular line outage are determined. The ranking of the outages are carried out based on the values of NCOSI, which reflects the impact/severity of particular line outage in the system. The results of the fuzzy contingency ranking are shown in Table 6, which are arranged in descending order of their NCOSI value/severity. As it can be seen from Table 6 that the line connected between bus 2 and 6 is of most severe contingency. This may lead to voltage instability/collapse, if proper preventive steps are not incorporated to protect the system.

In order to test the consistency of the proposed reactive power loss index, the weak buses identification is carried for the two different loading conditions in the system. The load bus voltage profile of the system under peak load and base-case load conditions is shown in Figure 13. The first five severe network contingencies as listed in Table 6 along with their respective normalized NCOSI values are considered in the identification process of weak buses in the system. The normalized RPLI value at each load buses is calculated for the system intact condition and five severe network contingencies cases under peak load and base-case load conditions in the system is shown in Tables 7 and 8, respectively. The summary of these results is shown in Table 9. It can be seen from Table 9 that the weak load buses identified from the proposed index under base-case and peak loading conditions are in agreement with each other.

The bus with lowest value of RPLI is assumed to be more stable and the bus with highest value of RPLI is considered to be the weakest bus in the system. From Table 9, bus 9 is the weakest bus and then followed by bus 8, bus 4, bus 10, bus 5, bus 7 and bus 6 for both load conditions. The occurrence of severe contingencies under peak/heavy load condition is the main cause for voltage instability/collapse problem in the power system (Ca˜nizares et al., 1996). Therefore, RPLI value at the load buses under peak load condition is considered for identification of weak buses in the system. Since the system un-

Table 6. Fuzzy logic based line outage/contingency ranking of sample 10-bus equivalent system

Sl No.	Outage Line		Overall Severity Index (OSI)			NCOSI	Normalized NCOSI
	From Bus	To Bus	Voltage Profile	L-Index	Line Loading		
0	Intact	condition	5.9573	5.7699	5.7663	17.494	–
1	2	6	6.6558	5.5209	6.8936	19.07	1
2	1	5	6.9158	5.9168	5.942	18.775	0.98449
3	1	4	5.9313	6.0258	6.0111	17.968	0.94221
4	6	9	5.523	6.2426	6.1256	17.891	0.93817
5	3	10	5.975	5.3709	6.2247	17.571	0.92136
6	3	7	5.6118	5.651	6.1175	17.38	0.91138
7	5	7	5.9444	5.6937	5.515	17.153	0.89947
8	5	8	5.9735	5.7852	5.265	17.024	0.89268
9	1	2	5.9188	5.7699	5.3213	17.01	0.89196
10	4	8	5.6092	5.8206	5.5333	16.963	0.8895
11	8	10	5.9796	5.3694	5.4969	16.846	0.88336
12	8	9	5.465	5	5.515	15.98	0.83795

Table 7. Normalized reactive power loss allocated at each load buses for different contingencies (multiplied with normalized NCOSI value of the respective outages): Peak load condition

Load Bus	Normalized Reactive Power Loss Allocation						Total
	Intact	Outage 1-4	Outage 1-5	Outage 2-6	Outage 3-10	Outage 6-9	
4	0.57551	0.90355	0.35249	0.21387	0.54737	0.26599	2.8588
5	0.47245	0.24949	0.93727	0.19937	0.32365	0.23646	2.4187
6	0.08944	0.02594	0.00616	0.35814	0.12781	-0.0659	0.5416
7	0.55986	0.32225	0.91271	0.27085	0.04299	0.30637	2.415
8	0.82225	0.58671	0.84111	0.53274	0.85658	0.53264	4.172
9	1.0000	0.58667	0.81327	1.0000	0.95655	0.89731	5.2538
10	0.50164	0.34267	0.59621	0.3013	0.75669	0.31582	2.8143

Table 8. Normalized reactive power loss allocated at each load buses for different contingencies (multiplied with normalized NCOSI value of the respective outages): Base-case condition

Load Bus	Normalized Reactive Power Loss Allocation						Total
	Intact	Outage 1-4	Outage 1-5	Outage 2-6	Outage 3-10	Outage 6-9	
4	0.6464	0.8792	0.4307	0.3456	0.6389	0.3499	3.2907
5	0.5036	0.2562	0.9477	0.2935	0.3807	0.2917	2.6734
6	0.093	0.0301	0.0211	0.3344	0.1379	-0.071	0.5457
7	0.5657	0.3023	0.8507	0.3595	0.0664	0.35	2.4944
8	0.8068	0.5074	0.7766	0.5912	0.8453	0.5606	4.0879
9	1	0.541	0.8004	1	0.988	0.9487	5.2781
10	0.5006	0.3109	0.5571	0.3632	0.7576	0.3467	2.8361

Table 9. Proposed reactive power loss index for 10-bus equivalent system under different loading conditions (considering the severe contingencies)

Severity Order	Load Bus	Base-Case Load	Load Bus	Peak Load
		RPLI		RPLI
1	9	5.2781	9	5.2538
2	8	4.0879	8	4.172
3	4	3.2907	4	2.8588
4	10	2.8361	10	2.8143
5	5	2.6734	5	2.4187
6	7	2.4944	7	2.415
7	6	0.54572	6	0.5416

Figure 13. Voltage profile of sample 10-bus equivalent system under base-case and peak load conditions

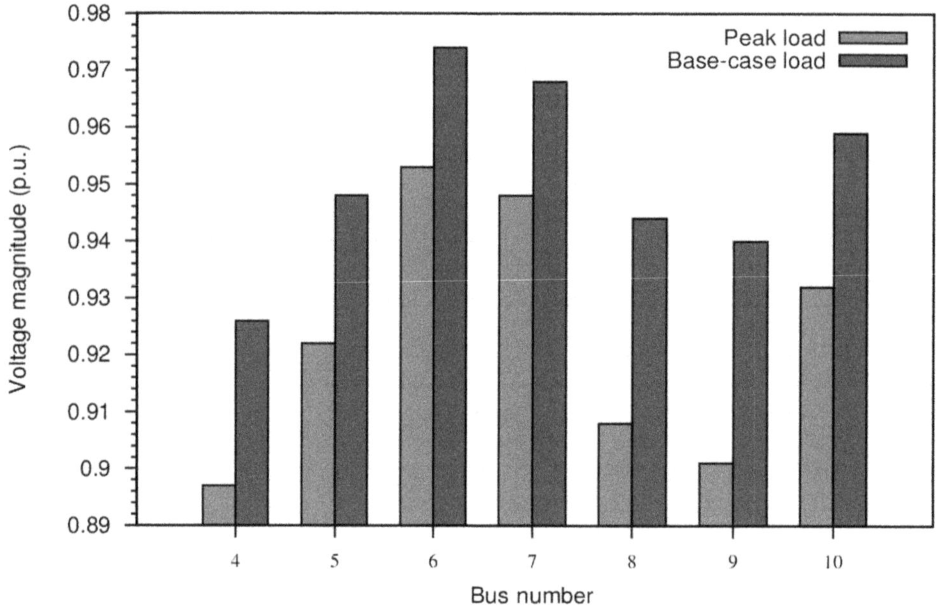

der consideration is small and for both load conditions with critical contingencies, the proposed index produced the same severity order of the weak load buses in the system. However, the severity order of weak load buses may differ in large interconnected system under different loading conditions.

The results from the proposed index is compared with that from other existing methods in the literature such as V-Q sensitivity based modal analysis (Gao et al., 1992), continuation power flow method (Ajjarapu & Christy, 1992), which have been used for identification of weak buses in the system. The results of the comparison are shown in Table 10. The V-Q sensitivity modal analysis is based on system Jacobian matrix near the point of voltage collapse and determining the bus participation factor (BPF) corresponding to the critical mode (least stable mode corresponds to minimum eigenvalue of reduced Jacobian matrix) of operation are used to determine the best site for placing the reactive compensation

Table 10. Identification of weak buses for sample 10-bus equivalent system: A comparison

Severity Order	Proposed Index		Modal Analysis (Gao et al., 1992)		CPF (Ajjarapu & Christy, 1992)	
	Bus	RPLI	Bus	BPF	Bus	Voltage (p.u.)
1	9	5.2538	9	0.3032	9	0.7793
2	8	4.172	8	0.2363	8	0.7936
3	4	2.8588	4	0.1404	4	0.8169
4	10	2.8143	10	0.134	10	0.8236
5	5	2.4187	5	0.0833	5	0.8379
6	7	2.415	6	0.0722	7	0.8555
7	6	0.5416	7	0.0307	6	0.8736

devices. The weak buses from the modal analysis method (Gao et al., 1992) are buses 9, 8, 4, 10, 5, 6 and 7. In the method based on continuation power flow (CPF), the real and reactive power at load buses are increased in proportion to their initial base-case load levels, step by step upto the critical loading point with the help of continuation parameter (loading parameter). The generator output is also increased in proportion to their initial base-case generations in order to meet the increased load. The weak buses from the continuation power flow method (Ajjarapu & Christy, 1992) are buses 9, 8, 4, 10, 5, 7 and 6.

From the results of the comparison, it can be seen that bus 9 is the weakest bus among the other load buses from the proposed index and as well as from other existing methods. The remaining buses 8, 4, 10, and 5 are in the same severity order produced by the proposed index and as well as from other existing methods. For improving steady state voltage stability, the best location for placing the reactive compensation is the weakest bus in the system.

Merits of Reactive Power Loss Index Over the Continuation Power Flow Method

As discussed in preceding Sections, from the proposed index the weak load buses are identified for the particular operating points/conditions in the system considering the few severe network contingencies without overloading the system up to the maximum loadability point as in case of other methods (Gao et al., 1992; Ajjarapu & Christy, 1992). In order to highlight the merits of reactive power loss index over the continuation power flow method, the loading on the system is increased step by step up to the maximum loadability point without considering the network contingencies. Correspondingly, the reactive power loss allocated to load buses is determined. The Figures 14 and 15 show the voltage profile and reactive power loss allocated to load buses for different loading conditions on the system for the intact condition. The two vertical lines in the Figures refer to two loading conditions in the system (these loading have been used for identification of weak load buses in the system. The red colour vertical line corresponds to base-case load condition and blue colour vertical line corresponds to peak load condition. It can be seen from Figures 14 and 15 that under both the load conditions (base-case as well as peak load conditions), the severity order of weak load buses produced from the proposed approach are buses 9, 8, 4, 7, 10, 5 and 6, and that from continuation power flow method are buses 4, 9, 8, 5, 10, 7 and 6. Further, when the same system under maximum loadability point, the severity order of weak load buses from the proposed approach are buses 9, 8, 7, 10, 4, 5 and 6, and that from continuation power flow method buses 9, 8, 4, 10, 5, 7 and 6.

The weak load buses identified in the system under base-case or peak load condition from the proposed approach are almost closely in agreement with that produced at the system maximum loadabilty point. In this case, the effect severe network contingencies is not considered, the actual severity order of the weak load buses are not identified from this type of analysis even for all loading conditions on the system. After considering the severe network contingencies in the system for the particular loading condition (either base-case or peak load condition), the severity order of actual weak load buses produced from the proposed approach are buses 9, 8, 4, 10, 5, 7 and 6 (refer Table 9). Therefore, the actual severity order of the weak load buses in the system from the proposed approach are identified only after considering the effect of severe network contingencies in the calculation of reactive power loss index. Hence, the effect of network contingencies plays an important role in the proposed approach for identifying the actual weak load buses in the system. However, in case of continuation power flow method, the severity order

Figure 14. Voltage profile at load buses for different loading conditions up to critical loading point of sample 10-bus equivalent system under intact condition

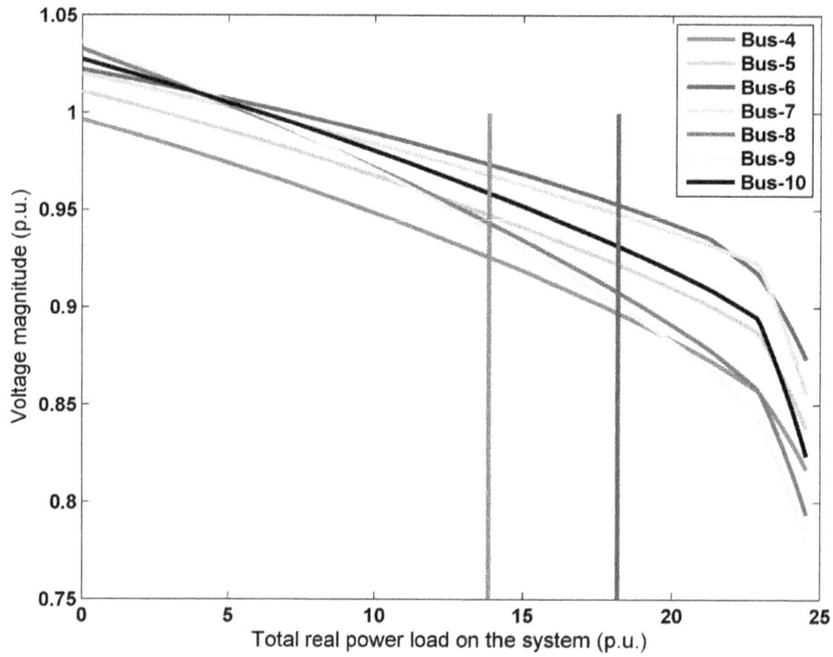

Figure 15. Reactive power loss (Qloss) allocated at load buses for different loading conditions up to critical loading point of sample 10-bus equivalent system under intact condition

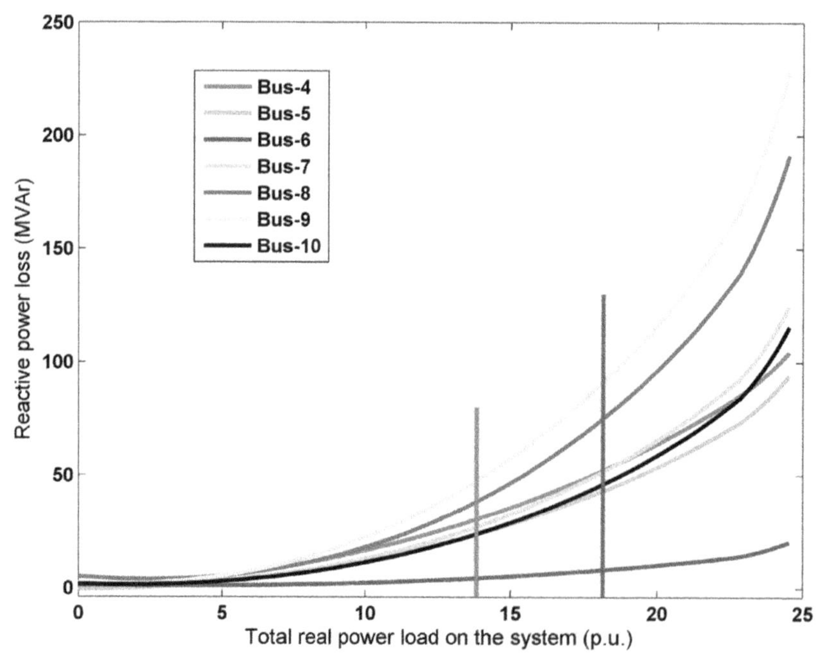

of weak load buses identified in the system under base-case/peak load condition are somewhat inconsistency with that produced at the system maximum loadability point as it can be seen from Figure 14.

From these discussions, it can be inferred that from the continuation power flow method the actual weak load buses are only identified when the system under maximum loadabilty point. It needs an iterative approach upto the critical loading points. Therefore, it seems to be computationally more expensive. However, from the proposed approach the actual weak load buses can be easily identified after considering the effect of severe network contingencies even under the base-case operating condition of the system as it clearly observed from Table 9.

Reactive Compensation Performance Studies

Once the identification of weak load buses in the system under consideration is completed, depending on the system size and reactive power reliability needs or requirements, the system operator can select few weakest buses for additional voltage support. These buses are considered as an optimal location for placing the reactive compensation devices to improve the steady state voltage stability. The objective of placing the reactive compensation is to bring the system back to its satisfactory secured operation from unsecured operation due to system disturbance caused due to either increase in heavy load demand or occurrence of severe network contingencies. For reliable and secure operation of the system for all anticipated operating conditions, the system operator may execute reactive power optimization algorithm, which is not the main scope of this Chapter. However, for the system performance studies/analysis, a short-term reactive power procurement/optimal reactive power dispatch (ORPD) analysis is carried out to determine the optimum size of the reactive compensation devices placed at the weak buses.

As reported in the literature (W. Zhang, Li, & Tolbert, 2007), the ORPD model may consider different objective functions with different constraints set depending on the system operator goal/objectives in the system, which are discussed detail in (W. Zhang et al., 2007). However, for system studies in this chapter, minimizing the sum of the squares of the voltage deviations from desired voltages at all load buses i.e., $\sum_{j=g+1}^{n} (V_j^d - V_j^a)^2$ is considered as an objective function subject to various system constraints. Where V_j^d and V_j^a are the desired and actual voltage at load bus j, respectively. Many algorithms can be used to solve the optimization based ORPD model (W. Zhang et al., 2007; W. Zhang & Tolbert, 2005). The ORPD model is solved by linear programming technique and the detailed mathematical formulation of the linear programming for the specified objective function is discussed in (Thukaram et al., 1984; Thukaram & Yesuratnam, 2006).

For system performance studies under different operating conditions (which may be caused due to either maximum/peak load or occurrence of severe network contingencies), the following cases are generated:

- Case-1: No compensation placed in the system.
- Case-2: Reactive compensation device placed at weakest load bus 9.

The optimum size of the reactive compensation device to be placed at the selected weak bus 9 for all anticipated operating conditions is determined from ORPD model (Thukaram & Yesuratnam, 2006) is given in Table 11. For each cases, system is evaluated with various performance indices or parameters viz., system power loss (real and reactive power losses) and load bus voltage profile parameters (V_{min},

Table 11. Optimal settings of the reactive compensation devices (in p.u.) for sample 10-bus equivalent system under different operating points/conditions

	Intact(Peak Load)	Outage (Buses 2-6)	Outage(Buses 1-5)
Case 1: No Comp.			
Case 2: Comp. at Bus 9	1.1	0.9	0.8

V_{max}, STDEV(V), $\Sigma(V_j^d - V_j^a)^2$ i.e., sum of the squares of the voltage deviations from desired voltages of buses, and voltage stability index at load buses (*L*-index)).The system performance parameters with and without compensation under different operating points/conditions in the system are presented in Table 12. The bus voltage profile for the different operating points/conditions in the system is shown in Figure 16. The voltage profile at load bus9 (which is the weakest among the all load buses) for various loading conditions upto the system maximum loadability for the different operating points/conditions is shown in Figures 17, 18 and 19.

It can be seen from Table 12 that the system with compensation, the real and reactive power losses are decreased by around 18 to 22% from its respective operating conditions and real power maximum loadability is increased by around 13 to 24% from its respective operating conditions of the system without compensation. Hence, placing the reactive compensation at the optimum locations in the system will enhance the system performance significantly as it can be observed from Table 12.

72-BUS EQUIVALENT SYSTEM OF INDIAN SOUTHERN REGION POWER GRID

A 72-bus, 400/220 kV level equivalent system of Indian southern region power grid is considered to demonstrate the effectiveness of the proposed index. The geographical map of Indian southern region power grid with zone-wise representation and single-line diagram are shown in Figures 20 and 21, respectively. The southern region power grid covers the electrical network off our south Indian states. The system data is taken from (Moger, 2016). The system comprising of 15 generators, which come under five generating utilities viz., Karnataka generating companies (K-Genco), Andhra Pradesh generating companies (AP-Genco), Tamil Nadu generating companies (TN-Genco), National Thermal Power Corporation (NTPC) and Nuclear Power Corporation (NPC) and 85 transmission lines including the transformers, which are of 400 kV and 200 kV lines in Karnataka, Andra pradesh and Tamil Nadu. The real and reactive loads are connected at 38 locations. The shunt reactors are connected at few buses for transient over voltage protection. The system has a peak load of 7044.36 MW and 3530.64 MVAr.

Weak Bus Identification and Validation

The peak load on the system is considered for identification of weak load buses in the system. Similar to the analysis on 10-bus equivalent system, the load flow analysis is carried out on the system under intact condition with peak load. The summary of the load flow results in shown in Table 13. Based on the load flow results, calculate the net reactive power support at each generator and load buses considering the line charging capacitances and other reactive sources/sinks at the respective buses. Then, the reactive power loss allocated to load buses under peak load condition without considering the network

Table 12. System performance parameters of sample 10-bus equivalent system under different operating points/conditions

Real and Reactive Power Losses (MW, MVAr)			
	Intact(Peak Load)	**Outage (Buses 2-6)**	**Outage(Buses 1-5)**
Case 1: No Comp.	46.159, 461.59	47.338, 473.38	34.889, 348.89
Case 2: Comp. at Bus 9	37.762, 377.621	37.088, 370.884	29.088, 290.883
Maximum Real Power Loadability (p.u.) from Continuation Power Flow Method (CPF)			
	Intact	**Outage (Buses 2-6)**	**Outage(Buses 1-5)**
Case 1: No Comp.	24.5322	16.0776	17.7408
Case 2: Comp. at Bus 9	30.3534	18.9882	20.097
Load Bus Voltage Profile Parameters (p.u.)			
Case 1: No Compensation			
Voltage Parameters	**Intact(Peak Load)**	**Outage(Buses 2-6)**	**Outage(Buses 1-5)**
V_{max}	0.943	0.9521	0.9572
V_{min}	0.8825	0.834	0.8832
STDEV(V)	0.0253	0.0459	0.0258
$\Sigma(V_d - V_a)^2$	0.061	0.1052	0.0492
L^{max}-index	0.2888	0.5069	0.2291
Case 2: Compensation at Bus 9			
Voltage Parameters	**Intact(Peak Load)**	**Outage(Buses 2-6)**	**Outage(Buses 1-5)**
V_{max}	1.0125	1.0141	1.0225
V_{min}	0.9503	0.9543	0.9502
STDEV(V)	0.0204	0.0222	0.0244
$\Sigma(V_d - V_a)^2$	0.005	0.0066	0.0046
L^{max}-index	0.2296	0.372	0.1948

contingency is shown in Table 14. The sum of reactive power loss allocated to load buses is in agreement with the total reactive power loss calculated by power flow method.

As discussed above, the contingency analysis plays an important role in identification of severe weak load buses in the system. Therefore, the contingency analysis on the system is carried out from the fuzzy logic approach to find out the severe network contingencies. The results of the fuzzy logic based contingency raking are shown in Table 15. Table shows first 20 severe network contingencies, and their respective overall severity index of each post contingent quantities and network composite overall severity index (NCOSI). These contingencies are arranged in descending order of their NCOSI value/severity. It can be seen from Table 15 that the line connected between bus 44 and 57 is of most severe contingency followed by the outage of line connected between bus 67 and 64. These severe contingencies and their respective normalized NCOSI values are considered in the identification of weak buses in the system.

The normalized RPLI value at each load buses is calculated for the system intact condition and 20 severe contingencies cases from the proposed approach and the summary of the results is shown in Table 16. From the results, it can be seen that bus 38 is the weakest bus and then followed by buses 53, 41, 36, 35 and so on.

Figure 16. Bus voltage profile of sample 10-bus equivalent system for different operating points/conditions: With and without compensation

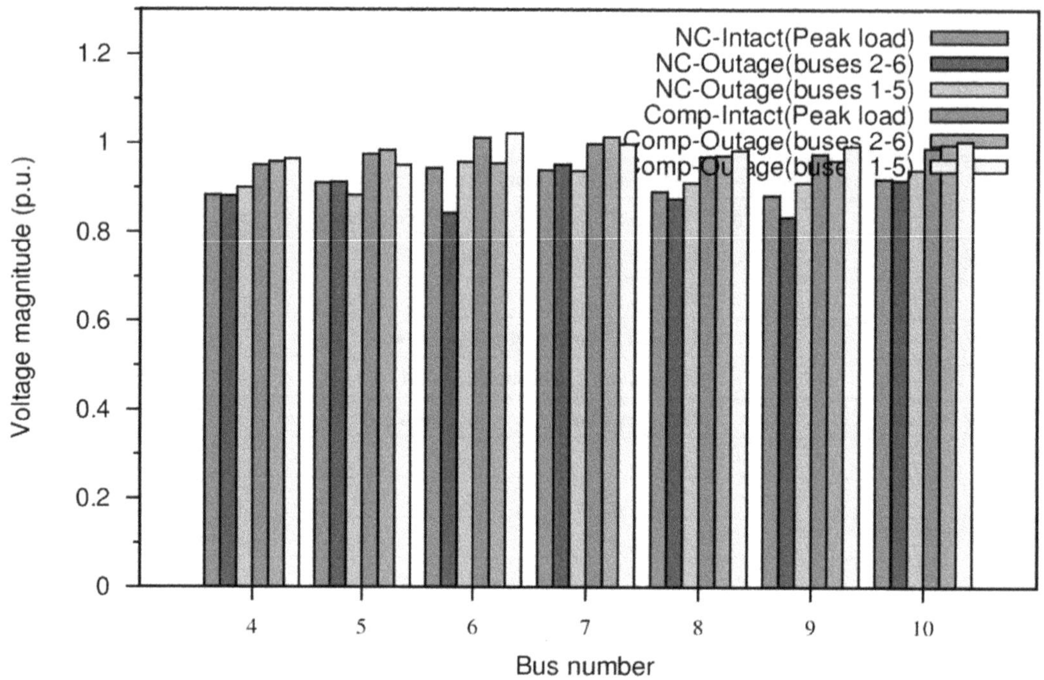

Figure 17. Voltage profile and maximum loadability at load bus 9 of sample 10-bus equivalent system under intact condition: With and without compensation at load bus 9

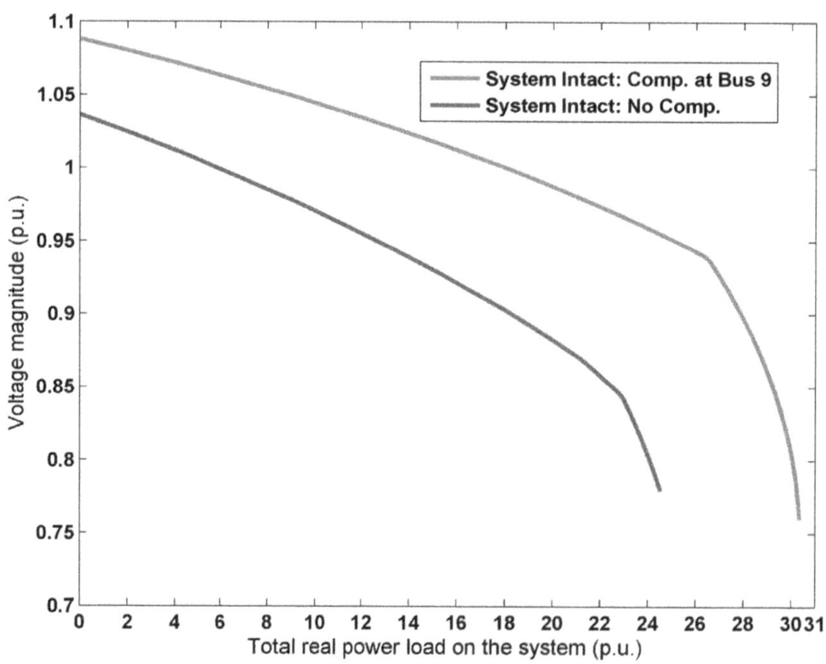

Figure 18. Voltage profile and maximum loadability at load bus 9 of sample 10-bus equivalent system under outage (buses 2-6) condition: With and without compensation at load bus 9

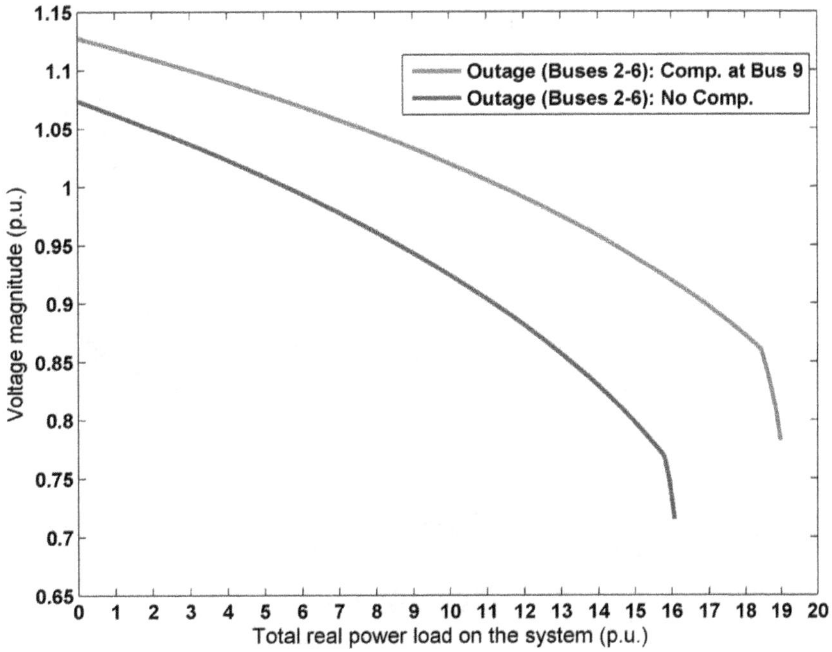

Figure 19. Voltage profile and maximum loadability at load bus 9 of sample 10-bus equivalent system under outage (buses 1-5) condition: With and without compensation at load bus 9

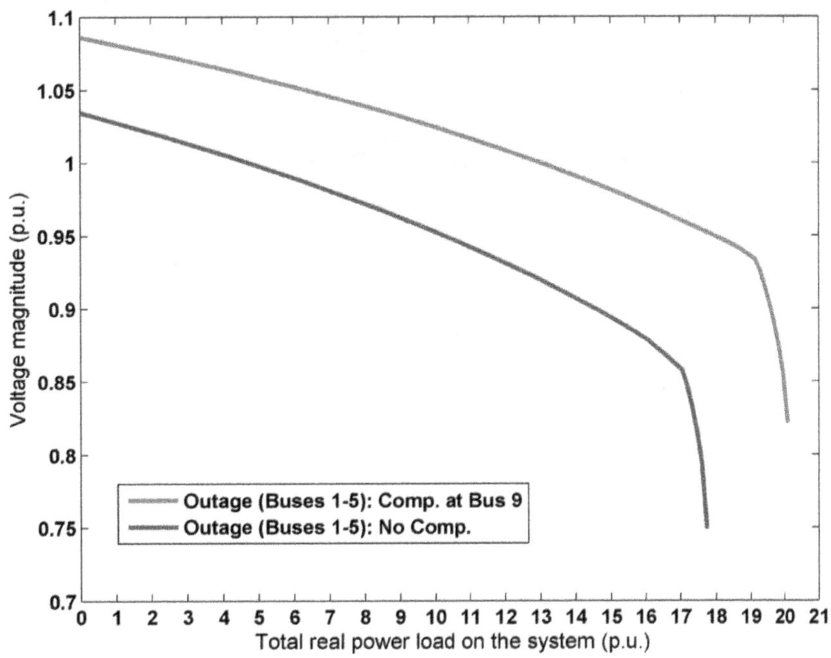

Figure 20. Zonal-wise grid map of 72-bus equivalent system of Indian southern region power grid

Figure 21. Single-line diagram of 72-bus equivalent system of Indian southern region power grid

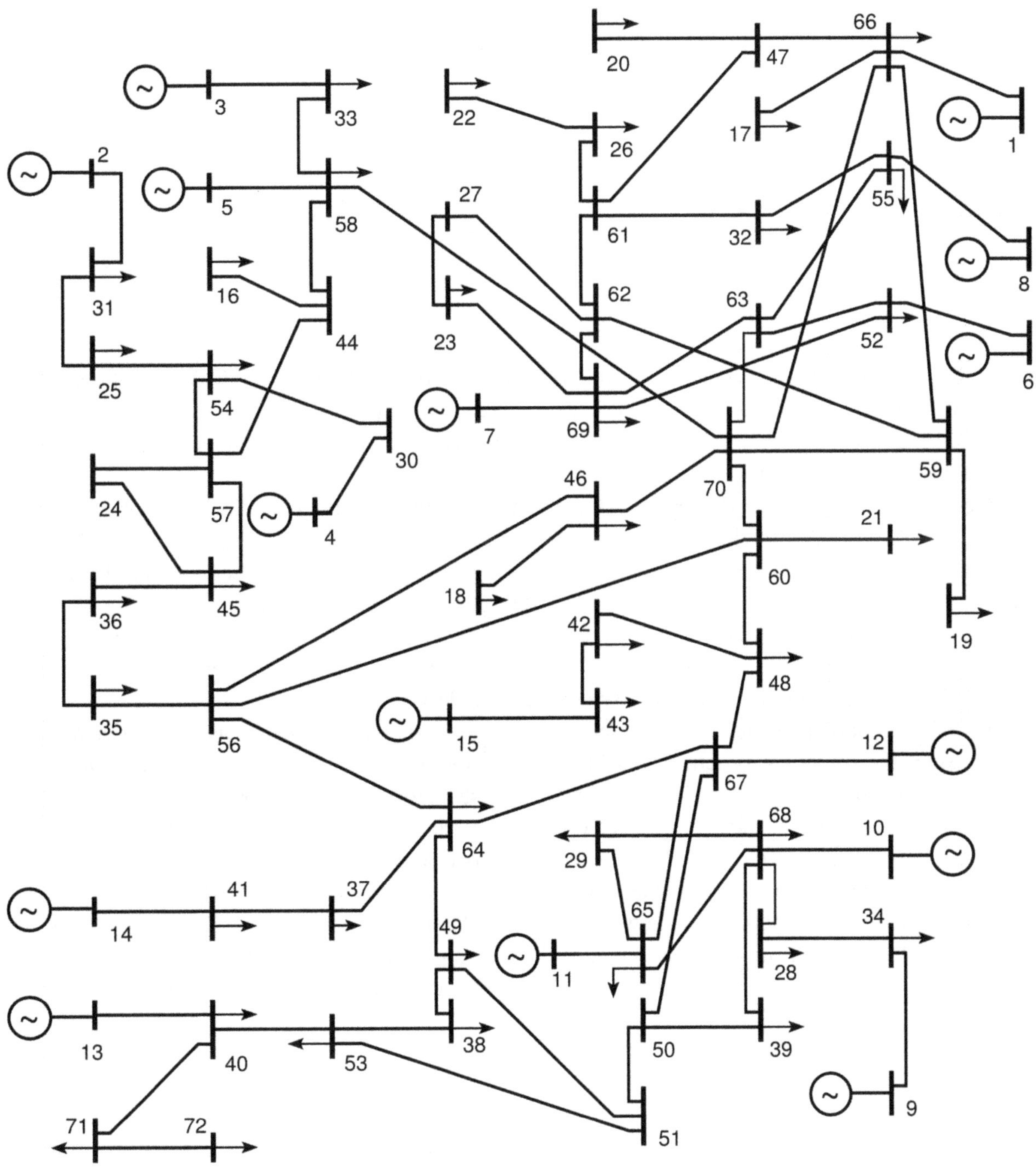

The bus with lowest value of RPLI is assumed to be more stable and the bus with highest value of RPLI is considered to be the most critical bus. Further, the system-wide top 15 weak load buses are selected as per their severity order in the system-wide (based on the values of RPLI) for validation studies. Out of 15 weak load buses, we can see that four buses are in Zone-1 (buses 36, 35, 31 and 54), one

Table 13. Load flow summary of SR 72-bus equivalent system

Load Flow Summary	
No. of generators:	15
No. of transmission links:	49
No. of transformers:	36
No. of shunt reactors:	17
Total generation:	7140.28 MW and 2450.02 MVAr
Total PQ load:	7044.36 MW and 3530.64 MVAr
Total active and reactive power losses:	95.91874 MW and 1364.7020 MVAr
Load bus voltage, V_{min}:	0.83669 (p.u.)

bus in Zone-2 (bus 19) and 10 buses are in Zone-3 (buses 38, 53, 41, 65, 40, 49, 39, 37, 43 and 68). Depending on the system topology and reactive power requirements, the system operator may select few weak load buses for placing the reactive compensation devices for additional voltage support to protect the system under severe disturbance.

Comparative analysis is carried out with the results from other existing methods (Ajjarapu & Christy, 1992; Gao et al., 1992) in the literature such as V-Q sensitivity based modal analysis (Gao et al.,1992), continuation power flow method (Ajjarapu & Christy, 1992), which have been used for identification of weak buses in the system. The results of system-wide comparison of weak load buses from the proposed index and other existing methods are shown in Table 17. From the proposed and other two existing methods, it can be seen that bus 38 (Zone-3) is one of the weakest/critical bus in the system. Similarly, the severity order of other load buses is also shown in Table 17.

Due to locational dependent nature of reactive power, system reactive power needs should be addressed locally. Since the size of the system is very large, there should be a sufficient amount of reactive power reserve available in the system, and it should be widely distributed across all 3-zonesto protect the system against the voltage collapse during unexpected severe disturbance like severe contingencies. Hence, the weak load buses identification is made on zone-wise as per their severity order in the respective zones. The comparative analysis of weak load buses identification in all three zones separately from the proposed index and other existing methods is shown in Table 18 (First five weak buses in each zone). It can be seen from Table 18 that from the proposed and other existing methods (Ajjarapu & Christy, 1992; Gao et al., 1992) the buses 36 and 35, and bus 38 are the weakest buses in Zone-1 and Zone-3, respectively.

Reactive Compensation Performance Studies

Similar to the analysis on 10-bus equivalent system, the reactive compensation performance analysis is carried out. For the studies, we considered the size of the reactive compensation device is 50MVAr for system performance analysis.

To highlight the features and robustness of the proposed index in comparison with other existing methods, the compensation performance analysis has been studied in two scenarios for different operating points/conditions caused due to either maximum/peak loading or occurrence of severe network contingencies (First 3 severe contingencies are considered as given in Table 15). Scenario-1 focuses on system-wide compensation selection andscenario-2 discuss on zone-wise compensation selection. In each

Table 14. Reactive power loss allocation to load buses of SR 72-bus equivalent system under intact condition with peak load

Load Bus	Net Demand		Total (MVAr)	Qloss (MVAr)	Load Bus	Net Demand		Total (MVAr)	Qloss (MVAr)
	PD (MW)	QD (MVAr)				PD (MW)	QD (MVAr)		
16	66.74	33.84	43.854	10.014	44	0	-28.938	-27.918	1.021
17	315.84	157.92	155.54	-2.379	45	0	-59.324	-58.815	0.509
18	9.4	6.58	7.768	1.188	46	0	-53.263	-52.309	0.954
19	304.56	152.28	178.62	26.337	47	0	-50.013	-44.079	5.933
20	58.28	29.14	28.276	-0.864	48	88.36	23.293	39.071	15.778
21	100.58	50.76	69.117	18.357	49	155.1	63.916	135.93	72.012
22	73.32	36.66	40.829	4.169	50	0	-114.6	-131.93	-17.325
23	136.3	59.927	70.778	10.851	51	0	-71.905	-84.579	-12.673
24	0	-86.848	-84.897	1.951	52	159.8	71.198	68.954	-2.244
25	65.8	19.441	32.486	13.045	53	236.88	93.717	200.92	107.2
26	0	-42.31	-37.073	5.237	54	152.28	41.963	71.766	29.803
27	0	-28.836	-25.961	2.875	55	423	193.65	184.8	-8.859
28	72.38	26.187	39.967	13.779	56	0	-77.691	-81.338	-3.647
29	38.54	13.401	21.176	7.775	57	0	-143.6	-139.39	4.215
30	211.5	75.118	93.534	18.416	58	0	-69.086	-65.31	3.776
31	464.36	222.55	294.37	71.82	59	0	-121.98	-110.94	11.033
32	42.3	20.258	19.743	-0.515	60	0	-184.93	-183.42	1.507
33	432.4	216.2	223.79	7.586	61	0	-176.28	-154.55	21.725
34	325.24	153.79	176.14	22.35	62	0	-159.22	-142.16	17.066
35	250.04	120.41	197.53	77.118	63	0	-34.192	-31.422	2.771
36	248.16	119.46	197.43	77.964	64	23.5	-29.097	-27.151	1.946
37	144.76	67.181	116.56	49.377	65	387.28	190.07	270.91	80.831
38	419.24	202.59	465.92	263.32	66	162.62	-83.981	-82.714	1.268
39	198.34	94.88	162.48	67.602	67	0	-193.36	-217.53	-24.171
40	216.2	87.034	164.29	77.256	68	115.62	47.956	71.353	23.397
41	306.44	147.83	251.52	103.69	69	295.16	119.41	128.81	9.403
42	37.6	8.647	14.986	6.339	70	0	-202.22	-187.52	14.694
43	290.46	135.45	184.14	48.692	71	6.58	-4.237	-3.235	1.002
					72	9.4	-1.498	0.925	2.423

scenario, the system performance in terms of reduction in real and reactive power losses, improvement in system voltage profile etc., have been studied.

Scenario-1: Looking into the system size, five weakest load buses are selected for placing the reactive compensation devices for system-wide analysis (Refer Table 17). The reactive compensation performance analysis is carried out for the following cases:

- Case-1: No compensation placed in the system.

Table 15. Fuzzy logic based line outage/contingency ranking of SR 72-bus equivalent system

Sl. No.	Outage Line		Overall Severity Index (OSI)			NCOSI	Normalized NCOSI
	From Bus	To Bus	Voltage Profile	L-Index	Line Loading		
0	Intact	condition	24.559	32.825	26.072	83.455	-
1	44	57	26.278	32.828	27.409	86.516	1
2	67	64	26.629	32.61	27.111	86.349	0.99808
3	56	60	26.334	32.28	27.039	85.653	0.99003
4	56	46	26.135	32.581	26.816	85.532	0.98863
5	66	70	26.278	32.105	26.984	85.367	0.98673
6	60	48	25.456	33.303	26.217	84.975	0.98219
7	57	24	26.749	32.253	25.926	84.928	0.98165
8	62	27	25.816	32.723	26.37	84.908	0.98142
9	57	45	26.902	32.033	25.937	84.872	0.981
10	56	64	24.387	34.04	26.324	84.751	0.9796
11	67	48	25.139	32.968	26.032	84.139	0.97253
12	59	62	25.659	32.604	25.84	84.102	0.9721
13	66	47	24.83	32.641	26.485	83.956	0.97041
14	45	24	25.078	32.668	25.904	83.651	0.96688
15	35	36	25.601	31.827	26.209	83.636	0.96672
16	58	70	25.136	32.587	25.912	83.634	0.96669
17	66	59	25.06	32.562	25.984	83.606	0.96637
18	62	61	25.353	32.533	25.677	83.563	0.96586
19	63	55	24.576	32.794	26.127	83.496	0.96509
20	47	61	24.707	32.629	26.138	83.474	0.96484

- Case-2: Reactive compensation devices placed at weak load buses 35, 36, 38, 41 and 53 using the proposed index (PI).
- Case-3: Reactive compensation devices placed at weak load buses 35, 36, 38, 49 and 53 using Q-V sensitivity modal analysis (QVS-MA) (Gao et al., 1992).
- Case-4: Reactive compensation devices placed at weak load buses 35, 38, 49, 51 and 53 using continuation power flow method (CPF) (Ajjarapu & Christy, 1992).

In a group of five weak buses, three weak buses are common in both from proposed index as well as other existing methods in the literature. For each case, system is evaluated with various performance indices or parameters as discussed in 10-bus equivalent system. The system voltage performance parameters for different operating points/conditions are summarized in Table 19. The real and reactive power losses of the system under different operating points/conditions for the various cases are shown in Figures 22 and 23. Similarly, load bus voltage magnitude of the system under different operating points/ conditions for the various cases is shown in Figures 24, 25, 26 and 27.

Table 16. Reactive power loss index for SR 72-bus equivalent system under peak loading conditions (considering the severe contingencies)

Severity Order	Load Bus	Zone	RPLI	Severity Order	Load Bus	Zone	RPLI
1	38	3	20.661	21	48	3	1.2746
2	53	3	8.3355	22	70	2	1.125
3	41	3	8.3292	23	25	1	1.0782
4	36	1	6.6372	24	28	3	1.0274
5	35	1	6.4066	25	23	2	1.0157
6	65	3	6.1082	26	69	2	0.9288
7	40	3	5.9744	27	59	2	0.8509
8	31	1	5.9092	28	33	1	0.8071
9	49	3	5.6572	29	16	1	0.7943
10	39	3	5.2104	30	29	3	0.5838
11	37	3	3.9689	31	42	3	0.4919
12	43	3	3.7805	32	47	2	0.4489
13	54	1	2.4653	33	26	2	0.4271
14	19	2	2.2037	34	22	2	0.3568
15	68	3	1.761	35	58	1	0.2957
16	61	2	1.6969	36	63	2	0.2273
17	30	1	1.619	37	27	2	0.2256
18	34	3	1.4479	38	64	3	0.2106
19	21	2	1.4405	39	57	1	0.2042
20	62	2	1.3031	40	72	3	0.1836

It can be seen that with compensation in the system, the real power and reactive power losses are decreased by around 8 to 19% from its respective system operating points/conditions. The load bus voltage profile parameters for all different cases from the proposed index are closely in agreement with that produced from other existing methods as it can be seen from Tables and Figures.

Scenario-2: As it can be observed from Table 17 that out of 15 weak load buses maximum weak buses from Zone-3 and few from Zone-1 and Zone-2. Therefore, in this scenario, selection of weak load buses is made zone-wise. Totally 9 weak load buses are selected for placing the compensation devices. Out of 9 weak buses, 3 weakest load buses from Zone-1, 2 weakest buses from Zone-2 and 4 weakest buses from Zone-3 are selected for reactive compensation devices placement (Refer Table 18). The reactive compensation performance analysis has been carried out for the following cases:

- Case-1: No compensation placed in the system.
- Case-2: Reactive compensation devices placed at the buses are Zone-1: buses 31, 35 and 36; Zone-2: buses 19 and 61; Zone-3: buses 38, 53, 41 and 65 from the proposed index.

Table 17. System-wide identification of weak buses for SR 72-bus equivalent system (First 15 weak buses): A comparison

Severity Order	Proposed Index			CPF (Ajjarapu & Christy, 1992)			Modal Analysis (Gao et al., 1992)		
	Load Bus	Zone	RPLI	Load Bus	Zone	Voltage	Load Bus	Zone	BPF
1	38	3	20.661	38	3	0.7587	38	3	0.1328
2	53	3	8.3355	49	3	0.8174	53	3	0.0665
3	41	3	8.3292	53	3	0.8396	49	3	0.0643
4	36	1	6.6372	51	3	0.8594	36	1	0.0594
5	35	1	6.4066	35	1	0.8769	35	1	0.0517
6	65	3	6.1082	36	1	0.878	41	3	0.0507
7	40	3	5.9744	37	3	0.8804	51	3	0.0406
8	31	1	5.9092	39	3	0.8838	37	3	0.0396
9	49	3	5.6572	41	3	0.8889	39	3	0.0389
10	39	3	5.2104	64	3	0.89	40	3	0.035
11	37	3	3.9689	56	1	0.8968	54	1	0.0344
12	43	3	3.7805	50	3	0.8988	48	3	0.0341
13	54	1	2.4653	72	3	0.9164	65	3	0.0319
14	19	2	2.2037	71	3	0.9177	21	2	0.027
15	68	3	1.761	40	3	0.9177	56	1	0.0267

Table 18. Zone-wise identification of weak buses for SR 72-bus equivalent system (First 5 weak buses in each zones): A comparison

Proposed Index			CPF (Ajjarapu & Christy, 1992)			Modal Analysis (Gao et al., 1992)		
Severity Order	Load Bus	RPLI	Severity Order	Load Bus	Voltage (p.u.)	Severity Order	Load Bus	BPF
Zone-1								
4	36	6.6372	5	35	0.8769	4	36	0.0594
5	35	6.4066	6	36	0.878	5	35	0.0517
8	31	5.9092	11	56	0.8968	11	54	0.0344
13	54	2.4653	18	45	0.9249	15	56	0.0267
17	30	1.619	21	25	0.9321	17	31	0.0207
Zone-2								
14	19	2.2037	16	21	0.9186	14	21	0.027
16	61	1.6969	17	17	0.9244	16	19	0.0263
19	21	1.4405	19	19	0.9298	30	17	0.0031
20	62	1.3031	23	18	0.9337	31	23	0.0019
22	70	1.125	25	46	0.937	33	59	0.0015
Zone-3								
1	38	20.661	1	38	0.7587	1	38	0.1328
2	53	8.3355	2	49	0.8174	2	53	0.0665
3	41	8.3292	3	53	0.8396	3	49	0.0643
6	65	6.1082	4	51	0.8594	6	41	0.0507
7	40	5.9744	7	37	0.8804	7	51	0.0406

Table 19. Voltage performance parameters of SR 72-bus equivalent system for different operating points/ conditions: System-wide compensation analysis

Voltage Parameters	Intact (Peak Load)	Outage (Buses 44-57)	Outage (Buses 67-64)	Outage (Buses 56-60)
Case 1: No Compensation				
V_{max}	1.012	1.0158	1.013	1.0186
V_{min}	0.81	0.82	0.8	0.819
STDEV (V)	0.0409	0.0371	0.0733	0.0423
$\Sigma(V_d - V_a)^2$	0.2484	0.1994	0.7109	0.2383
L_{max}-index	0.5427	0.484	0.6835	0.4959
Case 2: RPLI Index				
V_{max}	1.0161	1.0191	1.0205	1.0213
V_{min}	0.8573	0.8729	0.852	0.8688
STDEV (V)	0.028	0.0263	0.0365	0.0274
$\Sigma(V_d - V_a)^2$	0.0966	0.0841	0.1522	0.0932
L_{max}-index	0.4523	0.4162	0.4743	0.4207
Case 3: Q-V Sensitivity Modal Analysis (Gao et al., 1992)				
V_{max}	1.0161	1.0192	1.0202	1.0213
V_{min}	0.8665	0.8843	0.857	0.8785
STDEV (V)	0.0269	0.0246	0.0374	0.0261
$\Sigma(V_d - V_a)^2$	0.0933	0.0762	0.1655	0.0887
L_{max}-index	0.4403	0.4032	0.4654	0.409
Case 4: Continuation Power Flow Method (CPF) (Ajjarapu & Christy, 1992)				
V_{max}	1.0154	1.0185	1.0194	1.0208
V_{min}	0.8732	0.89	0.863	0.8841
STDEV (V)	0.025	0.023	0.0365	0.0257
$\Sigma(V_d - V_a)^2$	0.0951	0.0808	0.1744	0.0979
L_{max}-index	0.4325	0.4069	0.4584	0.4031

Figure 22. Real power loss for SR 72-bus equivalent system: System-wide compensation

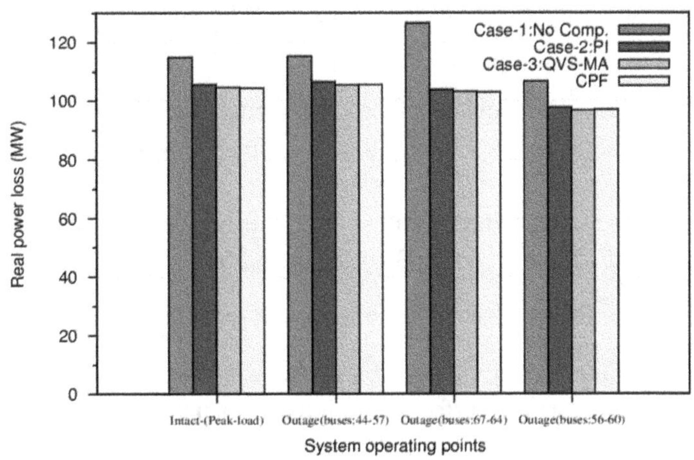

Figure 23. Reactive power loss for SR 72-bus equivalent system: System-wide compensation

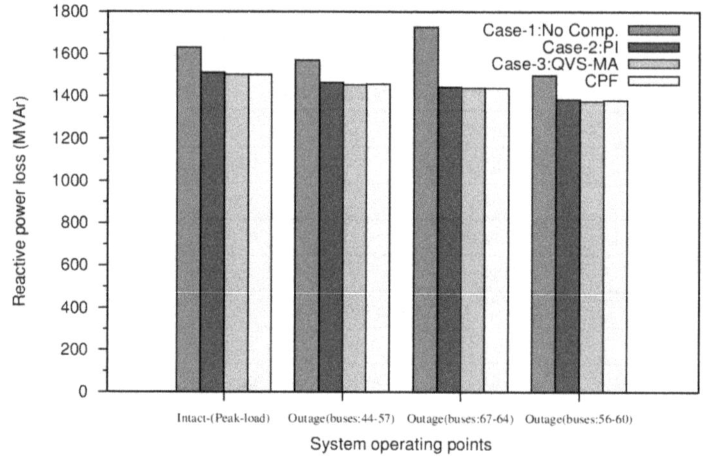

Figure 24. Voltage profile of SR 72-bus equivalent system under intact condition (peak load) for different cases: System-wide compensation

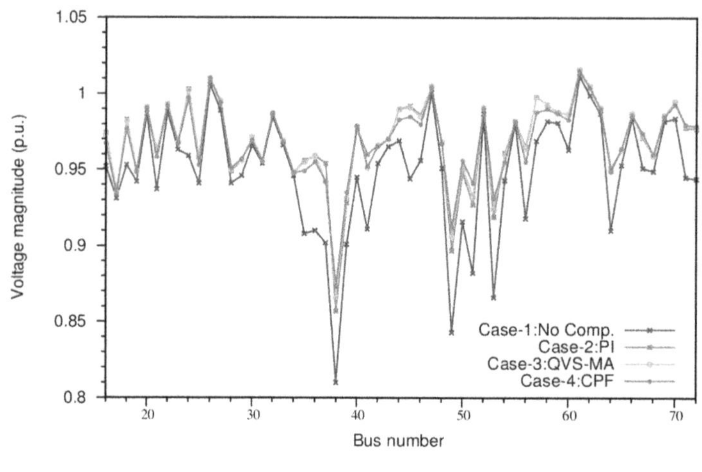

- Case-3: Reactive compensation devices placed at the buses are Zone-1: buses 36, 35 and54; Zone-2: buses 21 and 19; Zone-3: buses 38, 53, 49 and 41 from Q-V sensitivity modal analysis (Gao et al., 1992).
- Case-4: Reactive compensation devices placed at the buses are Zone-1: buses 35, 36 and 56; Zone-2: buses 21 and 17; Zone-3: buses 38, 49, 53 and 51 from continuation power flow method (Ajjarapu & Christy, 1992).

Similar to Scenario-1, the reactive compensation performance analysis is carried out. For each cases, system is again evaluated with various performance indices or parameters. The system voltage performance parameters for different operating points/conditions are summarized in Table 20. The real and reactive power losses of the system under different operating points/conditions for the various cases are

Figure 25. Voltage profile of SR 72-bus equivalent system under outage condition (outage buses44-57) for different cases: System-wide compensation

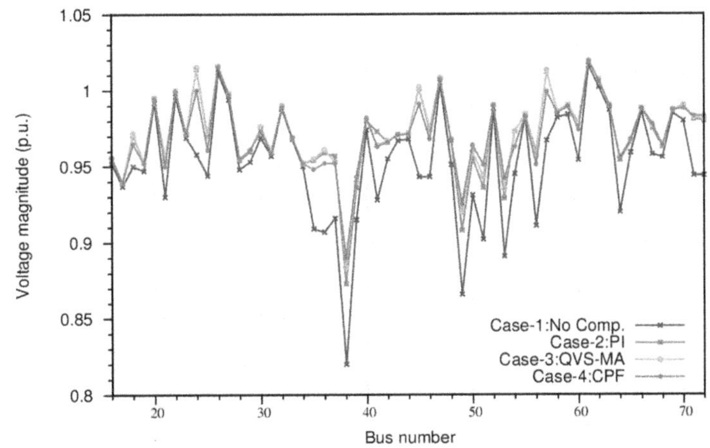

Figure 26. Voltage profile of SR 72-bus equivalent system under outage condition (outage buses 67-64) for different cases: System-wide compensation

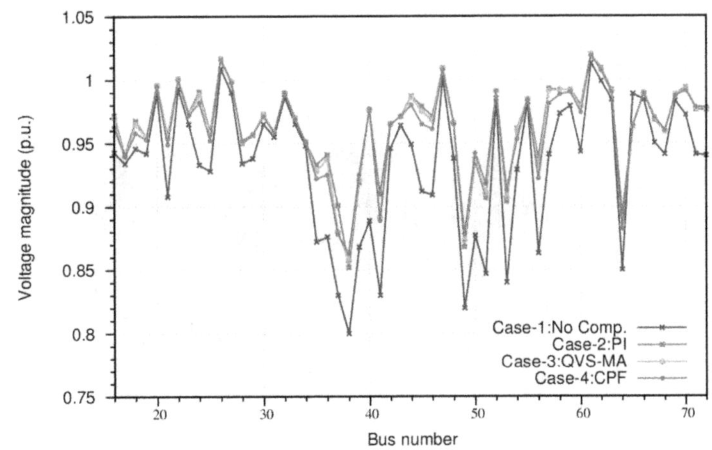

shown in Figures 28 and 29. Similarly, load bus voltage magnitude of the system under different operating points/conditions for the various cases is shown in Figures 30 and 31.

It can be seen from the Figures that with compensation in the system, real power and reactive power losses are decreased by around 9 to 22% from its respective system operating points/conditions. The load bus voltage profile parameters for all different cases from the proposed index are closely in agreement with that produced from other existing methods as it can be seen from Tables and Figures.

Figure 27. Voltage profile of SR 72-bus equivalent system under outage condition (outage buses 56-60) for different cases: System-wide compensation

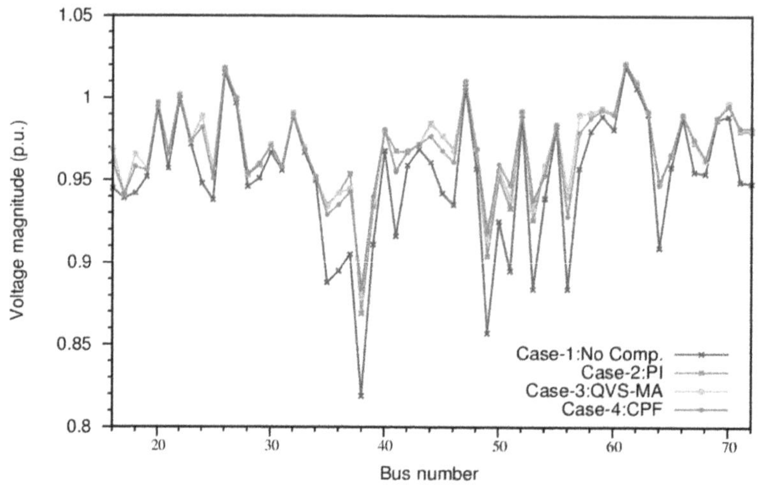

Figure 28. Real power loss for SR 72-bus equivalent system: Zone-wise compensation

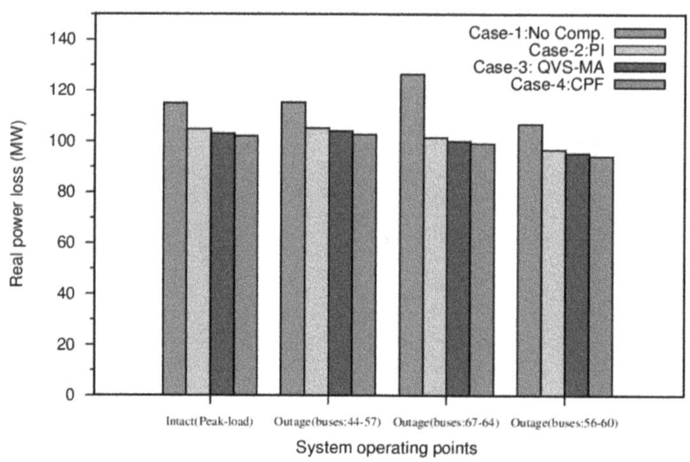

Inference

From the detailed system performance studies, it can be observed from the Figures and Tables that the system performance produced by the proposed index is closely in agreement with that produced by other existing methods (Ajjarapu & Christy, 1992; Gao et al., 1992) in the literature. One of the main advantages of the proposed index is that the weak load buses are identified in the system under particular loading condition (especially peak load) with incorporating the severe network contingencies. There is

Figure 29. Reactive power loss for SR 72-bus equivalent system: Zone-wise compensation

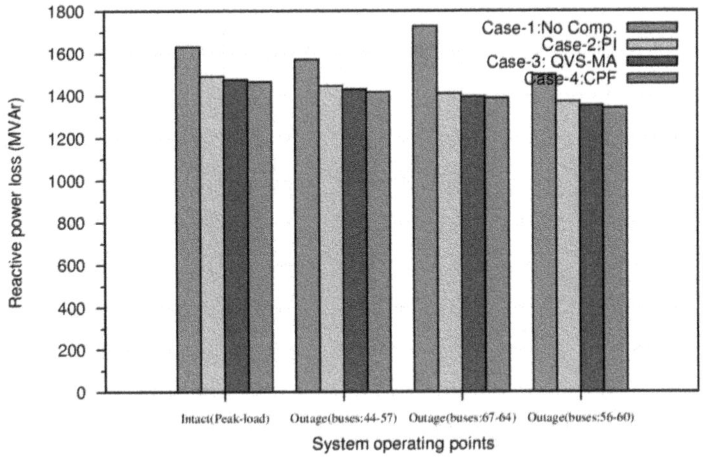

Table 20. Voltage performance parameters of SR 72-bus equivalent system for different operating conditions: Zone-wise compensation analysis

Voltage Parameters	Intact (Peak Load)	Outage (Buses 44-57)	Outage (Buses 67-64)	Outage (Buses 56-60)
Case 1: No Compensation				
V_{max}	1.012	1.0158	1.013	1.0186
V_{min}	0.81	0.818	0.805	0.81
STDEV (V)	0.0409	0.0371	0.0789	0.0423
$\Sigma(V_d - V_a)^2$	0.2484	0.1994	0.7109	0.2383
L_{max}-index	0.5427	0.484	0.6835	0.4959
Case 2: RPLI Index				
V_{max}	1.0311	1.034	1.0354	1.03
V_{min}	0.8596	0.879	0.845	0.875
STDEV (V)	0.03	0.0279	0.0409	0.0286
$\Sigma(V_d - V_a)^2$	0.0876	0.0762	0.1389	0.0857
L_{max}-index	0.4488	0.4129	0.4693	0.4176
Case 3: Q-V Sensitivity Modal Analysis (Gao et al., 1992)				
V_{max}	1.0218	1.0394	1.0267	1.0269
V_{min}	0.8749	0.8905	0.8587	0.8852
STDEV (V)	0.0276	0.0266	0.0364	0.0264
$\Sigma(V_d - V_a)^2$	0.0689	0.0598	0.0941	0.066
L_{max}-index	0.4302	0.3956	0.4411	0.4013
Case 4: Continuation Power Flow Method (CPF) (Ajjarapu & Christy, 1992)				
V_{max}	1.0201	1.0352	1.0247	1.0252
V_{min}	0.8842	0.9	0.8627	0.8956
STDEV (V)	0.0255	0.0239	0.0353	0.0236
$\Sigma(V_d - V_a)^2$	0.0615	0.0506	0.0977	0.0554
L_{max}-index	0.4189	0.3851	0.4343	0.3897

Figure 30a. Voltage profile of SR 72-bus equivalent system under different operating conditions: Zone-wise compensation (for peak load and outage buses 44-57): Peak load condition for various cases

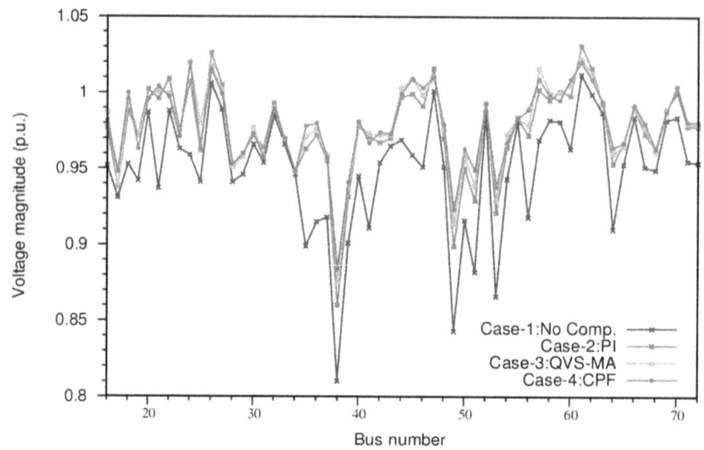

Figure 30b. Voltage profile of SR 72-bus equivalent system under different operating conditions: Zone-wise compensation (for peak load and outage buses 44-57): Outage condition (outage buses 44-57) for various cases

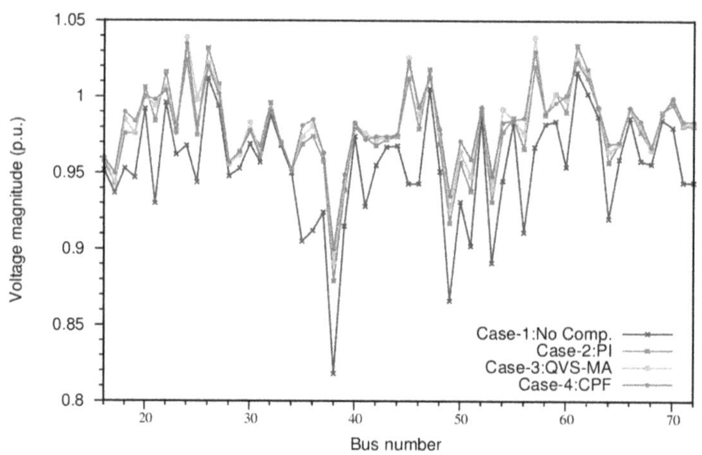

no need to load the system upto its critical loading point as done in other two existing methods. In the existing methods, the actual weak load buses can only be identified when the system is very close to diverging point. Hence, the system has to move from an initial operating point upto the collapse point by changing the loading factor. Thus, the whole process goes in an iterative manner and therefore, it is computationally more expensive. With the knowledge of system voltage profile by the system operator, the peak loading can easily simulated by taking 10-15% more load from the normal loading condition in

Figure 31a. Voltage profile of SR 72-bus equivalent system under different operating conditions: Zone-wise compensation (for outage conditions): Outage condition (outage buses 67-64) for various cases

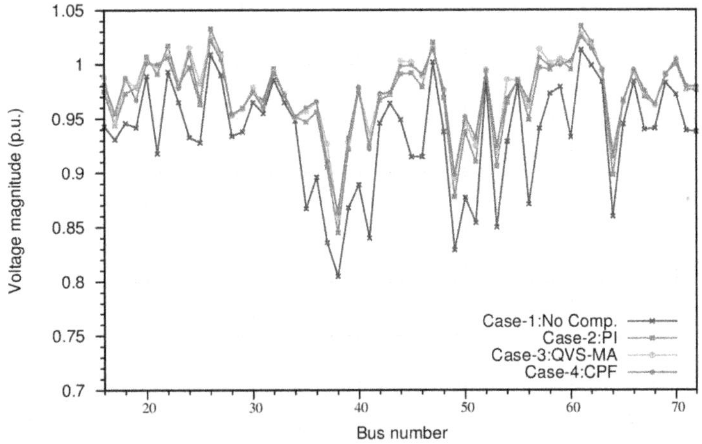

Figure 31b. Voltage profile of SR 72-bus equivalent system under different operating conditions: Zone-wise compensation (for outage conditions): Outage condition (outage buses 56-60) for various cases

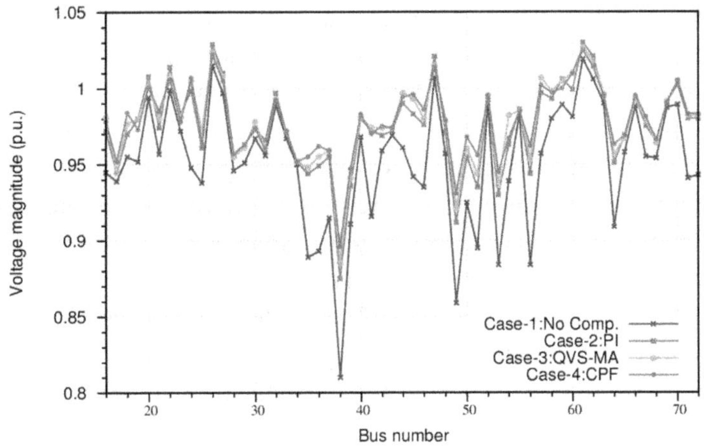

the system. The proposed index for identification of weak load buses is completely non-iterative manner, thus involving minimal computational efforts. The whole procedure will be completed by taking one set of results from the state estimator with simulating the peak load and readily available offline network contingency results. Hence, this may find application in control center for the control and monitoring of the system against voltage vulnerability/instability in real time. If necessary, the system operator may execute some preventive measures with the information obtained from the proposed index to protect the system.

FUTURE RESEARCH DIRECTIONS

A new index is determined based on the reactive power loss allocated at the load buses considering the steady state models of the power system components. There is a scope for defining reactive power loss index by incorporating machine dynamics, load dynamics and other network controls like FACTS and HVDC controllers, which give a better indication of the real-time monitoring of the practical systems. Further, it can be extended to grid integration of the renewable energy resources such as wind, solar, fuel cells, micro turbines, bio-gas and so on.

CONCLUSION

In this chapter, a new reactive power loss index is proposed to identify the weak load buses in the power system. This index is obtained from the reactive power support and loss allocation algorithm using Y-bus approach for the system under intact condition as well as severe network contingencies cases. The fuzzy logic approach is used to identify the most severe network line outages in the system. The proposed index is effectively demonstrated on sample 5-bus system for its significance to the system voltage stability. This index is also tested on the sample 10-bus equivalent system and 72-bus equivalent system of Indian southern region power grid. The test results demonstrate the effectiveness of the proposed index in terms of reducing the real and reactive power losses, improvement in system voltage profile and enhance the real power maximum loadability in the power system under different operating points/conditions. Comparative analysis of the proposed approach with the other existing methods such as Q-V sensitivity modal analysis and continuation power flow method is presented. The advantage of the proposed index is that the weak load buses are identified in the system under particular loading condition considering the severe network contingencies. However, the existing methods need an iterative approach upto the critical loading point. The whole procedure of identifying the weak buses from the proposed index is completely non-iterative, thus reducing the computational efforts. Hence, this can find application in control center where the control and monitoring of the system against voltage vulnerability/instability in real time. With the information obtained from the proposed index, the system operator may execute some preventive measures in order to ensure the reliable and secure operation of the power system.

REFERENCES

Abido, M., & Bakhashwain, J. (2005). Optimal VAR dispatch using a multiobjective evolutionary algorithm. *International Journal of Electrical Power & Energy Systems*, *27*(1), 13–20. doi:10.1016/j.ijepes.2004.07.006

Ajjarapu, V. (2007). *Computational techniques for voltage stability assessment and control*. Springer Science & Business Media. doi:10.1007/978-0-387-32935-2

Ajjarapu, V., & Christy, C. (1992). The continuation power flow: A tool for steady state voltage stability analysis. *IEEE Transactions on Power Systems*, *7*(1), 416–423. doi:10.1109/59.141737

Ajjarapu, V., & Lee, B. (1998). Bibliography on voltage stability. *IEEE Transactions on Power Systems*, *13*(1), 115–125. doi:10.1109/59.651622

Begovi'c, M. M., & Phadke, A. G. (1992). Control of voltage stability using sensitivity analysis. *IEEE Transactions on Power Systems*, *7*(1), 114–123. doi:10.1109/59.141694

Bijwe, P., Kothari, D., Nanda, J., & Lingamurthy, K. (1986). Optimal voltage control using a constant sensitivity matrix. *Electric Power Systems Research*, *11*(3), 195–203. doi:10.1016/0378-7796(86)90034-9

Ca˜nizares, C., & Faur, Z. T. (1999). Analysis of svc and tcsc controllers in voltage collapse. *IEEE Transactions on Power Systems*, *14*(1), 158–165. doi:10.1109/59.744508

Canizares, C. A. (2002). Voltage stability assessment: concepts, practices and tools. *Power System Stability Subcommittee Special Publication IEEE/PES, Final Document* (thunderbox.uwaterloo. ca/˜ claudio/ claudio. html# VSWG). IEEE.

Canizares, C. A., & Alvarado, F. L. (1993). Point of collapse and continuation methods for large ac/dc systems. *IEEE Transactions on Power Systems*, *8*(1), 1–8. doi:10.1109/59.221241

Ca˜nizares, C. A., De Souza, A. C., & Quintana, V. H. (1996). Comparison of performance indices for detection of proximity to voltage collapse. *IEEE Transactions on Power Systems*, *11*(3), 1441–1450. doi:10.1109/59.535685

Chebbo, A., Irving, M., & Sterling, M. (1992). Voltage collapse proximity indicator: Behavior and implications. *IEE Proceedings. Generation, Transmission and Distribution*, *139*(3), 241–252. doi:10.1049/ip-c.1992.0037

Chen, Y.-L. (1996). Weak bus-oriented optimal multi-objective VAR planning. *IEEE Transactions on Power Systems*, *11*(4), 1885–1890. doi:10.1109/59.544659

Chen, Y.-L., Chang, C.-W., & Liu, C.-C. (1995). Efficient methods for identifying weak nodes in electrical power networks. *IEE Proceedings. Generation, Transmission and Distribution*, *142*(3), 317–322. doi:10.1049/ip-gtd:19951728

CIGRE Task Force 38.02.11 Report, W. (1994). *Indices predicting voltage collapse including dynamic phenomena*. CIGRE.

Crisan, O., & Liu, M. (1994). Voltage collapse prediction using an improved sensitivity approach. *Electric Power Systems Research*, *28*(3), 181–190. doi:10.1016/0378-7796(94)90032-9

Crow, M. L. (2002). *Computational methods for electric power systems*. CRC Press. doi:10.1201/9781420058284

Da Costa, G. (2002). Modified newton method for reactive dispatching. *International Journal of Electrical Power & Energy Systems*, *24*(10), 815–819. doi:10.1016/S0142-0615(02)00013-3

de Souza, A. Z. (2000). Discussion on some voltage collapse indices. *Electric Power Systems Research*, *53*(1), 53–58. doi:10.1016/S0378-7796(99)00042-5

De Souza, A. Z., Canizares, C. A., & Quintana, V. H. (1997). New techniques to speed up voltage collapse computations using tangent vectors. *IEEE Transactions on Power Systems*, *12*(3), 1380–1387. doi:10.1109/59.630485

Deeb, N., & Shahidehpour, S. (1988). An efficient technique for reactive power dispatch using a revised linear programming approach. *Electric Power Systems Research*, *15*(2), 121–134. doi:10.1016/0378-7796(88)90016-8

Deeb, N., & Shahidehpour, S. (1990). Linear reactive power optimization in a large power network using the decomposition approach. *IEEE Transactions on Power Systems*, *5*(2), 428–438. doi:10.1109/59.54549

Dobson, I., & Lu, L. (1992). Computing an optimum direction in control space to avoid saddle node bifurcation and voltage collapse in electric power systems. *IEEE Transactions on Automatic Control*, *37*(10), 1616–1620. doi:10.1109/9.256397

Dommel, H. W., & Tinney, W. F. (1968). Optimal power flow solutions. *IEEE Transactions on Power Apparatus and Systems*, *PAS-87*(10), 1866–1876. doi:10.1109/TPAS.1968.292150

El-Keib, A., & Ma, X. (1995). Application of artificial neural networks in voltage stability assessment. *IEEE Transactions on Power Systems*, *10*(4), 1890–1896. doi:10.1109/59.476054

Elgerd, O. 1. (1982). *Electic energy systems theory-an introduction*. TMH.

FERC. (2005). *Principles for efficient and reliable reactive power supply and consumption*. FERC Staff Reports, USA, Docket No. AD05-1-000.

Galiana, F. (1984). Bound estimates of the severity of line outages in power system contingency analysis and ranking. *IEEE Transactions on Power Apparatus and Systems*, *PAS-103*(9), 2612–2624. doi:10.1109/TPAS.1984.318430

Gao, B., Morison, G., & Kundur, P. (1992). Voltage stability evaluation using modal analysis. *IEEE Transactions on Power Systems*, *7*(4), 1529–1542. doi:10.1109/59.207377

Hcbson, E. (1980). Network constrained reactive power control using linear programming. *IEEE Transactions on Power Apparatus and Systems*, *PAS-99*(3), 868–877. doi:10.1109/TPAS.1980.319715

He, T., Kolluri, S., Mandal, S., Galvan, F., & Rasigoufard, P. (2004). Identification of weak locations in bulk transmission systems using voltage stability margin index. In *Power engineering society general meeting, 2004* (pp. 1814–1819). IEEE. doi:10.1109/PES.2004.1373193

Hong, Y.-Y., & Gau, C.-H. (1994). Voltage stability indicator for identification of the weakest bus/area in power systems. *IEE Proceedings. Generation, Transmission and Distribution*, *141*(4), 305–309. doi:10.1049/ip-gtd:19949985

Hsu, Y.-Y., & Kuo, H.-C. (1992). Fuzzy-set based contingency ranking. *IEEE Transactions on Power Systems*, *7*(3), 1189–1196. doi:10.1109/59.207333

Kessel, P., & Glavitsch, H. (1986). Estimating the voltage stability of a power system. *IEEE Transactions on Power Delivery*, *1*(3), 346–354. doi:10.1109/TPWRD.1986.4308013

Kirschen, D., & Strbac, G. (1999). Tracing active and reactive power between generators and loads using real and imaginary currents. *IEEE Transactions on Power Systems*, *14*(4), 1312–1319. doi:10.1109/59.801890

Kumano, T., Yokoyama, A., & Sekine, Y. (1994). Fast monitoring and optimal preventive control of voltage instability. *International Journal of Electrical Power & Energy Systems*, *16*(2), 117–125. doi:10.1016/0142-0615(94)90048-5

Kundur, P., Balu, N. J., & Lauby, M. G. (1994). *Power system stability and control* (Vol. 7). McGraw-hill New York.

Kundur, P., Paserba, J., Ajjarapu, V., Andersson, G., Bose, A., & Canizares, C. (2004). Definition and classification of power system stability IEEE/CIGRE joint task force on stabilityterms and definitions. *IEEE Transactions on Power Systems*, *19*(3), 1387–1401. doi:10.1109/TPWRS.2004.825981

Li, H., & Song, Y. (2002). Identification of weak bus bars in large scale power system. In *International conference on Power system technology (Powercon)* (*Vol. 3*, pp.1700–1704). Academic Press. 10.1109/ICPST.2002.1067823

Lof, P.-A., Andersson, G., & Hill, D. (1993). Voltage stability indices for stressed power systems. *IEEE Transactions on Power Systems*, *8*(1), 326–335. doi:10.1109/59.221224

Lof, P.-A., Smed, T., Andersson, G., & Hill, D. (1992). Fast calculation of a voltage stability index. *IEEE Transactions on Power Systems*, *7*(1), 54–64. doi:10.1109/59.141687

Lomi, A., & Thukaram, D. (2012). Optimum reactive power dispatch for alleviation of voltage deviations. *TELKOMNIKA*, *10*(2), 257–264. doi:10.12928/telkomnika.v10i2.793

Mamandur, K., & Chenoweth, R. (1981). Optimal control of reactive power flow for improvements in voltage profiles and for real power loss minimization. *IEEE Transactions on Power Apparatus and Systems*, (7): 3185–3194. doi:10.1109/TPAS.1981.316646

Mangoli, M. K., Lee, K. Y., & Park, Y. M. (1993). Optimal real and reactive power control using linear programming. *Electric Power Systems Research*, *26*(1), 1–10. doi:10.1016/0378-7796(93)90063-K

Menezes, T., da Silva, L. C., Affonso, C., & Da Costa, V. F. (2004). Mvar management on the pre-dispatch problem for improving voltage stability margin. *IEE Proceedings. Generation, Transmission and Distribution*, *151*(6), 665–672. doi:10.1049/ip-gtd:20040802

Moger, T. (2016). *Reactive power planning and operation of power systems with wind farms for voltage stability improvement* (PhD dissertation). Department of Electrical Engineering, Indian Institute of Science, Bangalore, India.

Moger, T., & Dhadbanjan, T. (2013a). An improved approach for evaluation of reactive power sources contribution to reactive load and loss. *Annual India Conference (INDICON)*. 10.1109/INDCON.2013.6725961

Moger, T., & Dhadbanjan, T. (2013b). Reactive power loss allocation to consumers in a deregulated power system. *Fifth International Conference on Power and Energy Systems (ICPS)*.

Moger, T., & Dhadbanjan, T. (2015). A novel index for identification of weak nodes for reactive compensation to improve voltage stability. *IET Generation, Transmission & Distribution*, *9*(14), 1826–1834. doi:10.1049/iet-gtd.2015.0054

Moger, T., & Dhadbanjan, T. (2017). Evaluation of Reactive Power Support and Loss Allocation in a Pool Based Competitive Electricity Market. *International Journal of Emerging Electric Power Systems*, *18*(2), 1–19. doi:10.1515/ijeeps-2016-0201

Moger, T., Johnson, T., & Dhadbanjan, T. (2018). Significance of Reactive Power Loss and Its Application to System Voltage Stability. In *IEEMA Engineer Infinite Conference (eTechNxT)*. IEEE. 10.1109/ETECHNXT.2018.8385343

Momoh, J. A., El-Hawary, M., & Adapa, R. (1999). A review of selected optimal power flow literature to 1993.PART I: Nonlinear and quadratic programming approaches. *IEEE Transactions on Power Systems*, *14*(1), 96–104.

Musirin, I., & Rahman, T. A. (2002). Estimating maximum loadability for weak bus identification using FVSI. *IEEE Power Engineering Review*, *22*(11), 50–52. doi:10.1109/MPER.2002.1045568

Overbye, T. J., & DeMarco, C. L. (1991). Improved techniques for power system voltage stability assessment using energy methods. *IEEE Transactions on Power Systems*, *6*(4), 1446–1452. doi:10.1109/59.116988

Pudjianto, D., Ahmed, S., & Strbac, G. (2002). Allocation of VAR support using LP and NLP based optimal power flows. *IEE Proceedings. Generation, Transmission and Distribution*, *149*(4), 377–383. doi:10.1049/ip-gtd:20020200

Qiu, J., & Shahidehpour, S. (1987). A new approach for minimizing power losses and improving voltage profile. *IEEE Transactions on Power Systems*, *2*(2), 287–295. doi:10.1109/TPWRS.1987.4335121

Singh, G., Singh, S., Raju, G., & Thukaram, D. (1996). Methodology for alleviation of voltage excursions in large power systems. *International Journal of Electrical Power & Energy Systems*, *18*(3), 167–173. doi:10.1016/0142-0615(95)00018-6

Singh, J., Singh, S., & Srivastava, S. (2007). An approach for optimal placement of static VAR compensators based on reactive power spot price. *IEEE Transactions on Power Systems*, *22*(4), 2021–2029. doi:10.1109/TPWRS.2007.907374

Sode-Yome, A., & Mithulananthan, N. (2004). Comparison of shunt capacitor, svc and STATCOM in static voltage stability margin enhancement. *International Journal of Electrical Engineering Education*, *41*(2), 158–171. doi:10.7227/IJEEE.41.2.7

Srivastava, S., Velayutham, A., Agrawal, K., & Bakshi, A. (2012). *Report of the enquiry committee on grid disturbance in northern region on 30th July 2012 and in northern, eastern & northeastern region on 31st July 2012*. Academic Press.

Stagg, G. W., El-Abiad, A. H., & El-Abiad, A. (1968). *Computer methods in power system analysis* (Vol. 9). McGraw-Hill New York.

Tamura, Y., Mori, H., & Iwamoto, S. (1983). Relationship between voltage instability and multiple load flow solutions in electric power systems. *IEEE Transactions on Power Apparatus and Systems, PAS-102*(5), 1115–1125. doi:10.1109/TPAS.1983.318052

Taylor, C. W. (1994). *Power system voltage stability*. McGraw-Hill Companies.

Thukaram, D., Jenkins, L., Khincha, H., Yesuratnam, G., & Kumar, B. R. (2004).Monitoring the effects of on-load tap changing transformers on voltage stability. In *International conference on Power system technology* (*Vol. 1*, pp. 419–424). Academic Press.

Thukaram, D., Jenkins, L., & Visakha, K. (2006). Optimum allocation of reactive power for voltage stability improvement in ac-dc power systems. *IEE Proceedings. Generation, Transmission and Distribution*, *153*(2), 237–246. doi:10.1049/ip-gtd:20045210

Thukaram, D., & Lomi, A. (2000). Selection of static VAR compensator location and size for system voltage stability improvement. *Electric Power Systems Research*, *54*(2), 139–150. doi:10.1016/S0378-7796(99)00082-6

Thukaram, D., & Parthasarathy, K. (1996). Optimal reactive power dispatch algorithm for voltage stability improvement. *International Journal of Electrical Power & Energy Systems*, *18*(7), 461–468. doi:10.1016/0142-0615(96)00004-X

Thukaram, D., Parthasarathy, K., & Prior, D. (1984). Improved algorithm for optimum reactive power allocation. *International Journal of Electrical Power & Energy Systems*, *6*(2), 72–74. doi:10.1016/0142-0615(84)90011-5

Thukaram, D., & Yesuratnam, G. (2006). Comparison of optimum reactive power schedule with different objectives using lp technique. *International Journal of Emerging Electric PowerSystems*, *7*(3), 1–31.

Thukaram, D., & Yesuratnam, G. (2008). Optimal reactive power dispatch in a large power system with ac-dc and facts controllers. *IET Generation, Transmission & Distribution*, *2*(1), 71–81. doi:10.1049/iet-gtd:20070163

Tiranuchit, A., & Thomas, R. (1988). A posturing strategy against voltage instabilities in electric power systems. *IEEE Transactions on Power Systems*, *3*(1), 87–93. doi:10.1109/59.43177

Tomsovic, K. (1992). A fuzzy linear programming approach to the reactive power/voltage control problem. *IEEE Transactions on Power Systems*, *7*(1), 287–293. doi:10.1109/59.141716

Van Cutsem, T. (1991). A method to compute reactive power margins with respect to voltage collapse. *IEEE Transactions on Power Systems*, *6*(1), 145–156. doi:10.1109/59.131057

Venikov, V., Stroev, V., Idelchick, V., & Tarasov, V. (1975). Estimation of electrical power system steady-state stability in load flow calculations. *IEEE Transactions on Power Apparatus and Systems*, *94*(3), 1034–1041. doi:10.1109/T-PAS.1975.31937

Venkatesh, B., Arunagiri, A., & Gooi, H. (2003). Unified opf method for maximizing voltage stability margin using successive fuzzy LP. *Electric Power Systems Research*, *64*(2), 119–128. doi:10.1016/S0378-7796(02)00152-9

Visakha, K., Thukaram, D., & Jenkins, L. (2004a). Application of upfc for system security improvement under normal and network contingencies. *Electric Power Systems Research*, *70*(1), 46–55. doi:10.1016/j.epsr.2003.11.011

Visakha, K., Thukaram, D., & Jenkins, L. (2004b). An approach for real power scheduling to improve system stability margins under normal and network contingencies. *Electric Power Systems Research*, *71*(2), 109–117. doi:10.1016/j.epsr.2004.01.005

Yesuratnam, G., & Thukaram, D. (2007). Optimum reactive power dispatch and identification of critical on-load tap changing (OLTC) transformers. *Electric Power Components and Systems*, *35*(6), 655–674. doi:10.1080/15325000601139641

Zhang, W., Li, F., & Tolbert, L. M. (2007). Review of reactive power planning: Objectives, constraints, and algorithms. *IEEE Transactions on Power Systems*, *22*(4), 2177–2186. doi:10.1109/TPWRS.2007.907452

Zhang, W., & Tolbert, L. M. (2005). Survey of reactive power planning methods. In *Power engineering society general meeting* (pp. 1430–1440). IEEE.

Zhang, Y., & Ren, Z. (2005). Optimal reactive power dispatch considering costs of adjusting the control devices. *IEEE Transactions on Power Systems*, *20*(3), 1349–1356. doi:10.1109/TPWRS.2005.851920

Zhu, J., & Xiong, X. (2003). Optimal reactive power control using modified interior point method. *Electric Power Systems Research*, *66*(2), 187–192. doi:10.1016/S0378-7796(03)00078-6

Zobian, A., & Ilic, M. D. (1996). A steady state voltage monitoring and control algorithm using localized least square minimization of load voltage deviations. *IEEE Transactions on Power Systems*, *11*(2), 929–938. doi:10.1109/59.496177

KEY TERMS AND DEFINITIONS

Contingency: Contingency is an outage of a transmission line or transformer that may lead to over loads in other branches and/or sudden system voltage drop.

Electric Grid: An electric grid is a network of synchronized power providers and consumers that are connected by transmission and distribution lines and operated by one or more control centers.

Fuzzy System: It is a component of machine learning techniques which takes membership values within 0 to 1 unlike crisp sets.

Optimization: A mathematical method to find the solution of a problem towards achieving better performance either in form of minimum or maximum under one or more given constraint.

Reactive Power: In electric power transmission and distribution, volt ampere reactive (VAR) is a unit by which reactive power is expressed in an AC electric power system.

Standard Deviation (STDEV): The standard deviation of the load voltage is used to quantify the amount of variation or dispersion of a set of load voltages.

Voltage Stability: Voltage stability refers to the ability of a power system to maintain steady voltage at all buses in the system after being subjected to a disturbance from a given initial operating condition.

APPENDIX

Comparison Method

Q-V Sensitivity Based Modal Analysis

The voltage stability problem has a dynamic nature in general, but static analysis techniques are promising tools for predicting the problem characteristics (Kundur et al., 1994). A modal analysis, as a static approach, can discover the instability characteristics and can be used to find the best sites for reactive power compensation, generator re-dispatch, and load-shedding programs. This is accomplished by solving the linearized power flow equations.

$$\begin{bmatrix} \Delta P \\ \Delta Q \end{bmatrix} = \begin{bmatrix} J_{P\theta} & J_{PV} \\ J_{Q\theta} & J_{QV} \end{bmatrix} \begin{bmatrix} \Delta \theta \\ \Delta V \end{bmatrix} \tag{21}$$

System voltage stability is affected by both P and Q. However, at each operating point we keep P constant and evaluate voltage stability by considering the incremental relationship between Q and V. This is analogous to the Q-V curve approach. Although incremental changes in Pare neglected in the formulation, the effects of changes in system load or power transfer levels are taken into account by studying the incremental relationship between Q and V at different operating conditions.

To reduce (19), let $\Delta P = 0$, then the reduced Jacobian matrix J_R is obtained as (Gao et al.,1992):

$$\left[J_R\right] = \left[J_{QV} - J_{Q\theta} J^{-1}{}_{P\theta} J_{PV}\right] \tag{22}$$

and

$$\Delta Q = J_R \Delta V \tag{23}$$

$$\Delta V = J^{-1}{}_R \Delta Q \tag{24}$$

Where,

J_R is the reduced Jacobian matrix of the system

The elements of the inverse of the reduced Jacobian matrix J_R are the Q-V sensitivities. A system is voltage stable at a given operating condition if for every bus in the system, bus voltage magnitude increases as reactive power injection at the same bus is increased. A system is voltage unstable if, for at least one bus in the system, bus voltage magnitude decreases as the reactive power injection at the same bus is increased. In other words, a system is voltage stable if V-Q sensitivity is positive for every bus and unstable if V-Q sensitivity is negative for at least one bus. The main disadvantage of V-Q sensitivity analysis is that it does not give the degree of stability or information about the proximity to voltage stability. Also it cannot identify the individual voltage collapse modes.

Let

$$J_R = \xi \Lambda \eta \tag{25}$$

Where,

ξ right eigenvector matrix of J_R
η left eigenvector matrix of J_R
Λ diagonal eigenvalue matrix of J_R

Then, inverting (25) yields

$$J_R^{-1} = \xi \Lambda^{-1} \eta \tag{26}$$

and substituting (26) in (24) results in

$$\Delta V = \xi \Lambda^{-1} \eta \Delta Q \tag{27}$$

Or

$$\Delta V = \sum_i \frac{\xi_i \eta_i}{\lambda_i} \Delta Q \tag{28}$$

Where,

ξ right eigenvector matrix of J_R
η_i is the *i*th row of the left eigenvector and
ξ_i is the ith column of the right eigenvector.

The *i*th mode of the Q-V response is defined by the *i*th eigenvalue λ_i, and the corresponding right eigenvector (ξ_i) and left eigenvector (η_i).

The V-Q sensitivity at bus *k* is defined as,

$$\frac{\Delta V_k}{\Delta Q_k} = \sum_i \frac{\xi_{ki}\eta_{ik}}{\lambda_i} \tag{29}$$

Generally V-Q sensitivity is found for minimum eigenvalue for application to static voltage collapse. Thus, in voltage stability study, the minimum singular value of Jacobian becoming zero corresponds to the critical mode of the system. It can serve as stability index, which indicates the distance between the studied operating point and the steady state voltage stability limit. To get qualitative as well as quantitative information of voltage stability margin, the sensitivity is often found at critical point for any given change of parameters.

Since $\xi^{-1} = \eta$, (27) may be written as

$$\eta \Delta V = \Lambda^{-1} \eta \Delta Q \tag{30}$$

By defining $v = \eta\Delta V$ as the vector of modal voltage variation and $q = \eta\Delta Q$ as the vector of modal reactive power variation, one can write uncoupled first-order equations as

$$v = \Lambda^{-1} q \tag{31}$$

Where,

v is the vector of modal voltage variation
q is the vector modal reactive power variation

Thus, for the ith mode, we have

$$v_i = \frac{1}{\lambda_i} q_i \tag{32}$$

If $\lambda_i > 0$, the ith modal voltage and the ith modal reactive power variations move in the same direction, indicating voltage stability of the system; whereas $\lambda_i < 0$ refers to instability of the system. The magnitude of λ_i indicates a relative degree of instability of the ith modal voltage. The voltage collapses when $\lambda_i = 0$, because any change in the modal reactive power causes an infinite change in the modal voltage.

The relative contribution of the power at bus k in mode i is given by the bus participation factor (BPF)

$$P_{ki} = \eta_{ki}\, \xi_{ik} \tag{33}$$

Where,

P_{ki} indicates the contribution of the ith eigenvalue to the V-Q sensitivity at bus k.

The bus participation factors determine the most critical areas, which lead the system to instability. Usually, the higher the magnitude of the participation factor of a bus in a specific mode, the better the remedial action on that bus in stabilizing the mode.

Continuation Power Flow Method

The power flow is a useful tool for monitoring the system voltages as a function of load change. One common application is to plot the voltage at a particular bus as the load is varied from the base case to a point of maximum loadability. At the loadability limit, the system Jacobian of the power flow equations will become singular. Consequently, the traditional Newton-Raphson method of obtaining the load flow solution will show convergence problem. The continuation power flow method overcomes this problem by reformulating the power flow equations so that they remain well-conditioned at all possible loading conditions.

The continuation power flow method introduces an additional equation and unknown into the basic power flow equations (Stagg et al., 1968). The additional equation is chosen specially to ensure that the augmented Jacobian in no longer singular at the loadability limit. The additional unknown is often called the continuation parameter.

The power flow equations are reformulated to include a load parameter λ. This reformulation can be accomplished by expressing the load and generator at a bus as a function of the load parameter (λ). Thus, the general forms of the new equations for each bus i are

$$\Delta P_i = P_{Gi}(\lambda) - P_{Li}(\lambda) - P_i(V,\delta) = 0 \tag{34}$$

$$\Delta Q_i = Q_{Gi} - Q_{Li}(\lambda) - Q_i(V,\delta) = 0 \tag{35}$$

Where

$$P_i(V,\delta) = \sum_{j=1}^{n_b} V_i V_j Y_{ij} \cos(\delta_i - \delta_j - \theta_{ij}) \tag{36}$$

$$Q_i(V,\delta) = \sum_{j=1}^{n_b} V_i V_j Y_{ij} \sin(\delta_i - \delta_j - \theta_{ij}) \tag{37}$$

Where,

λ is the loading/continuation parameter: $0 \le \lambda \le \lambda_{critical}$; $\lambda = 0$ corresponds to the base case and $\lambda = \lambda_{critical}$ corresponds to critical case
P_{Gi} is the real power generation at bus i
Q_{Gi} is the reactive power generation at bus i

P_{Li} is the real power load at bus i
Q_{Li} is the reactive power load at bus i
P_i is the real power injection at bus i
Q_i is the reactive power injection at bus i
V_i is the voltage at bus i
Y_{ij} is the ijth elements of the bus admittance matrix

To simulate different loading on the system, the P_{Li} and Q_{Li} can be modified as

$$P_{Li}(\lambda) = P_{Lib} + \lambda[K_{Di} S_{\Delta base} \cos(\varphi_i)] \tag{38}$$

$$Q_{Li}(\lambda) = Q_{Lib} + \lambda[K_{Di} S_{\Delta base} \sin(\varphi_i)] \tag{39}$$

By replacing $S_{\Delta base} \cos(\varphi_i) = P_{Lib}$; $S_{\Delta base} \sin(\varphi_i) = Q_{Lib}$ and $Q_{Lib} = \tan(\varphi)$
The equations (38) and (39) can be rewritten as

$$P_{Li}(\lambda) = P_{Lib}[1 + \lambda K_{Di}] \tag{40}$$

$$Q_{Li}(\lambda) = P_{Lib} \tan(\Phi_i)[1 + \lambda K_{Di}] \tag{41}$$

Similarly, the active power generation can be modified to

$$P_{Gi}(\lambda) = P_{Gib}[1 + \lambda K_{Gi}] \tag{42}$$

Where,

P_{Lib} base case real power demand at bus i
Q_{Lib} base case reactive power demand at bus i
K_{Di} constant multiplier showing the rate of change in load at the ith bus
Φ is the power factor angle of load change at bus i
$S_{\Delta base}$ is the base case apparent power is chosen to provide appropriate scaling of λ
P_{Gib} base case active power generation at bus i
K_{Gi} constant multiplier showing the rate of change in generation at the ith bus

The predictor and corrector scheme is applied to get the power flow solution (Ajjarapu & Christy, 1992; Ajjarapu, 2007).

Chapter 8
Energy Management

Maheswari M.
Nalla Malla Reddy Engineering College, India

Gunasekharan S.
Lord's Institute of Engineering and Technology, India

Sumadeepthi Veeraganti
Malla Reddy Engineering College (Autonomous), India

ABSTRACT

Energy is described as the amount of work that can be done by force. There are various forms of energy such as kinetic energy, potential energy, thermal energy, light energy, sound energy, and electromagnetic energy. As per the law of conservation of energy, it is neither created nor destroyed. In this modern era, energy became an integral part of our life. The life without energy is not at all possible nowadays. The energy is not offered at free of cost and it comes at an affordable prize. The generation of energy requires natural resources which are exhaust day by day. At the same time, the usage of energy is increasing exponentially. Managing and reducing energy consumption not only saves money but also helps in mitigating climate change and enhancing corporate reputation. The organizations can achieve appreciable energy reduction by adopting simple measures. This chapter discuss about the present scenario of energy, need for energy management, energy management program, and its various steps involved.

INTRODUCTION

In this modern era, energy became an integral part of our life. The society which depends on energy is ours than before. Just thing about the equipments or utilities used in our day to day life either in home or work place which need energy for operation. The life without energy is unimaginable now a days. Also our productivity would drastically fall down, our society with computers cease to function and our gross domestic product (GDP) also reduced. Similarly, if oil supply stops then the fabric of our society also fall quickly. The example for the above said instants are power cut in California in the year 2001 and the strike of oil protesters in UK in September 2000. But in case of developing countries the scenario is just opposite to the developed countries. In such countries, the power is only supplied to the major

DOI: 10.4018/978-1-5225-8551-0.ch008

cities not to the rural areas. This situation will not affect the life of the people living there but it shows the change in low GDP of that country. The developing countries has majority of population but they consumes minority of energy. One - third of the world's population utilizes energy very easily but the remaining two-third population are unable to secure enough energy to grow economically. This concept can be highlighted by the example of USA consumes approximately 26% of all the world's energy but having 4.4% of the World's population.

The demand for energy will grow due to the developing economies of the world. This will indeed increase the pressure on the earth's dwindling fossil fuel supply and also increase atmospheric pollution due to excess green house gas emissions. The issues like climate change and atmospheric pollution has no national boundaries and impinging on the comfort and security of the developed world. The perceived threat of the global climate change is the driving force for all intergovernmental environmental summits of the latest years. Based on the resolution passed in the summits, many countries has implemented the alternatives. The most of the people are looking only the consumption of the energy but the energy supply itself is the large and important sector of the world's economy. Apart from all these summits and discussions, now the energy and its utilization became forefront of the public consciousness.

ENERGY

Energy is defined as the *"one joule of work done when a force of one Newton acts on an object and it moves to one metre distance in the direction of the force"*. There are various forms of energy such as kinetic energy, potential energy, thermal energy, light energy, sound energy and electromagnetic energy. As per the law of conservation of energy, it neither created nor destroyed. The generation of energy requires natural resources which are exhaust day by day. At the same time, the usage of energy is increasing exponentially.

ENERGY CONSUMPTION AND GDP

The GDP of any nation is related to its energy consumption. To illustrate this concept the energy consumption from an historical viewpoint is given in Table 1. It shows the average daily consumption of people in various societies from early period to recent. The increase in per capita energy consumption exponentially as the advancement and industrialization took place in the society is shown in Table1. In early years the humans were lived in forest and had fruits, nuts and vegetables. Then the people started to hunt the animals and they learnt to use fire for cooking and heating. After later stages, societies developed and the first energy usage applied for agricultural and then for industrial practices. Now a days the advancements took place in each and every aspect and all requires energy. It is clear from Table 1, the per capita energy consumption approximately increases from 2000 kilocalories to 21000 kilocalories from ancient days to late 1990's.

From the above Table and discussion it is clear that there is a strong relationship between the GDP and energy consumption of a nation. For the development of the nation, they should manage the energy consumption in proper way.

Table 1. Overview of per capita energy consumption from BC to AD

Period and Location	Type of Society	Characteristics	Estimated Daily per Capita Energy Consumption
Very early	Gatherers Gathered	wild fruit, nuts and vegetables	2000 kCal (8.2 MJ)
1000 000 BC	Hunter-gatherers	Gathered wild fruit etc., hunted and cooked food	4 000 kCal (16.4 MJ)
4000 ac Middle East	Settled farmers	Sowed crops and kept animals	12 000 kCal (49.2 MJ)
AD 1500 Europe	Agricultural with small scale Industry	Agricultural society with specialized industry	21 000 kCal (88.2 MJ)
AD 1900 Europe	Industrialized Society	Large Scale Industry and Mass production.	90 000 kCal (378 MJ)
AD 1500 USA, Europe and others	Advanced Industrialized Society	Consumer society, mass transport, many labour saving devices	250 000 kCal (1 GJ)

Huge challenges are put forth related to energy use. Many of the industries in the world have reduced the energy intensities by using and developing energy efficient techniques and management strategies. This decision makes them to contribute for high energy end use and contribution to energy related environmental problems. Due to these steps, industries have gained improvement in environmental protection as well as economic dividend. There are numerous researches are carried out to show case the benefits of the implementation of energy efficiency and management measures and also it shows the greater savings of the developing countries.

VALUE OF ENERGY MANAGEMENT

In the last few years, Government and Private Industries are under pressure to optimize the economic and environmental effects. To compete in the global market place economically and meet the increasing environmental standards to reduce pollution become major driving factors in most of the organizations. Energy Management is the important tool to help the organizations to meet these critical objectives for their success both in short term and long term. Significant energy and economic saving can be accomplished through energy management. The common facilities like schools, hospitals, office buildings and manufacturing plants etc., can able to save as per the following table

Table 2. Energy saving through energy management

Activities	Percentage of Energy Saving
Low cost activities first year or two	5 to 15%
Moderate cost, significant effort, three to five years	15 to 30%
Long-term potential, higher cost, more engineering	30 to 50%

Energy Management also does implementation of new energy efficient techniques, new materials, new manufacturing processes and new technologies in equipment for industries to improve the productivity and service quality. Energy saving alone is not the driving force for companies while implementing energy management, it also provides increase in productivity, quality, reduce environmental emissions and reduced energy costs. energy management plays a key role in helping the nation to move towards the energy efficient economy. This chapter proposes the fundamentals of an energy management program can be implemented in large or small organizations.

ENERGY MANAGEMENT PROGRAM

To implement an efficient energy management system, it is essential to develop an organizational structure. The major components of the energy management program is depicted in Figure 1. The components are organizational structure, a policy, plans for audits, education, reporting and strategy. Each component of energy management is discussed in detail in the succeeding section.

Organizational Structure

The generic organizational chart for energy management is shown in figure 1 and it can be adopted to fit into the available structure of each organization. For example, the presidential block may be replaced by general manager, VP block may be by division managers however the fundamental principle is same. It is very important that the location of the energy manager. actually this position is enough to access to key players in management and to have a knowledge on current events within in the organization. The position of the energy manager is also indicative and they should be able to understand the present energy projects, funding availability and other management priorities.

Energy Manager

The energy manager should have the top management support and he should secure this support. Energy manager should have a vision of energy management and its returns to the organization. To make energy management as a successful one, there will be a one person who takes care of things happen. The energy

Figure 1. Energy management program

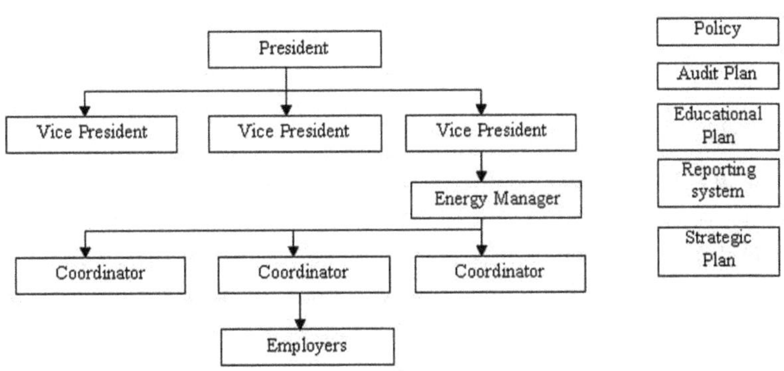

manager may be an energy engineer and should take efforts alone. But for a long term, the involvement of each and everyone at the facility will yield more productive and permanent. The major role of the energy manager is to develop a organizational structure. The major requirements of the energy management are,

- Set up an Energy Management Plan
- Establish energy records
- Identify outside assistance
- Assess future energy needs
- Identify financing sources
- Make energy recommendations
- Implement recommendations
- Provide liaison for the energy committee
- Plan communication strategies
- Evaluate program effectiveness

Initially, the energy management can be implemented in a division of a large organization. Many of the organizations which has facilities and responsibilities are initiating the program in corporate level. When the energy management program is initiated in corporate level there are some advantages like:

- There are more resources like budget, staff and facilities are available for implementation
- If top management is secured in corporate level then it will be easy to get support in the division level.
- Expensive equipments for testing can be purchased and can be used by divisions when needed.
- Reporting system can be followed combinable
- Financing should be creative and most needed
- Energy and environmental legislation impacts can be determined at corporate level
- Rates for electrical utilities and structures can be evaluated at corporate level

At the same time, the following precautions should be taken:

- At divisional level, many people have already done good job in energy saving and cautions to be taken to avoid the credits taken by the corporate staff.
- All divisions will not progress at the same speed. Initially work with people who are interested and report to the top management for their credit. Then others request for assistance.

Energy Team

The coordinators block shown in Figure 1 indicates the energy management team within one organizational structure. It is the heart of the program and the most interested people should be considered as members. The members should be from all divisions such as accounts, maintenance and representatives from each major department. The duration of the energy team may be for one or two years and new people can be given opportunity to get new innovative ideas. The coordinators should supplement the skills lacking in energy manager since it is not possible that one energy manager will have all qualifications. Hence the skills required for the team including energy manager are,

- Have enough technical knowledge to either understand the technology used by the organization or trainable in that technology
- Have knowledge on new technology that may be applicable to the program
- Have planning skills to establish structure, energy surveys, find energy needs and develop strategic management plan.
- Understand the economic evaluation system used by the organization like payback and life cycle cost analysis.
- Have good communication and motivational skills since energy management involves everyone within the organization.

The strengths of each team may be evaluated in the above said desired skills and their assignments should be made accordingly.

Employees

In Figure 1, employees are also shown as part of the organizational structure but they are the untapped resource in an energy management program. A structure method of expressing their ideas for more efficient use of energy will give most productive effort in the energy management program. A good energy manager should spent 20% of his total time with employees. Employees in manufacturing plants will know more about the equipment than anyone else. They definitely know how to operate it more efficiently. But there is no input mechanism available to get their views. To overcome these situations, employee involvement program can be conducted. The three key factors of motivational are,

- People already have motivation within them. The task of supervisor is to release it.
- The amount of energy and enthusiasm will vary person to person and should remember that all are not over achievers but all are not lazy.
- The amount of personal satisfaction to be derived determines the amount of energy an employee will invest in the job.

Achieving personal satisfaction is one of the most popular research area of industrial psychologists and they emerged some enlightening facts such as need for food, emotional need, need for acceptance, recognition or achievement. Also they highlighted that many efforts to motivate employees deal to satisfy physical needs, such as raises, bonuses or fringe benefits. These factors are effective only for short duration and we should look beyond these to other needs that may be sources of releasing motivation. Heresy and Blanchard has done a study in 1977 about the rank of job related factors among workers, they are

- Full appreciation for work done
- Feeling "in" on things
- Understanding of personal problems
- Job security
- Good wages
- Interesting work
- Promoting and growth in the company
- Management loyalty to workers

- Good working conditions
- Tactful discipline of workers

The raking of supervisors are just opposite to the raking of workers. Workers gave first priority to good wages. Hence job enrichment is the key to motivation. Hence the energy manager can plan a program involving employees that provide enrichment by some simple and inexpensive recognitions.

Energy Policy

A well written energy policy has to be formed by the management. It provides information to the energy manager about the authority to be involved in business planning, new facility created, selection of production equipment, purchase of equipment, energy reporting and training. The energy policy should be short and maximum two pages at most. One should not confuse the policy with the procedure manual. The energy policy should contain minimum the following items,

- *Objectives:* It should contain the standard flag and motherhood statements about energy. At the same time, the organization should incorporate energy efficiency into facilities and new equipment with emphasis on life cycle cost analysis rather than lowest initial cost.
- *Accountability:* It should establish the organizational structure and the authority for the energy manager, coordinators and other task groups.
- *Reporting:* The energy manager should have the authority from the top management to require others within the organization. Policy is the right place to establish it and provides legitimate for requesting funds to procure instrumentation to measure energy usage.
- *Training:* The training is the most important aspect among the organization at all levels. If it is included in the policy then it will become very easy to include training in budget.

Many organizations, instead of writing a comprehensive policy, it is described by simple statement including all aspects.

Planning

The most important aspect of the energy management is Planning. It consists of two major functions in the program. First, planning will acts as shield from disruptions and second, by scheduling the events throughout the year, will make the program very active. It is better to discuss with the top management and get the suggestions before planning to incorporate the suggestions also. The events such as training programs, audits, planning sessions, demonstrations, research projects, lectures etc., It is advisable to involve the people implementing the plan while planning process. This makes the plan workable and also people feel committed if they were a part of the design. At the same time, a management people should be there to prevent the dominating ideas from most outspoken members of a committee and rejecting ideas from less outspoken members. Andre Delbecq and

Andrea Van de Ven has developed new technique in the year 1980's and it consists of following steps:

- **Problem Definition:** The problem has to clearly defined to the members in the group.
- **Grouping:** Divide the large groups into 7 to 10 groups and elect secretary.

- **Silent Generation of Ideas:** The members can write many answers to the problem within a specified time.
- **Round-Robin Listing:** The secretary lists each idea individually until all have been recorded.
- **Discussion:** The ideas can be discussed for clarification, elaboration, evaluation and combining.
- **Ranking:** Each person ranks the five most important items and can find the first choice of the group.

Audit Planning

The audit should be planned before the actual audits and it includes types of audits to be performed, team makeup and dates. The audit should be specific instead of general for more energy saving. Some types of audits might be considered are,

- Tuning-Operation-Maintenance (TOM)
- Compressed air
- Motors
- Lighting
- Steam system
- Water
- Controls
- HVAC
- Employee suggestions

The individual audits will make easy to identify the proper team for the audit and the external persons for electric utility and natural gas representatives also added in the team for better outcome. Apart from scheduling, contribution to the audit will make the program more active. The audit can be done in two ways such as they can go for performance contactor or they can set up their own team . Each method has advantages as well as disadvantages also. The advantages of performance contract are,

- There is no investment is required for the company other than that involved in the contracting process which can be very time consuming.
- Only few in-house people like energy manager and financial people are involved in the process.

Similarly the disadvantages are,

- Technical resources are generally limited to the contracting organization
- Performance contracting is still maturing and many firms underestimate the required work.
- The contractor may not have the skills required for full spectrum
- The contractor will not show interest on low budget projects.

Advantages of setting own audit team are:

- The selection of team is based on the equipment to be audited and the members may be in-house personnel, outside specialist or combination of both.

- It is easy to carry out all level projects like low or high cost projects.
- The audit is an excellent training tool by involving others in the process.

Disadvantages of setting own audit team are:

- Financing identified projects become a separate issue for the energy manager
- It takes a well organized energy management structure to take full advantage of the work of the team

Educational Planning

The major role of energy manager is to educate the persons within the organization about the energy. Actually, there is a lot of ignorance among everybody about the energy management. The awareness among all persons about the energy will give big dividend. The energy management program will be more effective if the management understands the complexities of energy, economic benefits due to energy, use of latest technologies.

The training has to be provided for three groups namely, management, the energy team and employees.

Management Training

Finding time for the management is very difficult, hence delicate ways to be determined to get them up to speed. Conducting regular meetings to provide the updates on the program is one way. A periodical good concise report is a tool to educate management. Along with the report shot articles relevant to the respective organizational goals taken from magazines and news paper can be attached. The management can be added as a part of training program either the energy team or employees or both. Generally, the energy manager be a part of business planning for the organization and this makes a good contact with the management people to train and educate.

Energy Team Training

The energy team is the core group of the energy management program, hence proper and thorough training should be given. The training is available from many sources and in many forms:

- **Self-Study:** It requires good material related to energy and from which coordinators has to select.
- **In-House Training:** The in-house training can be planned periodically by the energy manager or the expert from the outside.
- Short-term courses offered by association such as Association of Energy engineers, by individual consultants or by colleges or universities.

For decentralized organizations having ten or more energy manager can plan annually two or three day seminar based on the educational program. The following suggestions should be incorporated into such a program:

- The quality speakers should be selected from both inside and outside the organization
- The top management people can be invited to give opening remarks.

- The participants can be involved in workshop activities to get the input for the seminar. Also, some practical tips on energy saving can be provided to them to implement immediately.
- A manual that includes a schedule of events, speaker information, list of attendees, information about each topic presented can be prepared.
- The logo for the program can be created and include in the favors of cups or carrier cases etc.,

Employee Training

A systematic approach has to be followed for employees. It should start from the basic in energy saving. It will produce much higher quality of ideas from them. Short training sessions like safety can be conducted and also include,

- Energy conservation at home
- Fundamentals of electric energy
- Energy saving fundamentals.

Strategic Planning

Strategic planning in energy management program constitutes objectives, strategies, programs and action items. It is the last and important step of the energy management. The name strategic planning is threatening to more technically inclined people. But by using simplified approach and involving the energy management team in the process, a plan in the form of flowchart can be made for the next five years. This plan should be protective and there should not any disturbance into the program, once it is established and approved. It provides the basis for resources such as funding and personnel for implementation. It projects strategic planning into overall planning by the organization. By involving the implementers in the planning process, there is a strong commitment to make it work.

Reporting

There is no specific form to prepare a report and there are too many variables such as organization size, product, project requirements and procedures already in existence. Many factors are there to influence any index, such as weather, production, expansion or contraction of facilities and new technologies. The basic need of any reporting system is to be customized to suit individual circumstances. It shows the effectiveness of the energy management program in an organization. The report is probably of most value to the one who prepares it. It should include all the information to be pulled together in a coherent manner. To make the preparation of the report more easier, the data should be collected and stored in computer periodically and finally combining production data and energy data to develop an energy index. The reporting requirements should be as simple as possible. The narratives should be short, and the supportive documents are kept in the file for deep discussion.

Ownership

The energy management program will be successful by a word "ownership". It extends to everyone within the organization. Employees who operate a machine should "own" that machine and any modification in that without their participation will not succeed. Because they have the knowledge to make or break the attempt. The tips for success that have been compiled from observing successful energy management programs:

- Have a plan: Plan dealing organization, surveys, training and strategic planning with events scheduled. It has the advantages like prevention of disruptions by non productive ideas and sets up scheduled events to make the program active.
- Give away: Share the ideas for saving energy. otherwise the project will get spoiled.
- Be aggressive: The energy team after some training will be the most energy knowledgeable group within the company. Too many management decisions are made with a meager knowledge of the effects on energy.
- Go with the winners: Not every department within a company will be enthused about the energy program. Make those who are look good through the reporting system to top management and all will follow.
- A final tip: The machine operator should be asked about the steps to reduce the energy. at the same time they should be rewarded for the ideas.

CONCLUSION

Energy management has developed to offer the best opportunities for the people willing to inest time and effort to learn the fundamentals. It requires technical and management skills which widen educational needs for both technical and management people desiring to enter into the field. Since it is the economic return to the top management, there is greater opportunities to the energy managers for recognition and advancement. Managing energy will be the continuous need for the development of the nation as well as worldwide and it will improve the environmental effects also.

REFERENCES

Delbecq, A. L., Van de Ven, A. H., & Gustafson, D. H. (1975). *Group techniques for program planning: A guide to nominal group and Delphi processes* Glenview, IL: Scott, Foresman.

Doty, S., & Turner, W. C. (2004). *Energy management handbook*. CRC Press.

Farmani, F., Parvizimosaed, M., Monsef, H., & Rahimi-Kian, A. (2018). A conceptual model of a smart energy management system for a residential building equipped with CCHP system. *International Journal of Electrical Power & Energy Systems*, *95*, 523–536. doi:10.1016/j.ijepes.2017.09.016

Hersey, P., & Blanchard, K. H. (1969). *Management of organizational behavior: Utilizing human resources*. Academic Press.

Mashburn, W. H. (1989). *Managing energy resources in times of dynamic change*. Academic Press.

Shafie-Khah, M., & Siano, P. (2018). A stochastic home energy management system considering satisfaction cost and response fatigue. *IEEE Transactions on Industrial Informatics*, *14*(2), 629–638. doi:10.1109/TII.2017.2728803

Thomas, D., Deblecker, O., & Ioakimidis, C. S. (2018). Optimal operation of an energy management system for a grid-connected smart building considering photovoltaics' uncertainty and stochastic electric vehicles' driving schedule. *Applied Energy*, *210*, 1188–1206. doi:10.1016/j.apenergy.2017.07.035

Wang, F., Zhou, L., Ren, H., Liu, X., Talari, S., Shafie-khah, M., & Catalão, J. P. (2018). Multi-Objective Optimization Model of Source–Load–Storage Synergetic Dispatch for a Building Energy Management System Based on TOU Price Demand Response. *IEEE Transactions on Industry Applications*, *54*(2), 1017–1028. doi:10.1109/TIA.2017.2781639

Chapter 9
Energy Efficient and Secure Localization in Wireless Sensor Networks:
An Approach Through Anchor Mobility Control

Rathindra Nath Biswas
A. J. C. Bose Polytechnic, India

Swarup Kumar Mitra
MCKV Institute of Engineering, India

Mrinal Kanti Naskar
Jadavpur University, India

ABSTRACT

This chapter describes the development of energy efficient and secure localization systems for wireless sensor networks (WSNs) based on anchor mobility control techniques. Towards improvement of energy efficiency and security over the network, sensors are assumed to broadcast messages in a periodic and coherent manner. Moreover, anchor is supposed to be location aware with GPS (global positioning system) receiver and capable of finding the directions of arrival (DoA) from intercepted signals using smart antennas. In each step along its trajectory, anchor communicates only with neighboring nodes having received signal strength (RSS) above a predefined threshold level. Mobility control schemes aim to explore few new nodes along with the existing ones in each subsequent anchor steps. Sensors would be able to localize themselves after receiving range (distance/angle) data from two distinct anchor positions. Accordingly, convergence speed of localization process is optimized. Simulation results corroborate its competency comparable to the existing methods.

DOI: 10.4018/978-1-5225-8551-0.ch009

INTRODUCTION

Currently, several important communication services both in civil and defense sectors, very often require simple and cost effective wireless network architectures (Akyildiz, Su, Sankarasubramaniam, & Cayirci, 2002). In these cases, the sink or base station (BS) also needs to be located at a convenient and safe place for uninterrupted operation throughout the networks. For example, monitoring of disaster-prone areas, military surveillance, or mobile target detection and tracking in the battlefield etc. are normally operated from a remote control room. Wireless sensor network infrastructures, consisting of a plentiful of low power micro sensor devices enabled with integrated sensing, processing and short distance communication capabilities, have been a good choice for such applications until now (Pazzi & Boukerche, 2008). However, these essentially require location based data from the sensors that needs to be relayed to the sink for realization of an event occurring within the harsh fields. Although, in practice, huge static tiny sensors are randomly deployed over the area of interest from an aircraft/vehicle, they remain scattered, unattended and unaware of their locations in most of the cases. Hence, localization systems are also required to be incorporated urgently in such networks for any location based application as mentioned earlier (Boukerche, Oliveira, Nakamura, & Loureiro, 2007).

Basically, the conventional localization systems work with their two constituent parts such as: (i) location estimation process and (ii) localization algorithm. Usually, the former functional segment performs position computation with/without measuring the range (distance/angle) information and hence it is classified into two categories as: (a) range-based and (b) range-free techniques. However, range-based schemes show more accuracy over its counterpart at the cost of system complexity (Biswas, Mitra, & Naskar, 2014). On the other hand, the latter section signifies operating principles of the process how it propagates throughout the entire networks. Again, it may take two forms as: (a) centralized and (b) distributed techniques. In fact, distributed systems often outperform the centralized one, reducing the amount of central coordination and thus, each sensor is capable of acting independently over the networks (Xiao, Chen, & Zhou, 2008). Accordingly, the network attributes like energy efficiency, scalability and robustness etc. are enhanced to some extent. Thus, most of the state-of-the-art localization systems are of distributed features. They also involve few location aware nodes (static/mobile), termed as anchors or beacons, as references in position computation (Zou & Chakrabarty, 2004; Ou, 2011).

However, lifetime of wireless sensor networks is usually dependent on power handling capacity of the battery driven tiny sensors. Hence, a control on proper utilization of the power is very crucial towards their sustained operation to fulfill a specific task. In this regard, reducing the consumption of energy at each sensor might be a strategic approach (Iguchi-Cartigny, Ruiz, Simplot-Ryl, Stojmenovic, & Yago, 2009). But it also appears to be a rather complicated and challenging task since the sensors function in a cooperative manner. Consequently, sensing a large volume of raw data and then communicating them to the neighbors involve a significant amount of power loss. Again, static sensor networks inherently possess some coverage problem after a random deployment because relocation of the sensors is not possible in such cases (Wang, Srinivasan, Wang, & Chua, 2008). Instead, mobile sensor networks can provide intelligent coverage over the networks. But they also become energy inefficient due to their additional power consumption in mobility assistance (Wu, Cho, D'Aurrio, & Lee, 2007). Hence, only tradeoff exists on including few mobile anchors along with static sensor networks making the entire network architectures as quasi-static in nature. Moreover, several malicious attacks (e.g., Sybil, replay and wormhole etc.) may occur on the localization systems degenerating one of its functional segments and misleading the decision about any event occurring within the networks (Boukerche, Oliveira, Nakamura, & Loureiro, 2008). Therefore, localization systems need to be made as energy efficient as possible along with increase in robustness and security.

This chapter is focused on the development of a new energy management technique in localization process to ensure prolonged network lifetime. This is actually accomplished here with an intelligent control on anchor mobility that makes localization process faster and reduces power consumption at sensors. For this purpose, a synchronous and static network model along with a mobile anchor is primarily considered. Anchor is supposed to be equipped with a GPS (Global Positioning System) receiver and smart antennas (Biswas, Mitra, & Naskar, 2014). These arrangements help to keep the track of its corresponding locations and determining angles from the incoming signals of neighboring nodes, having signal strength above a prescribed threshold value (also termed as clustering nodes). Such antenna array normally produces a narrow beam with higher directive gain. Thus, it can also guarantee trivial loss during data transmission between the anchor and the clustering nodes. In the networks, stationary nodes are configured to broadcast message in a periodic way. The anchor usually receives messages from all clustering nodes at each interval. Also, it controls its next step movements in such a way that it would explore more number of clustering nodes (i.e. making cluster size larger) keeping all nodes of the present cluster. Hence, clustering nodes must be localized after receiving messages from two distinct anchor positions along its trajectory. Two anchor mobility approaches such as: (i) two-step centroid / deterministic based method and (ii) Fuzzy based system are adopted here. Thus, such schemes improve energy efficiency attaining higher degree of convergence speed at the localization process. Towards enhancement of energy savings, localized clustering nodes are also switched to sleep mode till the entire process ends. Besides, two-step centroid scheme is found to perform well against Sybil and replay type attacks in WSN environments, making the localization systems robust and secure enough. Basically, security algorithm operates on the principles of blocking/filtering out malicious nodes. Malicious nodes are usually selected detecting misbehaviors/ anomalies of node IDs in two consecutive anchor steps. Therefore, such method is simple enough for its realization in wireless sensor network infrastructures.

The rest of this chapter is organized as follows. In the next section, the related works in the area of conventional localization schemes with mobile anchor are briefly summarized. The following section provides a brief overview of the network architectures, attack scenarios and radio propagation features and all assumptions made for developing the proposed localization method. The anchor mobility control schemes and localization process are described in detail in the next section. In the second to last section, the simulation results are discussed and the final section concludes the chapter and discusses future research.

RELATED WORKS

To reduce energy consumption, data should be relayed over the shortest route within the networks. For this purpose, it is important to have knowledge of locations in each constituent node. After a random deployment, sensors usually become familiar with their own locations through localization systems.

Hence, localization process needs to be made energy efficient so that an elongated network lifetime is ensured accomplishing its future goals. During past decades, several pioneering research works have been done in this regard. An overview of the existing localization methods using mobile anchors is presented in this section. Sichitiu and Ramadurai (2004) proposed a localization method using single mobile beacon. Using a Bayesian inference approach, each sensor node estimates its position inferring proximity constraints to received data packets from the mobile beacon. Trajectory is set in such a way that at least three non-collinear beacons are required to provide full coverage on all possible positions of a stationary node. It seems to be punier in terms of location estimation error (order of a few meters). Ou (2011) also introduced a range-free localization system taking few mobile anchors in the networks. Anchors are equipped with four directional antennas and sensors estimate their locations by statistical median process, receiving few beacon messages. A chessboard like anchor trajectory is considered to obtain lower energy consumption and least localization error in noisy environments. Similarly, Zhang and Yu (2008) presented an energy efficient localization algorithm with an anchor moving in a straight trajectory. Anchor is embedded with both RF (radio frequency) and ultrasonic transmitter whereas sensor nodes are containing both RF and ultrasonic receiver. Stationary node can estimate its location measuring distance from three non-collinear virtual anchor points with TDoA (time difference of arrival) method and solving nonlinear equations using Newton's iterative process. Although higher energy efficiency and an acceptable amount of accuracy in localization can be obtained, system complexity is increased to some extent. Ssu et al. (2005) also described a range-free localization scheme using few mobile anchor points. Each sensor node computes its location based on the perpendicular bisector of a chord conjecture principle where at least three end points (beacon points) on the circle (sensor radio range) are collected for making two chords. Sensors usually estimate the intersection point of two chord bisector using Cramer's rule as its location. Lee et al. (2005) proposed a new localization scheme to estimate sensor location from possible areas by using geometric constraints. Location of any node is determined as the midpoint of three selected beacon points among all received beacons and obtaining an intersection area from two of them using geometric constraints. The third beacon point is used to sort out correct location. Likewise, Xiao et al. (2008) described a range-free distributed localization method using a single moving beacon. Each sensor node estimates its position taking arrival and departure overlap (AOD) of beacon messages. An analytical model with three different beacon trajectories (e.g. sparse straight line, dense straight line and random) shows that the upper bound of localization error is determined by radio range and beacon interval. It also points out that beacon movement pattern play a pivotal role in the sensor localization. Kim and Kim (2010) proposed also a method of distance estimation with weighted least squares technique. It is suitable for range based localization with single mobile beacon traversing along linear tracks. Dong and Severance (2007) proposed a novel position estimation scheme with multiple mobile beacons. Each sensor node assumes its position as the center of circle formed by four non-collinear boundary beacon points. Then it establishes confident position estimation through an iterative refinement technique using Levenberg-Marquardt optimization. Zhenjie and Changjia (2006) presented a localization scheme with mobile beacon. Each sensor node determines the range differences among three different beacon positions, measuring TDoA of received RF signals and estimates its location using trilateration principle. Most of the algorithms mentioned above use random anchor mobility model and hence they ignored the conditions of preserving security against the malicious attacks as well as improving energy efficiency during localization process. However, authors acknowledge the innovative works of Xiao et al. (2008) to suit well acquiring greater energy efficiency but it fails under the adversarial attacks. In this perspective, proposed anchor mobility control schemes, selecting an appropriate trajectory, would initiate faster convergence in the localization process. Thus, optimum efficiency in energy consumption could be guaranteed over the networks. In addition, two-step deterministic model could make localization systems secure and robust with feasible infrastructures.

LOCALIZATION ASPECTS IN SENSOR NETWORKS

There is a huge demand of location based services in various sectors today. Towards fulfillment, all sensor nodes in WSNs need to be incorporated with GPS receivers or there must have suitable localization systems enabling an autonomous determination process of node location. However, installation of GPS system to each node might be an unrealistic approach because it makes sensors bulkier and more expensive. On the other hand, relaying of location based data to the sink is also essential in observing and detecting any event within the networks. Localization systems using references of few static/mobile anchors (with GPS receivers) might be a feasible solution in this regard (Vivekanandan & Wong, 2007; Rahman & Kleeman, 2009). They also need to be made as energy efficient as possible to ensure prolong network lifetime. Energy efficiency can be improved with reduction of localization time in each constituent node. As localization systems play crucial role in location based data communications, they are now becoming the tempting targets for various malicious attacks as well. Attacks may occur in different manners (i.e. direct/indirect and external/internal) but they aim at hampering normal operations of the localization systems in most cases. Thus, they should also be made resilient and secure enough both to sustain their uninterrupted functioning over the networks.

Wireless Sensor Networks Preliminaries

Wireless sensor networks are often deployed in remote and hostile environments for data gathering. Accordingly, adequate smart battery-driven microsensor devices are configured over the region of interest. In real-time, several issues such as constraints in energy efficiency, coverage, scalability and security etc. are very common to exist there. In most of the cases, these conditions are fulfilled by setting up suitable network architectures and protocols. For example, energy efficiency can be enhanced by selecting appropriate modes (e.g. active, sleep, idle etc.) in operation of the constituent nodes, reducing their power consumption over the networks (Heinzelman, Chandrakasan, & Balakrishnan, 2002). Coverage problems can be mitigated introducing some mobility on the nodes to facilitate their relocations in WSNs. Algorithms also need to be defined properly to hold scalability attributes. However, preserving node security against several malicious attacks in WSNs is rather a complicated and challenging task. Although several cryptographic schemes exist in literature, they may not be applicable to tiny sensors. Because the complex algorithm imposes huge computational overheads to the memory restrained nodes. Also, it consumes more power driving the nodes to active mode for a long period (Zheng et al., 2017).

This section deals with the network configurations, radio propagation characteristics and possible attacks on the localization systems. All necessary assumptions are introduced here to expedite the description and analysis of the proposed anchor mobility control schemes, making energy efficient and secure localization systems, in the subsequent sections.

Network Architectures and Protocols

To implement the proposed anchor mobility control schemes, several assumptions are made at network architectures and protocols (as shown in Figure 1). These are discussed as follows.

1. Node Deployment Attributes
   a. Sufficient number of stationary sensors along with a mobile anchor is deployed in random fashion over the region of interest. The sensor fields should be an obstacle free two dimensional (2-D) plane where sensors are scattered over the entire areas.
   b. Sensors remain unaware of their own locations during the deployment and become location aware using a localization system. They are activated to broadcast message using omnidirectional antenna in a synchronous manner and with a periodic interval of time (τ).

Figure 1. Trajectories of the mobile anchor in WSNs

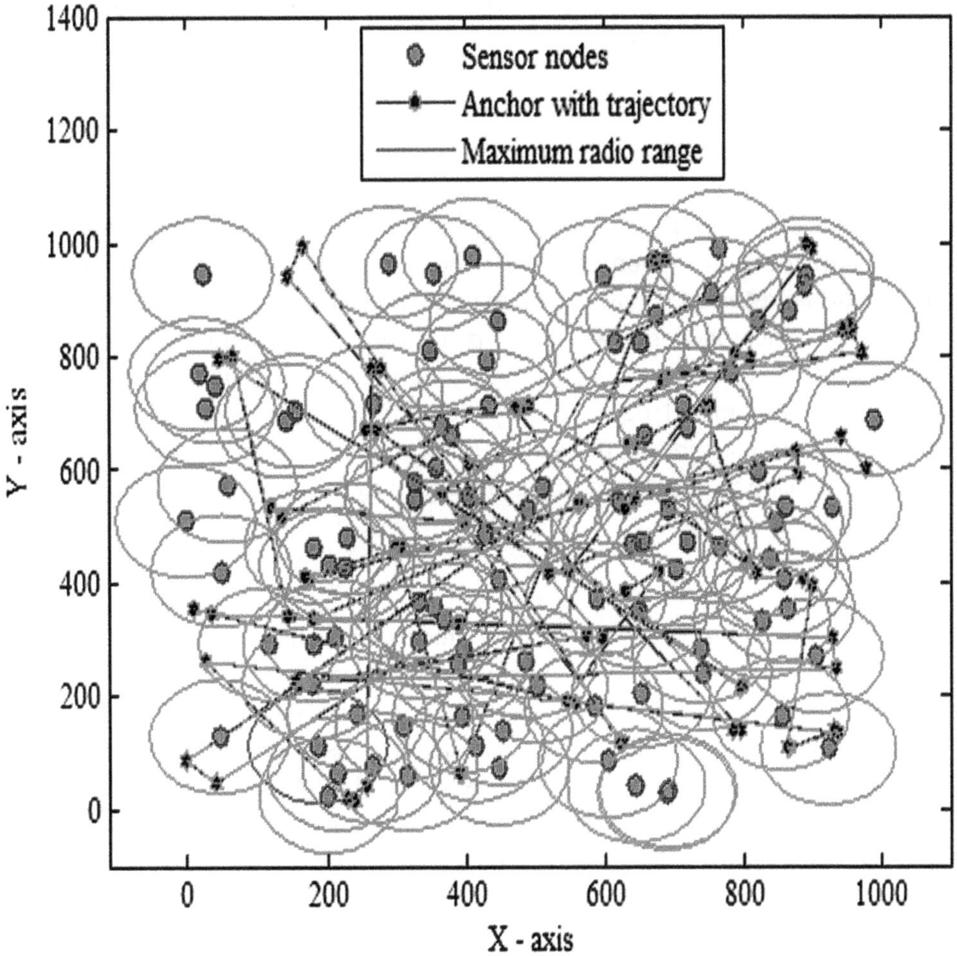

c. Anchors are equipped with both GPS receivers and smart antennas. The mobility features are also incorporated mounting them on wheels/vehicles. They have an extra capacity of memory and refilling facilities to their power source.

d. Towards having better network coverage and connectivity, the quality of services (QoS) in data communications is assumed to be constant within the maximum radio range (d_{max}).

2. Data Communication Capabilities

a. Sensors are homogeneous having similar capabilities in sensing, processing and relaying data packets over the networks. They are also configured with unique IDs to broadcast message.

b. Sensors are restrained by their capacity of energy consumption and memory usage. They are provided with a TPC (transmit power control) mechanism so that power is radiated in proportional to the square of distance during data transmission. Towards energy savings, they are also enabled with a switching system to toggle themselves into sleep mode immediately after their localization.

c. Anchors are capable of estimating the range information while the RSS exceeds the prescribed threshold limit on signal to noise ratio (SNR_{th}) using path loss model and ESPRIT (estimation of signal parameters via rotational invariance technique) algorithm.

d. Base station (BS) is fixed and located at the center of sensor fields so that it can communicate equally to all sensors in WSNs.

Radio Propagation Environments

While sensors transmit signals for relaying data packets over the networks, several propagation features such as reflection, refraction, diffraction and dispersion etc. are commonly exist in radio environments. Thus, mean power of the received signal can be expressed as an exponential decaying function of the distance travelled. This phenomenon is known as path loss and it is used for measuring the distance information in most of the RSS based localization systems (Wang & Xiao, 2007). However, suitable choice for signal propagation model, defining the path loss in a more realistic manner, is also important to develop the energy efficient and secure localization systems (Ramakrishnan & Thyagarajan, 2012). Some assumptions are made in this regard that is described as follows:

1. Path Loss Model

a. Signals flowing through the radio environments are usually composite form of large scale path loss component, medium scale slow varying component and small scale fast varying component. Medium scale slow varying component uses log-normal distributions and small scale fast varying component uses Rician/ Rayleigh distributions in LoS (line of sight)/ NLoS (non-line of sight) data communications respectively (Wang & Xiao, 2007).

b. Path loss is determined using small scale propagation characteristics with fast signal variations due to multipath fading over short distances (order of few wavelengths)/ short duration of time (order of few seconds).

In log-distance distributions (Liberti & Rappaport, 1999), large scale path loss component (L_p) can be defined by a function of distance (d) and usually expressed (in dB) as

$$L_p(d) = L_p(d_0) + 10\alpha \log\left(\frac{d}{d_0}\right) \quad (1)$$

where, α is known as path loss exponent. It indicates the rate of increase in path loss with distance. Also, d_0 denotes a small reference distance from the transmitter.

Now, considering variable cluttering effects of environments, the expression of path loss component is modified and defined with log-normal shadowing model (in dB) as

$$L_{ps}(d) = L_p(d) + X_\sigma \quad (2)$$

where, X_σ is a random variable of zero mean Gaussian distributions with standard deviation σ.

Again, LoS propagation can be described with Frii's free space equation and expressed (in dB) as

$$P_r = P_t + G_t + G_r - 32.45 - 20\log_{10}(d / KM) - 20\log_{10}(f / MHz) \quad (3)$$

where,

$$L_p = 32.45 + 20\log_{10}(d / KM) + 20\log_{10}(f / MHz)$$

is the path loss component. Also, P_t and G_t signify power output and gain of the transmitting antenna. P_r and G_r imply power intercepted and gain of the receiving antenna.

2. Link Budget Analysis

a. Link budget equations are formulated assuming equal energy dissipation per bit in the electronic circuits (ε_{elec}) of wireless sensor transceivers (as shown in Figure 2). Also, extra energy consumption in the power amplifier circuit (ε_{amp}) is considered during transmission of each data bit at unit distance over the free space.
b. Radio propagation model is considered to be compatible with MICAz mote (Crossbow, n.d.). It follows the IEEE 802.15.4 protocols in wireless personal area networks (WPAN). As per medium access control (MAC) protocols, each clustering node remains in the active mode prior to localization process and then switches to sleep mode.

Figure 2. Functional blocks of the Crossbow MICAz mote

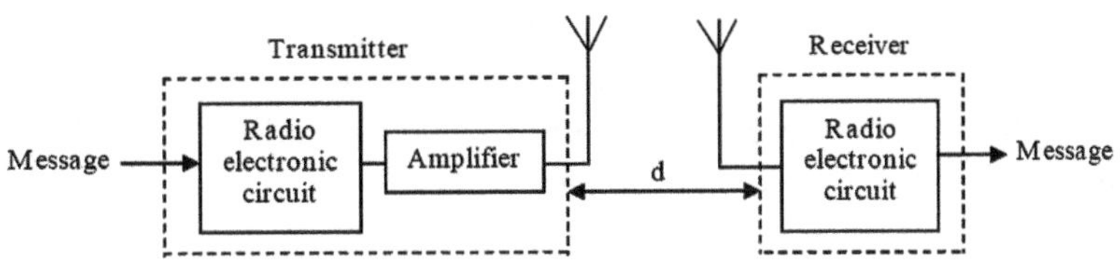

For any clustering node, power dissipation (in Joules per second) is calculated at its two radio links. In forward path: broadcasting signals (E_{TX}) into free space and in reverse link: receiving data packets from the anchor (E_{RX}).Thus, total energy consumption (E_{Total}) during localization of k-th clustering node cab be expressed (in Joules) as

$$E_{Total} = t_k.E_{TX} + 2.E_{RX} \tag{4}$$

where, t_k is the anchor steps required to localize the k-th clustering node.

Now, keeping analogy with the LEACH (low energy adaptive clustering hierarchy) algorithm where energy consumption in any clustering node during transmission/receiving n bit data packet is expressed (in Joules) as (Heinzelman, Chandrakasan, & Balakrishnan, 2002).

$$E_{TX}(n,d) = n.\left(\varepsilon_{elec} + \varepsilon_{amp}.d^2\right) \tag{5}$$

$$E_{RX}(n) = n.\varepsilon_{elec} \tag{6}$$

Low power dissipation model of Crossbow MICAz mote operating at 2.4 GHz ISM (Industrial, Medical and Scientific) band is adopted here. Thus, parametric settings for n = 250 Kbps, ε_{elec} = 50 nJ/bit, ε_{amp} = 10 pJ/bit/m^2 and d_{max} = 90 m are chosen.

Attack Scenarios

Attacks may take place in different form such as internal and external to the WSNs. By setting up passive links to control few benevolent nodes might be a strategic way. Thus, the adversaries could easily get access to their stored data. Otherwise, tampering the radio environments over specific areas, propagation characteristics of some benevolent nodes present there could be deteriorated. Such types of attacks are usually known as internal attacks and obviously, it causes the localization process very tricky.

1. Through Compromised Nodes
   a. Distance estimations could be made erroneous varying the transmission power or delaying the data packets transmission time of compromised nodes towards RSS based or ToA/TDoA based measurement techniques respectively (Chen, Yang, Trappe, & Martin, 2010). To make tampering of angle estimations through AoA based measurement methods, compromised nodes could send signals with reduced SNR. Besides, position computations could be made incorrect corrupting the anchor references.
2. Through compromised environments
   a. By introducing obstacles, smoke or noise etc. to change the physical medium over the WSNs, both RSS based or ToA/TDoA based measurement methods could be made inaccurate. However, AoA based measurement schemes could be compromised deploying magnets over the sensor fields. Also, position computations could be hampered jamming the GPS signals to make erroneous anchor references (Boukerche et al., 2008).

On the other hand, adversaries could install huge advanced electronic devices and circuits to eavesdrop at few nodes and capture MAC IDs from broadcast messages over the networks. Then they might appear with such IDs and communicate with the benevolent nodes. During communications, they could inject garbage/deceived information to these nodes, causing erroneous data routing throughout the entire networks. Such types of attacks are often termed as external attacks. This chapter is focused on to design the localization systems preserving security against the external attacks.

Conventional Localization Systems Issues

Localization systems could play an important role establishing reliable network architectures in location based data communications (Patwari et al., 2005). This can be viewed as regulating the power consumption to ensure longer network lifetime or keeping data integrity/confidentiality from unauthorized nodes during localization process. So, localization algorithms need to be made energy efficient and secured as far as possible.

Energy Efficiency Constraints

Mostly, the source of energy in micro-sensors is battery as mentioned earlier. Thus, it is almost impossible/ inconvenient to recharge their batteries once they are deployed in remote and hostile environments. Sensors are assumed to consume most of their energy in relaying data packets over the networks and negligible amount of energy is needed for sensing and processing raw data from their surroundings. Also, energy consumption of nodes is dependent on the size of relayed packets, required time and distance in data transmission. So, energy efficiency can be enhanced by reducing the localization time.

Security Challenges

Usually, the objectives of any attack could be viewed as to mislead important plans/decisions taken over the accumulated data in WSNs. As localization systems take major role to detect/track the events occurred in various location based services, preserving security on them is crucial. Attacks may occur in direct/indirect ways to malfunction any segment of the localization systems, leading to overall systems breakdown.

ALGORITHM IMPLEMENTATION STRATEGY

Development of energy efficient and secure localization systems is accomplished with an intelligent control on anchor mobility. A two-step anchor movement is considered in localization process. The strategic approach always efforts to keep the anchor on such trajectories that can initiate exploring more number of clustering nodes. Also, it must ensure same clustering nodes in its subsequent steps. Thus, the clustering nodes are able to determine their locations autonomously after receiving range and reference information from two consecutive anchor steps. However, there might be some cases when there is no clustering node present in anchor proximity. Otherwise, anchor could not find out any new clustering node in its next step movement and belongs to same cluster as in its earlier step. This actually happens when the localization process tends to be end and most of the nodes transit themselves into sleep mode after being localized in WSNs. In such cases, anchor is supposed to follow a random trajectory until/unless it discovers any clustering node. As the nodes are assumed to be homogeneous, better energy efficiency is guaranteed through the proposed anchor mobility scheme rather than that of the conventional random walk model. Because it emphasizes on selection of larger cluster size making higher convergence speed in the localization process. Similarly, detecting any misbehavior of the clustering nodes in two successive steps, anchor could be able to identify and block the malicious nodes through an iterative process.

Anchor Mobility Controls

Two anchor mobility controls such as (i) centroid/deterministic method and (ii) fuzzy logic system are proposed here. Basic objective of such schemes is to improve energy efficiency and preserving security in the localization process. In both algorithms, next step of the anchor movement is controlled on the basis of cluster size. Again, such movement must be made including at least all nodes in the existing cluster. Also, anchor relays data packets to each clustering node as per priority index given on them. Priority index is usually set in each step according to the RSS values from the clustering nodes. It normally varies as an inverse function with respect to this parameter. Thus, localization of any clustering node needs only two consecutive anchor steps in the networks. This chapter considers both Sybil and replay type attacks. So, adversaries are supposed to act either altering the transmission power in malicious nodes or repeating the broadcast message of any benign node in their neighborhood. In this way, they could mislead the anchor about their actual locations in WSNs.

Problem Definition

For localization systems to be made as energy efficient, it requires anchor to visit near the clustering nodes very frequently as per priority indices. So, proposed algorithms are based on controlling two factors as: (i) displacement (d_a) and (ii) direction (ϕ) in the next step of anchor movements. These parameters are varied with respect to the distance of clustering nodes (d_k), their priority indices (w_k) and cluster size (cs). For the k-th clustering node, distance (d_k) is to be estimated using received signal power (P_k). Now, applying log-normal shadowing conditions in equation (3), it takes the form as

$$P_k = P_t + G_t - L_p(d_0) - 10\alpha \log\left(\frac{d_k}{d_0}\right) - X_\sigma \quad (7)$$

Priority index is expressed by the following equation as

$$w_k = \frac{d_k}{d_{max}} \quad (8)$$

After estimating these two mobility parameters, the anchor reference (x_{an},y_{an}) for its next step can be determined as

$$\left.\begin{aligned} x_{an} &= x_a + d_a.\cos\varphi \\ y_{an} &= y_a + d_a.\sin\varphi \end{aligned}\right\} \tag{9}$$

where, (x_a,y_a) denotes anchor reference in the present cluster.

Likewise, making localization systems as secured against the attacks, anchor needs to be present among the same clustering nodes to monitor their behaviors in its two consecutive steps. So, finding any discrepancy in received MAC IDs during this course of time is primarily treated as suspicious nodes. In fact, suspicious nodes can be defined as: (i) if any ID appears twice or more in the present cluster (ii) if any ID of present cluster becomes absent in the next step and (iii) if any ID does not appear at all. Thus, suspicious nodes include both benevolent and malicious nodes of the networks. Certainly, they need to be considered again in the localization process and hence a selective tuning is done to filter out the malicious ones in an iterative manner. However, the anchor goes for a random walk while there is no discrepancy found in its two-step movements.

Algorithm Implementation

Towards saving energy of each node, it is important to make the localization process very fast. For this, anchor should move in a trajectory that would explore more number of nodes over the sensor fields. It also evaluates distances and angles of clustering nodes estimating their RSS values using path loss model and ESPRIT algorithm (discussed in the following sections). Then, priority indices are assigned as per the equation (7). Accordingly, it steers itself with a displacement (d_a) and angle (ϕ) that would determine its new reference at next step movement.

1. Centroid/deterministic method

In this method, displacement (d_a) and direction (ϕ) are calculated as

$$d_a = \min\left\{d_{\max} - d_k\right\} \tag{10}$$

where, k = 1,2,…..,cs

$$\varphi = \frac{\sum_{k=1}^{cs} w_k\left(\theta_k + \theta_{ik}\right)}{2\sum_{k=1}^{cs} w_k} \tag{11}$$

where, θ_{ik} is called the image angle of θ_k and they are evaluated as follows.

$$\theta_k = \begin{cases} \theta_k & if \quad \theta_k \geq 0 \\ 2\pi - \left|\theta_k\right| & otherwise \end{cases} \tag{12}$$

$$\theta_{ik} = \begin{cases} \pi + \theta_k & if\, \theta_k \geq 0 \\ \pi - \left|\theta_k\right| & otherwise \end{cases} \tag{13}$$

P_t and P_r are normally represented in dB/dBm. Also, G_t and G_r are represented in dBi. Parametric settings are made as per the Crossbow MICAz motes keeping compatibility with FCC limits. For low power tiny sensors, they are chosen as P_t = 30 dBm, G_t = 10 dBi, G_r = 0 dBi, d_0 = 1 m, SNR_{th} = -40 dBm, X_σ = 0 dB and $\alpha = 2$.

Pseudo code of two-step centroid based anchor mobility controls making secure localization system is shown in Figure 3. Also, generalized version of pseudo code to realize energy efficient localization system is shown in Figure 4.

Fuzzy Logic System

In fuzzy inference system, distance of clustering nodes (d_k) and cluster size (cs) both are chosen as linguistic variables for the crisp inputs. Again, displacement (d_a) is considered as linguistic variable for the crisp output. Both inputs and output are defined with five fuzzy levels as: very small (VS), small (S), medium (M), large (L) and very large (VL). In this method, fuzzy sets small (S), medium (M) and large (L) are represented by triangular membership functions and trapezoidal membership functions are used to represent very small (VS) and very large (VL) fuzzy sets (as shown in Figure 5, Figure 6 and Figure 7). Fuzzy rules are formulated as a collection of if-then statements (Mendel, 1995). For example, if the cs is VL and the d_k is VL then the d_a is VS. Thus, a total of 5×5 = 25 (twenty five) fuzzy rules are exist (as given in Table 1). For defuzzification, the centroid is estimated over a number of sample points. The formula for the aggregation of output membership function is

$$COG = \frac{\sum \mu(x).x}{\sum \mu(x)} \tag{14}$$

Table 1. Fuzzy rules for evaluating displacement (d_a)

	Fuzzy Sets	Node Distance (d_k)				
		VS	**S**	**M**	**L**	**VL**
Cluster Size (cs)	VS	VL	L	M	S	VS
	S	L	M	M	S	VS
	M	M	S	S	VS	VS
	L	S	S	VS	VS	VS
	VL	S	S	VS	VS	VS

where, μ(x) is the degree of membership function at fuzzy point x.

The control surface (as shown in Figure 8) obtained through simulation in MATLAB fuzzy logic toolbox (MathWorks, 2019), provides the unified decision of proposed fuzzy controller 'anchor.fis'. The displacement (d_a) parameter of anchor next-step movement is evaluated with the help of this control surface. To achieve this, MATLAB command can be written as

```
a = readfis('anchor.fis');

da = evalfis([cs dk], a);
```

The direction (ϕ) factor in anchor mobility can also be determined as in the earlier case.

Node Localization Process

Localization systems involve anchor references in position computation of sensor nodes. Proposed algorithm is based on the triangulation principle (Gezici, 2008). It usually requires at least two number of angle information in estimation of node location. Anchor is supposed to estimate such angle data of all received signals from the clustering nodes using smart antennas. It also sends message containing these angle information and its reference to each of them as per their priority indices. Thus, sensor nodes can estimate their individual locations taking messages in two-step anchor movements. The steps in node localization process are discussed as follows.

Estimation of Angle Information

Angle information is calculated using a typical high precision direction of arrival (DoA) technique known as ESPRIT in smart antennas literature (Roy & Kailath, 1989). ESPRIT algorithm exploits the rotational invariance property in the signal subspace. It normally utilizes a translational invariance structure that consists of two identical sub-arrays with a finite separation Δ (also known as doublets). Hence, assuming number of signal sources (D) is less than that of array elements (M), the signals induced on each of the sub-arrays can be written as

$$x_1(t) = A_1.s(t) + n_1(t) \tag{15}$$

and

$$x_2(t) = A_2.s(t) + n_2(t) = A_1.\Phi.s(t) + n_2(t) \tag{16}$$

where,

$$\Phi = diag\left\{e^{j\beta.\Delta\sin\theta_1}, e^{j\beta.\Delta\sin\theta_2}, \ldots\ldots, e^{j\beta.\Delta\sin\theta_D}\right\}$$

Figure 3. Pseudo-code for secure localization in WSNs

1: Initialization of the network parameters
1.1 Set network boundaries : $\{(x_{min}, y_{min}), (x_{max}, y_{max})\}$ and maximum iterations (t_{max}) for localization
1.2 Choose random positions for the stationary nodes (x,y) and the mobile anchor (x_a,y_a) in the networks
1.3 Assign threshold limit in signal-to-noise ratio (SNR_{th}) for an effective estimation of range data
1.4 Configure each stationary node with distinct MAC ID
1.5 Set a mod-2 down counter in each stationary node
1.6 Choose few random MAC IDs as malicious nodes
1.7 Set maximum rounds (τ_{max}) for filtering out the malicious nodes
2: Selection of the clustering nodes
2.1 **while** all the rounds $\tau \le \tau_{max}$ do
2.2 **while** all the iterations $t \le t_{max}$ do
2.3 **for** all received signals k do
2.4 **if** ($RSS_k > SNR_{th}$) then
2.5 Compute distances and angles from current anchor position using equation (7) and equation (21)
2.6 **end if**
2.7 Store all corresponding data along with node IDs into anchor memory
2.7 **end for**
2.8 Record the cluster size (cs)
3: Blocking of the malicious nodes
3.1 **for** all clustering nodes m do
3.2 Check whether any node appears twice or more in the cluster
3.3 Consider such nodes as suspicious and remove their contents from anchor memory
3.4 **end for**
4: Transfer of the range data
4.1 **if** (cs $\ne$ null) then
4.2 Set the priority indices in clustering nodes as per their RSS values
4.2 **for** all clustering nodes m do
4.3 Relay the range (distance/angle) data along with current anchor position to each clustering node
4.4 Decrement the counter
4.5 **end for**
5: Control on the anchor mobility
5.1 Determine displacement (d_a) for anchor movement using equation (10)
5.2 Estimate direction (ϕ) for anchor movement using equation (11)
5.3 Evaluate new anchor position using equation (9)
5.4 **else**
5.5 Choose a random value for new anchor position
5.6 **end if**
6. Estimation of the node locations
6.1 **for** all clustering nodes m do
6.2 **if** (counter = reset) then
6.3 Evaluate locations of the clustering nodes using equation (24)
7: Scheduling the mode of operation
7.1 Switch the clustering nodes to sleep mode
7.2 **end if**
7.3 **end for**
8: Conditions for the process termination
8.1 **if** (active nodes = null) then
8.2 Stop the localization process
8.3 **end if**
8.4 **end while**
8.5 **if** (suspicious nodes = null) then
8.6 Stop the filtration process
8.7 **end if**
8.8 **end while**

Figure 4. Pseudo-code for energy efficient localization in WSNs

1: Initialization of the network parameters
1.1 Set network boundaries : $\{(x_{min}, y_{min}), (x_{max}, y_{max})\}$ and maximum iterations (t_{max}) for localization
1.2 Choose random positions for the stationary nodes (x,y) and the mobile anchor (x_a,y_a) in the networks
1.3 Assign threshold limit in signal-to-noise ratio (SNR_{th}) for an effective estimation of range data
1.4 Configure each stationary node with distinct MAC ID
1.5 Set a mod-2 down counter in each stationary node
2: Selection of the clustering nodes
2.1 **while** all the iterations $t \leq t_{max}$ **do**
2.2 **for** all received signals k **do**
2.3 **if** ($RSS_k > SNR_{th}$) **then**
2.4 Compute distances and angles from current anchor position using equation (7) and equation (21)
2.5 **end if**
2.6 Store all corresponding data along with node IDs into anchor memory
2.7 **end for**
2.8 Record the cluster size (cs)
3: Transfer of the range data
3.1 **if** (cs ≠ null) **then**
3.2 Set the priority indices in clustering nodes as per their RSS values
3.2 **for** all clustering nodes m **do**
3.3 Relay the range (distance/angle) data along with current anchor position to each clustering node
3.4 Decrement the counter
3.5 **end for**
4: Control on the anchor mobility
4.1 Determine displacement (d_a) for anchor movement using equation (10)/ MATLAB command: evalfis([cs dk], a)
4.2 Estimate direction (ϕ) for anchor movement using equation (11)
4.3 Evaluate new anchor position using equation (9)
4.4 **else**
4.5 Choose a random value for new anchor position
4.6 **end if**
5. Estimation of the node locations
5.1 **for** all clustering nodes m **do**
5.2 **if** (counter = reset) **then**
5.3 Evaluate locations of the clustering nodes using equation (24)
6: Scheduling the mode of operation
6.1 Switch the clustering nodes to sleep mode
6.2 **end if**
6.3 **end for**
7: Conditions for the process termination
7.1 **if** (active nodes = null) **then**
7.2 Stop the localization process
7.3 **end if**
7.4 **end while**

is a diagonal unitary matrix with progressive phase shifts β. Also, A_1 and A_2 is called Vandermonde matrix of steering vectors for two sub-arrays.

Now, considering the contributions of both sub-arrays, total received signal can be expressed as

$$x(t) = \begin{bmatrix} A_1 \\ A_1 . \Phi \end{bmatrix} . s(t) + \begin{bmatrix} n_1(t) \\ n_2(t) \end{bmatrix} \tag{17}$$

Figure 5. Membership function of cluster size (cs)

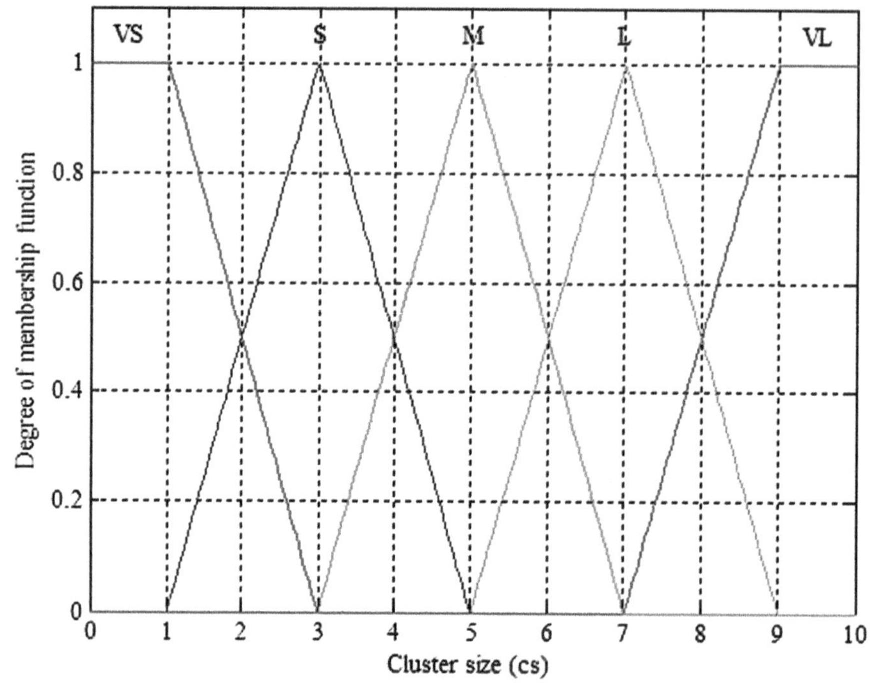

Figure 6. Membership function of node distance (d_k)

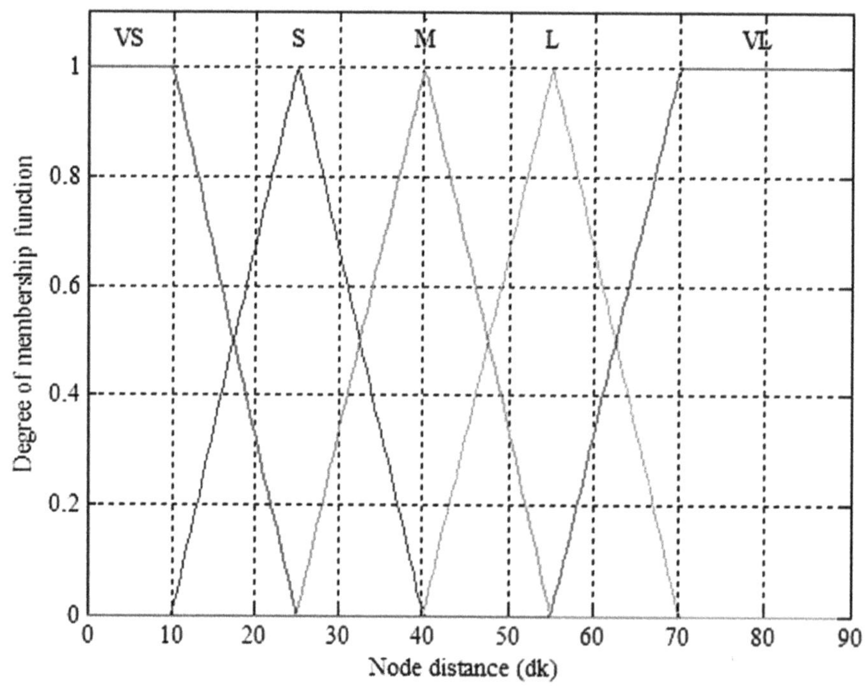

Figure 7. Membership function of displacement (d_a)

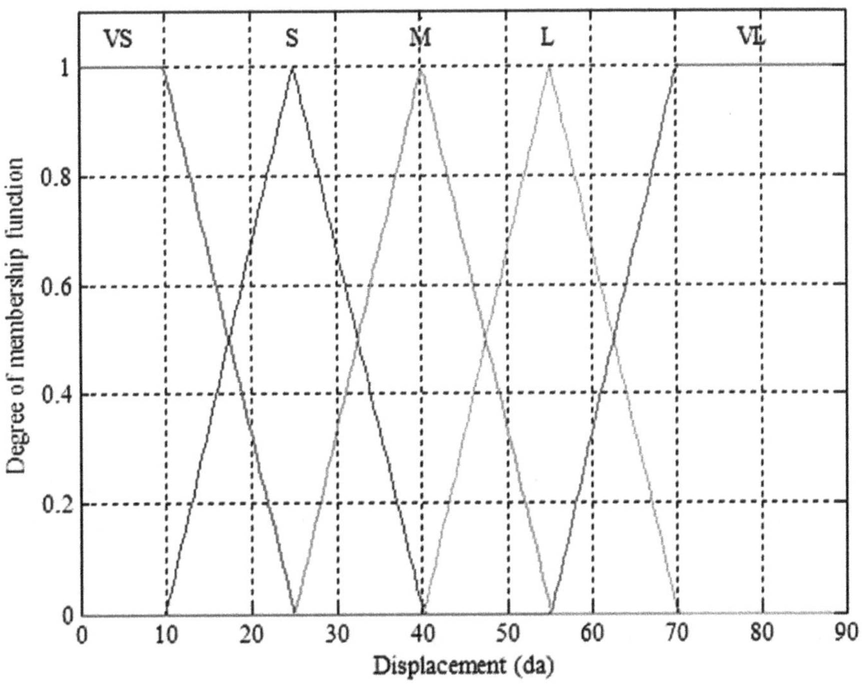

Figure 8. Control surface of fuzzy inference system

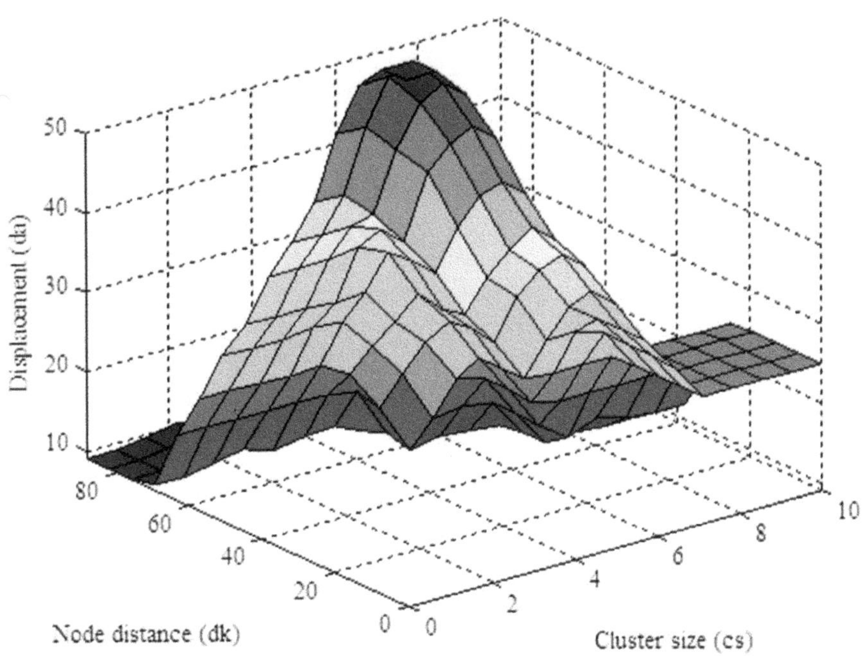

Thus, correlation matrix for the complete array is normally expressed as

$$R_{xx} = E[x.x^H] = A_1 R_{ss} A_1^H + \sigma_n^2 I \quad (18)$$

Again, correlation matrices for two sub-arrays are usually represented as

$$R_{11} = E[x_1.x_1^H] = A_1 R_{ss} A_1^H + \sigma_n^2 I \quad (19)$$

and

$$R_{22} = E[x_2.x_2^H] = A_1 R_{ss} A_1^H + \sigma_n^2 I \quad (20)$$

Due to the invariance array structure, signal subspace (E_x) can be decomposed into two subspaces: E_1 and E_2 whose columns include the D eigenvectors corresponding to the largest eigenvalues of R_{11} and R_{22}. Since these arrays are related with translational invariance features, E_1 and E_2 would be related by a unique non-singular transformation matrix Ψ such that $E_1 \Psi = E_2$.

Similarly, there must also be a unique non-singular transformation matrix Γ such that $E_1 = A_1 \Gamma$ and $E_2 = A_1 \Phi \Gamma$.

From the above relationships, it can be derived that $\Gamma \Psi \Gamma^{-1} = \Phi$.

Thus, the eigenvalues of Ψ must be equal to the diagonal elements of Φ such that

$$\lambda_1 = e^{j\beta.\Delta \sin\theta_1}, \lambda_2 = e^{j\beta.\Delta \sin\theta_2}, \ldots\ldots.., \lambda_D = e^{j\beta.\Delta \sin\theta_D}$$

Also, the columns of Γ must be the eigenvectors of Ψ.

Now, the angle information of signals arriving at k-th clustering node (θ_k) can be estimated as

$$\theta_k = \sin^{-1}\left(\frac{\arg\left(\lambda_k\right)}{\beta.\Delta}\right) \quad (21)$$

where, k = 1,2,….,D

Computation of Node Locations

In WSNs, anchor is aware of its own location through GPS receiver. As discussed earlier, position computation requires both angle and reference data from two consecutive anchor steps. Thus, equations for two straight lines are formulated using such data in triangulation method. For k-th clustering node, these are expressed as

$$\left.\begin{aligned} y_k - y_{1a} &= \tan\theta_{1k}.\left(x_k - x_{1a}\right) \\ y_k - y_{2a} &= \tan\theta_{2k}.\left(x_k - x_{2a}\right) \end{aligned}\right\} \tag{22}$$

where, θ_{nk} and (x_{na},y_{na}) denote corresponding estimated angles and anchor position in the n-th steps.

Now, the above equation can also be written, in matrix form, as

$$AX = B \tag{23}$$

where,

$$A = \begin{bmatrix} -\tan\theta_{1k} & 1 \\ -\tan\theta_{2k} & 1 \end{bmatrix},\ X = \begin{bmatrix} x_k \\ y_k \end{bmatrix} \text{ and } B = \begin{bmatrix} y_{1a} - \tan\theta_{1k}.x_{1a} \\ y_{2a} - \tan\theta_{2k}.x_{2a} \end{bmatrix}$$

So, location of k-th clustering node can be computed as

$$X = A^{-1}B \tag{24}$$

PERFORMANCE EVALUATION

As discussed earlier, proposed two-step anchor mobility control is based on centroid/deterministic and fuzzy logic. These methods aim to provide higher convergence speed filtering out the malicious nodes in localization process. Thus, the system would become robust enough maintaining energy efficiency and security. Their performances are studied under several benchmarks in real-time node deployment scenarios. This section also deals with an analytical discussion on various simulation results.

Performance Metrics

Performance of the proposed localization algorithm normally depends on how different anchor mobility control procedures work in WSNs. Hence, the metrics such as localization error, energy consumption and localization time etc. are considered for verification its competency. Also, the results so obtained are compared with that of a conventional random walk model.

(i) Localization error (δ): It is the amount of deviation in estimated location (x_e,y_e) from the actual one (x,y) for any sensor node in the network. For the k-th node, localization error (δ_k) is expressed as

$$\delta_k = \sqrt{\left(x_k - x_{ek}\right)^2 + \left(y_k - y_{ek}\right)^2} \tag{25}$$

Also, average localization error (δ_{av}) in the network can be represented as

$$\delta_{av} = \frac{1}{N}\sum_{k=1}^{N}\delta_k \tag{26}$$

where, N is the total number of stationary nodes in WSNs.

(ii) Energy consumption (ξ): It is the amount of energy consumed by any sensor node for transmission and reception of data packets during localization process in the network. For the k-th node, energy consumption (ξ_k) is expressed as

$$\xi_k = E_{Total} \tag{27}$$

Also, average energy consumption (ξ_{av}) in the network can be written as

$$\xi_{av} = \frac{1}{N}\sum_{k=1}^{N}\xi_k \tag{28}$$

(iii) Localization time (τ_c): It is the time required to complete localization process of all sensor nodes in the network. So, it can be measured with the maximum number of anchor steps needed in localization and expressed as

$$\tau_c = \max\left\{t_1, \ldots, t_N\right\} \tag{29}$$

Simulation Environments

Simulation environments are implemented on MATLAB software (version 7) platform. Several off-line PC (personal computer) generated data, making analogy with the real-time WSNs scenarios, are assumed here. A random deployment of 100 stationary sensor nodes over a two dimensional (2-D) field of 1000m×1000m area is considered as network architectures. Sensors are supposed to be configured maintaining a reasonable space for increasing network coverage and avoiding interferences among them. Anchors are also considered to move along the prescribed trajectories over the sensor field and relay necessary data packets via private links as per the priority indices of clustering nodes. Adversaries are assumed to act either capturing few nodes making them compromised or participating directly in data communications over the networks. As radio propagation over free space, log-normal shadowing model is assumed here.

Simulation Parameters

Simulation parameters are set in keeping similarity with common wireless narrowband transmission systems such as WPAN IEEE 802.15.4 (Benevent, n.d.). These are given as in the Table 2.

Table 2. Simulation parameters

Parameters	Values
Network size	1000m x 1000 m
Number of nodes	100
Number of malicious nodes	5,10,15
Energy consumption in electronic circuit (ε_{elec})	50 nJ/bit
Maximum radio range (d_{max})	90 m
Energy consumption in amplifier circuit (ε_{amp})	10 pJ/bit/m^2
Threshold level in SNR (SNR_{th})	−40 dBm
Path loss exponent (α)	2
Shadowing noise variance (σ^2)	1

Simulation Results

Proposed localization algorithm is tested under extensive simulations on its performance parameters as mentioned above. Such parameters are typically dependent on the anchor movements in WSNs. However, empirical cumulative distribution functions (ECDF) of the results are made, by taking 30 runs of the program, towards proper explanations in most cases.

1. Localization error, energy consumption and localization time in each node

Estimation of localization error, energy consumption and localization time etc. are made in each node, assuming several anchor mobility models over WSNs. They are illustrated as in Figure 9, Figure 10 and Figure 11 respectively. It shows that the least error can be achieved with random model, whereas centroid method seems to have more error than others. On the other hand, centroid method appears to be more effective in terms of energy consumption and localization time over its counterparts. However, random model is found to be consistent in both cases. Fuzzy scheme is apparently more likely to a random model.

Considering few malicious nodes in the networks, simulation of these parameters is also made and their results are shown as in Figure 12, Figure 13 and Figure 14 respectively.

2. Average of localization error, energy consumption and localization time in WSNs

Taking an average of localization error, energy consumption and localization time etc. over the networks, empirical cumulative probability distribution functions (ECDF) are defined with 30 runs of the program in each case. They are also shown in Figure 15, Figure 16 and Figure 17 respectively. Again, assuming few malicious nodes in the networks, simulation of these parameters is also made and their results are shown as in Figure 18, Figure 19 and Figure 20 respectively.

3. Convergence speed and cluster size in localization process

Figure 9. Localization error in each sensor node

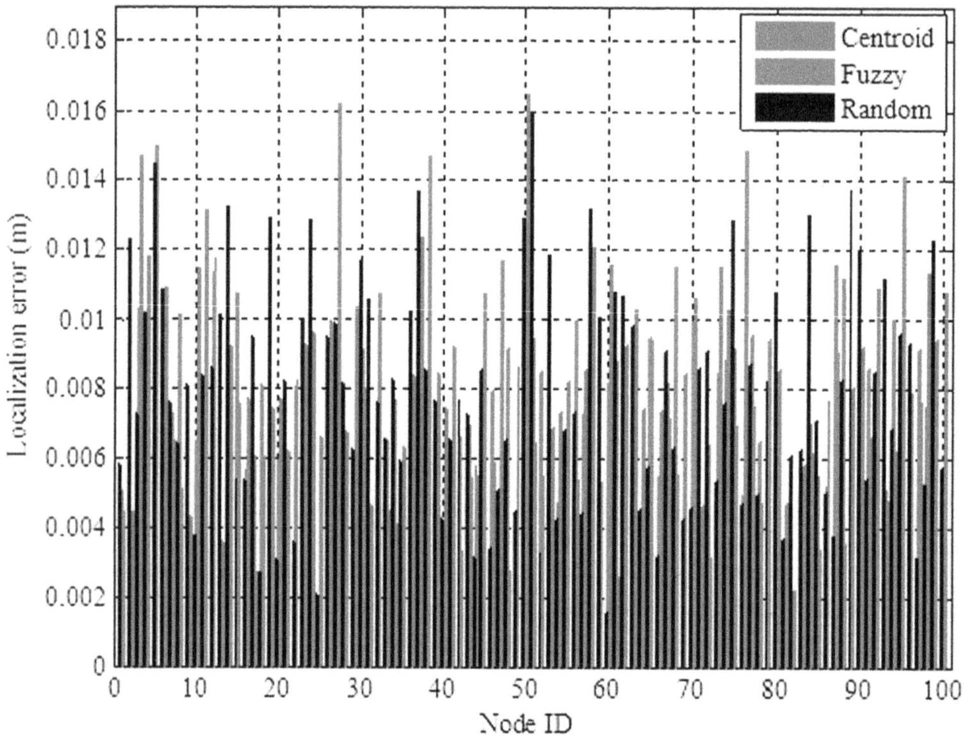

Figure 10. Energy consumption in each sensor node

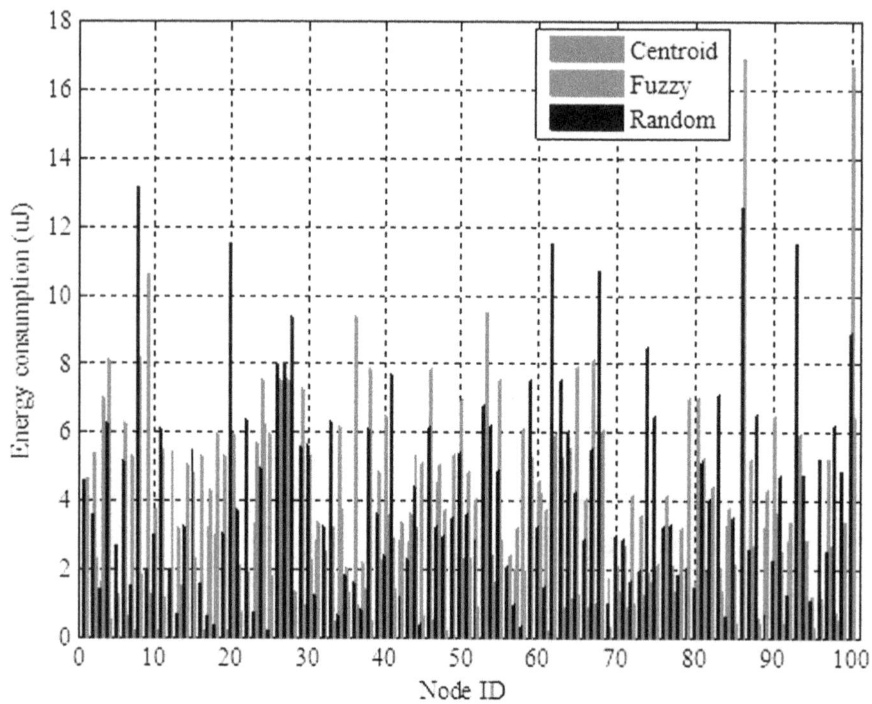

Figure 11. Localization time in each sensor node

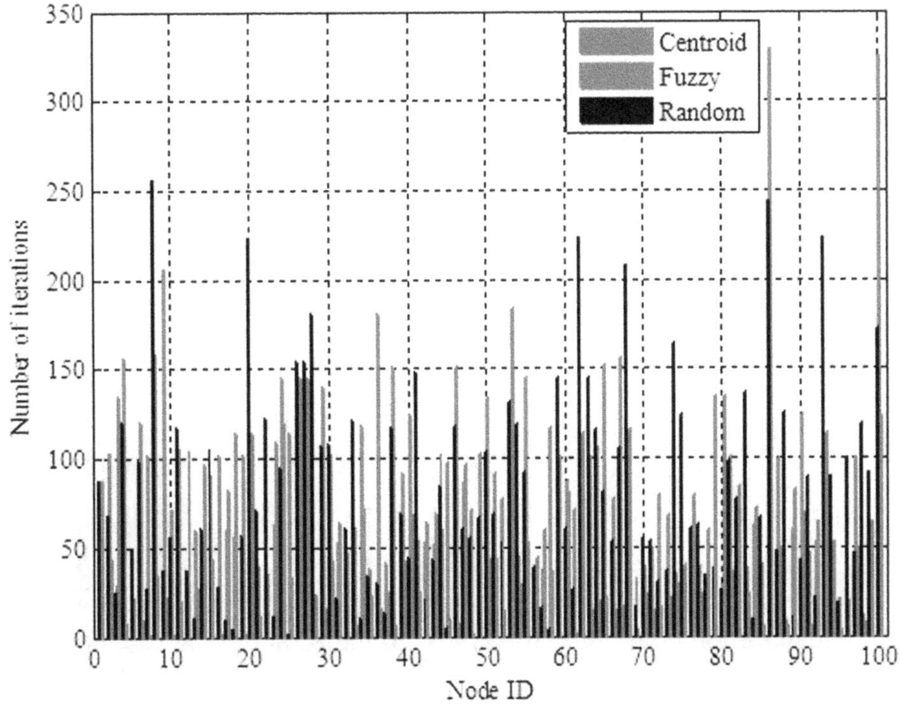

Figure 12. Localization error in each benign node

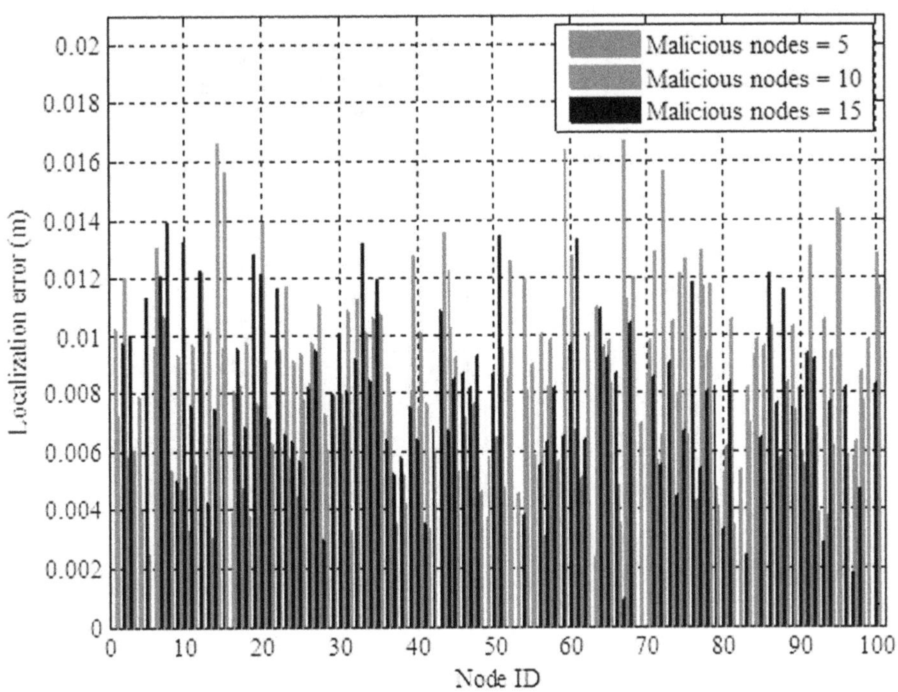

Figure 13. Energy consumption in each benign node

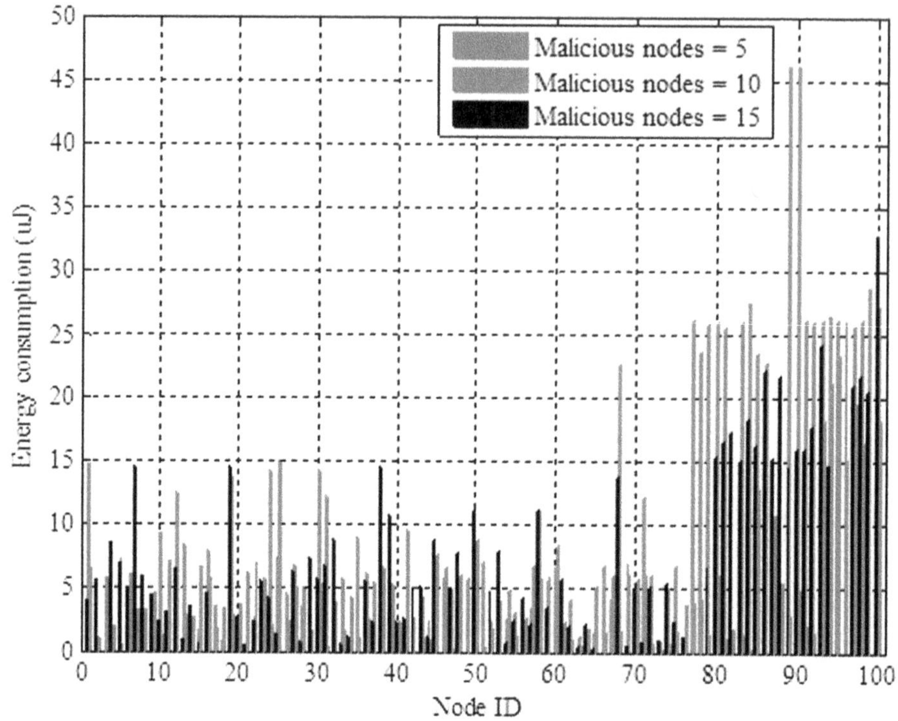

Figure 14. Localization time in each benign node

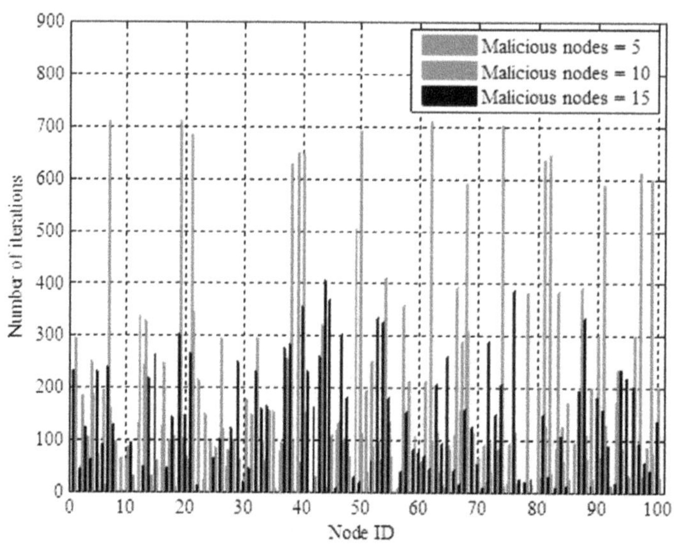

Figure 15. Average localization error in WSNs

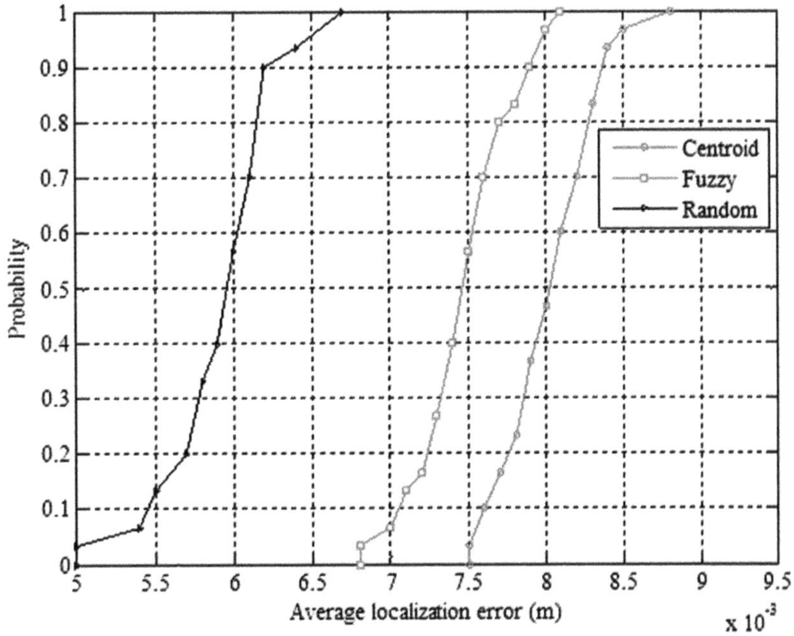

Figure 16. Average energy consumption in WSNs

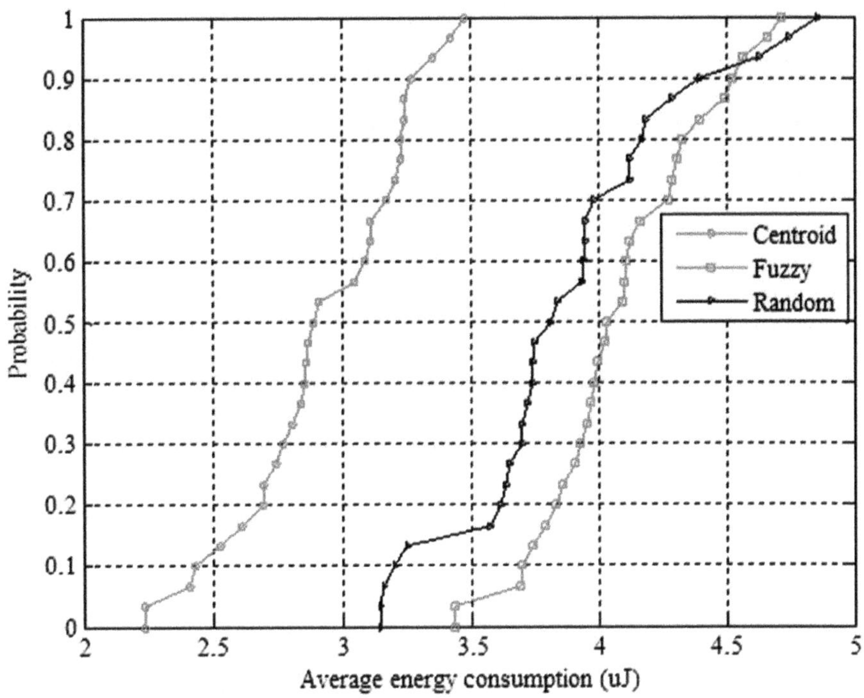

Figure 17. Average localization time in WSNs

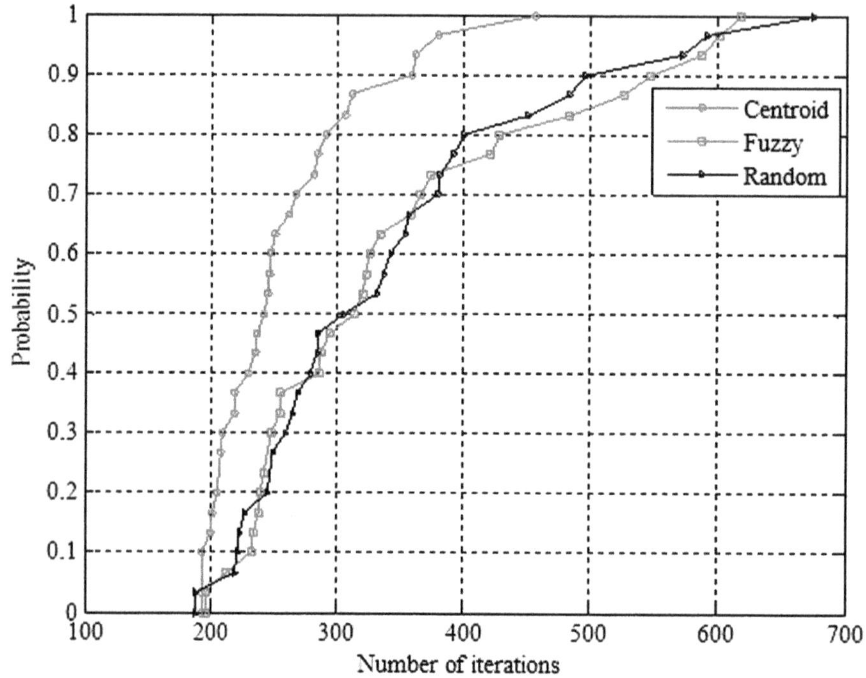

Figure 18. Average localization error under attacks

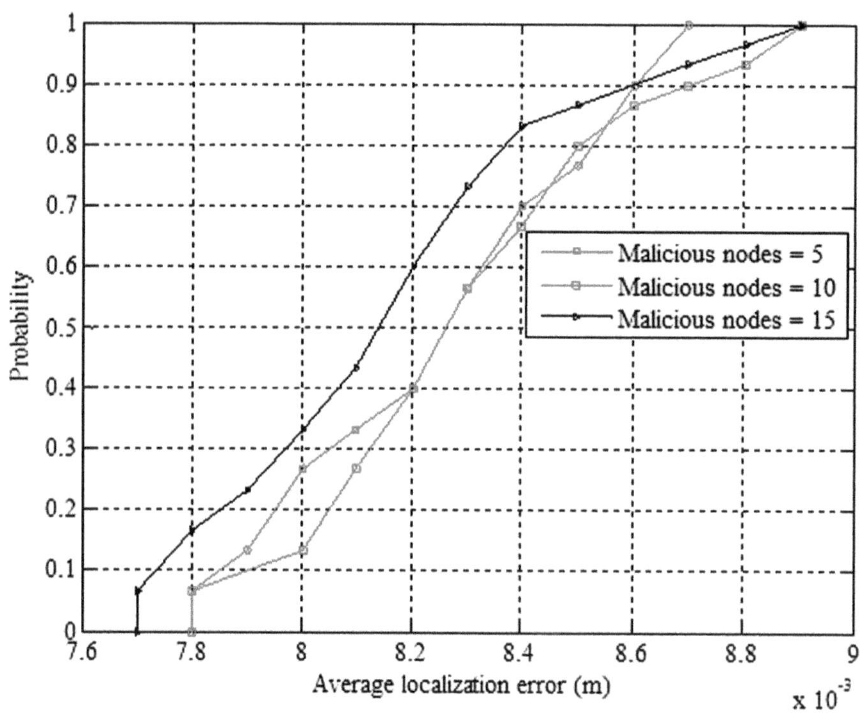

Figure 19. Average energy consumption under attacks

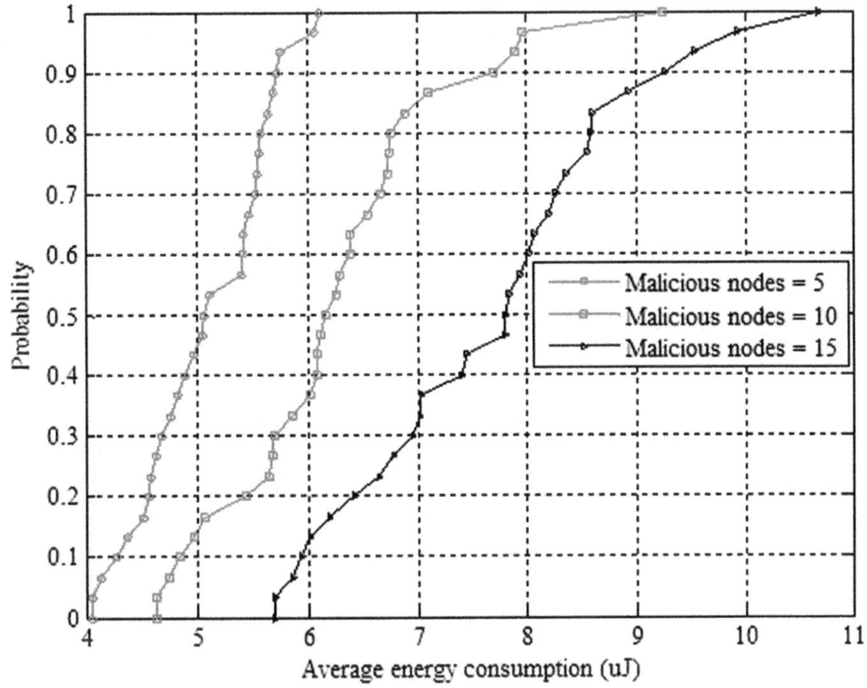

Figure 20. Average localization time under attacks

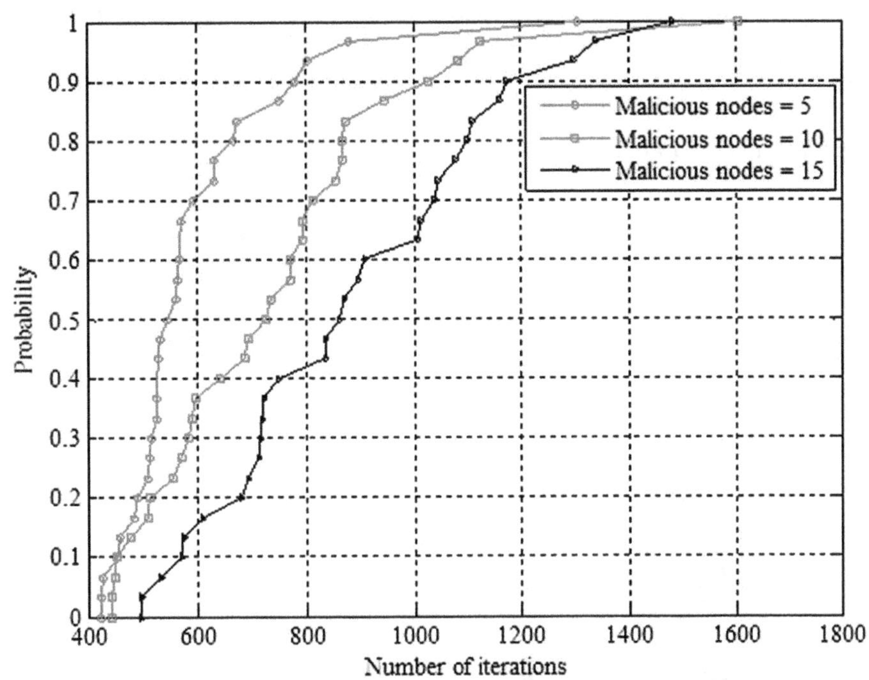

Convergence speed usually indicates number of anchor steps required completing entire localization process. It also depends on the cluster size formed in each anchor step over the networks. They are shown in Figure 21 and Figure 22 respectively. It is obvious that nodes are being localized faster in centroid method acquiring larger cluster size than others. Also, it is evident that both fuzzy and random schemes are comparable to each other in this case.

Figure 21. Convergence speed in localization process

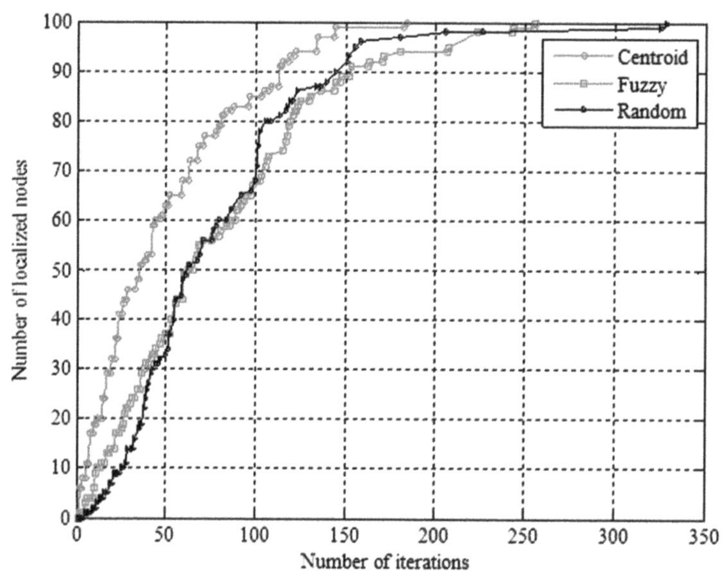

Figure 22. Cluster size formed in each anchor step

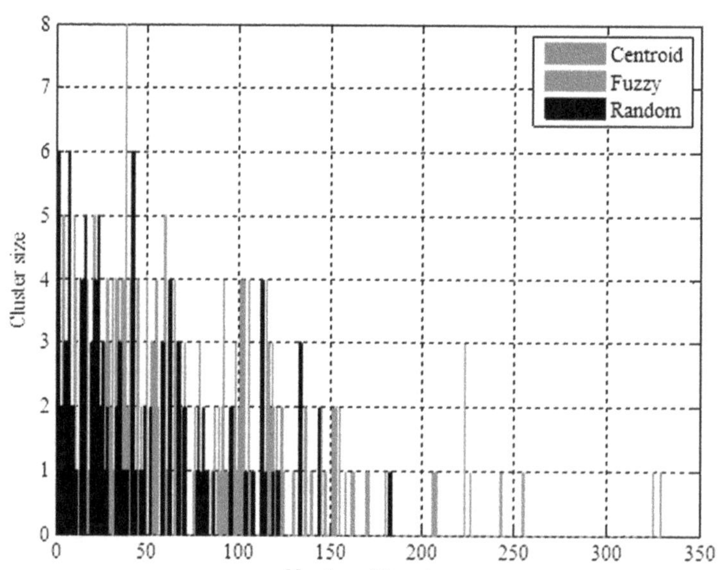

Performance Analysis

The proposed localization scheme is developed to satisfy two key aspects in WSNs environment. These are mentioned as to enhance energy efficiency and preserving security in the networks. Such attributes are examined here through three dissimilar anchor mobility models on a common node deployment platform. The network architectures are based on homogeneous nodes and its protocols utilize a synchronous and periodic broadcasting of signals. Considering transmission of radio waves to cover maximum range, equal energy consumption is always estimated on them. Hence, a control on the anchor movements, keeping it to appropriate trajectories, localization time could be reduced in each case. This methodology, in turn, guarantees least amount of energy consumption over the networks. This is actually made possible enforcing frequent visits of mobile anchor to the regions with maximum number of clustering nodes. Also, switching of each node to sleep mode after its localization would make the system as energy efficient. But, it also introduces some delay in the localization process. In fact, anchor undergoes haphazard movements on several clusters finding zero number of nodes although there may exist some nodes physically. In practice, this occurs while the localization process tends to be end.

On the other hand, mobility strategy might also be helpful to mitigate the effects of malicious attacks in WSNs. Complexity of various attack models, however, must have a great impact on its performance. Attackers are assumed to be present making some compromised nodes in the networks. Such malicious nodes usually broadcast messages with fake IDs or IDs captured from benevolent nodes at their neighborhood. In both cases, they might appear with some clustering nodes although they do not actually belong to them. This would mislead the localization process of those particular clustering nodes. Hence, identifying the misbehaviors of compromised nodes, it is possible to block/ filter out their effects, making the localization process secure as well.

CONCLUSION

Two anchor mobility controls are presented towards design of energy efficient and secure localization systems in WSNs. Amongst them, one is deterministic technique based on the centroid principle and other is of a fuzzy logic method. Performance of the proposed methods is tested and verified with extensive simulations on several metrics. Effectiveness is justified with various results compared to a common random mobility model. Deterministic approach is found to outperform its counterparts yielding both least energy consumption and localization time. But, it seems slightly weaker in measurement of localization error. Also, it is apparent to be very effective against the Sybil and replay type attacks. Fuzzy based system, on the other hand, could also perform more likely to a random mobility model. Implementation of such algorithms is simple enough because it does not impose any requirement of extra hardware and computational overheads on the existing WSN infrastructures.

In this work, only one mobile anchor is considered in the networks. Hence, it would limit speed of convergence in the localization process and violate the criterion of saving maximum energy in the system. So, all mobility models should be studied further with multiple anchors emphasizing the enhancement of energy efficiency during localization. Also, considering complex attack models and higher node density in WSNs, its security aspects must be verified in future.

REFERENCES

Akyildiz, I. F., Su, W., Sankarasubramaniam, Y., & Cayirci, E. (2002). A survey on sensor networks. *IEEE Communications Magazine*, *40*(8), 102–114. doi:10.1109/MCOM.2002.1024422

Benevent, E. (n.d.). *IEEE 802.15.4 -2006 standard.* Retrieved September 13, 2018, from https://www.unirc.it/documentazione/materiale_didattico/599_2009_192_7229.pdf

Biswas, R. N., Mitra, S. K. & Naskar, M. K. (2014). A robust mobile anchor based localization system for wireless sensor networks using smart antenna. *International Journal of Ad-Hoc and Ubiquitous Computing*, *15*(1/2/3), 23-37.

Boukerche, A., Oliveira, H. A. B. F., Nakamura, E. F., & Loureiro, A. A. F. (2007). Localization systems for wireless sensor networks. *IEEE Wireless Communications*, *14*(6), 6–12. doi:10.1109/MWC.2007.4407221

Boukerche, A., Oliveira, H. A. B. F., Nakamura, E. F., & Loureiro, A. A. F. (2008). Secure localization algorithms for wireless sensor networks. *IEEE Communications Magazine*, *46*(4), 96–101. doi:10.1109/MCOM.2008.4481347

Chen, Y., Yang, J., Trappe, W., & Martin, R. P. (2010). Detecting and localizing identity-based attacks in wireless and sensor networks. *IEEE Transactions on Vehicular Technology*, *59*(5), 2418–2434. doi:10.1109/TVT.2010.2044904

Crossbow. (n.d.). *Data sheet of Crossbow MICAz mote*. Retrieved September 13, 2018, from http://edge.rit.edu/edge/P08208/public/Controls_Files/MICaZ-DataSheet.pdf

Dong, L., & Severance, F. L. (2007). Position estimation with moving beacons in wireless sensor networks. In *Proceedings of International Conference on Wireless Communications and Networking* (pp. 2317–2321). Kowloon, Hong Kong: IEEE. 10.1109/WCNC.2007.433

Gezici, S. (2008). A survey on wireless position estimation. *Wireless Personal Communications*, *44*(3), 263–282. doi:10.100711277-007-9375-z

Heinzelman, W. B., Chandrakasan, A. P., & Balakrishnan, H. (2002). An application-specific protocol architecture for wireless microsensor networks. *IEEE Transactions on Wireless Communications*, *1*(4), 660–670. doi:10.1109/TWC.2002.804190

Iguchi-Cartigny, J., Ruiz, P. M., Simplot-Ryl, D., Stojmenovic, I., & Yago, C. M. (2009). Localized minimum-energy broadcasting for wireless multihop networks with directional antennas. *IEEE Transactions on Computers*, *58*(1), 120–131. doi:10.1109/TC.2008.125

Kim, E., & Kim, K. (2010). Distance estimation with weighted least squares for mobile beacon-based localization in wireless sensor networks. *IEEE Signal Processing Letters*, *17*(6), 559–562. doi:10.1109/LSP.2010.2047169

Lee, S., Kim, E., Kim, C., & Kim, K. (2009). Localization with a mobile beacon based on geometric constraints in wireless sensor networks. *IEEE Transactions on Wireless Communications*, *8*(12), 5801–5805. doi:10.1109/TWC.2009.12.090319

Liberti, J. C., & Rappaport, T. S. (1999). *Smart antennas for wireless communications: IS-95 and third generation CDMA applications*. Prentice Hall.

MathWorks. (2019). *Fuzzy logic toolbox*. Retrieved September 13, 2018, from https://www.mathworks.com/help/pdf_doc/fuzzy/fuzzy.pdf

Mendel, J. M. (1995). Fuzzy logic systems for engineering: A tutorial. *Proceedings of the IEEE, 83*(3), 345–377. doi:10.1109/5.364485

Ou, C.-H. (2011). A localization scheme for wireless sensor networks using mobile anchors with directional antennas. *IEEE Sensors Journal, 11*(7), 1607–1616. doi:10.1109/JSEN.2010.2102748

Patwari, N., Ash, J. N., Kyperountas, S., Hero, A. O., Moses, R. L., & Correal, N. S. (2005). Locating the nodes: Cooperative localization in wireless sensor networks. *IEEE Signal Processing Magazine, 22*(4), 54–69. doi:10.1109/MSP.2005.1458287

Pazzi, R. W. N., & Boukerche, A. (2008). Mobile data collector strategy for delay-sensitive applications over wireless sensor network. *Computer Communications, 31*(5), 1028–1039. doi:10.1016/j.comcom.2007.12.024

Rahman, M. Z., & Kleeman, L. (2009). Paired measurement localization: A robust approach for wireless localization. *IEEE Transactions on Mobile Computing, 8*(8), 1087–1102. doi:10.1109/TMC.2008.173

Ramakrishnan, S., & Thyagarajan, T. (2012). Fuzzy logic-based transmission power control algorithm for energy efficient MAC protocol in wireless sensor networks. *International Journal of Communication Networks and Distributed Systems, 9*(3/4), 247–265. doi:10.1504/IJCNDS.2012.048873

Roy, R., & Kailath, T. (1989). ESPRIT-estimation of signal parameters via rotational invariance techniques. *IEEE Transactions on Acoustics, Speech, and Signal Processing, 37*(7), 984–995. doi:10.1109/29.32276

Sichitiu, M. L., & Ramadurai, V. (2004). Localization of wireless sensor networks with a mobile beacon. In *Proceedings of International Conference on Mobile Ad-hoc and Sensor Systems* (pp. 174-183). Fort Lauderdale, FL: IEEE. 10.1109/MAHSS.2004.1392104

Ssu, K.-F., Ou, C.-H., & Jiau, H. C. (2005). Localization with mobile anchor points in wireless sensor networks. *IEEE Transactions on Vehicular Technology, 54*(3), 1187–1197. doi:10.1109/TVT.2005.844642

Vivekanandan, V., & Wong, V. W. S. (2007). Concentric anchor beacon localization algorithm for wireless sensor networks. *IEEE Transactions on Vehicular Technology, 56*(5), 2733–2744. doi:10.1109/TVT.2007.899962

Wang, C., & Xiao, L. (2007). Sensor localization under limited measurement capabilities. *IEEE Network, 21*(3), 16–23. doi:10.1109/MNET.2007.364254

Wang, W., Srinivasan, V., Wang, B., & Chua, K.-C. (2008). Coverage for target localization in wireless sensor networks. *IEEE Transactions on Wireless Communications, 7*(2), 667–676. doi:10.1109/TWC.2008.060611

Wu, X., Cho, J., D'Aurio, B. J., & Lee, S. (2007). Mobility-assisted relocation for self-deployment in wireless sensor networks. *IEICE Transactions on Communications, E90-B*(8), 2056–2069. doi:10.1093/ietcom/e90-b.8.2056

Xiao, B., Chen, H., & Zhou, S. (2008). Distributed localization using a moving beacon in wireless sensor networks. *IEEE Transactions on Parallel and Distributed Systems, 19*(5), 587–600. doi:10.1109/TPDS.2007.70773

Zhang, B., & Yu, F. (2008). An energy efficient localization algorithm for wireless sensor networks using a mobile anchor node. In *Proceedings of International Conference on Information and Automation* (pp. 215-219). Zhangjiajie, China: IEEE.

Zheng, K., Wang, H., Li, H., Xiang, W., Lei, L., Qiao, J., & Shen, X. S. (2017). Energy- efficient localization and tracking of mobile devices in wireless sensor networks. *IEEE Transactions on Vehicular Technology, 66*(3), 2714–2726. doi:10.1109/TVT.2016.2584104

Zhenjie, X., & Changjia, C. (2006). A localization scheme with mobile beacon for wireless sensor networks. In *Proceedings of 6th International Conference on ITS Telecommunications* (pp. 1017-1020). Chengdu, China: IEEE. 10.1109/ITST.2006.288725

Zooghby, A. E. (2005). *Smart antenna engineering*. Boston: Artech House.

Zou, Y., & Chakrabarty, K. (2004). Sensor deployment and target localization in distributed sensor networks. *ACM Transactions on Embedded Computing Systems, 3*(1), 61–91. doi:10.1145/972627.972631

Chapter 10
Power System Voltage Stability

Hemanthakumar Chappa
Maulana Azad National Institute of Technology, India

Tripta Thakur
Maulana Azad National Institute of Technology, India

ABSTRACT

Understanding the voltage instability phenomenon and its effects in detail facilitates the research community to develop methodologies that can detect instability in a timely manner. Traditionally voltage instability in the system is identified through P-V and Q-V curves that are plotted using repetitive runs of load flow programs. It is observed that voltage stability is affected by the load dynamics, voltage control devices like OLTC, and hitting of over excitation limiters of the synchronous generators. In the following sections of this chapter, the concept of voltage instability with P-V and Q-V curves, load restoration mechanism with on load tap changer (OLTC), and with different types of loads are briefly presented.

INTRODUCTION

Increased load demand and deregulation of the power system makes the transmission lines operations close to the stability limits. This enforces increased reactive power demand on the system. Insufficient reactive power in the system leads to voltage problems. A system may be transiently stable but may not be able to maintain the voltage profile, which leads to voltage instability. Voltage instability incidents worldwide are due to insufficient reactive power in the system.

According to Prabha Kundur (1998) Voltage stability is the ability of the system to maintain acceptable voltages at all nodes in the system before and after the occurrence of a disturbance. The maintenance of acceptable voltages depends on the ability of the system to restore stable equilibrium between load demand and load supplied. Voltage instability leads to voltage collapse. Voltage collapse is progressive decline in voltage magnitude at electric power system load buses leading to complete or partial blackout as per Byung Ha Lee (1991). The occurrence of blackouts are rare but the repercussions are very severe. Prominent examples of blackouts due to voltage problems include North East of US in 2003

DOI: 10.4018/978-1-5225-8551-0.ch010

A. Atputharajah (2009), Athens blackout in 2004 Machowski (2008), Brazil blackout in 2009 Ordacgi Filho (2010), Indian grid blackout in July 2012 Vaishali Rampurkar (2016), Turkey blackout in 2015 Project Group Turkey (2015). Various blackout reports in W. R Lachs(1992), C. W Taylor (1997) and other literature Hemanthakumar Chappa (2018) confirms that voltage collapse occurs due to increased load demand or contingency under peak loaded conditions.

Understanding the voltage instability phenomenon and its effects in detail facilitates the research community to develop methodologies that can detect instability in a timely manner. Traditionally voltage instability in the system is identified through P-V and Q-V curves that are plotted using repetitive runs of load flow programs. It is observed that voltage stability is affected by the load dynamics, voltage control devices like OLTC and hitting of over excitation limiters of the synchronous generators.

In the following sections of this chapter the concept of voltage instability with P-V and Q-V curves, load restoration mechanism with on load tap changer (OLTC) and with different types of loads are briefly presented.

CONCEPT OF VOLTAGE INSTABILITY

Consider the two bus system shown in Figure 1, where E is generator terminal voltage and δ is the phase angle between generatorvoltage (E) and load voltage (V).

To compute the voltage instability condition, critical voltages and critical powers are to be calculated. For the Figure 1 the apparent power flow at the receiving end node is given as Carson W. Taylor (1994)

$$S = VI^* \tag{1}$$

$$I = \left[\frac{E\angle\delta - V}{jX}\right] \tag{2}$$

$$S = V\left[\frac{E\angle\delta - V}{jX}\right]^* \tag{3}$$

Figure 1. Simple two bus system

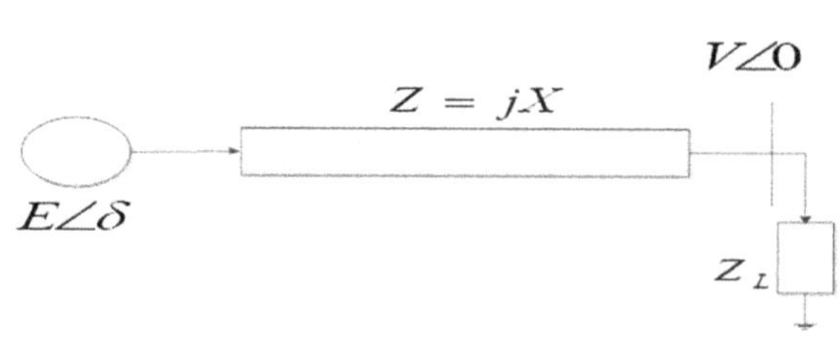

$$S = P + jQ \tag{4}$$

Simplifying the above equations by separating out the real and imaginary terms will result in

$$P = \frac{EV}{X}\sin\delta \tag{5}$$

$$Q = \frac{EV}{X}\cos\delta - \frac{V^2}{X} \tag{6}$$

From (1) and (2), $\sin\delta = \dfrac{PX}{EV}$ and $\cos\delta = \dfrac{QX + V^2}{EV}$. Considering $\cos^2\delta + \sin^2\delta = 1$ and substituting the suitable terms, it then results in

$$\left(\frac{PX}{EV}\right)^2 + \left(\frac{QX + V^2}{EV}\right)^2 = 1$$

$$P^2X^2 + Q^2X^2 + V^4 + 2V^2QX = E^2V^2$$

$$V^4 + V^2(2QX - E^2) + X^2(P^2 + Q^2) = 0 \tag{7}$$

Let

$$v = \frac{V}{E}, p = \frac{PX}{E^2}, q = \frac{QX}{E^2}$$

then (7) can be written as

$$v^4 + v^2(2q - 1) + p^2 + q^2 = 0 \tag{8}$$

From Figure 2 $q = p\,\mathrm{Tan}\,\varphi$. Plug in the value of q in (8) and the solution for voltage can be expressed as

$$v^2 = \frac{-(2p\,\mathrm{Tan}\,\varphi - 1) \pm \sqrt{(2p\,\mathrm{Tan}\,\varphi - 1)^2 - 4p^2\,\mathrm{Sec}^2\,\varphi}}{2} \tag{9}$$

Figure 2. Phasor relationship of p,q,v

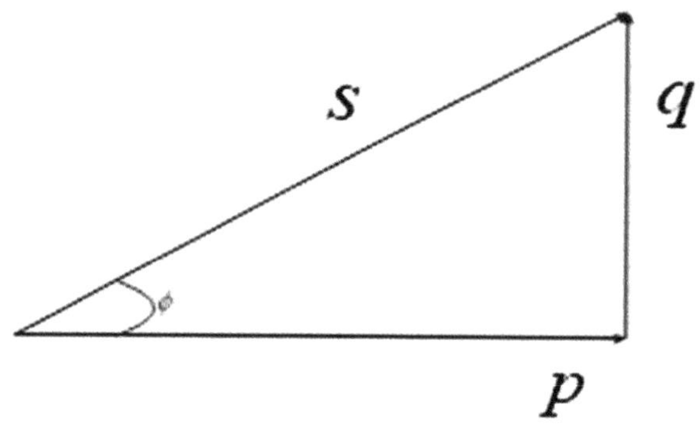

Now v has four solutions, consider only the two meaningful solutions, they are called as high voltage solution and low voltage solution Therry Van Cutsem (1998) and Venkataramana Ajjarapu (2007). The two meaningful solutions will coalesce when the square root term in (9) becomes equal to zero. The operating point at which both the solutions meet is called as bifurcation point. At this point voltage is called critical voltage and the power at which this critical voltage occurs is critical power.

P-Q-V CURVES

Voltage stability surface can be drawn by considering the equation (9) as per Therry Van Cutsem(1998). P-Q-V surface has been plotted by using this equation in matlab software and it is shown in Figure 3. All the coloured lines in the figure show high voltage solutions and all the blue lines shows low voltage solutions at different power factors. The black line is drawn at a power factor angle of -30 degrees and towards right of the black curve each curve is plotted with increasing power factor angle of 10 degrees till power factor angle reaches 80 degrees.

Figure 3. P-Q-V plot

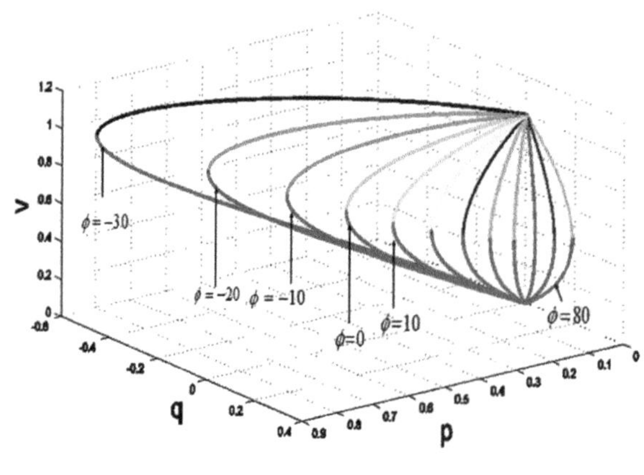

The points at which two solutions become equal are called maximum power points. By utilizing the above plot the meridians can be projected to on P-V plane to obtain the famous P-V curves or the network P-V curves. The plot of the network P-V curve is shown in Figure 4.

From Figure 4 for a unity power factor and for real power equal to zero the two voltage solutions are 1 and 0. The critical voltage and critical power can be obtained by considering the square root term in (9) equal to zero.

$$\left(2p\,\mathrm{Tan}\,\varphi - 1\right)^2 - 4p^2\,\mathrm{Sec}^2\,\varphi = 0$$

Consider p as p_c (critical power)

$$(1 - 2p_c\,\mathrm{Tan}\,\varphi)^2 = (2p_c\,\mathrm{Sec}\,\varphi)^2$$

$$1 - 2p_c\,\mathrm{Tan}\,\varphi = 2p_c\,\mathrm{Sec}\,\varphi$$

Simplifying the above equation for p_c will result in

$$p_c = \frac{\cos\varphi}{2(1 + \sin\varphi)} \qquad (10)$$

Consider now the (9) for critical voltage and replace p with p_c and v with v_c ,then

$$v_c^{\,2} = \frac{-(2p_c\,\mathrm{Tan}\,\varphi - 1)}{2}$$

Plug in the value of p_c in above expression

Figure 4. P-V curves

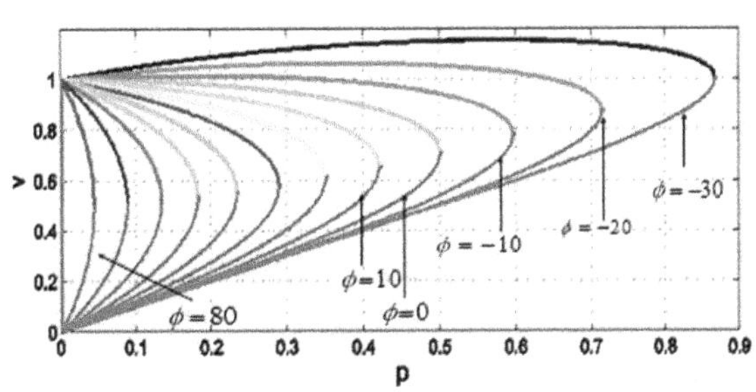

$$v_c^{\ 2} = \frac{1 - 2\left[\dfrac{\cos\varphi}{2(1+\sin\varphi)}\operatorname{Tan}\varphi\right]}{2}$$

$$v_c^{\ 2} = \frac{1 - \dfrac{\sin\varphi}{1+\sin\varphi}}{2}$$

$$v_c = \frac{1}{\sqrt{2(1+\sin\varphi)}} \tag{11}$$

The Phasor angle (δ) between sending end node and receiving end node at critical condition can be obtained by considering (5)

Multiply both sides by $\dfrac{X}{E^2}$ to (5) then

$$p_c = v_c \sin\delta$$

$$\sin\delta = \frac{p_c}{v_c} \tag{12}$$

Using the trigonometric relationship $\sin^2\delta + \cos^2\delta = 1$, then

$$\cos\delta = \sqrt{1 - \sin^2\delta}$$

$$\cos\delta = \sqrt{1 - \left[\frac{p_c}{v_c}\right]^2} \tag{13}$$

Substituting (10) and (11) in (13), then

$$\cos\delta = \sqrt{1 - \frac{\cos^2\varphi}{2(1+\sin\varphi)}}$$

$$\cos\delta = \sqrt{\frac{2 + 2\sin\varphi - \cos^2\varphi}{2(1+\sin\varphi)}}$$

$$\cos\delta = \sqrt{\frac{1+2\sin\varphi+\sin^2\varphi}{2(1+\sin\varphi)}}$$

$$\cos\delta = \sqrt{\frac{(1+\sin\varphi)}{2}}$$

$$\delta = \cos^{-1}\left[\sqrt{\frac{(1+\sin\varphi)}{2}}\right] \tag{14}$$

if power factor angle φ is equal to zero, then the critical power is 0.5 as per (10) and critical voltage is 0.707 as per (11). For different power factor angles, the values of the critical variables are tabulated in table1. From this table it is observed that the angle δ is maintained a secure value for critical power transfer. So voltage instability may be initiated much before the initiation of angle stability Venkataramana Ajjarapu (2007).

The negative power factor angle means load is compensated. For a compensated load the power transfer increases but critical voltages are nearly equal to the nominal voltages. So voltage magnitude is not a reliable indicator for monitoring voltage instability because maintaining an acceptable voltage profile does not guarantee voltage stability and at the same time voltage instability need not be associated with low voltage profiles as per M.K Pal (1992).

Table 1. Critical variables for different power factors

φ	p_c	v_c	q	δ
90	0.0000	0.5000	0.2500	0
80	0.0437	0.5019	0.2481	5.0000
70	0.0882	0.5077	0.2422	10.0000
60	0.1340	0.5176	0.2321	15.0000
50	0.1820	0.5321	0.2169	20.0000
40	0.2332	0.5517	0.1956	25.0000
30	0.2887	0.5774	0.1667	30.0000
20	0.3501	0.6104	0.1274	35.0000
10	0.4195	0.6527	0.0740	40.0000
0	0.5000	0.7071	0	45.0000
-10	0.5959	0.7779	-0.1051	50.0000
-20	0.7141	0.8717	-0.2599	55.0000
-30	0.8660	1.0000	-0.5000	60.0000

QV CURVES

These curves are plotted by considering a constant power in (9) Venkataramana Ajjarapu (2007). The Q-V curves represent the reactive power requirement at a load to maintain desirable voltage at a real power demand. For the curves shown in figure 5 for a real power load of 0.6 pu, the reactive power requirement is 0.083pu to maintain voltage at 1pu.

The negative reactive power means inductive reactive power and positive reactive power means capacitive reactive power.

LOAD RESTORATION WITH ON-LOAD TAP CHANGER (OLTC)

Load restoration depends on the voltage dependent load Therry Van Cutsem (1998). For the load tap changer shown in figure 6 with tap ratio (a:1) connected to restore the voltage and load at the distribution side whereas primary side of the transformer is connected to the transmission side. The distribution side voltage is assumed to be voltage dependent load. The system is initially operating at load (P2 and Q2) as shown if figure 7. If due to disturbance in the system if the post disturbance network P-V curve is shifted from the pre disturbance network P-V curve as shown is figure 7 then OLTC will change its tap setting in the primary side so that secondary voltage in distribution side will be improved. If pre disturbance voltage is V_1^0 on the primary side and V_L^0 is the pre disturbance voltage on the distribution side or load side, then load real and reactive powers can be expressed as

Figure 5. Q-V curves

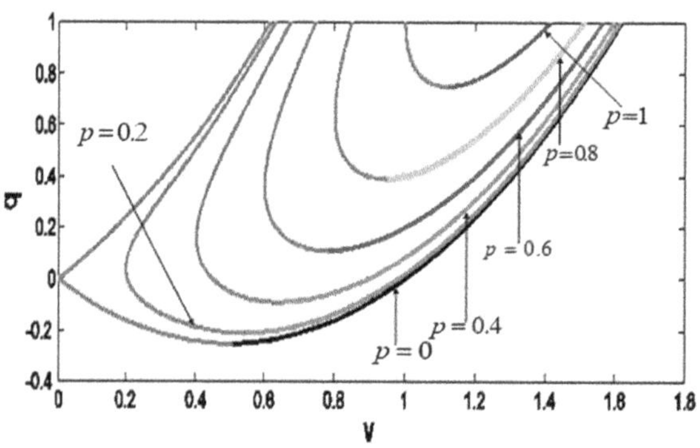

Figure 6. System with OLTC

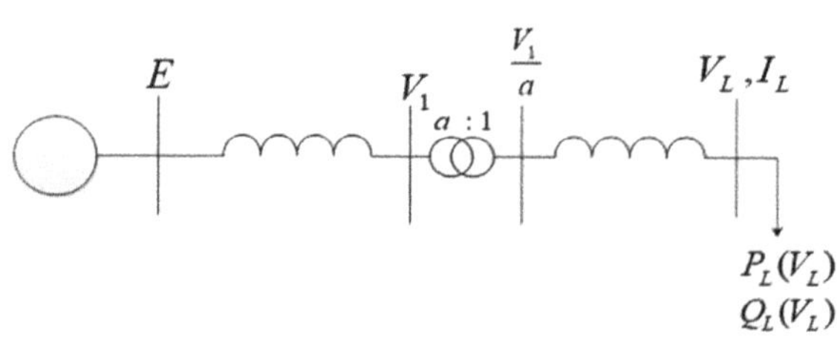

Figure 7. OLTC restoration

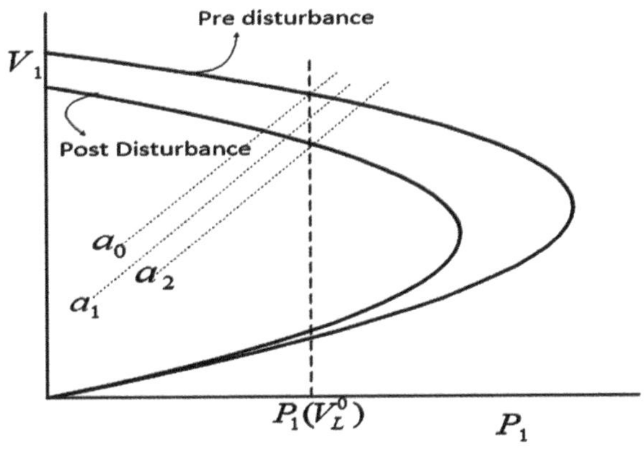

$$P_L = P_L(V_L^0) \text{ and } P_1 = P_1\left(\frac{V_1^0}{a}\right) = P_L(V_L^0) \tag{15}$$

$$Q_L = Q_L(V_L^0) \text{ and } Q_1 = Q_1\left(\frac{V_1^0}{a}\right) \tag{16}$$

If the tap changer initial position is at a_0 then from figure 7 then post disturbance load power is less than the pre disturbance load power and tap will be reduced to increase the load voltage. At tap position a_0, the voltage on the high voltage(HV) side ($V_1^{'}$) is less than pre disturbance voltage on HV side(V_1^0) and the post disturbance voltage on the load side ($V_L^{'}$) is alsoless than the pre disturbance distribution side voltage (V_L^0). The load power is $P_L(V_L^{'})$ less than $P_L(V_L^0)$.

As the voltage on load side is less than V_L^0, the tap setting on HV side reduces and becomes equal to a_1.This causes further reduction inHV side voltage and a little improvement in load side voltage. In this condition the load side voltage ($V_L^{"}$) is greater than $V_L^{'}$ but still less than V_L^0. Load power ($P_L(V_L^{"})$) is greater than $P_L(V_L^{'})$ but still less than $P_L(V_L^0)$. This again causes a change in tap position a_2. In this position HV side voltage further reduces and load side voltage improves and is equal to pre disturbance voltage (V_L^0). Therefore the power at the load bus is restored to pre disturbance load power. At each tap position the voltage on the HV side reduces, so the current on the HV side increases. The restoration of distribution voltage using OLTC actually increases losses in the system. This condition is analogous to inserting a fictitious inductor on HV side and afictitious capacitor on distribution side.

The voltage and load restoration with OLTC is helpful only if the load characteristic is having intersection point with network characteristic at upper portion of the P-V curve Prabha Kundur (1994). If the intersection point is at lower portion of the P-V curve and OLTC is used for restoration of load power and voltage then reverse action of OLTC causes voltage instability. This is the reason that OLTC's are blocked after reaching certain tap position in real time power system operations.

VOLTAGE INSTABILITY MECHANISM WITH DIFFERENT LOADS

Voltage Dependent Load

The voltage dependent real and reactive power at any load bus can be represented as:

$$P_L = zP_0 \left(\frac{V}{V_0} \right)^{\alpha} \tag{17}$$

$$Q_L = zQ_0 \left(\frac{V}{V_0} \right)^{\beta} \tag{18}$$

Where z is the amount of load demand, V_0 is the reference voltage. α and β are voltage impact on real and reactive power respectively. For this load model the network and load PV curves are shown in Figure 8 Therry Van Cutsem (1998).

The initial operating point or pre disturbance operating point at a load demand of LD_1 is a_1. The disturbance in system causes a shift in P-V curve and the operating point moves to $a_1^{!}$, which is the intersection point of network P-V curve and load P-V curve.

If the load demand increases to LD_2 then the operating point shifts to b_1 in pre disturbance period and $b_1^{!}$ in post disturbance period. If load demand further increases to LD_3 then the operating point shifts to c_1 in pre disturbance period and the load P-V curve has no intersection point with the network P-V curve in post disturbance period and therefore system will collapse. For this load demand the load P-V curve has become tangent to the network P-V curve which indicates the stability limit of the system. Any further load increment beyond LD_3 will cause instability even at pre disturbance condition as there will not be any intersection between network and load P-V curves. From this it can be understood that voltage instability is a local problem and heavy load demand alone can initiate the problem and therefore it is also called as load instability.

Figure 8. PV curves with voltage dependent loads

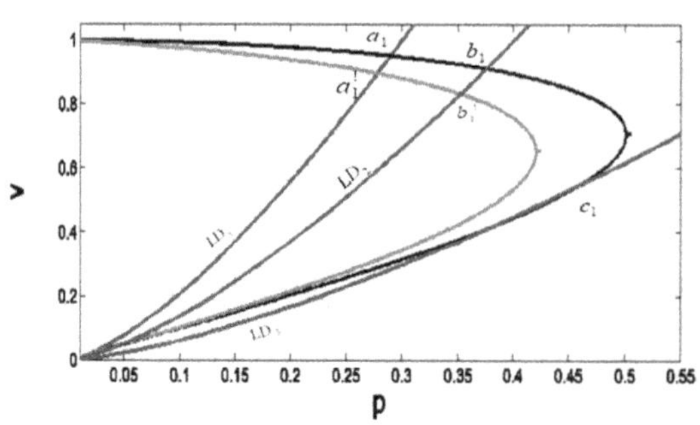

With Constant Power Load

Constant power load or constant MVA load has been considered as the most stringent load from voltage stability point of view Ordacgi Filho (2010). It may because of nature of load or nature of network voltage restoration mechanism that tries to maintain the load voltage. So it is very important to consider the dynamics of the load because constant MVA load is not a static load. If the load demand changes to a new value the load will not change instantaneously to new demand. It changes according to the instantaneous load characteristics. The instantaneous load characteristic may be constant current load or constant impedance load. A change in demand causes instantaneously adjust the load current from system until the load demand satisfies at final system voltage. Same is the case with sudden change in system voltage. However this process is not instantaneous. A time lag is involved in this process. So the dynamic behaviour of the constant MVA load can be modelled by using first order delay model and is represented as

$$T_L \frac{dG}{dt} = P_0 - V_L^2 G \tag{19}$$

Where V_L is load voltage, G is load conductance, P_0 is power set point and T_L is load time constant.

At steady state

$$P_0 = V_L^2 G \tag{20}$$

For the example considered in figure 9 load current(I_L) is represented as

$$I_L = V_L G \tag{21}$$

and

$$E = V_L + I_L jX \tag{22}$$

$$E^2 = V_L^2 + V_L^2 G^2 X^2 \tag{23}$$

Figure 9. Two bus system with constant MVA load

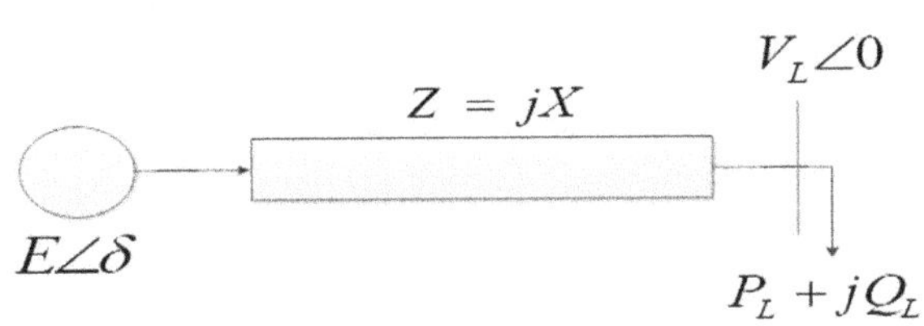

$$V_L^2 = \frac{E^2}{1 + G^2 X^2} \tag{24}$$

Substituting (24) in (20) then

$$P_0 = \frac{E^2}{1 + G^2 X^2} G$$

$$P_0 (1 + G^2 X^2) = E^2 G \tag{25}$$

divide both sides of the above expression by $P_0 X^2$ then

$$G^2 - G \frac{E^2}{X^2 P_0} + \frac{1}{X^2} \tag{26}$$

$$G = \frac{\frac{E^2}{X^2 P_0} \pm \sqrt{\left(\frac{E^2}{X^2 P_0}\right)^2 - \frac{4}{X^2}}}{2} \tag{27}$$

Assume

$$G_H = \frac{\frac{E^2}{X^2 P_0} + \sqrt{\left(\frac{E^2}{X^2 P_0}\right)^2 - \frac{4}{X^2}}}{2}$$

and

$$G_L = \frac{\frac{E^2}{X^2 P_0} - \sqrt{\left(\frac{E^2}{X^2 P_0}\right)^2 - \frac{4}{X^2}}}{2}$$

The G_H and G_L correspond to low voltage and high voltage solution respectivelyin the P-V curve. This can be verified by considering (8) at upf.

From (19)

Figure 10. PV curves with constant MVA load

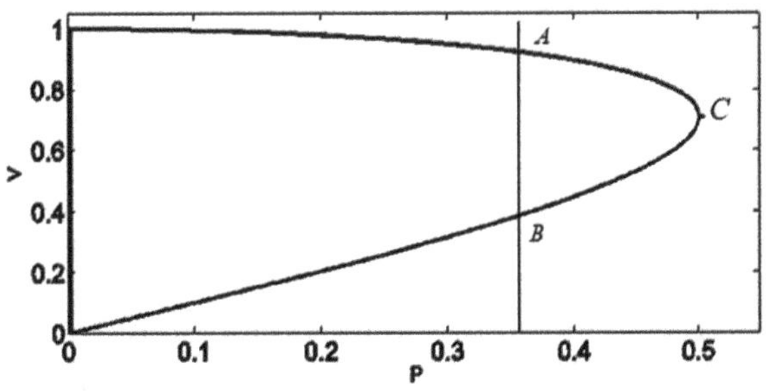

$$T_L \frac{dG}{dt} = \frac{P_0(1+G^2X^2) - E^2 G}{(1+G^2X^2)} \tag{28}$$

$$T_L \frac{dG}{dt} = \frac{P_0 X^2 \left[(G-G_H)(G-G_L)\right]}{(1+G^2X^2)} \tag{29}$$

If the initial operating condition is on the left side of point A on p-V curve of figure 10then $G < G_L < G_H$. As per (29), $\frac{dG}{dt}$ is positive means G is increasing and therefore the operating point moves towards A and settle at A. If the initial operating point is anywhere between ACB then $G_L < G < G_H$ and $\frac{dG}{dt}$ is negative, means G is reducing and final operating point will again be A. This is because at the operating point anywhere between ACB the power delivered is more than the set power. This causes the load restoration mechanism reduces the current drawn from the system. This leads to improvement in voltage and rate of change in voltage is dominant in this region up to point C and power drawn will increase. But from point C onwards the power reduces till the operating point settles at A. Similarly if the initial operating point is towards the left of point B then $G_L < G_H < G$ and $\frac{dG}{dt}$ is positive, so G increases and operating point moves away from B. The upper portion of the P-V curve and the region to the right of point B is the region of attraction of A which is stable equilibrium point.

The load and network P-V curves for constant power load are shown in Figure 11. If load demand increases then the load P-V curves will shift to the right (vertical lines in figure 11). If the load demand increases beyond the stability limit then the equilibrium point suddenly disappears and voltage instability takes place. On the other hand if demand is not increasing but a disturbance in system can cause the network P-V curve to shrink and to this post disturbance P-V curve may not have an intersection with load P-V curve. This condition again causes voltage instability. Usually network voltage recovery takes place faster than the load recovery. In the figure shown above if the load level is P_1 then the pre distur-

Figure 11. P-V curves with instantaneous load characteristics and with different load demands

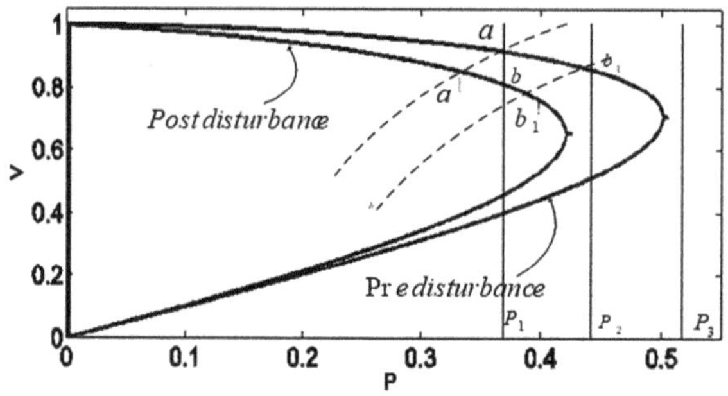

bance operating point is a which is the intersection of load P-V curve (vertical line) and network P-V curve. The transient load characteristic is shown with dotted lines. Immediately after the disturbance load exhibits either constant impedance or constant current load characteristic (dotted lines) the operating point will become $a^|$ and this point is within the region of attraction of final operating point b on post disturbance P-V curve. If the system is operating at an initial load demand of P_2 then the operating point is b_1 on pre disturbance P-V curve. However the load P-V curve does not have any intersection with post disturbance network P-V curve and therefore voltage instability takes place. If the load demand further increased and becomes P_3, this curve does not have any intersection with any of the curves and voltage instability occurs. A heavy demand on the system alone can cause voltage instability or disturbance on the system such that post disturbance P-V curve does not intersect with load P-V curve can cause voltage instability.

With Both Static and Dynamic Load

The dynamics of the load and the dynamics of the system voltage control devices(Excitation control) has appreciable effects on stability. The total load at upf with constant power and static (resistive) load cab be represented as M.K Pal (1992)

$$P = V_L^2(G + G_L) \tag{30}$$

where G is the conductance of the constant MVA load and its dynamics are given in (19)and G_L is the conductance of the static load.

The real and reactive power flow from sending end to receiving end are given as

$$V_L(G + G_L)X = E\sin\delta \tag{31}$$

$$V_L = E\cos\delta \tag{32}$$

Linearizing (19), (31) and (32) for small change in δ, G and V_L, neglecting the smaller terms then (19) becomes

$$T_L \frac{d\Delta G}{dt} = -V_L^2 \left[1 + \frac{2G\Delta V_L}{\Delta G V_L}\right] \Delta G \tag{33}$$

(31) becomes

$$V_L \Delta G + \Delta V_L (G + G_L) X = E \cos\delta \Delta\delta \tag{34}$$

(32) becomes

$$\Delta V_L = -V_L \operatorname{Tan}\delta \Delta\delta \tag{35}$$

With suitable mathematical manipulations of (34) and (35) and substituting for ΔG in (33) yields

$$T_L \frac{d\Delta G}{dt} = -V_L^2 \left[1 - \frac{2GX \operatorname{Tan}\delta}{1 + (G + G_L) X \operatorname{Tan}\delta}\right] \Delta G \tag{36}$$

obtain expression for $(G + G_L)X$ using (31) and (32) and substitute in (36), this results in

$$T_L \frac{d\Delta G}{dt} = -V_L^2 \left(\frac{G \cos 2\delta + G_L}{G + G_L}\right) \Delta G \tag{37}$$

The above expression is stable if $G \cos 2\delta + G_L \geq 0$ or

$$\begin{aligned} \cos 2\delta &\geq \frac{-G_L}{G} \\ &\geq \frac{-G_L V^2}{G V^2} \\ &\geq \frac{-G_L V^2}{P_0} \end{aligned} \tag{38}$$

The same condition for stability can be verified by considering the power flow Jacobian matrix. If the determinant of the power flow Jacobian made equal to zero and with suitable mathematical manipulations the same condition in (38) will emerge. So for detecting voltage instability for large systems Jacobian matrix singularity can be used as an indicator for voltage instability. Many voltage instability detection methodologies are based on this condition.

EFFECT OF SYNCHRONOUS GENERATOR EXCITATION SYSTEM

Synchronous generators are the conventional power sources in power system. Synchronous generator is normally represented with a voltage source in series with synchronous reactance as shown in figure 12. Modern alternators have a synchronous reactance of 2 pu.

The network voltage restoration usually takes place by changing the field current of the synchronous generators so that the excitation voltage changes and terminal voltage of the synchronous machine maintains constant value. However, the synchronous machine is provided with excitation limits. The field current cannot exceed certain value for long time due rotor heating effects. So the automatic voltage regulator (AVR) when hits its limits the terminal voltage will no longer maintained constant. If the synchronous generator operating within the reactive power limits then terminal voltage V_S will be maintained constant and excitation voltage E various continuously depends on the field current. If the over excitation limiter hits its limits then terminal voltage V_S cannot be maintained constant whereas the excitation voltage E will become constant. The power system will visualize this condition as a sudden injection of synchronous reactance X_s in series with the transmission line reactance X. The overall system reactance will increase and therefore the network P-V curve shifts left. If the system is operating at a particular load demand with stable equilibrium point and at this condition if the exciter hits its limits then the operating point will shift to the new network P-V curve and this operating point may or may not be in the region of attraction of the final operating point. If the operating point will shift to a point which is not the region of attraction of final operating point then voltage instability will be occur.

From the above discussion it is observed that voltage stability is a local phenomena and it is affected by the system initial operating condition, type of connected load and load demand, operation of OLTC and generator reactive power limits. In this chapter the effect of all the mentioned apparatus and conditions are studied for understanding the voltage stability problem. This knowledge is very much useful for understanding the literature in voltage stability area and novel methodologies may be developed for detecting the voltage instability in power system.

Figure 12. Two bus system with generator exciter

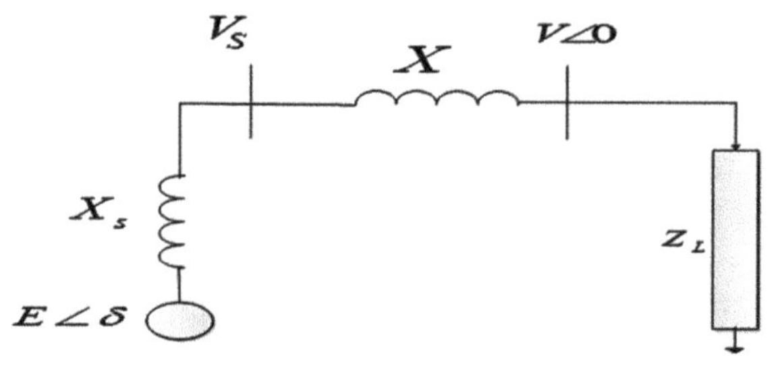

REFERENCES

Ajjarapu, V. (2007). *Computational Techniques for Voltage Stability Assessment and Control*. Springer. doi:10.1007/978-0-387-32935-2

Atputharajah & Saha. (2009). Power system blackouts - Literature review. *Proc. in ICIIS*, 460-465.

Carson, W. (1994). *Taylor, Power System Voltage Stability*. McGraw-Hill.

Chappa & Thakur. (2018, July). A Fast Online Voltage Instability Detection in Power Transmission System Using Wide-Area Measurements. *Iranian Journal of Science and Technology, Transactions of Electrical Engineering*.

Kundur. (1994). Power System Stability and Control. New York: McGraw-Hill.

Lachs, W. R., & Sutanto, D. (1992). Voltage instability in interconnected power systems: A simulation approach. *IEEE Transactions on Power Systems*, *7*(2), 753–761. doi:10.1109/59.141782

Lee, B. H., & Lee, K. Y. (1991). A study on voltage collapse mechanism in electricpower systems. *IEEE Transactions on Power Systems*, *6*(3), 966–973. doi:10.1109/59.119236

Machowski, J., Bialek, J. W. & Bumby. (2008). *Power System Dynamics*. John Wiley and Sons.

Ordacgi Filho, J. M. (2010). *Brazilian blackout 2009: Blackout watch*. Protection, Automation and Control World.

Pal, M. K. (1992, February). Voltage Stability Conditions Considering Load Characteristics. *IEEE Transactions on Power Systems*, *7*(1), 243–249. doi:10.1109/59.141710

Project Group Turkey. (2015). *Report on Blackout in Turkey on 31st March 2015*. European Network of Transmission System Operators for Electricity.

Rampurkar, V., Pentayya, P., Mangalvedekar, H. A., & Kazi, F. (2016). Cascading Failure Analysis for Indian Power Grid. *IEEE Transactions on Smart Grid*, *7*(4), 1951–1960. doi:10.1109/TSG.2016.2530679

Taylor & Erickson. (2005). Recording and analyzing the July 2 cascading outage Western USA power system. *IEEE Computer Applications in Power*, *10*(1), 26–30.

Van Cutsem. (1998). *Costas Vournas*. Voltage Stability of Electric Power Systems.

Chapter 11
Load Frequency Control in Multi-Area Interconnected Power Systems Using Second Order Sliding Mode

Ark Dev
National Institute of Technology Manipur, India

Mrinal Kanti Sarkar
National Institute of Technology Manipur, India

ABSTRACT

The chapter focuses on load frequency control (LFC) problems in multi area power systems using nonlinear second order sliding mode control (NL-SOSMC) under load disturbances and parameter uncertainties. A sudden load disturbance can causes deviation in frequency and tie line power from their schedule value. The main objective of the chapter is to give knowledge about the application of robust control technique mainly sliding mode control (SMC) for load frequency problems. The designed controller ensures finite time convergence of frequency and tie line power deviations with chattering free control signal. The proposed controller confirms better transient and steady state behavior. Furthermore, the controller is validated under matched uncertainty, random step load disturbances, parameter uncertainties, and with nonlinearities in power system like generation rate constraints (GRC) and governor dead band (GDB). The stability of the controller is theoretically proved using Lyapunov candidate function and verified using simulations in MATLAB R2015a.

INTRODUCTION

Electric power systems are an inevitable part of an industrial economy. Power systems extend over almost all developing and developed nations and represent one of the most expensive and largest man-made systems on the earth. Economic soundness of any country depends on how good and quickly its

DOI: 10.4018/978-1-5225-8551-0.ch011

power sector responds to any transience and disturbance in the line. Discovering new sources of energy, transforming one source of energy into another, transmitting the energy from resource centric region to load centric region also constitute an integral part of power systems. Growing demand of energy in every sphere of life has resulted in increased complexity, functionality and interconnection of power systems. Thus, a special attention is required to be directed towards electric power systems of any nation. A sudden disturbance in power systems may have an indirect impact on the lives and economy of the nation. Thus, power system stability is of an utmost importance and should be dealt with modern state of art techniques.

In power systems, frequency stability is an important index of power quality. A stable frequency symbolises the balance between the generation and demand. A sudden load disturbance can cause deviations in frequency and tie line power from their schedule value. In this chapter three area interconnected non-reheated power system is considered. Interconnected power systems are more reliable and economical. A sudden increase in load in one area can be met by borrowing power from adjoining interconnected area. Interconnected power systems have many advantages like reliability, effective use of generators, reduction in the reserve generation capacity of any area etc. However, interconnection requires proper management because control operation of frequency and tie line power deviations under load disturbance can become very challenging. A sudden change of load in one area can cause deviations in frequency and tie line power in the neighboring area. A schematic diagram of a three-area interconnected power system model is illustrated in figure 1.

Load frequency control is an important aspect in an interconnected power system. In multi area power systems, frequency is dependent on active power whereas voltage is on reactive power. This combination of active power and frequency control is known as load frequency control (LFC) and is also known as power frequency control or automatic generation control (Pandey et al., 2013). LFC has made the operation and control of interconnected power systems possible. In the case of interconnected power systems, decentralized control is preferable over centralized control because it uses local area state information to reduce the frequency deviation (Mi et al., 2013). Thus, reducing the computational task and making control more practical and simpler. Decentralized control is an effective way of physically breaking complex control action into smaller and easy control task. As the size of power systems increases, it leads to increase in the complexity of the system that requires precise control technique and

Figure 1. Schematic diagram of interconnected of power systems

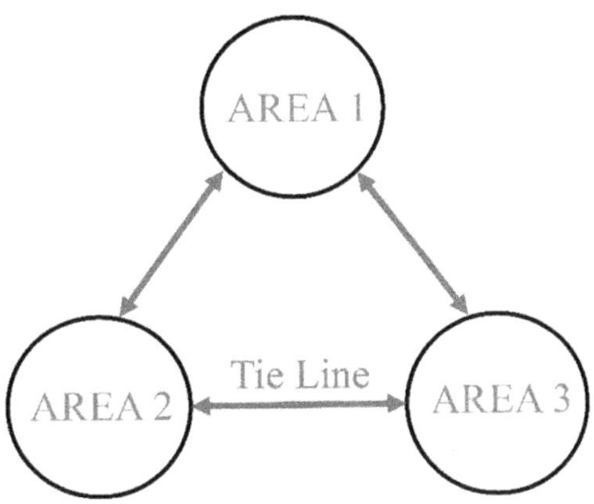

mechanism to have healthier and reliable power supply. Therefore, the LFC of an interconnected power systems have the following two main aspects:

1. maintenance of stable frequency.
2. regulation of tie line flow deviation between the interconnected power systems at desired values.

LFC has been exhaustively studied in the past using different control techniques. Many control techniques have stood the test of time and few have failed under heavy uncertain conditions. In this chapter, LFC is discussed using robust controller known as sliding mode control (SMC). SMC, a particular type of variable structure control (VSC), is another method to solve load frequency problems in multi area power systems. SMC is well known for its robustness, finite time convergence and insensitivity to disturbances and parameter uncertainties. However, the main disadvantage of SMC is the problem of chattering. First order SMC is affected by discontinuous nature of control signal which results in high but finite oscillations also known as “chattering”. Chattering can be one of the main operational problems during practical implementation of controller under heavy uncertain conditions. High frequency oscillations (chattering) have many disadvantages such as low control accuracy, high wear and tear of moving mechanical parts, and high heat losses in power circuits etc. Chattering in control signal can be overcome by close linear approximation of signum function by high gain saturation function, sigmoid function etc. However, these methods reduce control accuracy. Another approach is to design a higher order sliding mode control (HOSMC) that significantly eliminates chattering from control signal and at the same time retains the inherent properties of robustness and high control accuracy.

The Main objective of using nonlinear second order sliding mode control (NL-SOSMC), for load frequency problem is to achieve fast settling time and low overshoot. Also, compared to first order SMC, the proposed design has an advantage of chattering free control with greater control accuracy. The performance of the proposed controller is compared to that of linear and nonlinear SMC under same system parameters and various nonlinearities in multi area power system. The proposed controller results in low overshoot and fast settling time compared to that of linear and nonlinear SMC. Performance indices like integral square error (ISE), integral absolute error (IAE) and integral time absolute error (ITAE) have also been considered and values of the performance indices are found to be lesser than that with linear and nonlinear SMC. In addition to this, proposed controller is also validated under random varying load disturbance, parameter variations and with nonlinearities in power system like generation rate constraints (GRC) and governor deadband (GDB). The stability of sliding surface is theoretically proved by selecting an appropriate Lyapunov candidate function and verified through simulations in MATLAB R2015a.

The remaining chapter give a brief background about the studies proposed by researchers and engineers in the field of LFC followed by detail description of the problem formulation and controller design. The controller design is then validated using simulation. Simulation results are discussed based on different case studies. Finally, the chapter ends with conclusion and scope for research and future work in the field of LFC.

BACKGROUND

Load frequency problems in multi area power systems which are inherently nonlinear uncertain systems fail using conventional controllers. Conventional controllers are often tuned for only one operating condition and they fail under variable operating conditions. Many classical control approaches have been reported however they result in relatively large overshoot and transient frequency deviation (Elgard and Fosha, 1970; Bechert and Chen, 1977). Khodabakhshian and Edrisi (2008) reported a systematic tuning method with a new structure to design a Proportional-Integral-Derivative (PID) control for LFC problems in multi area power systems on the basis of linearized model. Many researches also proposed robust and intelligent techniques to optimize PI controller for improved system performance under parameter uncertainties and load disturbances. Fosha and Elgerd (1970) proposed a state variable frequency control model based on a new optimal feedback control law for improved dynamic response and stability margin for two area interconnected non reheat type thermal power systems. Calovic Milan (1972) reported a linear regulator design for the LFC based on optimal linear regulator theory. A review of recent efforts in applying optimal linear regulator theory with intent to clarify the objectives of LFC, particularly as regard to the application of modern control theory was reported by Kwatny et al., in 1975. A systematic approach to design an optimal variable structure controller (VSC) for the LFC problem in an interconnected power system was reported by Hsu and Chan in 1984. Since all the system states are not available for feedback, considering state reconstruction, observer-based control design for load frequency problems in multi area power systems have already been considered. Moorthi and Aggarwal (1972) presented a modified optimal and sub optimal control based on modern control theory namely optimal observer using nonlinear transformation and reduced order models with local observer. Optimal LFC based on differential approximation, Luenberger observer, adaptive observer for estimation of unmeasured states and unknown deterministic demands are present in literature (Yamashita and Taniguchi, 1986; Feliachi A, 1987; Velusami and Ramar, 1997). Number of research works in LFC in multi area power systems based on adaptive and self-tuning approach under varying operating condition are reported in (Liaw, 1994; Pan and Liaw, 1989). Intelligent control schemes with the use of soft computing techniques for LFC in multi area power systems such as fuzzy logic, artificial neural network (ANN), genetic algorithm (GA), particle swarm optimization (PSO) algorithm are present in literature (Chaturvedi et al., 1999; Indulkar and Raj, 1995; Ghoshal, 2003; Sinha et al., 2010; Aditya & Das, 2003; Juang and Lu, 2002). Ray et al. (1999) reported a robust control for the load frequency problems for interconnected power systems with uncertain parameters. A '*N*' interlinked riccati equation separated by a decoupling technique for '*N*' local robust LFC for *N* area power systems is reported in Lim et al., 1996. Rerkpreedapong et al. (2003) proposed a robust H_∞ control design using linear matrix inequality (LMI) techniques against parameter uncertainties. Robust decentralized PI control design based on H_2 / H_∞ for LFC is reported by Bevrani and Hiyama in 2009.

However, focus of the chapter is to give a background study of the extent of research and application of robust SMC in the field of load frequency problem in multi area power system. Given below is the list of research work related to LFC in multi area power system using sliding mode.

Ray et al. (2004) proposed a decentralized VSC approach to the LFC problem using proportional plus integral type switching function. The technique effectively reduces the effect of constant disturbances in the input channel and specifically parameter perturbation that satisfies the matching conditions. However, the limitations of VSC is not discussed and also frequency deviation under nonlinearities is not considered.

Tsay T. S. (2011) presented LFC of interconnected power system with governor backlash uncertainty. Governor nonlinearities are responsible for oscillations in area frequency and tie line power transient response. In practice, nonlinearities in power system cannot be ignored.

Jiang et al. (2012) introduced the concept of delay dependent stability for LFC with constant and time varying delay. Constant delay exists in the dedicated communication channels while time varying delay is introduced by open communication channels. Presence of time delay degrade the performance of the controller and at worse can cause system instability. Here, concept of delay margin is used to tune the PI controller.

Mi et al. (2013) reported a PI type first order switching surface-based SMC for load frequency problem in power system with matched and unmatched uncertainty. In this article, nonlinearities in the form of generation rate constraints and governor deadband is considered. However, the first order SMC is affected by the problem of chattering which could be solved by using higher order SMC.

Mi et al. (2016) introduced the penetration of wind and solar power in multi area power systems. In this article, sliding mode LFC for hybrid power systems based on disturbance observer is reported. LFC of multi-source power systems, also commonly known as hybrid power system is very challenging. The problem of chattering due to the first order sliding mode is not properly investigated.

Tan et al. (2017) presented LFC of power systems with typical nonlinearities like GRC and GDB with an objective to add anti-windup scheme for any designed load frequency controller. In this article, modelling of GRC and GDB is discussed in details.

Prasad et al. (2017) formulated a nonlinear sliding surface-based SMC with matched and unmatched uncertainties for LFC application in three-area interconnected power system. The reported work results in better transient and steady state response under load disturbance however, the limitations of SMC which is the problem of chattering is not investigated.

Mi et al. (2017) reported a sliding mode LFC for multi area time delay power system with wind power integration. Output power fluctuation of wind turbine is dependent on the speed of wind, and thus load frequency problem become challenging with the integration of wind energy resource. In this article also, the problem of chattering is not stated and solved.

Liao K. & Xu Y. (2017) proposed a new robust LFC scheme using a super twisting SMC technique and extended disturbance observer for multi area power systems. Super twisting control is one type of continuous control technique where control law acts only on first time derivative of sliding surface to obtain chattering free control signal.

Prasad et al. (2018) reported a generalized extended state observer and non-linear SMC for frequency deviation problem in multi-area power systems. Generalized extended state observer is used not only for accurate state and disturbance estimation but also for disturbance rejection.

Sarkar et al. (2018) presented a robust chattering free adaptive integral higher order SMC for load frequency problems in three area power systems focusing on robustness and chattering free control. In this article, a PI type second order sliding mode LFC is proposed in the presence of heavy uncertainties. In this article, the problem of chattering is investigated using second order SMC.

Dev et al. (2019) reported an event triggered based adaptive integral higher order SMC for load frequency problems in multi area power system where triggering of control signal is based on a prespecified condition which is independent of system parameters. In this article also, the problem of chattering is solved using second order SMC.

Dev et al. (2019) reported a nonlinear sliding surface based super twisting load frequency control in three area interconnected power systems using estimated system states. In this work, the problem of chattering is addressed using a super twisting controller which is a continuous control action.

Recent studies in load frequency problems in last few years, suggest the presence of time delays in the system. Time delays in power systems can be present in the sensor loop or control loop or both. Modernization of power systems have made the transmission and control unit more computer dependent. This introduces time delays in the system. So far, authors have either ignored the presence of time delays or considered a controller that is robust against time delays. Presence of time delays can deteriorate the performance of the controller or at worse can cause system instability. Ignorance of time delays can have a catastrophic effect on the controller design. Thus, time delays in power systems cannot be ignored. Limited works are present in this field and can referred from Li J. et al. (2018); Sargolzaei A. et al. (2016); Sun Y. et al. (2018) and Zhang C. K. et al. (2013). LFC in the presence of time delays is not in the scope of this chapter and hence has not been included.

SYSTEM MODELLING

The foremost step in the design of any feedback control scheme is the accurate mathematical modelling of the system. The two most common way of describing a system are transfer function model and state space model. The state space approach is used to describe the linear as well as nonlinear systems and is more accurate compared to that of transfer function approach. However, transfer function is the simplest and need linearization based on proper approximation and assumptions. Conventional power systems are very complex, dynamic and non-linear. Linearized model is permissible in LFC because only small changes in load can be expected during its normal operation. Each area of multi area power system is governed by its own linearized model of governor, turbine and load. Linearized modelling can be referred from any standard text book. Figure 2 illustrates the block diagram of multi area power systems.

The state space dynamic equation of ith area power system can be formulated in the following way,

$$\Delta \dot{f}_i(t) = -\frac{1}{T_{P_i}}\Delta f_i(t) + \frac{K_{P_i}}{T_{P_i}}\Delta P_{g_i}(t) - \frac{K_{P_i}}{T_{P_i}}\Delta P_{d_i}(t) - \frac{K_{P_i}}{2\pi T_{P_i}}\sum_{\substack{j\in N\\ j\neq i}}^{N} K_{sij}\{\Delta\delta_i(t) - \Delta\delta_j(t)\} \tag{1}$$

$$\Delta \dot{P}_{g_i}(t) = -\frac{1}{T_{T_i}}\Delta P_{g_i}(t) + \frac{1}{T_{T_i}}\Delta X_{g_i}(t) \tag{2}$$

$$\Delta \dot{X}_{g_i}(t) = -\frac{1}{R_i T_{G_i}}\Delta f_i(t) - \frac{1}{T_{G_i}}\Delta X_{g_i}(t) - \frac{1}{T_{G_i}}\Delta E_i(t) + \frac{1}{T_{G_i}}u_i(t) \tag{3}$$

Figure 2. Block diagram of interconnected power systems
(Mi et al., 2013)

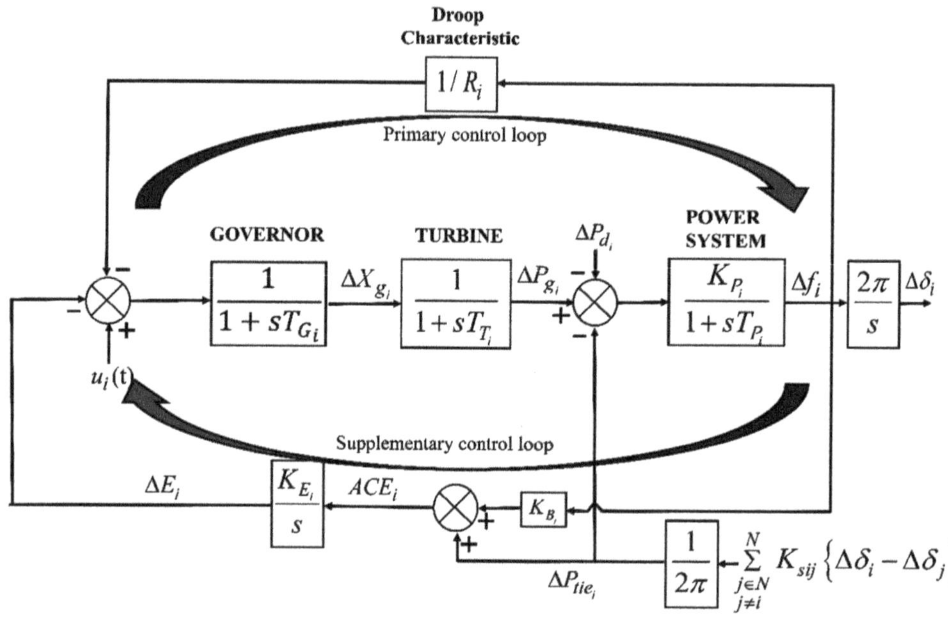

$$\Delta \dot{E}_i(t) = K_{E_i}[K_{B_i}\Delta f_i(t) + \frac{1}{2\pi}\sum_{\substack{j\in N \\ j\neq i}}^{N} K_{sij}\{\Delta\delta_i(t) - \Delta\delta_j(t)\}] \tag{4}$$

$$\Delta\dot{\delta}_i(t) = 2\pi\Delta f_i(t) \tag{5}$$

where i=1, 2…… N and N denote number of areas. The matrix form representation of the equations (1)-(5) can now be given as,

$$\dot{x}_i(t) = A_i x_i(t) + B_i u_i(t) + \sum_{\substack{j\in N \\ j\neq i}}^{N} E_{ij} x_j(t) + F_i \Delta P_{d_i}(t) \tag{6}$$

where matrices $A_i \in \Re^{n_i \times n_i}$, $B_i \in \Re^{n_i \times m_i}$, $C_i \in \Re^{r_i \times n_i}$, $E_{ij} \in \Re^{n_i \times n_j}$ and $F_i \in \Re^{n_i \times k_i}$ are the constant matrices of suitable dimensions.

$$A_i = \begin{bmatrix} -\frac{1}{T_{P_i}} & \frac{K_{P_i}}{T_{P_i}} & 0 & 0 & \frac{-K_{P_i}}{2\pi T_{P_i}} \sum_{\substack{j \in N \\ j \neq i}}^{N} K_{sij} \\ 0 & -\frac{1}{T_{T_i}} & \frac{1}{T_{T_i}} & 0 & 0 \\ -\frac{1}{R_i T_{G_i}} & 0 & -\frac{1}{T_{G_i}} & -\frac{1}{T_{G_i}} & 0 \\ K_{E_i} K_{B_i} & 0 & 0 & 0 & \frac{K_{E_i}}{2\pi} \sum_{\substack{j \in N \\ j \neq i}}^{N} K_{sij} \\ 2\pi & 0 & 0 & 0 & 0 \end{bmatrix}$$

$$B_i = \begin{bmatrix} 0 & 0 & \frac{1}{T_{G_i}} & 0 & 0 \end{bmatrix}^T$$

$$C_i = \begin{bmatrix} 1 & 0 & 0 & 0 & 0 \end{bmatrix}$$

$$F_i = \begin{bmatrix} -\frac{K_{P_i}}{T_{P_i}} & 0 & 0 & 0 & 0 \end{bmatrix}^T$$

$$E_{ij} = \begin{bmatrix} 0 & 0 & 0 & 0 & \frac{K_{P_i}}{2\pi T_{P_i}} K_{sij} \\ 0 & 0 & 0 & 0 & 0 \\ 0 & 0 & 0 & 0 & 0 \\ 0 & 0 & 0 & 0 & -\frac{K_{E_i}}{2\pi} K_{sij} \\ 0 & 0 & 0 & 0 & 0 \end{bmatrix}$$

$$x_i(t) = \begin{bmatrix} \Delta f_i(t) \\ \Delta P_{g_i}(t) \\ \Delta X_{g_i}(t) \\ \Delta E_i(t) \\ \Delta \delta_i(t) \end{bmatrix}$$

Vector $x_i(t) \in \Re^{n_i}$ is the state vector ; $u_i(t) \in \Re^{m_i}$ is the control vector ; $x_j(t) \in \Re^{n_j}$ is the adjacent state vector of $x_i(t)$; $\Delta P_{d_i}(t) \in \Re^{k_i}$ is the vector of load disturbance (p.u. MW). Variables $\Delta f_i(t)$ is the change in frequency (Hz); $\Delta P_{g_i}(t)$ is the change in power output (p.u. MW); $\Delta X_{g_i}(t)$ is the change in governor valve position (p.u. MW); $\Delta E_i(t)$ is the change in integral control (p.u. MW) ; $\Delta\delta_i(t)$ is the change in rotor angle deviation (radian). Parameters T_{G_i}, T_{T_i} and T_{P_i} are the time constants of governor, turbine and sub system model respectively with a standard unit of 'seconds'; K_{P_i}, R_i, K_{E_i} and K_{B_i} are power system gain (Hz/p.u. MW), speed regulation coefficient (Hz/p.u. MW), integral control gain (Hz/p.u. MW) and frequency bias factor (p.u. MW/Hz) respectively. K_{sij} (p.u. MW) is the interconnection gain between area *i* and *j (i≠j)* and if there is no exchange of power between *i* and *j*, $K_{sij} = 0$.

The dynamics of ith area is transformed in the regular form using a transformation matrix T_{r_i} (Prasad et al., 2017),

$$z_i(t) = \begin{bmatrix} z_{1_i}(t) \\ z_{2_i}(t) \end{bmatrix} = T_{r_i} x_i(t),$$

where the transformation matrix satisfies, $T_{r_i} T_{r_i}^T = I$, and

$$T_{r_i} B_i = \begin{bmatrix} 0_{(n_i - p_i)\times p_i} \\ \left[B_{2_i}\right]_{(p_i \times p_i)} \end{bmatrix},$$

where *I* is the identity matrix and B_{2_i} is the non-singular matrix. Thus, the transformed states of ith area is obtained as,

$$\begin{cases} z_{1_i}(t) = \left[\Delta f_i(t) \quad \Delta P_{g_i}(t) \quad -\Delta\delta_i(t) \quad \Delta E_i(t)\right]^T \\ z_{2_i}(t) = \left[\Delta X_{g_i}(t)\right]^T \end{cases} \tag{7}$$

Equation (6), in regular form can be written as,

$$\begin{bmatrix} \dot{z}_{1_i}(t) \\ \dot{z}_{2_i}(t) \end{bmatrix} = \begin{bmatrix} A_{11_i} & A_{12_i} \\ A_{21_i} & A_{22_i} \end{bmatrix} \begin{bmatrix} z_{1_i}(t) \\ z_{2_i}(t) \end{bmatrix} + \begin{bmatrix} 0 \\ B_{2_i} \end{bmatrix} u_i(t) + \sum_{\substack{j=1 \\ j\neq i}}^{N} \begin{bmatrix} E_{11_{ij}} & E_{12_{ij}} \\ E_{21_{ij}} & E_{22_{ij}} \end{bmatrix} \begin{bmatrix} z_{1_j}(t) \\ z_{2_j}(t) \end{bmatrix} + \begin{bmatrix} F_{1_i} \\ F_{2_i} \end{bmatrix} \Delta P_{d_i}(t) \tag{8}$$

where

$$z_{1_j} = \left[0 \quad 0 \quad -\Delta\delta_j \quad 0\right]^T, \; z_{2_j} = \left[0\right]^T.$$

Matrix sub block in equation (8) is given as,

$$T_{r_i} A_i T_{r_i}^{-1} = A_{reg_i} = \begin{bmatrix} A_{11_i} & A_{12_i} \\ A_{21_i} & A_{22_i} \end{bmatrix}, T_{r_i} F_i = F_{reg_i} = \begin{bmatrix} F_{1_i} \\ F_{2_i} \end{bmatrix},$$
$$T_{r_i} E_{ij} T_{r_i}^{-1} = E_{reg_{ij}} = \begin{bmatrix} E_{11_{ij}} & E_{12_{ij}} \\ E_{21_{ij}} & E_{22_{ij}} \end{bmatrix}, T_{r_i} B_i = B_{reg_i} = \begin{bmatrix} 0 \\ B_{2_i} \end{bmatrix} \quad (9)$$

To achieve asymptotic stability of system, switching surface and control signal for each area are designed separately based on the following assumptions.

Assumption 1: Each subsystem is completely controllable for i^{th} area interconnected system i.e. each matrix pair (A_i, B_i) is controllable. This implies that the pair,

$$\left(A_{11_i} + \sum_{\substack{j=1 \\ j\neq i}}^{N} E_{11_{ij}}, A_{12_i} + \sum_{\substack{j=1 \\ j\neq i}}^{N} E_{12_{ij}} \right)$$

is also controllable in the regular form (Prasad et al., 2017).

The state equation of ith area in regular form (upon transformation) can now be written as,

$$\dot{z}_{1_i}(t) = A_{11_i} z_{1_i}(t) + \sum_{\substack{j=1 \\ j\neq i}}^{N} E_{11_{ij}} z_{1_j}(t) + A_{12_i} z_{2_i}(t) + \sum_{\substack{j=1 \\ j\neq i}}^{N} E_{12_{ij}} z_{2_j}(t) + F_{1_i} \Delta P_{d_i}(t) \quad (10)$$

$$\dot{z}_{2_i}(t) = A_{21_i} z_{1_i}(t) + \sum_{\substack{j=1 \\ j\neq i}}^{N} E_{21_{ij}} z_{1_j}(t) + A_{22_i} z_{2_i}(t) + \sum_{\substack{j=1 \\ j\neq i}}^{N} E_{22_{ij}} z_{2_j}(t) + B_{2_i} u_i(t) + F_{2_i} \Delta P_{d_i}(t) \quad (11)$$

Considering system parameter uncertainties in (10) and (11),

$$\dot{z}_{1_i}(t) = (A_{11_i} + \Delta A_{11_i}) z_{1_i}(t) + \sum_{\substack{j=1 \\ j\neq i}}^{N} E_{11_{ij}} z_{1_j}(t) + (A_{12_i} + \Delta A_{12_i}) z_{2_i}(t)$$
$$+ \sum_{\substack{j=1 \\ j\neq i}}^{N} E_{12_{ij}} z_{2_j}(t) + (F_{1_i} + \Delta F_{1_i}) \Delta P_{d_i}(t) \quad (12)$$

$$\dot{z}_{2_i}(t) = (A_{21_i} + \Delta A_{21_i})z_{1_i}(t) + \sum_{\substack{j=1\\ j\neq i}}^{N} E_{21_{ij}} z_{1_j}(t) + (A_{22_i} + \Delta A_{22_i})z_{2_i}(t)$$
$$+\sum_{\substack{j=1\\ j\neq i}}^{N} E_{22_{ij}} z_{2_j}(t) + (B_{2_i} + \Delta B_{2_i})u_i(t) + (F_{2_i} + \Delta F_{2_i})\Delta P_{d_i}(t) \tag{13}$$

In general, A_i, B_i and F_i are the matrices of the nominal parameters and $\Delta A_i, \Delta B_i$ and ΔF_i are the matrices of parameter uncertainties.

Defining,

$$g_{1_i}(t) = \Delta A_{11_i} z_{1_i}(t) + \Delta A_{12_i} z_{2_i}(t) + (F_{1_i} + \Delta F_{1_i})\Delta P_{d_i}(t), \tag{14}$$

$$g_{2_i}(t) = \Delta A_{21_i} z_{1_i}(t) + \Delta A_{22_i} z_{2_i}(t) + \Delta B_{2_i} u_i(t) + (F_{2_i} + \Delta F_{2_i})\Delta P_{d_i}(t) \tag{15}$$

as the aggregated uncertainties. Equations (12) & (13) can now be rewritten as,

$$\dot{z}_{1_i}(t) = A_{11_i} z_{1_i}(t) + \sum_{\substack{j=1\\ j\neq i}}^{N} E_{11_{ij}} z_{1_j}(t) + A_{12_i} z_{2_i}(t) + \sum_{\substack{j=1\\ j\neq i}}^{N} E_{12_{ij}} z_{2_j}(t) + g_{1_i}(t) \tag{16}$$

$$\dot{z}_{2_i}(t) = A_{21_i} z_{1_i}(t) + \sum_{\substack{j=1\\ j\neq i}}^{N} E_{21_{ij}} z_{1_j}(t) + A_{22_i} z_{2_i}(t) + \sum_{\substack{j=1\\ j\neq i}}^{N} E_{22_{ij}} z_{2_j}(t) + B_{2_i} u_i(t) + g_{2_i}(t) \tag{17}$$

SLIDING MODE CONTROL

SMC is a particular type of VSC. VSC is a discontinuous control where the dynamics of a system changes based on a high frequency switching control. The state feedback control law varies from one smooth control to another possibly at a very high frequency. Introduced in late 1970s and still continuing today, SMC is well known for its robustness, finite time convergence and insensitivity to disturbances and uncertainties. Design of a SMC involves two fundamental steps: -

1. Design of a stable sliding surface.
2. Design of control law that drives the system state trajectories from any initial condition to the sliding surface in finite time and forces it to remain on it thereafter.

The design of sliding surface which is normally linear combination of system states (not always true) should satisfy all the constraints and design specification and thus, it should be designed optimally. Sliding surface is also known as switching surface because the control signal has two gains based on whether the state trajectories are above the surface or below the surface. To understand the notion of SMC, a simple first order linear time invariant (LTI) system is considered (Bandyopadhyay et al., 2009),

$$\dot{x}(t) = ax(t) + bu(t) + d(x,t) \tag{18}$$

where $x(t) \in \Re$ represent system states, $u(t) \in \Re$ represent the control input, $d(x,t) \in \Re$ represent the uncertainty and only the bounds of uncertainty are known. Parameters '*a*' and '*b*' are known nonzero constants.

A sliding surface can be designed as,

$$S(x,t) = \sum_{i=1}^{n} c_i x_i = \sum_{i=1}^{n-1} c_i x_i + x_n \tag{19}$$

where,

$$c_i = \begin{bmatrix} c_1 & \cdots & c_{n-1} & 1 \end{bmatrix}^T.$$

In SMC, the parameters $c_1, c_2, \ldots, c_{n-1}$ are selected such that

$$s^{n-1} + c_{n-1} s^{n-2} + \ldots\ldots + c_1$$

is Hurwitz polynomial, where '*s*' is a Laplace operator.

So, for the first order LTI system, the sliding surface is $S(x,t) = x(t)$. Thus, formulating a control law based on the sliding surface,

$$u(t) = -\frac{1}{b}\left[ax(t) + \beta sign\left(S(x,t)\right)\right] \tag{20}$$

where, $\beta > 0$ and is known as switching constant and selected such that $\left|d(x,t)\right| \le \beta$ and

$$sign\left(S(x,t)\right) = \begin{cases} 1, S(x,t) > 0 \\ 0, S(x,t) = 0 \\ -1, S(x,t) < 0 \end{cases}.$$

Using $u(t)$ from (20) and substituting it in (18) results,

$$\dot{x}(t) = -\beta sign(S(x,t)) + d(x,t) \tag{21}$$

Let us understand (21) in detail by considering three cases.

1. When $S(x,t) > 0$, $\dot{x}(t) < 0$. This conclude that $x(t)$ is decreasing and moving towards $S(x,t) = 0$.
2. When $S(x,t) < 0$, $\dot{x}(t) > 0$. This conclude that $x(t)$ is increasing and moving towards $S(x,t) = 0$.
3. When $S(x,t) = 0$, the trajectories $x(t)$ is forced back to $S(x,t) = 0$ from either direction.

The phase when the trajectories are moved towards $S(x,t)$ is called the reaching phase. And, the phase when the trajectories start sliding over the sliding surface is called sliding phase. Once the trajectories are on the sliding surface, the system is governed by the properties of the sliding surface and become independent of the disturbance. Figure 3 show the motion of state trajectory on the sliding surface.

SLIDING MODE CONTROL: REACHING LAW

As discussed earlier, the concept of sliding mode can be described easily with the help of reaching phase and sliding phase. Classical reaching laws for SMC from literature is described briefly (Bandyopadhyay et al., 2009).

1. 1. Constant rate reaching law

$$\dot{S}(x,t) = -\beta sign(S(x,t));\ \beta > 0$$

Figure 3. State trajectory in the vicinity of sliding surface
(Bandyopadhyay et al., 2009)

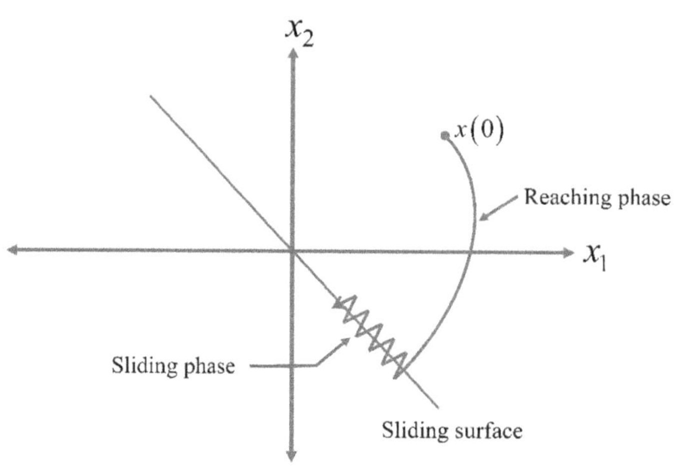

This law enables the state variable to reach the sliding surface at a constant rate β. The merits of constant rate reaching law is its design simplicity. However, large value of β will result in high frequency oscillations (chattering) which will be discussed latter in this section.

2. 2. Exponential reaching law

$$\dot{S}(x,t) = -\beta sign\left(S(x,t)\right) - \lambda S(x,t);\ \beta > 0, \lambda > 0$$

The addition of extra proportional rate term $-\lambda S(x,t)$, enables the state trajectory to reach the sliding surface faster when $S(x,t)$ is large.

3. 3. Power rate reaching law

$$\dot{S}(x,t) = -\lambda \left|S(x,t)\right|^{\alpha} sign\left(S(x,t)\right);\ \lambda > 0, 1 > \alpha > 0$$

This law increases the reaching speed when the trajectories are far away from the sliding surface and reduces the speed when the trajectories are near the sliding surface. The advantage of the power rate reaching law is fast convergence and low chattering effect.

STABILITY CONDITION

The existence condition of SMC is given by classical Lyapunov stability theorem. For the finite time convergence of state trajectories on to the sliding surface, the sign of $S(x,t)$ and $\dot{S}(x,t)$ should be opposite. Considering a positive definite function,

$$V = 0.5\left(S(x,t)\right)^2 \tag{22}$$

and ensuring the first time derivative of function (22) to be negative. This would confirm finite time convergence of system state trajectories on the sliding surface.

$$\dot{V} = S(x,t)\dot{S}(x,t) < 0 \tag{23}$$

LIMITATIONS OF SLIDING MODE CONTROL: CHATTERRING

In practice, SMC is affected by oscillations having finite frequency and amplitude also known as 'chattering'. Chattering is caused due to discontinuous nature of control signal (signum function is discontinuous in nature). The problem of chattering reduces the control accuracy, causes wear and tear of moving mechanical parts, high heat loss in power circuits etc. and at worse can cause system instability. Active research is going on to find the solutions to the problem of chattering. The most practiced approaches include close linear approximation of discontinuous signum function by using saturation function, sigmoid function etc. However, in doing so, the control accuracy of SMC is affected to a greater extent. Also, the insensitivity to disturbance property of SMC is also affected. Another effective approach to suppress the problem of chattering is the use of second and higher order SMC. Unlike first order SMC which acts on the first time derivative of sliding surface, HOSMC acts on higher order time derivative of sliding surface to suppress the problem of chattering. In this chapter, both first order SMC and second order SMC are considered for load frequency problems in multi area power systems. The simulation results are also compared which confirm that second order sliding mode guarantees low overshoot and fast settling time with chattering free control.

Design of Linear Sliding Surface and Control Signal

Let us consider a linear sliding surface (Fulwani et al., 2011),

$$S_i\left(z_i,t\right)=\left[c_i \quad 1\right]z_i\left(t\right) \tag{24}$$

$$S_i\left(z_i,t\right)=\left[c_i \quad 1\right]\begin{bmatrix} z_{1_i}\left(t\right) \\ z_{2_i}\left(t\right)\end{bmatrix},$$

where $c_i = K_i$ and K_i is chosen such that the poles of $\left(A_{11_i} - A_{12_i}K_i\right)$ have negative real part.

Differentiating equation (24),

$$\dot{S}_i\left(z_i,t\right)=c_i\dot{z}_{1_i}\left(t\right)+\dot{z}_{2_i}\left(t\right) \tag{25}$$

Substituting (17) in (25) and solving further for $u_i(t)$,

$$u_i(t)=-B_{2_i}^{-1}\begin{pmatrix} K_i\dot{z}_{1_i}\left(t\right)+A_{21_i}z_{1_i}\left(t\right)+\sum_{\substack{j=1\\ j\neq i}}^{N}E_{21_{ij}}z_{i_j}\left(t\right)+A_{22_i}z_{2_i}\left(t\right)+\sum_{\substack{j=1\\ j\neq i}}^{N}E_{22_{ij}}z_{2_j}\left(t\right) \\ +\beta\,\mathrm{sign}\left(\mathrm{S}_i(\mathrm{z}_i,\mathrm{t})\right)+\mathrm{g}_{2_i}\left(t\right)\end{pmatrix} \tag{26}$$

where β is a positive constant.

Stability of Sliding Surface During Sliding Mode

Theorem 1: The control signal (26) is robust against aggregated uncertainty and drives the system state trajectories given by equations (16) & (17) from any initial condition to the sliding surface and forces it to remain on it thereafter provided $\left(A_{11_i} - A_{12_i}K_i\right)$ is stable.

During the sliding mode (i.e. $S(z_i, t) = 0$), equation (24) becomes, (Fulwani et al., 2011),

$$z_{2_i}\left(t\right) = -c_i z_{1_i}\left(t\right) \tag{27}$$

$$\dot{z}_{1_i}\left(t\right) = \left(A_{11_i} - A_{12_i}K_i\right)z_{1_i}\left(t\right) + \sum_{\substack{j=1\\j\neq i}}^{N} E_{11_{ij}} z_{1_j}\left(t\right) + \sum_{\substack{j=1\\j\neq i}}^{N} E_{12_{ij}} z_{2_j}\left(t\right) + g_{1_i}\left(t\right) \tag{28}$$

Considering the Lyapunov equation for i^{th} area,

$$V_i\left(t\right) = z_{1_i}^T\left(t\right)P_i z_{1_i}\left(t\right) \tag{29}$$

where P_i is a positive definite matrix satisfying the Lyapunov equation for some value of W_i which is symmetric positive definite,

$$\left(A_{11_i} - A_{12_i}K_i\right)^T P_i + P_i(A_{11_i} - A_{12_i}K_i) = -W_i.$$

$$\dot{V}_i\left(t\right) = \dot{z}_{1_i}^T\left(t\right)P_i z_{1_i}\left(t\right) + z_{1_i}^T\left(t\right)P_i \dot{z}_{1_i}\left(t\right) \tag{30}$$

Using (28) in the above equation,

$$\begin{aligned}\dot{V}_i\left(t\right) &= z_{1_i}^T((A_{11_i} - A_{12_i}K_i)^T + \sum_{\substack{j=1\\j\neq i}}^{N}(E_{11_{ij}} z_{1_j}(t))^T + \sum_{\substack{j=1\\j\neq i}}^{N}(E_{12_{ij}} z_{2_j}(t))^T + g_{1_i}^T(t))P_i z_{1_i}(t)\\ &+ z_{1_i}^T P_i\{(A_{11_i} - A_{12_i}K_i)z_{1_i}(t) + \sum_{\substack{j=1\\j\neq i}}^{N} E_{11_{ij}} z_{1_j}(t) + \sum_{\substack{j=1\\j\neq i}}^{N} E_{12_{ij}} z_{2_j}(t) + g_{1_i}(t)\}\end{aligned} \tag{31}$$

$$\dot{V}_i(t) = z_{1_i}^T((A_{11_i} - A_{12_i}K_i)^T P_i + P_i(A_{11_i} - A_{12_i}K_i))z_{1_i}(t)$$
$$+z_{1_i}^T(\sum_{\substack{j=1\\j\neq i}}^{N}(E_{11_{ij}}z_{1_j}(t))^T + \sum_{\substack{j=1\\j\neq i}}^{N}(E_{12_{ij}}z_{2_j}(t))^T + g_{1_i}^T(t))P_i z_{1_i}(t) \tag{32}$$
$$+z_{1_i}^T P_i\{\sum_{\substack{j=1\\j\neq i}}^{N}E_{11_{ij}}z_{1_j}(t) + \sum_{\substack{j=1\\j\neq i}}^{N}E_{12_{ij}}z_{2_j}(t) + g_{1_i}(t)\}$$

Substituting

$$\left(A_{11_i} - A_{12_i}K_i\right)^T P_i + P_i(A_{11_i} - A_{12_i}K_i) = -W_i$$

in (32), we get,

$$\dot{V}_i(t) = z_{1_i}^T\{-W_i + (\sum_{\substack{j=1\\j\neq i}}^{N}(E_{11_{ij}}z_{1_j}(t))^T + \sum_{\substack{j=1\\j\neq i}}^{N}(E_{12_{ij}}z_{2_j}(t))^T + g_{1_i}^T(t)P_i\}z_{1_i}(t)$$
$$+z_{1_i}^T P_i\{\sum_{\substack{j=1\\j\neq i}}^{N}E_{11_{ij}}z_{1_j}(t) + \sum_{\substack{j=1\\j\neq i}}^{N}E_{12_{ij}}z_{2_j}(t) + g_{1_i}(t)\} \tag{33}$$

Since,

$$\sum_{\substack{j=1\\j\neq i}}^{N}\left(E_{11_{ij}}z_{1_j}(t)\right) = 0, \sum_{\substack{j=1\\j\neq i}}^{N}E_{12_{ij}}z_{2_j}(t) = 0$$

and assuming $g_{1_i}(t)$ to be small, gives $\dot{V}_i(t) < 0$. The negative derivative of Lyapunov function indicates finite time convergence of system states (Fulwani et al., 2011).

Design of Nonlinear Sliding Surface and Control Signal

Considering nonlinear sliding surface reported in (Prasad et al., 2017; Fulwani et al., 2011; Mondal and Mahanta, 2011),

$$S_i(z_i, t) = \begin{bmatrix} c_i & 1 \end{bmatrix} z_i(t) \tag{34}$$

$$S_i(z_i, t) = \begin{bmatrix} c_i & 1 \end{bmatrix} \begin{bmatrix} z_{1_i}(t) \\ z_{2_i}(t) \end{bmatrix} \tag{35}$$

$$S_i\left(z_i,t\right)=\left[K_i-\psi_i(y_i)A_{12_i}^T P_i \quad 1\right]\begin{bmatrix} z_{1_i}\left(t\right) \\ z_{2_i}\left(t\right)\end{bmatrix} \tag{36}$$

where K_i is chosen such that the poles of $\left(A_{11_i}-A_{12_i}K_i\right)$ has negative real part and dominant poles have low damping ratio. $\Psi_i(y_i)$ is the non-positive differentiable function in y_i used to change the damping ratio. P_i is a positive definite matrix satisfying the Lyapunov equation for some value of W_i which is symmetric positive definite.

$$\left(A_{11_i}-A_{12_i}K_i\right)^T P_i+P_i(A_{11_i}-A_{12_i}K_i)=-W_i \tag{37}$$

The choice of nonlinear function, $\Psi_i(y_i)$ for i^{th} area is not fixed. The nonlinear function $\Psi_i(y_i)$ should have the following two properties:

1. It should change from 0 to $-\alpha_i$ as the output approaches the set point from initial value, where $\alpha_i>0$.
2. It should be differentiable in y_i to ensure the existence of sliding mode.

For the above case, nonlinear function $\Psi_i(y_i)$ is selected as,

$$\Psi_i(y_i)=\frac{-\alpha_i}{1-e^{-1}}\left[e^{-(1-((y_i-y_{i0})/(r_i-y_{i0}))^2)}-e^{-1}\right],\text{ where } y_{i0}=y_i(0). \tag{38}$$

Differentiating the sliding surface (35) for the computation of control signal,

$$\dot{S}_i\left(z_i,t\right)=\dot{c}_{1_i}z_{1_i}\left(t\right)+c_{1_i}\dot{z}_{1_i}\left(t\right)+\dot{z}_{2_i}\left(t\right) \tag{39}$$

Substituting the value of $\dot{z}_{2_i}\left(t\right)$ in the above equation and solving further for $u_i(t)$,

$$\begin{aligned} u_i(t)=-B_{2_i}^{-1}\{&K_i\dot{z}_{1_i}(t)-\psi_i(y_i)A_{12_i}^T P_i\dot{z}_{1_i}(t)+A_{21_i}z_{1_i}(t)+\sum_{\substack{j=1\\ j\neq i}}^{N}E_{21_{ij}}z_{1_j}(t)+A_{22_i}z_{2_i}(t)\\ &+\sum_{\substack{j=1\\ j\neq i}}^{N}E_{22_{ij}}z_{2_j}(t)-\frac{d\psi_i(y_i)}{dt}A_{12_i}^T P_i z_{1_i}(t)+\beta sign\left(S_i(z_i,t)\right)+g_{1_i}(t)\}\end{aligned} \tag{40}$$

where β is a positive constant. Equation (40) governs the nonlinear sliding surface based first order SMC and is affected by chattering. In order to design second order sliding surface, let us consider the second order time derivative of sliding surface (39) (Mondal and Mahanta, 2011; Mondal et al., 2012),

$$\ddot{S}_i\left(z_i,t\right)=c_{1_i}\ddot{z}_{1_i}\left(t\right)+\dot{c}_{1_i}\dot{z}_{1_i}\left(t\right)+\dot{c}_{1_i}\dot{z}_{1_i}\left(t\right)+\ddot{c}_{1_i}z_{1_i}\left(t\right)+\ddot{z}_{2_i}\left(t\right) \tag{41}$$

In SOSMC, objective is to bring the sliding surface $S_i\left(z_i,t\right)$ and $\dot{S}_i\left(z_i,t\right)$ to zero in finite time by using a discontinuous control signal $\dot{u}_i(t)$, such that actual control signal $u_i(t)$ obtained by integrating discontinuous control signal $\dot{u}_i(t)$ is continuous.

Using $\ddot{z}_{2_i}\left(t\right)$ in above equation (41) and solving further for $\dot{u}_i(t)$,

$$\begin{aligned}\dot{u}_i(t)=&-B_{2_i}^{-1}(K_i\ddot{z}_{1_i}(t)+A_{21_i}\dot{z}_{1_i}(t)+A_{22_i}\dot{z}_{2_i}(t)+\sum_{\substack{j=1\\j\neq i}}^{N}E_{21_{ij}}\dot{z}_{1_j}(t)+\sum_{\substack{j=1\\j\neq i}}^{N}E_{22_{ij}}\dot{z}_{2_j}(t)\\&-2\frac{d\psi_i(y_i)}{dt}A_{12_i}^TP_i\dot{z}_{1_i}(t)-\frac{d^2\psi_i(y_i)}{dt^2}A_{12_i}^TP_iz_{1_i}(t)+\dot{g}_{2_i}(t)+\lambda sign\left(S_i(z_i,t)\right)+\mu(\dot{S}_i(z_i,t)))\end{aligned} \tag{42}$$

where λ,μ are the positive constants such that, $\lambda>\mu$. Unlike equation (26) & (40), equation (42) acts on the second order time derivative of sliding surface and is free from the problem of chattering. It should be noted that the integration of any discontinuous function results in continuous function, and thus the above control signal is free from any discontinuous part and thus chattering in the control signal is reduced considerably.

Stability of Nonlinear Sliding Surface During Sliding Mode

Theorem 2: Control signals (40) for nonlinear SMC is robust against uncertainties and drives the system state trajectories from any initial condition to the sliding surface and forces it to remain on it thereafter provided $(A_{11_i}-A_{12_i}K_i)$ is stable and nonlinear function $\psi_i(y_i)$ of the form is given by equation (38).

During sliding mode, $S_i\left(z_i,t\right)=0$, equation (36) becomes,

$$z_{2_i}(t)=-K_iz_{1_i}(t)+\psi_i(y_i)A_{12_i}^TP_iz_{1_i}(t) \tag{43}$$

$$\dot{z}_{1_i}(t)=(A_{11_i}-A_{12_i}K_i)z_{1_i}(t)+\psi_i(y_i)A_{12_i}A_{12_i}^TP_iz_{1_i}(t)+\sum_{\substack{j=1\\j\neq i}}^{N}\{(E_{11_{ij}}-E_{12_{ij}}K_i)+\psi_i(y_i)E_{12_{ij}}A_{12_i}^TP_i\}z_{1_j} \tag{44}$$

Assumption 2: The following matching condition is satisfied, $g_i(t)=B_ig_i'(t)$, it is assumed that the aggregated disturbance $g_i'\left(t\right)$ is bounded. This implies, $\left\|g_i'(t)\right\|\leq d_i=0$.

Proof: Considering the Lyapunov stability theorem, the system state trajectories in the sliding mode is proved based on following assumption (Prasad et al., 2017).

$$V_i\left(t\right)=z_{1_i}^T\left(t\right)P_iz_{1_i}\left(t\right) \tag{45}$$

$$\dot{V}_i(t) = \dot{z}_{1_i}^T(t) P_i z_{1_i}(t) + z_{1_i}^T(t) P_i \dot{z}_{1_i}(t) \tag{46}$$

$$\begin{aligned}
&= z_{1_i}^T(t)[(A_{11_i} - A_{12_i}K_i) + \psi_i(y_i)A_{12_i}^T P_i z_{1_i}(t)]^T P_i z_{1_i}(t) + z_{1_i}^T(t) P_i[(A_{11_i} - A_{12_i}K_i) \\
&+\psi_i(y_i)A_{12_i}^T P_i]z_{1_i}(t) + z_{1_j}^T\{\sum_{\substack{j=1 \\ j\neq i}}^{N}(E_{11_{ij}} - E_{12_{ij}}K_i) + \psi_i(y_i)E_{12_{ij}}A_{12_i}^T P_i\}P_i z_{1_i}(t) \\
&+z_{1_i}^T(t)P_i\{\sum_{\substack{j=1 \\ j\neq i}}^{N}(E_{11_{ij}} - E_{12_{ij}}K_i) + \psi_i(y_i)E_{12_{ij}}A_{12_i}^T P_i\}z_{1_j}(t)
\end{aligned} \tag{47}$$

$$\begin{aligned}
&= z_{1_i}^T(t)\left[-W_i + 2\psi_i(y_i)P_i A_{12_i} A_{12_i}^T P_i\right] z_{1_i}(t) + z_{1_i}^T\{\sum_{\substack{j=1 \\ j\neq i}}^{N}(E_{11_{ij}} - E_{12_{ij}}K_i) + \psi_i(y_i)E_{12_{ij}}A_{12_i}^T P_i\}P_i z_{1_i}(t) \\
&+z_{1_i}^T(t)P_i\{\sum_{\substack{j=1 \\ j\neq i}}^{N}(E_{11_{ij}} - E_{12_{ij}}K_i) + \psi_i(y_i)E_{12_{ij}}A_{12_i}^T P_i\}z_{1_j}(t)
\end{aligned} \tag{48}$$

Thus,

$$= z_{1_i}^T(t)\left[-W_i + 2\psi_i(y_i)P_i A_{12_i} A_{12_i}^T P_i\right] z_{1_i}(t) + z_{1_j}^T M_i^T P_i z_{1_i} + z_{1_i}^T M_i P_i z_{1_j}(t) \tag{49}$$

Nonlinear function varies from zero (0) to a very low negative value, $-\alpha_i$.

$$= z_{1_i}^T(t)\left[-W_i - 2\alpha_i P_i A_{12_i} A_{12_i}^T P_i\right] z_{1_i}(t) + z_{1_j}^T M_i^T P_i z_{1_i} + z_{1_i}^T M_i P_i z_{1_j}(t) \tag{50}$$

where,

$$M_i = \sum_{\substack{j=1 \\ j\neq i}}^{N}\left(\left(E_{11_{ij}} - E_{12_{ij}}K_i\right) - \beta_i E_{12_{ij}} A_{12_i}^T P_i\right)^T$$

is a positive definite matrix. Thus, $\dot{V}_i(t) < 0$, this proves that system states are asymptotically stable.

Stability of Nonlinear Second Order Sliding Surface

Let us consider a Lyapunov function for the stability of second order sliding surface (Mondal and Mahanta, 2011),

$$V_i\left(t\right)=0.5S_i^{\,2}\left(z_i,t\right)+0.5\dot{S}_i^{\,2}\left(z_i,t\right) \tag{51}$$

$$\dot{V}_i\left(t\right)=S_i\left(z_i,t\right)\dot{S}_i\left(z_i,t\right)+\dot{S}_i\left(z_i,t\right)+\ddot{S}_i\left(z_i,t\right) \tag{52}$$

$$\dot{V}_i\left(t\right)=S_i\left(z_i,t\right)\dot{S}_i\left(z_i,t\right)+\dot{S}_i\left(z_i,t\right)\left(-\lambda sign\left(\left|S_i\left(z_i,t\right)\right|\right)-\mu sign\left(\left|\dot{S}_i\left(z_i,t\right)\right|\right)\right) \tag{53}$$

$$\dot{V}_i\left(t\right)=S_i\left(z_i,t\right)\dot{S}_i\left(z_i,t\right)-\lambda\dot{S}_i\left(z_i,t\right)sign\left(\left|S_i\left(z_i,t\right)\right|\right)-\mu\dot{S}_i\left(z_i,t\right)sign\left(\left|\dot{S}_i\left(z_i,t\right)\right|\right) \tag{54}$$

If the choice of λ and μ are such that, $\lambda>\mu$ and $\mu>\dot{g}_{2_i}\left(t\right)$, then,

$$\leq S_i\left(z_i,t\right)\dot{S}_i\left(z_i,t\right)-\lambda\dot{S}_i\left(z_i,t\right)\frac{S_i\left(z_i,t\right)}{\left|S_i\left(z_i,t\right)\right|}-\left(\mu-\dot{g}_{2_i}\left(t\right)\right)\left|\dot{S}_i\left(z_i,t\right)\right| \tag{55}$$

$$\leq\left|S_i\left(z_i,t\right)\right|\left|\dot{S}_i\left(z_i,t\right)\right|\left(1-\frac{\lambda}{\left|S_i\left(z_i,t\right)\right|}\right)-\left(\mu-\dot{g}_{2_i}\left(t\right)\right)<0 \tag{56}$$

The above equation ensures that the sliding surface is stable and therefore SOSMC converges to zero in finite time with matched uncertainties and disturbances.

SIMULATION AND RESULTS: CASE STUDIES

To illustrate the effectiveness of the said design, a three-area interconnected power system is considered. The basic system parameters of three area power system can be referred from Table 1 (Mi et al., 2013).

Each area in a three-area power system is studied individually. Nonlinear sliding surface and controller is designed for each control area separately. The closed loop poles considered for the calculation of K_i are $-2.0020\pm2.0389j$; $-1.9979\pm1.5023j$. Step load disturbances $\Delta P_{d_1}(t)=0.02$ p.u. MW, $\Delta P_{d_2}(t)=0.015$ p.u. MW and $\Delta P_{d_3}(t)=0.01$ p.u. MW (Mi et al., 2013) which can also be referred as unmatched uncertainties in three control areas and are applied at *t=0* seconds simultaneously. Based on the system parameter given in table 1, system matrix, input matrix, disturbance matrix and interconnection gain matrix are formulated for each area.

For Area 1:

Table 1. Parameter values used for three area interconnected power system

Area	T_{p_i}	K_{P_i}	T_{T_i}	T_{G_i}	R_i	K_{E_i}	K_{B_i}	K_{sij}
1	20	120	0.3	0.08	2.4	10	0.41	0.55
2	25	112.5	0.33	0.072	2.7	9	0.37	0.65
3	20	115	0.35	0.07	2.5	7.1	0.4	0.545

$$\bar{A}_1 = \begin{bmatrix} -0.05 & 6 & 0 & 0 & -0.955 \\ 0 & -3.472 & 3.472 & 0 & 0 \\ -5.878 & 0 & -13.021 & -13.021 & 0 \\ 4 & 0 & 0 & 0 & 1.592 \\ 6.283 & 0 & 0 & 0 & 0 \end{bmatrix}$$

$$\bar{B}_1 = \begin{bmatrix} 0 & 0 & 13.021 & 0 & 0 \end{bmatrix}^T$$

$$\bar{F}_1 = \begin{bmatrix} -6 & 0 & 0 & 0 & 0 \end{bmatrix}^T$$

$$E_{12} = \begin{bmatrix} 0 & 0 & 0 & 0 & 0.478 \\ 0 & 0 & 0 & 0 & 0 \\ 0 & 0 & 0 & 0 & 0 \\ 0 & 0 & 0 & 0 & -0.796 \\ 0 & 0 & 0 & 0 & 0 \end{bmatrix}$$

$$E_{13} = \begin{bmatrix} 0 & 0 & 0 & 0 & 0.478 \\ 0 & 0 & 0 & 0 & 0 \\ 0 & 0 & 0 & 0 & 0 \\ 0 & 0 & 0 & 0 & -0.796 \\ 0 & 0 & 0 & 0 & 0 \end{bmatrix}$$

For Area 2:

$$\bar{A}_2 = \begin{bmatrix} -0.04 & 4.5 & 0 & 0 & -0.716 \\ 0 & -3.157 & 3.157 & 0 & 0 \\ -5.805 & 0 & -14.468 & -14.468 & 0 \\ 4 & 0 & 0 & 0 & 1.592 \\ 6.283 & 0 & 0 & 0 & 0 \end{bmatrix}$$

$$\bar{B}_2 = \begin{bmatrix} 0 & 0 & 14.468 & 0 & 0 \end{bmatrix}^T$$

$$\bar{F}_2 = \begin{bmatrix} -4.5 & 0 & 0 & 0 & 0 \end{bmatrix}^T$$

$$E_{21} = \begin{bmatrix} 0 & 0 & 0 & 0 & 0.358 \\ 0 & 0 & 0 & 0 & 0 \\ 0 & 0 & 0 & 0 & 0 \\ 0 & 0 & 0 & 0 & -0.796 \\ 0 & 0 & 0 & 0 & 0 \end{bmatrix}$$

$$E_{23} = \begin{bmatrix} 0 & 0 & 0 & 0 & 0.358 \\ 0 & 0 & 0 & 0 & 0 \\ 0 & 0 & 0 & 0 & 0 \\ 0 & 0 & 0 & 0 & -0.796 \\ 0 & 0 & 0 & 0 & 0 \end{bmatrix}$$

For Area 3:

$$\bar{A}_3 = \begin{bmatrix} -0.05 & 5.75 & 0 & 0 & -0.915 \\ 0 & -2.976 & 2.976 & 0 & 0 \\ -6.448 & 0 & -14.881 & -14.881 & 0 \\ 4 & 0 & 0 & 0 & 1.592 \\ 6.283 & 0 & 0 & 0 & 0 \end{bmatrix}$$

$$\bar{B}_3 = \begin{bmatrix} 0 & 0 & 14.881 & 0 & 0 \end{bmatrix}^T$$

$$\bar{F}_3 = \begin{bmatrix} -5.75 & 0 & 0 & 0 & 0 \end{bmatrix}^T$$

$$E_{31} = \begin{bmatrix} 0 & 0 & 0 & 0 & 0.458 \\ 0 & 0 & 0 & 0 & 0 \\ 0 & 0 & 0 & 0 & 0 \\ 0 & 0 & 0 & 0 & -0.796 \\ 0 & 0 & 0 & 0 & 0 \end{bmatrix}$$

$$E_{32} = \begin{bmatrix} 0 & 0 & 0 & 0 & 0.458 \\ 0 & 0 & 0 & 0 & 0 \\ 0 & 0 & 0 & 0 & 0 \\ 0 & 0 & 0 & 0 & -0.796 \\ 0 & 0 & 0 & 0 & 0 \end{bmatrix}$$

Case 1: In this case, plant is designed with nominal system parameters under normal load disturbances. A comparative study is done between controllers designed with linear sliding surface, nonlinear sliding surface and nonlinear second order sliding surface under normal load disturbances. Controller designed with nonlinear second order sliding surface ensures better finite time convergence of frequency and tie line power deviations which can be seen from figure 4. From figure 5, it can be concluded that finite time convergence of frequency and tie line power deviations using NL-SOSMC is obtained with chattering free control signal whereas linear SMC and nonlinear SMC is affected by the problem of chattering. In practice, chattering free control signal is always advantageous as it reduces wear and tear of moving mechanical parts, heat losses in power circuits etc. The said design ensures better transient and steady state response compared to that of linear SMC and nonlinear SMC.

Case 2: In this case, robustness of the proposed controller is validated for plant designed with matched uncertainties (Mi et al., 2013). The designed controller confirms finite time convergence of frequency deviation under load disturbances with matched uncertainty. Here, the matched uncertainty ΔA_i is compensated by NL-SOSMC. From figure 6, it can be seen that the response with matched uncertainties has comparatively more undershoot and large settling time compared to that of plant considered without matched uncertainties. This shows that the said design can work even with matched uncertainty without any major deviations in frequency. It should be kept in mind that matched uncertainties lie in the range space of input matrix and enter the system along with input.

$$\Delta A_i = \begin{bmatrix} 0 & 0 & 0 & 0 & 0 \\ 0 & 0 & 0 & 0 & 0 \\ -2.26\cos(t) & 2\cos(t) & -2.604\cos(t) & 3\cos(t) & 0 \\ 0 & 0 & 0 & 0 & 0 \\ 0 & 0 & 0 & 0 & 0 \end{bmatrix}$$

Figure 4. Response under step load disturbance using linear SMC, NL-SMC and NL-SOSMC

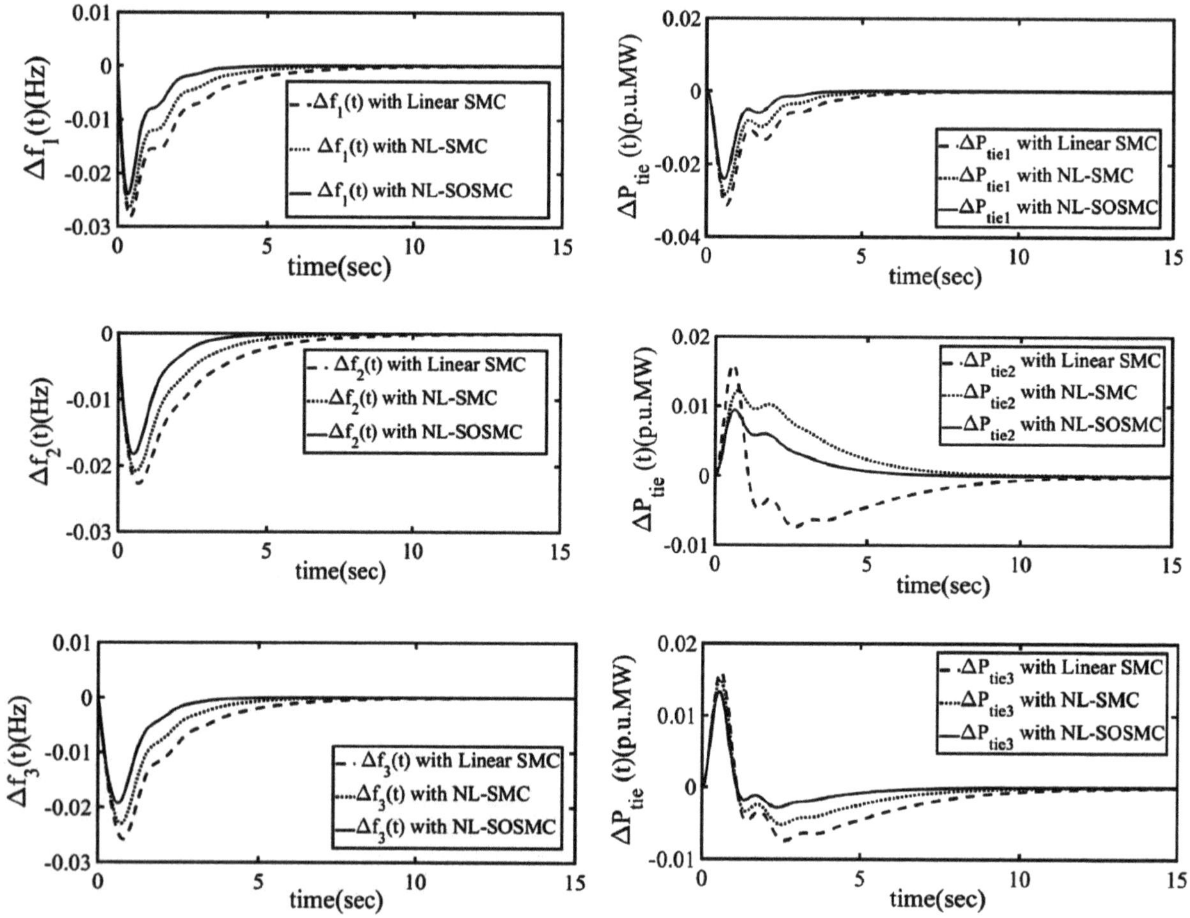

a) Frequency deviation $\left(\Delta f_i(t)\right)$ b) Tie line power deviation $\left(\Delta P_{tie_i}(t)\right)$

Furthermore, the robustness of the proposed controller is also validated with parameter variations. The following range of parameter variation is considered which is obtained by varying the values of system parameters by 20% from their nominal values. Figure 7 confirm that the proposed controller guarantees finite time convergence of frequency with system parameter variations. The proposed controller can also be validated with variation of parameters up to (35-40) % from their nominal value.

$$\frac{1}{T_{T_1}} \in [2.778\ 4.167],$$

$$\frac{1}{T_{G_1}} \in [10.417\ 15.625],$$

Figure 5. Control signals for linear SMC, nonlinear SMC and nonlinear SOSMC

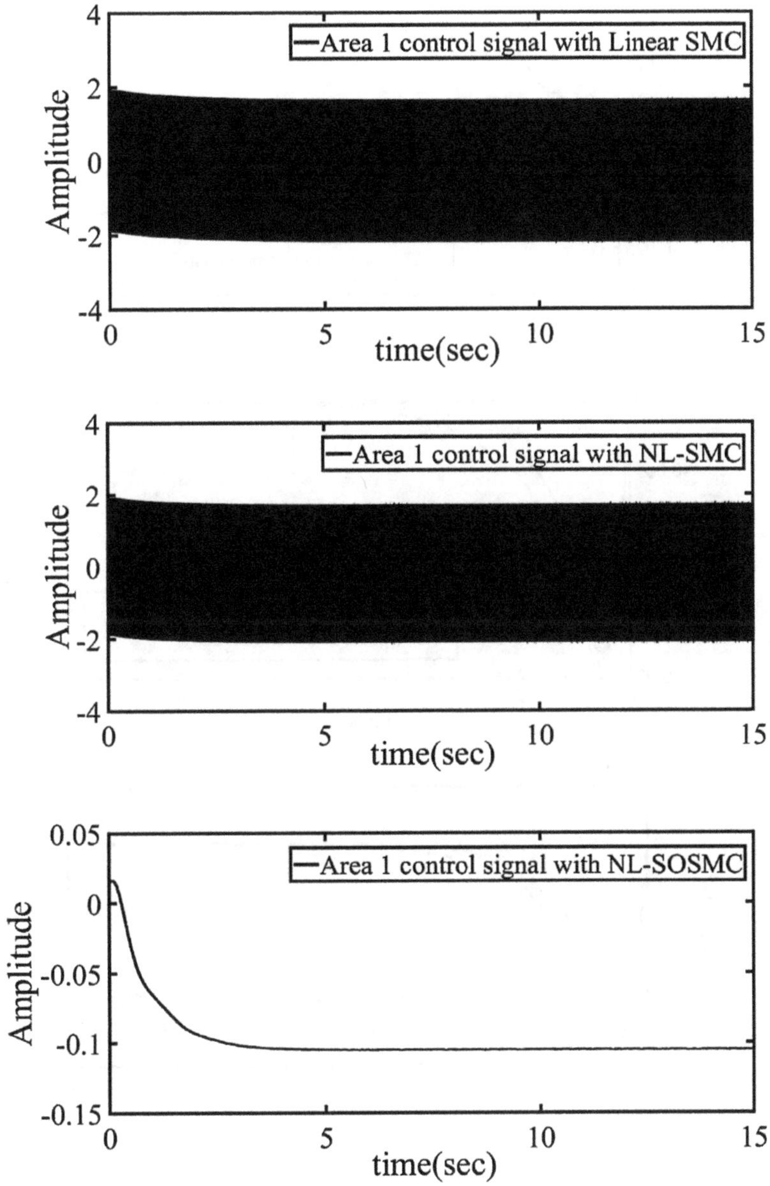

$$\frac{1}{R_1 T_{G_1}} \in [3.617\ 8.138],$$

$$\frac{1}{T_{T_2}} \in [2.525\ 3.788],$$

Figure 6. $\Delta f_i(t)$ *using nonlinear SOSMC without and with matched uncertainty*

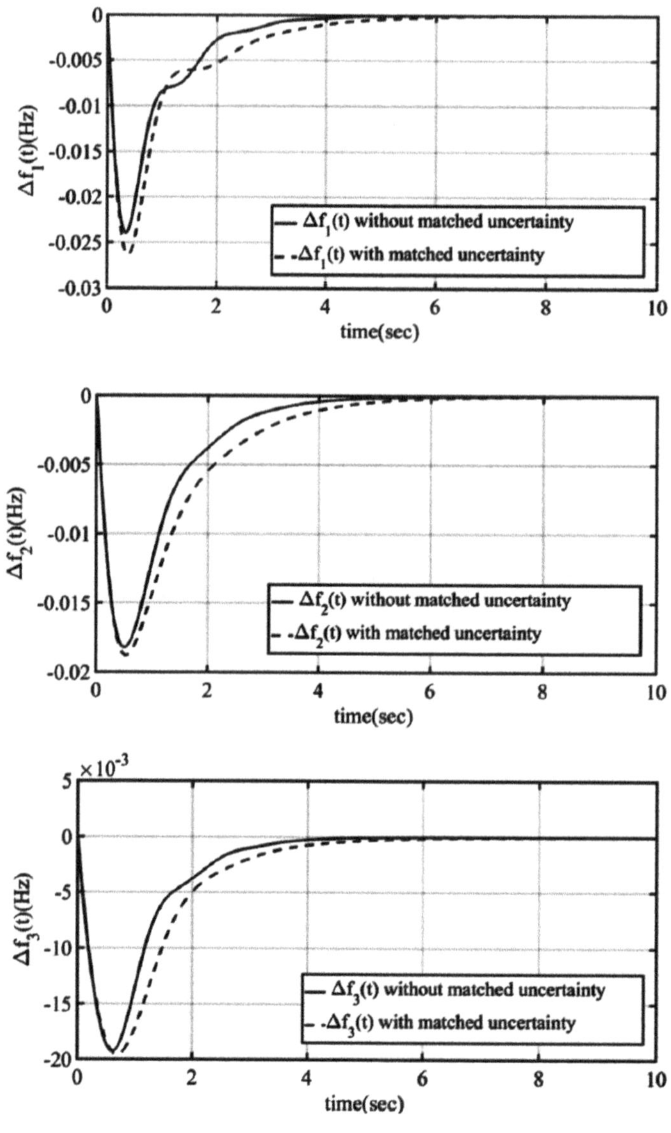

$$\frac{1}{T_{G_2}} \in [11.574\ 17.361],$$

$$\frac{1}{R_2 T_{G_2}} \in [3.968\ 8.929],$$

$$\frac{1}{T_{T_3}} \in [2.381\ 3.571],$$

Figure 7. Frequency deviation $\left(\Delta f_i\left(t\right)\right)$ *using nonlinear SOSMC under parameter variation*

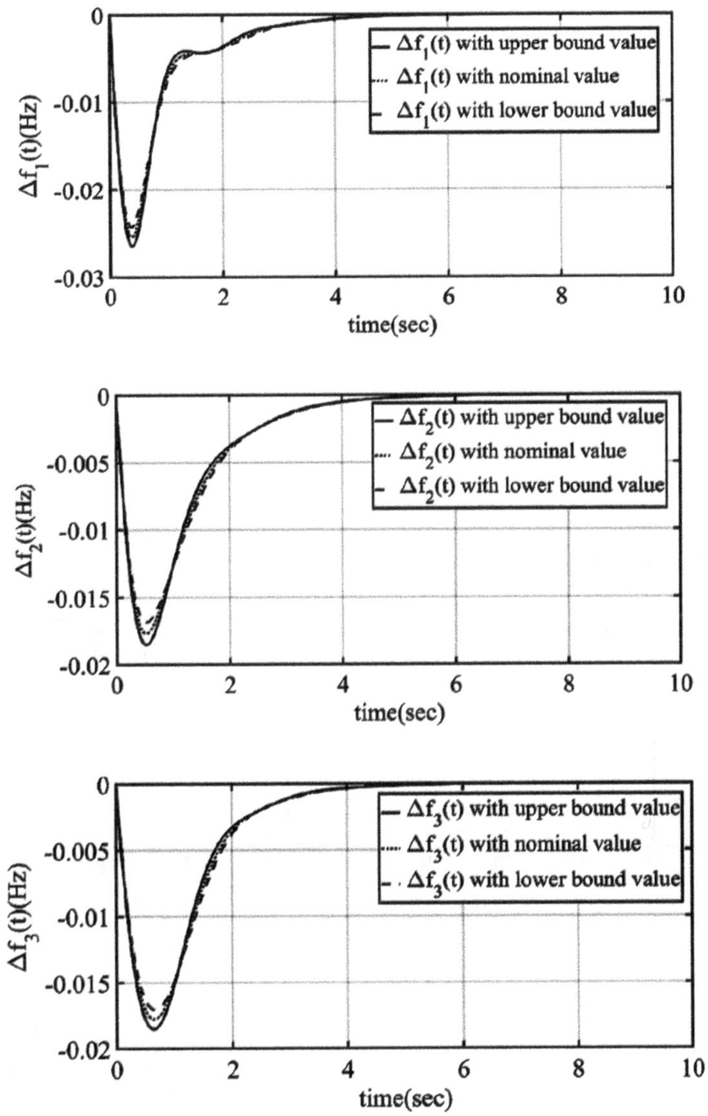

$$\frac{1}{T_{G_3}} \in [11.905\ 17.857],$$

$$\frac{1}{R_3 T_{G_3}} \in [3.968\ 8.929]$$

Case 3: In this case, robustness of the proposed controller is tested under random step load disturbances. Random step load disturbances shown in figure 8 are given in all the three areas of power systems. The frequency deviations using nonlinear second order SMC under the random load disturbance confirms better transient and steady state performance compared to that of linear and nonlinear SMC. Figures 8 confirm the robustness and insensitivity property of SMC under random step load disturbances.

Case 4: In this case, performance indices like ISE, IAE and ITAE are compared for each controller only for area 1 under normal load disturbance. It is observed from table 2 as well as from simulation results in figure 9 that ISE, IAE and ITAE for the proposed controller is significantly small compared to that of linear SMC and nonlinear SMC. It can now be concluded that the said design has lower value of ISE, IAE and ITAE.

$$\text{ISE}=\int \left(\Delta f\right)^{2} dt \tag{57}$$

Figure 8. Random load disturbances and frequency deviation $\left(\Delta f_i\left(t\right)\right)$

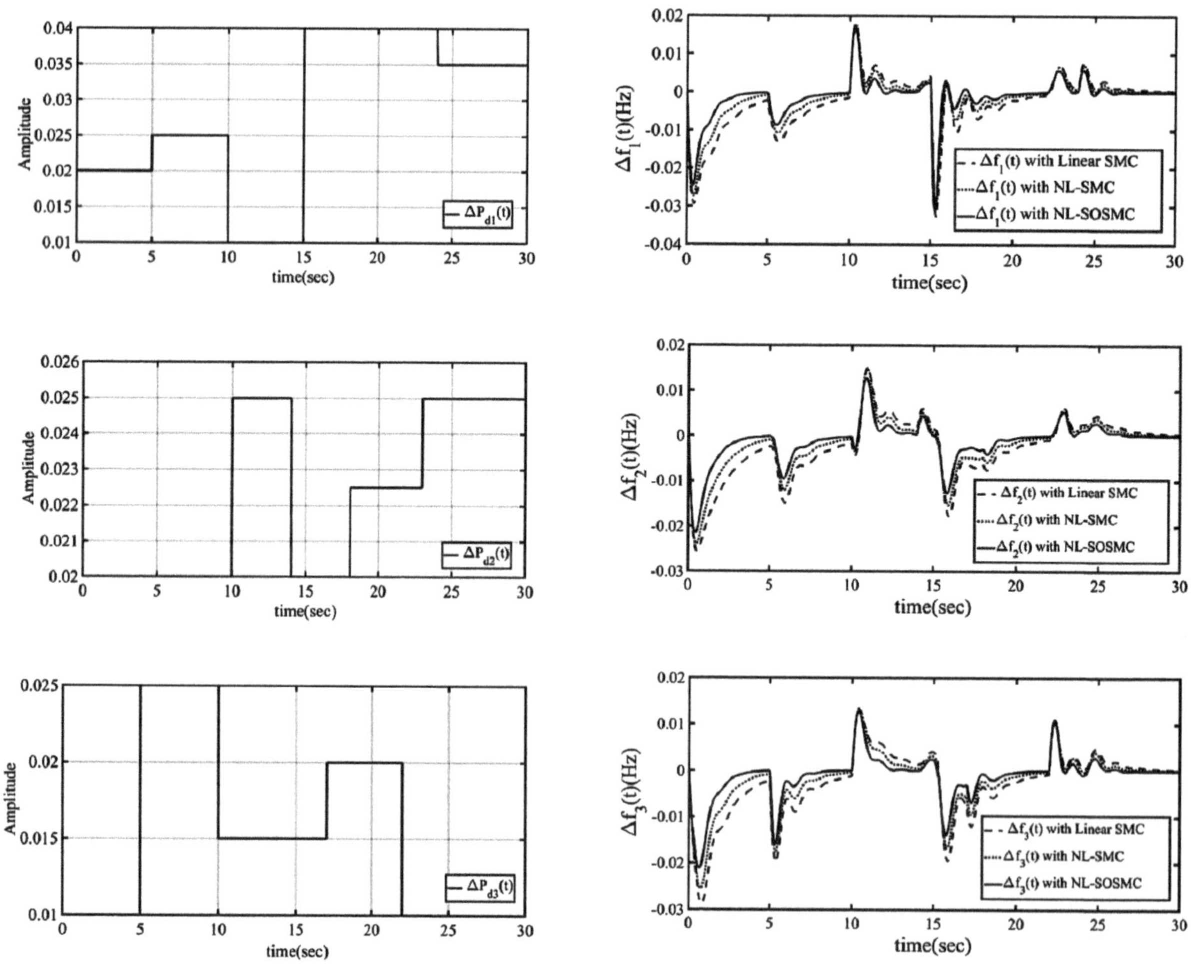

Table 2. Measure of performance indices

Controller	Integral Square Error	Integral Absolute Error	Integral Time Absolute Error
Linear SMC	0.00076	0.0534	0.104
Nonlinear SMC	0.00054	0.0375	0.054
Nonlinear SOSMC	**0.00032**	**0.0242**	**0.023**

Figure 9. ISE, IAE and ITAE of area 1 using linear SMC, nonlinear SMC and nonlinear SOSMC under normal load disturbance

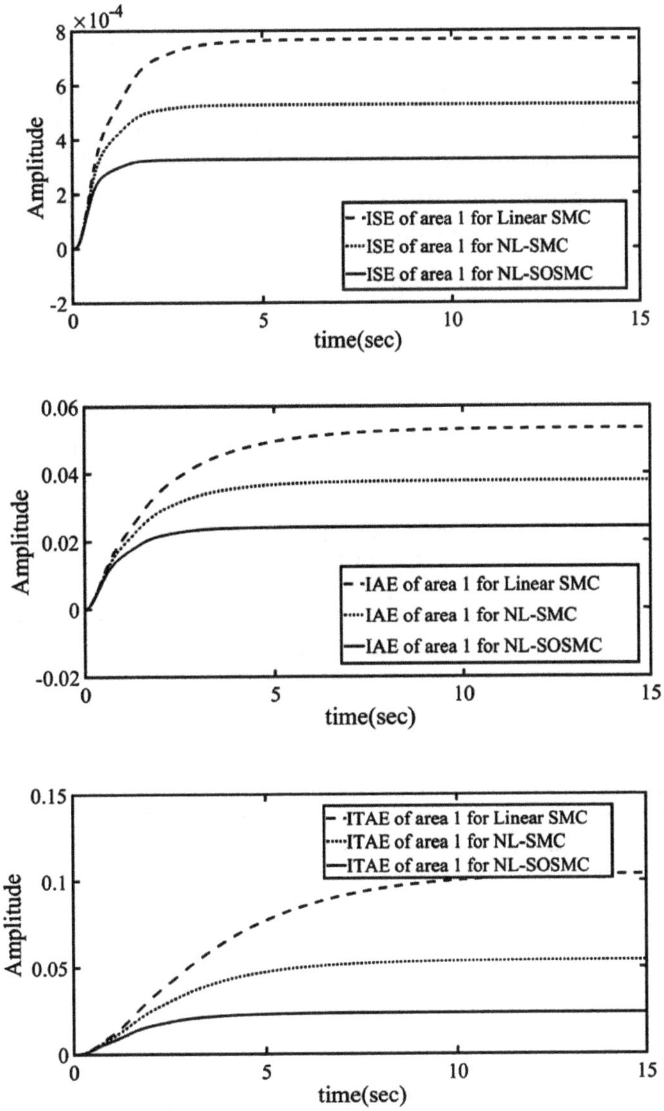

$$\text{IAE}=\int |\Delta f|\,dt \tag{58}$$

$$\text{ITAE}=\int t\,|\Delta f|\,dt \tag{59}$$

Case 5: In practice, nonlinearities in power systems cannot be avoided. In this case plant is designed with typical nonlinearities in power systems like GRC and GDB. GRC is a maximum limit on the rate of the change in the generating unit due to physical limitations of turbine (Tan et al., 2017). There are several ways of modelling GRC and it can be refereed from literature. In this chapter, GRC is modelled using closed loop modelling method as depicted in the figure 10. GDB is an interval of a signal band where the output of the system is zero. The values for GRC and GDB considered are ± 0.2 p.u. MW/sec and ± 0.05 respectively. Figure 11 illustrates comparative study of the proposed controller with and without nonlinearities. Plant considered with nonlinearities have comparatively higher overshoot and more settling time however the proposed controller still guarantees stable frequency deviation under GRC and GDB. This also confirms robustness of the proposed design with hard nonlinearities. The response obtained with the proposed controller for all the areas are better compared to that of linear and nonlinear SMC in the presence of nonlinearities.

Figure 10. Block diagram of interconnected power systems with GRC and GDB
(Mi et al., 2013)

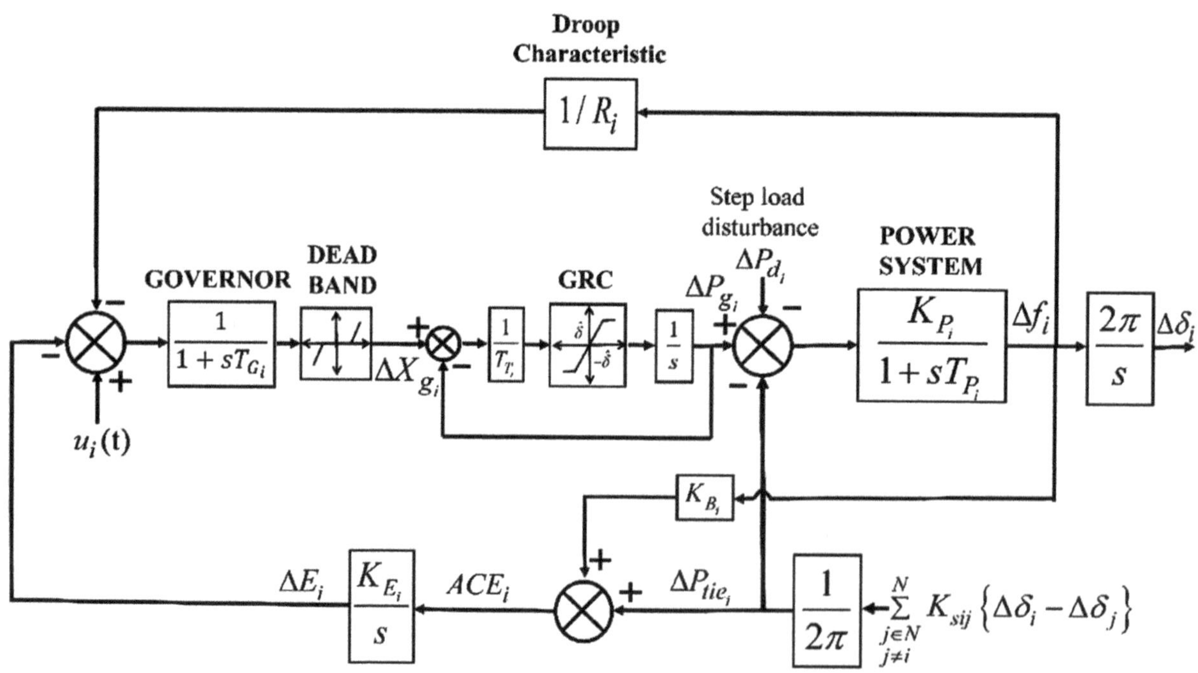

Figure 11. $\Delta f_i(t)$ *using linear SMC, nonlinear SMC and nonlinear SOSMC with and without GRC and GDB*

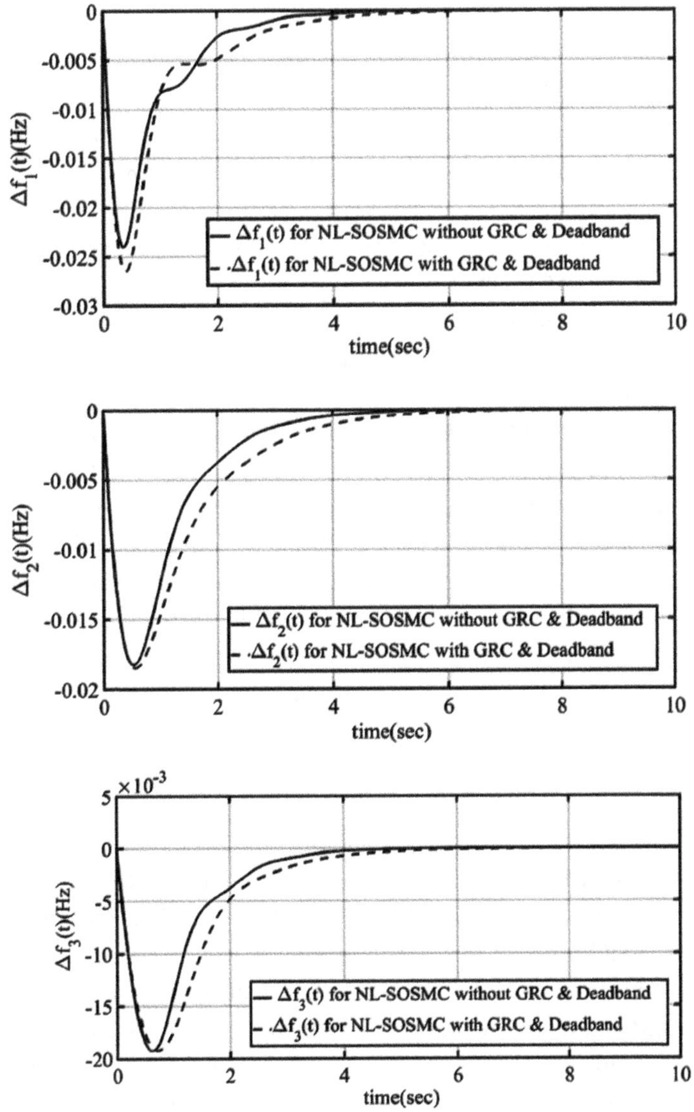

FUTURE RESEARCH DIRECTIONS

Industrial control systems are prone to cyber-attacks. LFC in power systems is one type of industrial control system. Power systems of any country can be an easy target for hackers and adversaries as it can have a major impact on the lives and economy of the country. Time delay switch attacks (TDS) can be one type of cyber-attacks. Time delays can be introduced in the sensing loop or control loop of the system or both. Introduction of time delays can degrade the performance of the controller or at worse can cause system instability. Modern power grids rely heavily on computers and multipurpose communication channels and can also be an easy target for introducing time delays. Time delays can be constant or time varying in nature. LFC of multi area power systems in the presence of constant or time varying delays can be very challenging. Thus, security of power systems from such attacks would ensure optimum performance of industrial and critical infrastructure. Load frequency problem in the presence of time delays using sliding mode is still open for research. This can be done by designing a delay estimator or predictor which can detect the time delays and can track it thus reducing the negative impact of time delays on load frequency problems.

Another aspect of future work can be integration of renewable energy resources. Depletion of fossil fuels have made way for renewable energy resources. Renewable energy resources are eco-friendly, cheap and abundant in nature. Another prospect of future research in this field can be LFC of hybrid power systems. Hybrid power systems include additional power sources like power from wind turbines, photovoltaic cells, geothermal sources, micro turbines etc. LFC become more challenging with more than one type of power sources. This require robust controllers like SMC to ensure system stability under uncertainties caused due to integration of renewable energy resources.

From control system point of view, observer-based control techniques for load frequency problems can also be a good area of research. Observer based systems are more practical compared to their counterparts. Design of any state feed back controller require information of all the system states. However, not all the system states are available or are easily measurable. In that case, observers are designed for accurate estimation of system states. State observers reduces the cost of sensors used to estimate the states directly. Also, observer-based system is less prone to external noise.

CONCLUSION

In this chapter, importance of LFC and the challenges associated with it is addressed. The extent of research in the field of LFC using different control techniques and in particular sliding mode is also briefly discussed. NL-SOSMC for load frequency problems in three area power system is designed to ensure finite time convergence of frequency and tie line power deviations under load disturbances. The proposed controller is validated under normal load disturbance (case 1), with matched and unmatched uncertainties (case 2), random load disturbances (case 3) and power systems nonlinearities like GRC and GDB (case 5). Simulation results confirm finite time convergence of frequency and tie line power deviations under load disturbance with chattering free control signal. In addition to this, the controller also ensures better transient and steady state behavior compared to that of linear SMC and nonlinear SMC in all the cases. The proposed controller also guarantees better performance indices compared to that of linear and nonlinear SMC (figure 9). Frequency deviations with power system nonlinearities like GRC and GDB is also found to be within acceptable range (figure 11). In future, LFC in hybrid power systems can be under taken. Hybrid power systems are dependent on more than one source (in general, renewable energy resources) for the production of power. Stochastic nature of renewable energy resources can be considered to model the power systems using stochastic differential equations. Along with this, LFC of time delay power systems can also be undertaken.

ACKNOWLEDGMENT

This research received no specific grant from any funding agency in the public, commercial, or not-for-profit sectors.

REFERENCES

Aditya, S. K., & Das, D. (2003). Design of load frequency controllers using genetic algorithm for two area interconnected hydropower system. *Electric Power Components and Systems*, *31*(1), 81–94. doi:10.1080/15325000390112071

Bandyopadhyay, B., Deepak, F., & Kim, K. S. (2009). Sliding Mode Control Using Novel Sliding Surfaces. *Lecture Notes in Control and Information Science*.

Bechert, T. E., & Chen, N. (1977). Area automatic generation control by multi-pass dynamic programming. *IEEE Transactions on Power Apparatus and Systems*, *96*(5), 1460–1468. doi:10.1109/T-PAS.1977.32474

Bevrani, H., & Hiyama, T. (2009). On load–frequency regulation with time delays: Design and real-time implementation. *IEEE Transactions on Power Systems*, *24*(1), 292–300.

Calovic, M. (1972). Linear regulator design for a load and frequency control. *IEEE Transactions on Power Apparatus and Systems*, *91*(6), 2271–2285. doi:10.1109/TPAS.1972.293383

Chaturvedi, D. K., Satsangi, P. S., & Kalra, P. K. (1999). Load frequency control: A generalized neural network approach. *Electrical Power & Energy Systems*, *21*(6), 405–415. doi:10.1016/S0142-0615(99)00010-1

Dev, A., & Sarkar, M. K. (2019). Robust Higher Order Observer Based Non-linear Super Twisting Load Frequency Control for Multi Area Power Systems via Sliding Mode. *International Journal of Control, Automation, and Systems*, *17*(7), 1814–1825. doi:10.100712555-018-0529-4

Dev, A., Sarkar, M. K., Asthana, P., & Narzary, D. (2019). Event-triggered adaptive integral higher order sliding mode control for load frequency problems in multi area power systems. *Iranian Journal of Science and Technology. Transaction of Electrical Engineering*, *43*(1), 137–152. doi:10.100740998-018-0078-0

Elgerd, O. I., & Fosha, C. (1970). Optimum megawatt frequency control of multi-area electric energy systems. *IEEE Transactions on Power Apparatus and Systems*, *89*(4), 556–563. doi:10.1109/TPAS.1970.292602

Feliachi, A. (1987). Load frequency control using reduced order models and local observers. *International Journal of Electrical Energy Systems*, *7*(2), 72–75.

Fosha, C. E., & Elgerd, O. I. (1970). The megawatt frequency control problem: A new approach via optimal control theory. *IEEE Transactions on Power Apparatus and Systems*, *89*(4), 563–577. doi:10.1109/TPAS.1970.292603

Fulwani, D., Bandyopadhyay, B., & Fridman, L. (2011). Non-linear sliding surface: Towards high performance robust control. *IET Control Theory & Applications*, *6*(2), 235–242. doi:10.1049/iet-cta.2010.0727

Ghoshal, S. P. (2003). Multi area frequency and tie line power flow control with fuzzy logic based integral gain scheduling. *Journal of Institute of Engineers*, *84*, 135–141.

Hsu, Y. Y. & Chan, W. C. (1984). Optimal variable structure controller for load-frequency control of interconnected hydro-thermal power systems. *Electrical Power & Energy Systems*, *6*(4), 221–229.

Indulkar, C. S., & Raj, B. (1995). Application of fuzzy controller to automatic generation control. *Electric Machines and Power Systems*, *23*(2), 209–220. doi:10.1080/07313569508955618

Jiang, L., Yao, W., Wu, Q. H., Wen, J. Y., & Cheng, S. J. (2012). Delay- dependent stability for load frequency control with constant and time varying delays. *IEEE Transactions on Power Systems*, *27*(2), 932–941. doi:10.1109/TPWRS.2011.2172821

Juang, C. F., & Lu, C. F. (2002). Power system load frequency control with fuzzy gain scheduling designed by new genetic algorithms. IEEE International Conference on Fuzzy Systems, 1, 64–68.

Khodabakhshian, A., & Edrisi, M. (2008). A new robust PID load frequency controller. *Control Engineering Practice*, *16*(9), 1069–1080. doi:10.1016/j.conengprac.2007.12.003

Kwatny, H. G., Kalnitsky, K. C., & Bhatt, A. (1975). An optimal tracking approach to load frequency control. *IEEE Transactions on Power Apparatus and Systems*, *94*(5), 1635–1643. doi:10.1109/T-PAS.1975.32006

Li, J., Liu, X., & Su, X. (2018). Sliding mode observer-based load frequency control of multi area power systems under delayed inputs attack. *Chinese Control and Decision Conference*. 10.1109/CCDC.2018.8407768

Liao, K., & Xu, Y. (2017). A robust load frequency control scheme for power systems based on second order sliding mode and disturbance observer. *IEEE Transactions on Industrial Informatics*, *14*(7), 3076–3086. doi:10.1109/TII.2017.2771487

Liaw, C. M. (1994). Design of a reduced-order adaptive LFC for an interconnected hydrothermal power system. *International Journal of Control*, *60*(6), 1051–1063. doi:10.1080/00207179408921510

Lim, K. Y., Wang, Y., & Zhou, R. (1996). Robust decentralized load-frequency control of multi-area power systems. *IEE Proceedings. Generation, Transmission and Distribution*, *143*(5), 377–386. doi:10.1049/ip-gtd:19960452

Mi, Y., Fu, Y., Li, D., Wang, C., Loh, P. C., & Weng, P. (2016). The sliding mode load frequency control for hybrid power system based on disturbance observer. *International Journal of Electrical Power & Energy Systems*, *74*, 446–452. doi:10.1016/j.ijepes.2015.07.014

Mi, Y., Fu, Y., Wang, C., & Wang, P. (2013). Decentralized Sliding Mode Load Frequency Control for Multi-Area Power Systems. *IEEE Transactions on Power Systems*, *28*(4), 4301–4309. doi:10.1109/TPWRS.2013.2277131

Mi, Y., Hua, X., Liu, Y., Fu, Y., Wang, C., Wang, P., & Loh, P. C. (2017). Sliding mode load frequency control for multi area time-delay power system with wind power integration. *IET Generation, Transmission & Distribution*, *11*(18), 4644–4643. doi:10.1049/iet-gtd.2017.0600

Mondal, S., Gokul, T. V., & Mahanta, C. (2012). Chattering Free Sliding Mode Controller for Mismatched Uncertain System. *6th International Conference on Industrial and Information Systems*. 10.1109/ICIInfS.2012.6304790

Mondal, S., & Mahanta, C. (2011). Nonlinear Sliding Surface based Second Order Sliding Mode Controller for Uncertain Linear Systems. *Communications in Nonlinear Science and Numerical Simulation, Elsevier*, *16*(9), 3760–3769. doi:10.1016/j.cnsns.2010.12.020

Moorthi, V. R., & Aggarwal, R. P. (1972). Suboptimal and near optimal control of a load frequency control system. *Proceedings of the Institution of Electrical Engineers*, *119*(11), 1653–1660. doi:10.1049/piee.1972.0329

Pan, C. T., & Liaw, C. M. (1989). An adaptive controller for power system and load frequency control. *IEEE Transactions on Power Systems*, *4*(1), 122–1228. doi:10.1109/59.32469

Pandey, S. K., Mohanty, S. R., & Kishor, N. (2013). A literature survey on load–frequency control for conventional and distribution generation power systems. *Elsevier Renewable and Sustainable Energy Reviews*, *25*, 318–334. doi:10.1016/j.rser.2013.04.029

Prasad, S., Purwar, S., & Kishor, N. (2017). Non-linear sliding mode load frequency control in multi area power system. *Control Engineering Practice*, *61*, 81–92. doi:10.1016/j.conengprac.2017.02.001

Prasad, S., Purwar, S., & Kishor, N. (2018). Load frequency regulation using observer based nonlinear sliding mode control. *Electrical Power and Energy Systems*, *104*, 178–193. doi:10.1016/j.ijepes.2018.06.035

Ray, G., Prasad, A. N., & Prasad, G. D. (1999). A new approach to the design of robust load-frequency controller for large scale power systems. *Electric Power Systems Research*, *51*(1), 13–22. doi:10.1016/S0378-7796(98)00125-4

Ray, G., Sitansu, D., & Bhattacharyya, T. K. (2004). Multi-area load Frequency control of power systems: A decentralized variable structure Approach. *Electric Power Components and Systems*, *33*(3), 315–331. doi:10.1080/15325000590474672

Rerkpreedapong, D., Hasanovic, A., & Feliachi, A. (2003). Robust load frequency control using genetic algorithms and linear matrix inequalities. *IEEE Transactions on Power Systems*, *18*(2), 855–861. doi:10.1109/TPWRS.2003.811005

Sargolzaei, A., Yen, K. K., & Abdelghani, M. N. (2016). Preventing time delay switch attacks on load frequency control in distributed power systems. *IEEE Transactions on Smart Grid*, *7*(2), 1176–1185.

Sarkar, M. K., Dev, A., Asthana, P., & Narzary, D. (2018). Chattering free robust adaptive integral higher order sliding mode control for load frequency problems in multi area power systems. *IET Control Theory & Applications*, *12*(9), 1216–1227. doi:10.1049/iet-cta.2017.0735

Sinha, S. K., Patel, R. N., & Prasad, R. (2010). Application of GA and PSO tuned fuzzy controller for AGC of three area thermal-hydropower system. *International Journal of Computer Theory and Engineering*, *2*(2), 1793–8201.

Sun, Y., Wang, Y., Wei, Z., Sun, G., & Wu, X. (2018). Robust H_∞ load frequency control of multi area power system with time delay: A sliding mode control approach. *IEEE/CAA Journal of Automatica Sinic*, *5*(2), 610-617.

Tan, W., Chang, S., & Zhou, R. (2017). Load frequency control of power system with nonlinearities. *IET Generation, Transmission & Distribution*, *11*(17), 4307–4313. doi:10.1049/iet-gtd.2017.0599

Tsay, T. S. (2011). Load frequency control of interconnected power system with governor backlash nonlinearities. *International Journal of Electrical Power & Energy Systems*, *33*(9), 1542–1549. doi:10.1016/j.ijepes.2011.06.005

Velusami, S., & Ramar, K. (1997). Design of observer-based decentralized load-frequency controllers for interconnected power systems. *International Journal of Power and Energy Systems*, *17*(2), 152–160.

Yamashita, K., & Taniguchi, T. (1986). Optimal observer design for load frequency control. *International Journal of Electrical Power & Energy Systems*, *8*(2), 93–100. doi:10.1016/0142-0615(86)90003-7

Zhang, C. K., Jiang, L., Wu, Q. H., He, Y., & Wu, M. (2013). Delay dependent robust load frequency control for time delay power systems. *IEEE Transactions on Power Systems*, *28*(3), 2192–2201. doi:10.1109/TPWRS.2012.2228281

ADDITIONAL READING

Bandyopadhyay, B., Deepak, F., & Kim, K. S. (2009). *Sliding Mode Control Using Novel Sliding Surfaces, Verlag Berlin Heidelberg*. Springer. doi:10.1007/978-3-642-03448-0

Elgerd, O. I. (1971). *Electric Energy Systems Theory: An Introduction*. New York, USA: The McGraw-Hill Book Company Inc.

Liu, J., & Wang, X. (2012). *Advance Sliding Mode Control for Mechanical Systems*. Beijing, China: Tsinghua University Press.

Nise, N. S. (2011). *Control Systems Engineering*. John Wiley & Sons, Inc.

Pandey, S. K., Mohanty, S. R., & Kishor, N. (2013). A literature survey on load–frequency control for conventional and distribution generation power systems. *Elsevier Renewable and Sustainable Energy Reviews*, *25*, 318–334. doi:10.1016/j.rser.2013.04.029

Pappachen, A., & Fathima, P. (2017). Critical research area on load frequency control issues in a deregulated power system: A state-of-the-art-of-review. *Elsevier Renew. Sust. Energy Rev*, *72*, 163–177. doi:10.1016/j.rser.2017.01.053

Saadat, H. (1999). *Power System Analysis*. New York, USA: The McGraw-Hill Book Company Inc.

Stevenson, W. D. (1955). *Elements of Power System Analysis*. New York, USA: The McGraw-Hill Book Company Inc.

Chapter 12
An Energy Storage System: Experimental Proposal for the Efficiency Improvement of the Electrical Network Management

Juan Aurelio Montero-Sousa
University of A Coruña, Spain

Tomás González-Ayuso
https://orcid.org/0000-0001-8146-4496
CIEMAT, Spain

Xosé Manuel Vilar Martínez
University of A Coruña, Spain

Luis Alfonso Fernandez-Serantes
FH Joanneum University of Applied Sciences, Austria

Esteban Jove
University of A Coruña, Spain

Héctor Quintián
University of A Coruña, Spain

José-Luis Casteleiro-Roca
https://orcid.org/0000-0001-9740-6477
University of A Coruña, Spain

Jose Luis Calvo Rolle
University of A Coruna, Spain

ABSTRACT

The increasing greenhouse emissions have led us to take advantage of renewable sources. The intermittency of these sources can be mitigated using energy storage systems. The present work shows three different strategies depending on the power management and other technical factors, such as energy quality, each one with a specific goal. The first strategy tries to improve the electricity quality, the second tries to reduce the penalties imposed by the grid manager to the power plant, and the third one tries to improve significantly the final economic profit of the generation companies. To achieve the above strategies, an intelligent model approach is explained with the aim to predict the energy demand and generation. These two factors play a key role in all cases. In order to validate the three proposed strategies, the data from a real storage/generation system consisting on an electrolyzer, a hydrogen tank, and a fuel cell were analyzed. In general terms, the three methods were checked, obtaining satisfactory results with an acceptable performance of the created system.

DOI: 10.4018/978-1-5225-8551-0.ch012

INTRODUCTION

Currently, the concern for climate change has led many countries to take measures to prevent global warming and mitigate its consequences. The effects of such warming can jeopardize the survival of the human race itself. This concern for the environment has led to the introduction of legislation not only internally in many countries, but also to signing international treaties.

Within these international treaties, it is worth to mention the 2015 Paris Agreement, which is expected to replace the Kyoto Protocol in the year 2020. This agreement has been achieved within the United Nations, the Framework Convention on Climate Change. Basically, it is intended to maintain the increase in the global average temperature below 2 °C, and reduce the emission of greenhouse gases.

To control the emission of greenhouse gases, the companies have emission limits. If a company exceeds the amount of gases allowed, it can buy from other companies what it is called emission rights. In Europe, the European Union Emissions Trading System (EU ETS) operates to regulate these emissions.

The energy demand usually increases with the level of development of a society and with the time (Figure 1). Then, it is necessary to improve the efficiency of the electrical systems and to cover the increase of the demand while taking care of the environment.

Two additional factors that must be mentioned in relation to the current trend and that determine the energy source in an electrical system are:

- Limited traditional sources. Factors like the price makes them less promising (Rao, 2011).
- Nuclear power is difficult to control and represents a high risk in case of accident, for example what happened in Chernobyl or Fukushima power plants (Ferguson, 2011).

Due to the previous factors, renewable energy is suitable for power generation. Although these sources have been known for many years, their use in energy generation is still arising. One significant issue that these energy sources have to deal with is the interruptible generation, the energy might be generated when there is not demand from the consumers. For instance, the solar energy is higher in the middle of

Figure 1. Electricity demand in Spain (GWh)

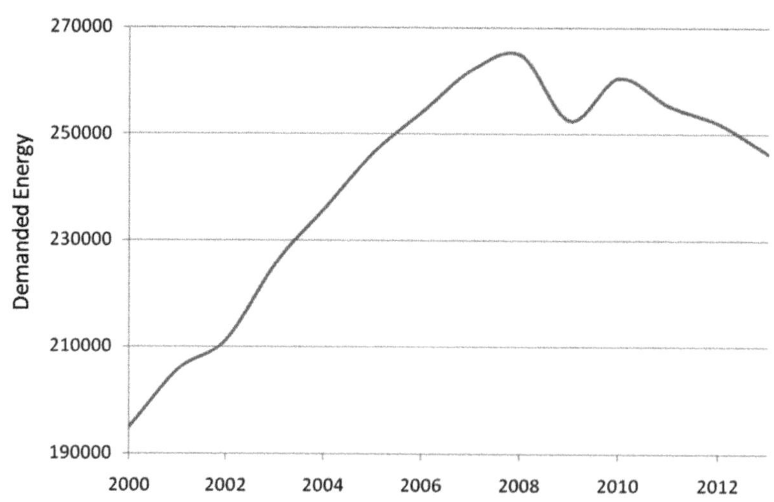

the day, where the heater is not necessary in a house; or the wind can blow when the energy demand decreases, such as what happens during the night (MacKay, 2009). This fact can be seen in Figure 2, where the wind speed and energy production registered at Sotavento, Galicia is represented.

With the aim of coordinating the generation and the demand, different storage systems have been implemented in the industry. For example, one way of energy storage consists on pumping water at reservoirs when there is an excess of energy generation and to turbine it when there is a demand (MacKay, 2009). Even though these energy storage systems are profitable and proved, their performance is remarkably low (Huggins, 2010). Therefore, improving the existing storage systems and coming up with new techniques that fits to the needs is justified.

The energy storage systems can add many different benefits in the electrical distribution system, depending on the type of technology of the storage system:

- They could be installed anywhere regardless of the location of the generation plant. In addition, the power and the stored energy could be of any size. The storage systems could be adapted to each specific use.
- For fast variations in the load, energy storage systems allow adapting the power generation in very short time; these systems could react faster than the power plants.
- If energy storage systems are used in conjunction with generation plants and electrical networks, it is possible to make transfer of energy from different points to others where the demand increases based on prediction systems.

Energy storage systems allow the improvement in the energy demand management. Several researches deals with the use of storage systems for peak shaving and valley filling. Moreover, the quality of the electrical system can be improved in terms of voltage and frequency.

One of the most known energy storage systems is the one based on hydrogen. In these kind of systems, a fuel cell generates electrical energy using the hydrogen as fuel. In an isolated fuel cell system, the hydrogen is stored in a tank and it is normally generated in another place. However, to consider it as an energy storage system, the hydrogen generation has to be integrated into the same facility, as part

Figure 2. Comparison of wind speed (blue - m/s) and production of a wind farm (red - kWh)

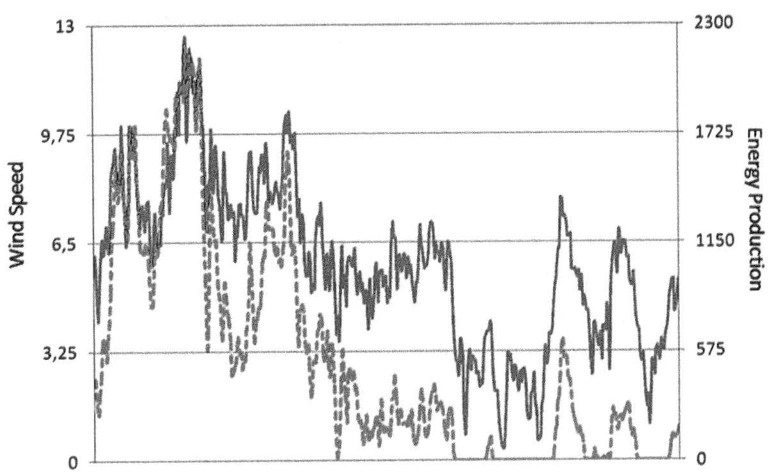

of the storage system. Usually, an electrolyzer is the equipment used to generate the hydrogen, which is stored in a pressurized tank for later use.

A proposal for the classification of the different existing techniques for managing the electricity demand is shown in this work. These systems are based on energy storage to use the energy later depending on the chosen strategy. The classification is made according to technical factors related to power distribution and economic factors.

With the aim to solve the present problem, a good model of the system performance is needed. Due to the high nonlinearity of the generation and consumption, the traditional regression techniques are not capable of solving this challenge. The random generation of renewable energies represents the main nonlinearity of the power system Moreover, the different response times to change the production from one power plant to other (to minimized cost), is, sometimes, a complicate task that need to take into account (for example, in smart grids). To manage the energy distribution in a right and efficient way, a model has to be created by using novel regression methods to predict the necessary variables (J. L. Casteleiro-Roca et al., 2013; J.-A. Montero-Sousa, Fernandez-Serantes, Casteleiro-Roca, Vilar-Martnez, & Calvo-Rolle, 2017; J. A. Montero-Sousa, Fernández-Serantes, Casteleiro-Roca, Vilar-Mart\'\inez, & Calvo-Rolle, 2017).

After this introduction, a brief summary of energy economy will be discussed. Then, strategies for interconnection between the renewable energy generation systems and the storage systems will be analyzed in detail. The different power management strategies will be explained and a new concept to model the consumption and the generation is presented. Then, a novel generic model approach to accomplish the new strategies is presented. Finally, a real experimental storage system is analyzed and the conclusions are shown.

THE NEED OF ENERGY STORAGE FROM THE ECONOMIC POINT VIEW

Economic Characteristics of Electricity

Nowadays, the electric service is a duty in modern societies and its access has to be guaranteed to every citizen. In the last years, the conservation of the environment, quality and cost have become issues that the supplies must conciliate with (Energía y sociedad, n.d.-a).

In comparison with other markets, the commercialization of electricity differs that the supply and demand cannot be just balanced by the quantity and price. The electricity is a joint and two folded economic service:

- The user has to pay for it and it is provided only if there is demand, so in this sense it is a private service.
- On the other hand, it behaves as a public service as it must have a minimum quality and security of supply. Furthermore, the safety of the system is not private and the government ensures the coverage rates (Utray Fabra, n.d.). Thus, regulatory mechanisms, stablished by the administration, must supplement the markets (Fabra, 2007).

In both cases, the system must be reliable, meaning that it is a non-excludable service (the improvement of the supply benefit all the consumers as they are all connected to the same network) and a non-rival service (the supply to other costumers is not stopped or reduced by the consumption that a person makes).

One problem of the electric system is how to remunerate the different generation systems, as the units of the service are indistinguishable for the different agents. In this way, the electricity can be defined as a homogeneous property while the ways to produce it are not homogenous (Utray Fabra, n.d.).

On the other hand, in the electricity market there are so call flat rates, where the consumer pays a fixed price for the unit of consumed energy independently of the time of the day or the location. In addition, it does not promote the transfer demand to the time of the day in which the price should be lower, or the transfer of large industrial consumer plants to areas with high potential energy generation. Large investments often have to be made in electricity transmission networks (Benavides, n.d.) (Fernandez-Serantes, Berger, Stocksreiter, & Weis, 2016; Fernández Serantes, 2014).

Evolution of the Electricity Market

At the end of the 19th century, the use of electricity for commercial purposes began. Firstly, electricity was obtained from small generators located in the cities where it was needed. The appearance of the transformers allowed to transport the electricity to large distances and to take advantage of the dams created in faraway rivers. For this purpose, large corporations were needed to contribute large amounts of capital, producing a concentration process of the sector (SOUSA, ROCA, & ROLLE, 2017a).

After the Second World War, there was a process of nationalization of the sector in most countries of the Western world (SOUSA, ROCA, & ROLLE, 2017b). Hence, natural monopolies were created. However, in Spain the sector remained in private hands but controlled by the government; the state also entered the sector with the creation of a company called ENDESA.

This period is characterized by searching new sources for electricity production and the improvement of the existing ones. For example, the first nuclear power plants are put into operation and the performance of thermal power plants is improved.

After the energy crisis that began in 1973, in many countries there was a deregulation process of the electricity market, eliminating natural monopolies and establishing competitive markets. In all countries belonging to the European Union, the community legislation enhanced this deregulation process; companies must have separate generation, transport and marketing activities. In addition, the power lines of different European have been interconnected.

More recently, the European Union has approved a package of climate and energy measures whose objectives must be met in 2020. The main objectives are:

- Reduction of 20% in greenhouse gas emissions (in relation to 1990).
- A 20% of renewable energy production.
- Improvement of 20% of energy efficiency.

The EU countries have assumed binding national objectives to increase until 2020 the percentage of renewable energies generation. The objectives vary according to the starting situations of renewable energy production in each country; the goal is to reach 20% in the whole European Union in 2020 and to achieve a 10% share of renewable energy in the transport sector (Blanco-Rodriguez, Fernández-Serantes, Otero-Pazos, Calvo-Rolle, & de Cos Juez, 2017).

The EU's main instrument for reducing greenhouse gas emissions is the emissions trading system. It is used for the aviation sector and the large installations of the electrical and industrial sectors. Apart from the emissions trading system, an emission reduction target has been set for the housing sector, agriculture, waste and transport (excluding air transport).

The current electrical systems are characterized by equalizing generation and demand at any time. The use of electrical energy storage systems (batteries, hydrogen…) would allow to improve the energy efficiency and the management of the energy obtained by wind turbines, solar panels and other renewable energy systems; with the storage, the excess energy generated would be saved and then, used when a greater demand arises.

The Influence of Renewable Energy in the Electricity Market

In recent years, environmental issues have led to a great development of renewable energies, obtained from inexhaustible sources (wind, geothermal, solar…). This type of energy does not cause damage to the environment, since they do not produce emissions. But the generation of wind energy has the characteristics of intermittency and unpredictability. For this reason, the implementation of wind energy as an energy generation instrument involves demand management problems. Electricity is generated only when the wind blows at a certain intensity (Figure 2). In times of high demand, the wind may not blow; on the contrary, it may be that in moments of low demand the wind blows with intensity. In this last case, since the demand and the generation have to be balanced at all times, we would need to reduce the production of conventional power plants or stop the production of wind turbines (wind spills). If the wind decreases or ceases, the conventional power plants have to take over the demand. In any case, the conventional power plants must be available to manage these lags.

The use of storage systems would manage better these differences between the supply and the demand, storing in the case of greater supply and delivering energy to the network in the case of higher demand. This fact would also allow not having to maintain the conventional support plants. Due to the wind dependence, a reserve power (hot standby) is needed to manage the wind power, which must reach, at least, 92.5% of the installed renewable power (Energía y sociedad, n.d.-b).

STRATEGIES OF INTERCONNECTION BETWEEN GENERATION SYSTEMS AND ENERGY STORAGE SYSTEMS

As the renewable energy is inherently variable due to the availability of the energy source (MacKay, 2009), it may be the case that the production of electricity occurs when the demand is low. In contrary, it could happen that when the demand is high, there is no production of energy from the renewable supplies, so it would be necessary to supplement this lack of energy using the power basis plants to cover the whole demand.

In order to balance the offer and the demand, the managers of the network make an estimation of the curve of energy demand of a country, which can approximate quite accurate the demand for the following 24 hours (Egido Cortés, Rouco Rodríguez, Alonso Perez, Porras Muñoz, & Ruiz Mendoza, n.d.) (Figure 3). The estimations are used by those managers to interlock the different power plants of the system with the objective of meeting the demand. The possible deviations from the estimation can be of two types:

Figure 3. Comparison between estimated (blue-continuous line) and real energy demand (red-dashed line)

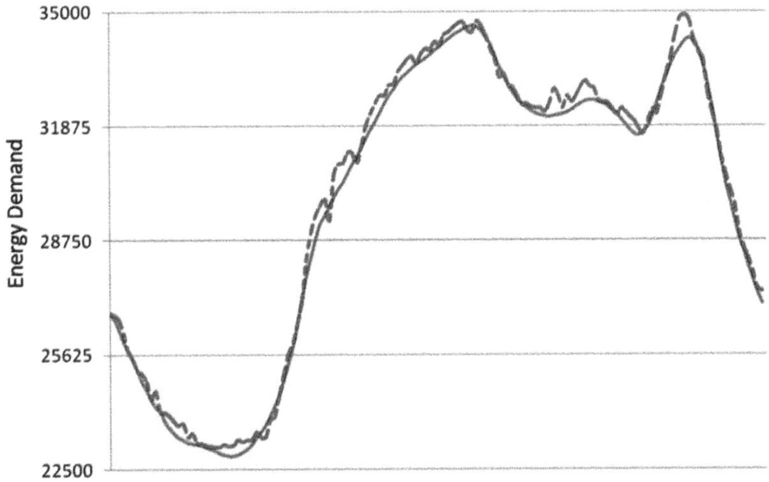

- The energy production is greater than the demand: in this case, some of the power plants will stop the energy production.
- The demand is greater than the generated energy at a certain time: the managers of the power network will connect several types of power plants depending on the time to engage the system (Hot Book) (Energía y sociedad, n.d.-b).

In this way, due to the interrupt generation of the renewable energy, a power network just based on this type of energy is not feasible. If the production is higher than the demand, some energy will be lost and if there is a higher demand than the production, it cannot be covered and intermittences will occur in the electric system.

In order to regulate the demand and the production in a system based on renewable energy, storage systems represent a solution. These systems store energy when the demand is lower than the production and supply or return this energy when the renewable energy is not available.

Stability of a power system is one of the most important parts to be controlled. This stability can be measured in an easy way taking into account that the frequency of a power system should be constant all time, and equal to 50 Hz (in Spain). An energy storage system could be used in a future to control the frequency, as they can change from "generator" to "consumer" depending on the system needs. Normally, storage system are used to have an additional energy to use if it is necessary, but from the stability point of view, that storage system should be used as load, to charge instead of turn off a power plant (for example). The system proposed in the next section could be used to predict the power consumption, and to adapt these storage systems to ensure the stability and to increase the efficiency of the power system.

This chapter presents three different strategies for energy management within a production system based on renewable energy. The classification is done depending on two main factors: the first refers to the storage system management, either the system is in load mode (accumulation) or download mode (generation). The second factor refers to a purely technical point of view of the stability and the quality, or taking into account an economic point of view.

Strategy Controlled by the System Manager

The system manager is in charge of controlling the optimal use of the generated energy in order to cover the demand, improving in this way factors such as the quality and the cost of electricity. The objective of this strategy is to accumulate the energy in the storage systems when there is more generation than demand. In addition, when the demand is higher than the produced energy, the demand would be covered by the storage system, releasing the previously accumulated energy.

The storage systems are controlled by the managers of the transmission system, who decide when the loading and unloading occurs. These storage systems are not associated or located nearby a generating plant, and their operation is independently of those.

Strategy Controlled Generation Plant: Avoid Penalties

In many power networks, the administrator of a power plant provides the manager of the electric network an estimation of the energy that is going to be generated at each time and with a certain deadline, competing in this way for the energy market (Egido Cortés et al., n.d.) (GaleanoGonzález & Botero Castro, n.d.). The predictions must be accurate and with small deviations. On the other hand, the manager can impose a financial penalty to the generator, either for deficit or excess of supplied energy (Salmerón, 2010).

In renewable energy, this assumption only makes sense when there is an available storage system that compensate the interruptible energy generation. Consequently, the storage system will regulate the energy provided to the electric network: it will accumulate the energy when the production is higher than the demand or otherwise, when the demand is higher than the production of electricity, it will be supplied by the storage system, matching in both cases with the provided estimation and avoiding penalties from the system manager.

Strategy Based on Independent Storage Plants or Related to Plant Generation

The price of the energy is set by the energy market at each country at any time by the amount of energy injected to the power network. Furthermore, in many countries, another factor that influences the price of the energy is the variation of the demand at a given time (Lucia López & Meneu Ferrer, 2005). For example, the price for energy generation when there is a peak of demand or in emergency situations is greater than in other cases, like when the demand is very low (Figure 4).

In situations when the price varies, the storage systems can be used with a different purpose than avoiding penalties. In this way, a third strategy can be defined: the use of the energy storage to supply when the cost of the electricity increases. In this case, the margin between the cost of storing energy and the profit to deliver it when the price of electricity is high, is used to increase the benefits of a power plant. In this manner, when the price of the energy is low, the storage system will accumulate the generated energy from the power plant and, when the energy price increases, the storage system will supply it. This strategy implies that selling the energy when the price is higher, is more profitable than the loss for not selling the produced energy at the time and storing it.

Figure 4. Evolution of the daily cost of energy

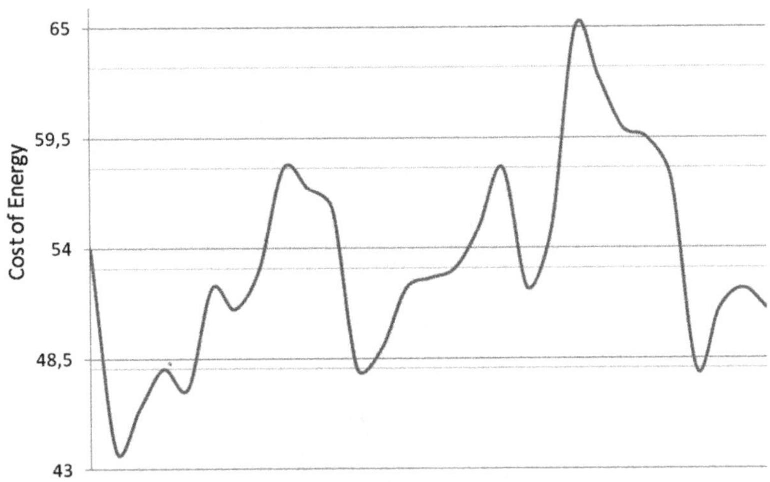

INTELLIGENT ENERGY MANAGEMENT SYSTEM

In this section, a novel approach to create the model used to manage the above explained strategies is described. In addition, due to the high nonlinearity of the system, the novel regression methods to accomplish the challenge are presented.

Energy Management System (EMS)

An Energy Management System (EMS) centralizes all the energy management of a specific location. In the case of a normal generation plant, the EMS controls the power generated at each moment. If the system uses renewable energies, the EMS should have the primary renewable energy predictions (the solar radiation, the wind speed…) for each moment. The EMS is very important when the generation and energy storage system are joined. It decides the different uses of the system according to the explained strategies.

An Intelligent Energy Management Systems would also have demand information, apart from all the variables of the generation and storage systems. In this way, the system can decide not only storing or generating energy, it could combine the generation and the "discharge" of the storage systems. In generation systems based on renewable energies, the intelligent EMS would also take into account the prediction of primary energies as it is mentioned (J.-L. Casteleiro-Roca, Gómez-González, et al., 2018; J.-L. Casteleiro-Roca et al., 2017; José-Luis Casteleiro-Roca, Barragán, Segura, Calvo-Rolle, & Andújar, 2019; José-Luis Casteleiro-Roca, Gómez-González, et al., 2019).

The Novel Approach

In this subsection, the procedure to create the models is described. Figure 5 shows the common steps to follow (J.-L. Casteleiro-Roca, Jove, et al., 2018; E. Jove, Blanco-Rodríguez, Casteleiro-Roca, Moreno-Arboleda, López-Vázquez, et al., 2018; E. Jove, Gonzalez-Cava, Casteleiro-Roca, Pérez, et al., 2018; Esteban Jove, Gonzalez-Cava, Casteleiro-Roca, Méndez-Pérez, Antonio Reboso-Morales, et al., 2018;

Figure 5. Generic approach for the model creation

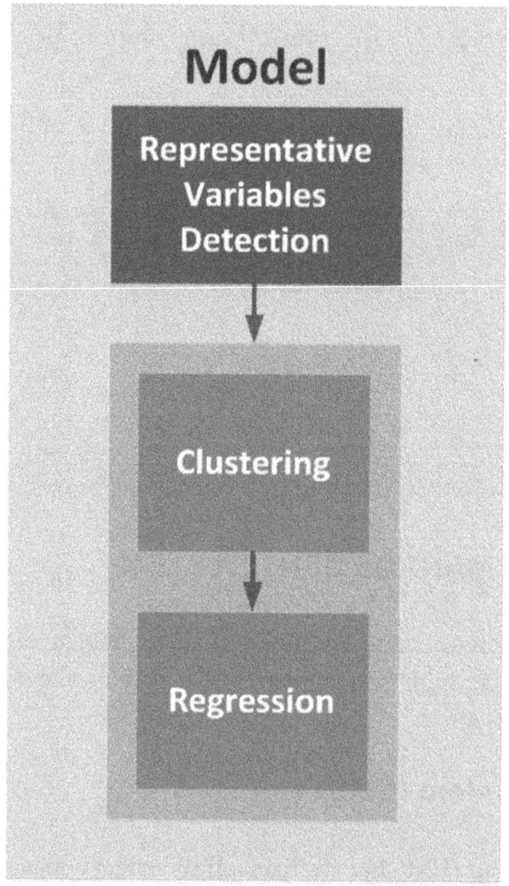

Esteban Jove, López, Fernández-Ibáñez, Casteleiro-Roca, & Calvo-Rolle, 2018): first, the representative variables have to be detected and selected from the dataset. Then, using the previously variables, the model is created, which is done in two different steps. Initially, the data is grouped in different clusters depending on its behavior and then, the regression for each cluster is made.

Figure 6 shows how the prediction of the energy demand from the created model is done. The inputs to the second block are the prediction of the demand and the energy production. This block allows obtaining an output to control the stock management, as explained above, from a function or a rule based system.

Obtaining the Intelligent Model

Systems with a high nonlinear component can be modeled using intelligent techniques (J. L. Calvo-Rolle, Quintián, Corchado, & Ferreiro-Garc\'\ia, 2013; Gonzalez-Cava et al., 2017; Manuel Vilar-Martinez, Aurelio Montero-Sousa, Luis Calvo-Rolle, & Luis Casteleiro-Roca, 2014). Due to the nonlinearity and complexity of the power management system described in this research, the process followed to obtain the intelligent model is divided in two main steps: clustering of the dataset and then, applying regression techniques to each cluster.

Figure 6. Energy stock management predicted with the model created

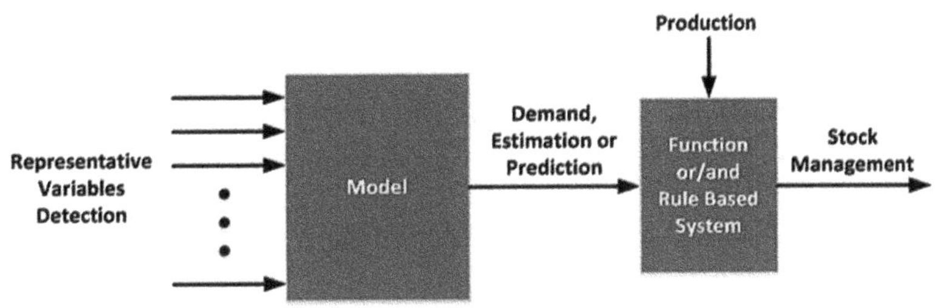

First, the selection of representative variables must be done for the system. These variables are used as inputs to the model. Furthermore, they can vary depending on the time of the year, geographical location…

These variables can be subject of specific and significant changes. For example, a sport event in winter, when all the light of the stadium are turned on plus the people watching the event at home or at a bar, or a summer music festival that congregates thousands of people (Alaiz-Moretón et al., 2018; Esteban Jove et al., 2017; Esteban Jove, Casteleiro-Roca, Quintián, Méndez-Pérez, & Calvo-Rolle, n.d.).

After detecting the representative variables, the clustering is done by making groups of the input variables and, afterwards, performing the regression of each cluster.

Detection of Representative Variables

The detection of the variables can be done with algorithms such as PCA (Principal Component Analysis) (Quintián, Calvo-Rolle, & Corchado, 2014), or SOM (Self Organization Maps) (López García & Machón González, 2004). They can be used to detect the variables that are more significant to calculate the outputs of the system, thus creating the desired model.

The SOM algorithm also has the ability to detect the correlation between variables. Thus, aims to reduce the number of input variables to the model and achieving an easier model (Esteban Jove, Aláiz-Moretón, Casteleiro-Roca, Corchado, & Calvo-Rolle, 2014).

Clustering

In nonlinear systems, the clustering techniques are used to perform a model. The variables with similar behavior are grouped and divided into local models. In this way, the model is improved by reducing the error between the predicted and the real outputs.

Algorithms such as SOM (Corchado & Baruque, 2012) and K-means algorithm (Fernández-Serantes, Estrada Vázquez, Casteleiro-Roca, Calvo-Rolle, & Corchado, 2014) are used for clustering, leading to successful results in non linear problems (Jose-Luis Casteleiro-Roca, Pérez, PIÑÓN-PAZOS, Calvo-Rolle, & Corchado, 2019; Garcia Iglesias et al., 2017; Gonzalez-Cava, Reboso, Casteleiro-Roca, Calvo-Rolle, & Méndez Pérez, 2018; E. Jove, Gonzalez-Cava, Casteleiro-Roca, Quintián, et al., 2018).

The SOM algorithms could be used to represent the behavior of the input variables in a low dimensional representation, allowing the visual detection of the number of clusters. In contrary, the K-means algorithm does not allow to detect the clusters, but it provides an efficient data division when the right number of clusters are selected (J. Calvo-Rolle et al., 2012; Quintián, Casteleiro-Roca, Perez-Castelo, Calvo-Rolle, & Corchado, 2016; Quintián & Corchado, 2017; Vega Vega, Quintián, Calvo-Rolle, Herrero, & Corchado, 2018).

Normally, these two algorithms are used concurrently. Using the SOM to detect the number of optimum clusters and then the K-means to divide the dataset. After, the modeling of each cluster can be done.

Regression

Some algorithms used to perform this step are the Polynomial Regression (J. Casteleiro-Roca, Quintián, Calvo-Rolle, Corchado, & del Carmen Meizoso-López, 2013) (Fernández-Serantes et al., 2014), the Artificial Neural Networks (ANN) (J. Casteleiro-Roca et al., 2013) (Fernández-Serantes et al., 2014), the Support Vector Regression (SVR) (J. Casteleiro-Roca et al., 2013) (Fernández-Serantes et al., 2014), and the Ensembles (Fernández-Serantes et al., 2014).

The Polynomial Regression is one of the most used regression algorithms, even though it has a lot of restrictions in nonlinear system modelling. However, when the clusters have a linear behavior, they have a good performance. The most important advantage is the low computational cost.

On contrary, the ANNs are very powerful algorithms that can be used in a wide variety of cases. The MLP (Multi-Layer Perceptron) is one of the most used ANN (Esteban Jove, Casteleiro-Roca, Quintián, Méndez-Pérez, & Calvo-Rolle, 2019, 2018; Esteban Jove, Gonzalez-Cava, et al., 2019; Padillo, Luna, Cano, & Ventura, 2016). This algorithm achieves very satisfactory results but only when the number of neurons, layers and activation functions are well selected. Hence, the efficiency depends on the internal net architecture.

The SVM (Support Vector Machines), which are classifications algorithms, can be used with the SVR modification. The SVR uses nonlinear transformation of inputs to perform regression in a high dimensional space. Then, the data is retransformed again. Moreover, the LS-SVR (Least Squared Support Vector Regression) is a modified version of the SVR that does not need any adjustment to tune the algorithm.

When the training data corresponds to all the operating range of the system, the Ensemble regression can be used. This algorithm uses a decision tree that is normally used for classification (Baruque, Porras, Jove, & Calvo-Rolle, 2019). It provides lower computational cost than other algorithms but needs to use more memory to achieve good results (Alaiz-Moretón et al., 2018).

Efficiency Measurement

The MSE (Mean Squared Error) (Fernández-Serantes et al., 2014) (Esteban Jove et al., 2014) allows the comparison and, therefore, the selection of the best model. This value compare the results obtained with the previous algorithms both for global and local models, with or without clustering. The MSE is calculated as a comparison of the model output and the real value of the system output.

This procedure could be used to compare algorithms, or to select the right number of clusters, when this number is not clear after the analysis previously explained.

EXPERIMENTAL ENERGY STORAGE SYSTEM

The experimental system shown in this work is located in the "Centro de Investigaciones Energéticas, Medioambientales y Tecnológicas" (CIEMAT) in Spain. In this specific case, the system is used to store the energy generated by two renewable sources: wind and solar. A schematic of the system is shown in Figure 7, where the local power bus is exposed to the generation and the consumption connections of the storage system.

Both, solar and wind generation, and the fuel cell are the power input points to the power bus, while the hydrogen generation equipment is the only consumption point shown. The energy storage system, in this case, consists on an electrolyzer, a hydrogen tank and a fuel cell. The Energy Management System decides the operation of the whole system: how much energy is provided to the power network or to the storage system. In the periods of high generation and low demand, it would be better to store the energy so it can be used when it is necessary.

Storaged System Analysis

The energy storage system is analyzed to calculate the global performance of the energy storage system.

Electrolyzer

The electrolyzer produces the hydrogen that feeds the fuel cell, which is in charge of producing the electric power when it is necessary.

The equipment uses H_2O; firstly, the water goes through a deionizer. The deionized water is mixed with potassium hydroxide - KOH (around 20 – 30%), creating the electrolyte. After, by applying an electric current between the negative and positive electrodes in the electrolyte, the H_2 and O_2 are obtained. The operating temperature should be between 70 - 80 °C. The O_2, in this case, is not stored and it is emitted directly to the atmosphere. The H_2 is filtered to eliminate the O_2 that may remain mixed. The hydrogen is stored in a tank to use it later in the fuel cell. If the pressure increases, a safety valve opens to reduce the pressure.

Figure 7. Energy Management System (EMS)

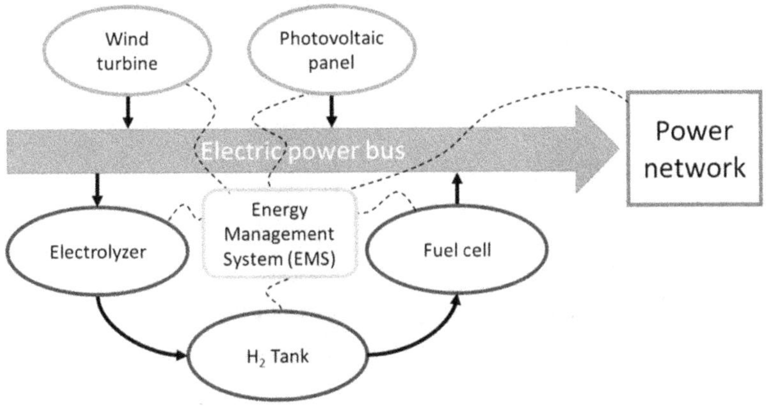

To increase the efficiency, the voltage should be slightly above 1.48 V, depending on the flow rate of H_2. It is used nitrogen in the electrolyzer to compensate the operating pressures.

H_2 Tank (Compressor)

Hydrogen storage is done in tanks at 10 bar of pressure. The size of the tanks depends exclusively on the amount of energy to store; the more hydrogen is stored, the more electrical energy can be produced.

As the system is experimental, the size of the deposit is not relevant, but it needs a compressor to store the hydrogen at the specific pressure. This compressor uses electrical energy so, it has to be taken into account to calculate the overall performance.

Fuel Cell

The fuel cell generates energy from H_2, which is obtained in the electrolyzer. The energy obtained is represented by the chemical reaction of the equation (1).

$$\left\{\begin{array}{ll} Anode & H_2 \rightarrow 2H^+ + 2e^- \\ Cathode & \frac{1}{2}O_2 + 2H^+ + 2e^- \rightarrow H_2O \end{array}\right\} \Rightarrow H_2 + \frac{1}{2}O_2 \rightarrow H_2O + Energy \quad (1)$$

This reaction occurs in each cell. If some cells are connected in series, the accumulated voltage is the sum of the individual voltages of each cell. The reaction, represented in the equation (1), takes place when each cell is filled with H_2 and O_2 (obtained from the air). A simplified scheme of the fuel cell is shown in Figure 8.

Figure 8. Working schematic of a fuel cell

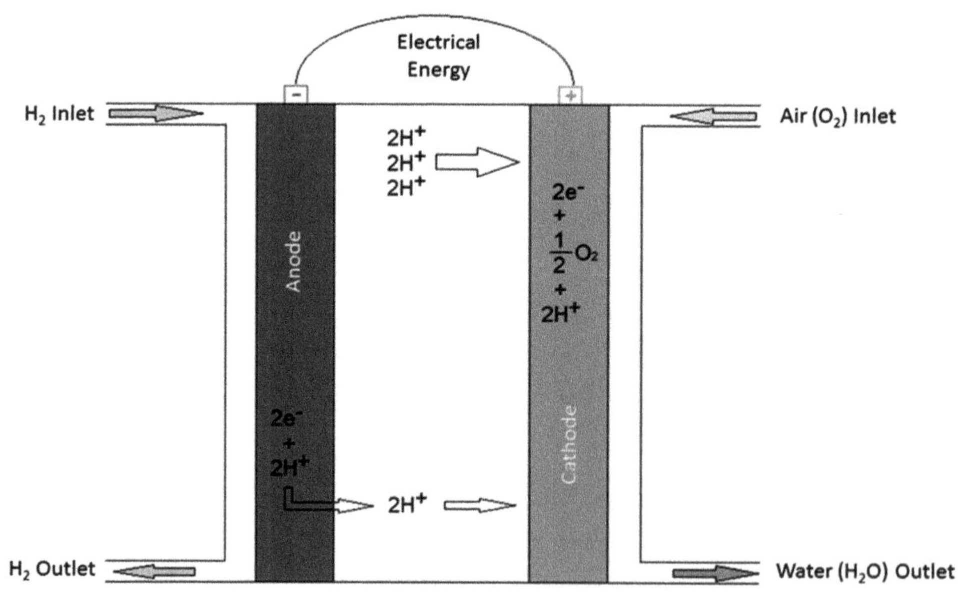

The hydrogen enters in the anode where a catalyst accelerates the reaction, as shown in equation (1). The H_2 releases an electron of each atom and a molecule of $2H^+$ is created. This molecule will react with ½ O_2 from the air that enters in the cathode. The result of this reaction is a water molecule.

For each molecule of water formed, two electrons are released, which are the ones that provide the electrical energy. These reactions occur as long as H_2 is supplied.

Global Performance of the Storage System

After analyzing the data provided by the CIEMAT, the following performance conclusions can be reach for each equipment in the storage system:

- The performance of the electrolyzer depends on the temperature, but it is between 18.7% and 28.5%.
- The compressor performance is around 32.4%.
- The fuel cell performance is 50.3%.

With these data, an overall system performance is nearly 3.5%. The system recovers only 3 kWh every 100 kWh used in the storage system. Although the low performance for an "only storage system", if we take into account that the application is based on renewable energies, the system can take advantage of the energy produced when there is no electricity demand in the network.

In spite of the low performance of the whole storage system, a lot of new researches continue with the study of these systems, but they center only in part of the system. For example, it is very interesting to study the fuel cell alone, as a generator. As Figure 8 shows, there is a H_2 outlet in the fuel cell, but this output should be reduced to the minimum to increase the efficiency. There are papers where models are created to predict the behavior of the fuel cells when their operation point changes. These models allow anticipating a change in the H_2 flow inlet when the output power was changed (to minimize the time response and the H_2 outlet, for example).

CONCLUSION

During last decades, the increase of greenhouse gases emissions has led the governments to promote the use of renewable sources for electric generation. However, the intermittency on the availability of renewable sources consists an important disadvantage. Due to this fact, it is possible that this renewable sources, such as solar or wind energy, produce more energy than the demanded or vice versa.

As mentioned in the present study, in this context, the use of energy storage systems can play a key role. It is possible to store energy when there is more generation than demand and then, this system can supply the power grid when there is an energy lack.

Depending on the management of the storage system and other technical factors, three different strategies for energy management within a production system based on renewable energy were presented in this chapter. In the first strategy, the system manager is in charge of controlling the optimal use of the generated energy. Then, the storage systems are controlled by the manager of the transmission system, who decide when the loading and unloading occurs. In the next strategy, the manager of a power plant decides to store or supply the energy depending on the estimation provided to the power manager with the objective of avoiding penalties and fitting the prediction. The last strategy occurs in power systems were the electricity price varies throughout the day. When the price is low, the power plant can store energy, and then release it when it is higher, increasing the benefits of the power plant.

Due to the complexity of the demand profile during one day and the different points of generation, the balance between the offer and the demand must be managed. To achieve this goal, the concept of an Intelligent Energy Management Systems that centralizes all the energy management of a specific place was presented in this chapter. To model the demand of the power grid, the use of intelligent techniques for clustering and regression are used.

To evaluate the different strategies, the data from a storage system consisting on an electrolyzer, a hydrogen tank and a fuel cell was analyzed, showing a low overall performance. Given the fact that the energy is generated from renewable sources (solar and wind energies), the achieved performance of the storage system is justified.

The contribution of this study is that different strategies have been applied to the presented plant successfully; despite the low performance of the system.

REFERENCES

Alaiz-Moretón, H., Casteleiro-Roca, J. L., Robles, L. F., Jove, E., Castejón-Limas, M., & Calvo-Rolle, J. L. (2018). Sensor Fault Detection and Recovery Methodology for a Geothermal Heat Exchanger. In *International Conference on Hybrid Artificial Intelligence Systems* (pp. 171–184). Academic Press. 10.1007/978-3-319-92639-1_15

Baruque, B., Porras, S., Jove, E., & Calvo-Rolle, J. L. (2019). Geothermal heat exchanger energy prediction based on time series and monitoring sensors optimization. *Energy*, *171*, 49–60. doi:10.1016/j.energy.2018.12.207

Benavides, J. (n.d.). *Precios, inversión y economía política de la energía eléctrica*. Academic Press.

Blanco-Rodriguez, P., Fernández-Serantes, L., Otero-Pazos, A., Calvo-Rolle, J., & de Cos Juez, F. (2017). Radon mitigation approach in a laboratory measurement room. *Sensors (Basel)*, *17*(5), 1090. doi:10.339017051090

Calvo-Rolle, J., Corchado, E., Quintian-Pardo, H., García, R., Román, J., & Hernández, P. (2012). A novel hybrid intelligent classifier to obtain the controller tuning parameters for temperature control. *Hybrid Artificial Intelligent Systems*, 677–689.

Calvo-Rolle, J. L., Quintián, H., Corchado, E., & Ferreiro-Garcia, R. (2013). Intelligent model to obtain current extinction angle for a single phase half wave controlled rectifier with resistive and inductive load. In *Soft Computing Models in Industrial and Environmental Applications* (pp. 249–256). Springer.

Casteleiro-Roca, J., Quintián, H., Calvo-Rolle, J., Corchado, E., & del Carmen Meizoso-López, M. (2013). Intelligent Model to Obtain Initial and Final Conduction Angle of a Diode in a Half Wave Rectifier with a Capacitor Filter. In *International Joint Conference SOCO'13-CISIS'13-ICEUTE'13*. Springer International Publishing.

Casteleiro-Roca, J.-L., Barragán, A. J., Segura, F., Calvo-Rolle, J. L., & Andújar, J. M. (2019). Fuel Cell Output Current Prediction with a Hybrid Intelligent System. *Complexity*, 2019.

Casteleiro-Roca, J. L., Fernández-Serantes, L. A., Calvo-Rolle, J. L., Machón-González, I., Crespo-Ramos, M. J., & López-Garcia, H. (2013). Study of the effect of a geothermal heat exchanger over the ground. In *2013 International Conference on New Concepts in Smart Cities: Fostering Public and Private Alliances (SmartMILE)* (pp. 1–6). Academic Press. 10.1109/SmartMILE.2013.6708197

Casteleiro-Roca, J.-L., Gómez-González, J. F., Calvo-Rolle, J. L., Jove, E., Quintián, H., Gonzalez Diaz, B., & Mendez Perez, J. A. (2019). Short-Term Energy Demand Forecast in Hotels Using Hybrid Intelligent Modeling. *Sensors (Basel), 19*(11), 2485. doi:10.339019112485

Casteleiro-Roca, J.-L., Gómez-González, J. F., Calvo-Rolle, J. L., Jove, E., Quintián, H., Martín, J. F. A., … Méndez-Perez, J. A. (2018). Prediction of the energy demand of a hotel using an artificial intelligence-based model. Lecture Notes in Computer Science, 10870. doi:10.1007/978-3-319-92639-1_49

Casteleiro-Roca, J.-L., Jove, E., Gonzalez-Cava, J. M., Méndez Pérez, J. A., Calvo-Rolle, J. L., & Blanco Alvarez, F. (2018). Hybrid model for the ANI index prediction using Remifentanil drug and EMG signal. *Neural Computing & Applications*. doi:10.100700521-018-3605-z

Casteleiro-Roca, J.-L., Jove, E., Sánchez-Lasheras, F., Méndez-Pérez, J.-A., Calvo-Rolle, J.-L., & De Cos Juez, F. J. (2017). Power Cell SOC Modelling for Intelligent Virtual Sensor Implementation. *Journal of Sensors, 2017*, 1–10. doi:10.1155/2017/9640546

Casteleiro-Roca, J.-L., Pérez, J. A. M., Piñón-Pazos, A. J., Calvo-Rolle, J. L., & Corchado, E. (2019). Intelligent Model for Electromyogram (EMG) Signal Prediction During Anesthesia. *Journal of Multiple-Valued Logic & Soft Computing, 32.*

Corchado, E., & Baruque, B. (2012). WeVoS-ViSOM: An ensemble summarization algorithm for enhanced data visualization. *Neurocomputing, 75*(1), 171–184. doi:10.1016/j.neucom.2011.01.027

Egido Cortés, I., Rouco Rodríguez, L., Alonso Perez, J., Porras Muñoz, E., & Ruiz Mendoza, J. L. (n.d.). Impacto en regulación frecuencia-potencia de los cambios de programa en escalón de las unidades generadoras. *Revista de La Asociación de Ingenieros Del ICAI.*

Energía y sociedad. (n.d.a). *Energía y sociedad.* Author.

Energía y sociedad. (n.d.b). *Energías renovables: tecnología, economía, evolución e integración en el sistema eléctrico.* Author.

Fabra, N. (2007). Ajustes regulatorios en el sector eléctrico español, La electricidad: mercado, inversiones y garantía de suministro. *ENEconomía Industrial*, 364.

Ferguson, C. D. (2011). *Nuclear Energy: What Everyone Needs to Know.* Oxford University Press.

Fernández Serantes, L. A. (2014). *Deseño e implementación dun emulador de fontes de corrente continua a partir dun convertedor DC/DC.* Academic Press.

Fernandez-Serantes, L. A., Berger, H., Stocksreiter, W., & Weis, G. (2016). Ultra-High Frequent Switching with GaN-HEMTs using the Coss-Capacitances as non-dissipative Snubbers. In *PCIM Europe 2016; International Exhibition and Conference for Power Electronics, Intelligent Motion, Renewable Energy and Energy Management* (pp. 1–8). Academic Press.

Fernández-Serantes, L. A., Estrada Vázquez, R., Casteleiro-Roca, J., Calvo-Rolle, J., & Corchado, E. (2014). Hybrid Intelligent Model to Predict the SOC of a LFP Power Cell Type. In M. Polycarpou, A. P. L. F. de Carvalho, J.-S. Pan, M. Wozniak, H. Quintian, & E. Corchado (Eds.), *Hybrid Artificial Intelligence Systems*. Springer International Publishing. doi:10.1007/978-3-319-07617-1_49

GaleanoGonzález, D. A., & Botero Castro, H. A. (n.d.). Representación mediante grafcet del accionamiento de generadores sincrónicos en una central hidroeléctrica modernizada. *Dyna, Revista de La Facultad de Minas - Universidad de Colombia.*

Garcia Iglesias, D., Jove Perez, E., Penyafiel Verdu, P., Valverde, I., Gonzalez Vasserot, M., & Rubin Lopez, J. M. (2017). P1034 Frequency domain analysis of surface ECG enables distinction of patients with brugada syndrome. *EP Europace, 19*(3), iii222--iii223.

Gonzalez-Cava, J. M., Arnay, R., Pérez, J. A. M., León, A., Martin, M., Jove-Perez, E., … de Cos Juez, F. J. (2017). A machine learning based system for analgesic drug delivery. In *International Joint Conference SOCO17-CISIS17-ICEUTE17 León, Spain, September 6--8, 2017, Proceeding* (pp. 461–470). Academic Press.

Gonzalez-Cava, J. M., Reboso, J. A., Casteleiro-Roca, J. L., Calvo-Rolle, J. L., & Méndez Pérez, J. A. (2018). A novel fuzzy algorithm to introduce new variables in the drug supply decision-making process in medicine. *Complexity.*

Huggins, R. A. (2010). *Energy Storage*. Springer. doi:10.1007/978-1-4419-1024-0

Jove, E., Aláiz-Moretón, H., Casteleiro-Roca, J., Corchado, E., & Calvo-Rolle, J. (2014). Modeling of Bicomponent Mixing System Used in the Manufacture of Wind Generator Blades. In E. Corchado, J. Lozano, H. Quintián, & H. Yin (Eds.), *Intelligent Data Engineering and Automated Learning-IDEAL.* Springer International Publishing.

Jove, E., Alaiz-Moretón, H., Garcia-Rodriguez, I., Benavides-Cuellar, C., Casteleiro-Roca, J. L., & Calvo-Rolle, J. L. (2017). PID-ITS: an intelligent tutoring system for PID tuning learning process. In *International Joint Conference SOCO17-CISIS17-ICEUTE17 León, Spain, September 6--8, 2017, Proceeding* (pp. 726–735). Academic Press.

Jove, E., Blanco-Rodríguez, P., Casteleiro-Roca, J. L., Moreno-Arboleda, J., López-Vázquez, J. A., de Cos Juez, F. J., & Calvo-Rolle, J. L. (2018). *Attempts prediction by missing data imputation in engineering degree* (Vol. 649). Advances in Intelligent Systems and Computing. doi:10.1007/978-3-319-67180-2_16

Jove, E., Casteleiro-Roca, J.-L., Quintián, H., Méndez-Pérez, J. A., & Calvo-Rolle, J. L. (2018). A New Approach for System Malfunctioning over an Industrial System Control Loop Based on Unsupervised Techniques. In M. Graña, J. M. López-Guede, O. Etxaniz, Á. Herrero, J. A. Sáez, H. Quintián, & E. Corchado (Eds.), *International Joint Conference SOCO'18-CISIS'18-ICEUTE'18* (pp. 415–425). Cham: Springer International Publishing.

Jove, E., Casteleiro-Roca, J.-L., Quintián, H., Méndez-Pérez, J. A., & Calvo-Rolle, J. L. (2019). Outlier Generation and Anomaly Detection Based on Intelligent One-Class Techniques over a Bicomponent Mixing System. In *International Workshop on Soft Computing Models in Industrial and Environmental Applications* (pp. 399–410). Academic Press.

Jove, E., Casteleiro-Roca, J.-L., Quintián, H., Méndez-Pérez, J. A., & Calvo-Rolle, J. L. (n.d.). A fault detection system based on unsupervised techniques for industrial control loops. *Expert Systems*, e12395.

Jove, E., Gonzalez-Cava, J. M., Casteleiro-Roca, J.-L., Méndez-Pérez, J.-A., Antonio Reboso-Morales, J., Javier Pérez-Castelo, F., ... Luis Calvo-Rolle, J. (2018). Modelling the hypnotic patient response in general anaesthesia using intelligent models. *Logic Journal of the IGPL*, *27*(2), 189–201. doi:10.1093/jigpal/jzy032

Jove, E., Gonzalez-Cava, J. M., Casteleiro-Roca, J. L., Pérez, J. A. M., Calvo-Rolle, J. L., & de Cos Juez, F. J. (2018). *An intelligent model to predict ANI in patients undergoing general anesthesia* (Vol. 649). Advances in Intelligent Systems and Computing. doi:10.1007/978-3-319-67180-2_48

Jove, E., Gonzalez-Cava, J. M., Casteleiro-Roca, J.-L., Quintián, H., Méndez-Pérez, J. A., Calvo-Rolle, J. L., ... Reboso, J. (2018). Remifentanil dose prediction for patients during general anesthesia. Lecture Notes in Computer Science, 10870. doi:10.1007/978-3-319-92639-1_45

Jove, E., Gonzalez-Cava, J. M., Casteleiro-Roca, J.-L., Quintián, H., Méndez-Pérez, J. A., & Calvo-Rolle, J. L. (2019). Anomaly Detection on Patients Undergoing General Anesthesia. In *International Joint Conference: 12th International Conference on Computational Intelligence in Security for Information Systems (CISIS 2019) and 10th International Conference on EUropean Transnational Education (ICEUTE 2019)* (pp. 141–152). Academic Press.

Jove, E., López, J. A. V., Fernández-Ibáñez, I., Casteleiro-Roca, J. L., & Calvo-Rolle, J. L. (2018). Hybrid intelligent system topredict the individual academic performance of engineering students. *International Journal of Engineering Education*, *34*(3), 895–904.

López García, H., & Machón González, I. (2004). Self-organizing map and clustering for wastewater treatment monitoring. *Engineering Applications of Artificial Intelligence*, *17*(3), 215–225. doi:10.1016/j.engappai.2004.03.004

Lucia López, J., & Meneu Ferrer, V. (2005). Los precios en los mercados reestructurados de electricidad: algunas lecciones básicas para la negociación derivada. *CICE Nº 69.*

MacKay, D. J. C. (2009). *Sustainable Energy: Without the Hot Air.* Uit Cambridge Limited.

Manuel Vilar-Martinez, X., Aurelio Montero-Sousa, J., Luis Calvo-Rolle, J., & Luis Casteleiro-Roca, J. (2014). Expert system development to assist on the verification of" TACAN" system performance. *Dyna (Bilbao)*, *89*(1), 112–121.

Montero Sousa, J. A., Casteleiro Roca, J. L., & Calvo Rolle, J. L. (2017a). El sector eléctrico desde su inicio hasta la segunda guerra mundial. *Dyna (Bilbao)*, *92*(1), 43–47. doi:10.6036/7947

Montero-Sousa, J. A., Fernández-Serantes, L. A., Casteleiro-Roca, J.-L., Vilar-Martinez, X. M., & Calvo-Rolle, J. L. (2017). Energy Management Strategies to Improve Electrical Networks Using Storage Systems. In Renewable and Alternative Energy: Concepts, Methodologies, Tools, and Applications (pp. 1500–1514). IGI Global.

Montero-Sousa, J.-A., Fernandez-Serantes, L.-A., Casteleiro-Roca, J.-L., Vilar-Martnez, X.-M., & Calvo-Rolle, J.-L. (2017). *Energy storage management for generation-distribution facilities*. Federacion Asociaciones Ingenieros Industriales Espana Alameda De Mazarredo.

Padillo, F., Luna, J. M., Cano, A., & Ventura, S. (2016). A data structure to speed-up machine learning algorithms on massive datasets. In *International Conference on Hybrid Artificial Intelligence Systems* (pp. 365–376). Academic Press. 10.1007/978-3-319-32034-2_31

Quintián, H., Calvo-Rolle, J. L., & Corchado, E. (2014). A Hybrid Regression System Based on Local Models for Solar Energy Prediction. *Informatica*, *25*(2), 265–282. doi:10.15388/Informatica.2014.14

Quintián, H., Casteleiro-Roca, J.-L., Perez-Castelo, F. J., Calvo-Rolle, J. L., & Corchado, E. (2016). Hybrid intelligent model for fault detection of a lithium iron phosphate power cell used in electric vehicles. In *International Conference on Hybrid Artificial Intelligence Systems* (pp. 751–762). Academic Press. 10.1007/978-3-319-32034-2_63

Quintián, H., & Corchado, E. (2017). Beta scale invariant map. *Engineering Applications of Artificial Intelligence*, *59*, 218–235. doi:10.1016/j.engappai.2017.01.002

Rao, K. R. (2011). *Energy and Power Generation Handbook: Established and Emerging Technologies*. American Society of Mechanical Engineers. doi:10.1115/1.859551

Salmerón, J. (2010). *Las renovables en el mercado de la electricidad: perspectivas de futuro*. Academic Press.

Sousa, J. A. M., Roca, J. L. C., & Rolle, J. L. C. (2017b). Evolución del sector eléctrico tras la Segunda Guerra Mundial. *Dyna (Bilbao)*, *92*(3), 280–284.

Utray Fabra, J. (n.d.). *El mercado de la electricidad: Algunas reflexiones y algunas propuestas*. Academic Press.

Vega Vega, R., Quintián, H., Calvo-Rolle, J. L., Herrero, Á., & Corchado, E. (2018). Gaining deep knowledge of Android malware families through dimensionality reduction techniques. *Logic Journal of the IGPL*, *27*(2), 160–176. doi:10.1093/jigpal/jzy030

Compilation of References

Aalami, H., Yousef, G. R., & Moghadam, M. P. (2010). Demand response modelling considering interruptible/curtailable loads and capacity market programs. *Applied Energy*, *87*(1), 243–250. doi:10.1016/j.apenergy.2009.05.041

Aalami, H., Yousef, G., & Moghadam, M. (2008, 21 - 24 April). Demand response model considering EDRP and TOU. In *IEEE PES*. Chicago, IL: Transmission and Distribution Conference and Exposition. doi:10.1109/TDC.2008.4517059

Abido, M., & Bakhashwain, J. (2005). Optimal VAR dispatch using a multiobjective evolutionary algorithm. *International Journal of Electrical Power & Energy Systems*, *27*(1), 13–20. doi:10.1016/j.ijepes.2004.07.006

Acharjee, P. (2012). Identification of maximum loadability limit and weak buses using security constraintgenetic algorithm. *International Journal of Electrical Power and Energy Systems*, *36*(1), 40–50. doi.org/10.1016/j.ijepes.2011.10.021

Adefarati, Bansal, & Jackson. (2017). Reliability and economic evaluation of a microgrid power system. *Proceedings of 9th International Conference on Applied Energy.*

Aditya, S. K., & Das, D. (2003). Design of load frequency controllers using genetic algorithm for two area interconnected hydropower system. *Electric Power Components and Systems*, *31*(1), 81–94. doi:10.1080/15325000390112071

Aghaei, J., Niknam, T., Azizipanah-Abarghooee, R., & Arroyo, J. M. (2013). Scenario-based dynamic economic emission dispatch considering load and wind power uncertainties. *International Journal of Electrical Power & Energy Systems*, *47*, 35167. doi:10.1016/j.ijepes.2012.10.069

Ajjarapu, V. (2007). *Computational techniques for voltage stability assessment and control.* Springer Science & Business Media. doi:10.1007/978-0-387-32935-2

Ajjarapu, V. (2009). *Computational Techniques for Voltage Stability Assessment and Control.* Springer Science and Business Media.

Ajjarapu, V., & Christy, C. (1992). The continuation power flow: A tool for steady state voltage stability analysis. *IEEE Transactions on Power Systems*, *7*(10), 304–311. doi:10.1109/59.141737

Ajjarapu, V., & Lee, B. (1992). Bifurcation theory and its application theory and its application to nonlinear dynamical phenomena in an electrical power system. *IEEE Transactions on Power Systems*, *7*, 424–431. doi:10.1109/59.141738

Ajjarapu, V., & Lee, B. (1998). Bibliography on voltage stability. *IEEE Transactions on Power Systems*, *13*(1), 115–125. doi:10.1109/59.651622

Akyildiz, I. F., Su, W., Sankarasubramaniam, Y., & Cayirci, E. (2002). A survey on sensor networks. *IEEE Communications Magazine*, *40*(8), 102–114. doi:10.1109/MCOM.2002.1024422

Alaiz-Moretón, H., Casteleiro-Roca, J. L., Robles, L. F., Jove, E., Castejón-Limas, M., & Calvo-Rolle, J. L. (2018). Sensor Fault Detection and Recovery Methodology for a Geothermal Heat Exchanger. In *International Conference on Hybrid Artificial Intelligence Systems* (pp. 171–184). Academic Press. 10.1007/978-3-319-92639-1_15

Ali, E. S., & Abd-Elazim, S. M. (2013). BFOA based design of PID controller for two area Load Frequency Control with nonlinearities. *Electrical Power and Energy Systems*, *51*(1), 224–231. doi:10.1016/j.ijepes.2013.02.030

Ali, M. H., Wu, B., & Dougal, R. A. (2010). An overview of SMES applications in power and energy systems. *IEEE Trans Sustain Energy*, *1*(1), 38–47. doi:10.1109/TSTE.2010.2044901

Alvarado, F., Dobson, I., & Hu, Y. (1994). Computation of closest bifurcations in power systems. *IEEE Transactions on Power Systems*, *9*(2), 918–928. doi:10.1109/59.317655

Anita, J. M., Babulal, C. K., Kannan, P. S., & Venkatesh, B. (2008). Determination of voltage Stability analysis limit using fuzzy logic. *International Journal of Power Energy Systems*, *28*(1), 3651–3692. doi:10.2316/Journal.203.2008.1.203-3651

Arya, L. D., Choube, S. C., Kothari, D. P., & Shrivastava, M. (2010). Loadability margin enhancement using co-ordinated aggregation based particle swarm optimization (CAPSO). *International Journal of Electrical Power & Energy Systems*, *32*(9), 975–984. doi:10.1016/j.ijepes.2010.02.007

Asma, A., Shafiullah, G. M., Amanullah, M. T. O., & Alex, S. (2010). Automatic Generation Control in Wind Integrated Power System: New Perspectives and Challenges. *Int. J. of Thermal & Environmental Engineering*, *12*(1), 27–38.

Atputharajah & Saha. (2009). Power system blackouts - Literature review. *Proc. in ICIIS*, 460-465.

Ayyarao, Tummala, Alluri, & Ramanarao. (2017). Optimal load frequency control of electric vehicle-based micro-grid using Grey Wolf Optimization. *Proceedings of International Conference on Computing of Power, Energy & Communication, 6*, 11-17.

Azmy, El-Sehiemy, & Sakr. (2017). Adaptive differential evolution algorithm for efficient reactive power management. *Applied Soft Computing*, *59*, 336–351. doi:10.1016/j.asoc.2017.01.004

Bae, I. S., Kim, J. O., & Kim, J. C. (2007). Reliability evaluation of distributed generation based on operation mode. *IEEE Transactions on Power Systems*, *22*(2), 785–790. doi:10.1109/TPWRS.2007.894842

Bae, I. S., Kim, J. O., Kim, J. C., & Singh, C. (2004). Optimal operating strategy for distributed generation considering hourly reliability worth. *IEEE Transactions on Power Systems*, *19*(1), 287–292. doi:10.1109/TPWRS.2003.818738

Balamourougan, V., Sidhu, T. S., & Sachdev, M. S. (2004). Technique for online prediction of voltage collapse. *IEE Proceedings. Generation, Transmission and Distribution*, *151*(4), 453–460. doi:10.1049/ip-gtd:20040612

Bandyopadhyay, B., Deepak, F., & Kim, K. S. (2009). Sliding Mode Control Using Novel Sliding Surfaces. *Lecture Notes in Control and Information Science.*

Barker, P. P., & De Mello, R. W. (2000). Determining the impact of distributed generation on power systems: Part 1 - Radial distribution systems. *Proceedings of the IEEE Power Engineering Society Transmission and Distribution Conference*, 3, 1645-1656. 10.1109/PESS.2000.868775

Baruque, B., Porras, S., Jove, E., & Calvo-Rolle, J. L. (2019). Geothermal heat exchanger energy prediction based on time series and monitoring sensors optimization. *Energy*, *171*, 49–60. doi:10.1016/j.energy.2018.12.207

Basu, M. (2007). Dynamic economic emission dispatch using evolutionary programming and fuzzy satisfied method. *Int J Emerg Electr Power Syst.*, *485*(8), 1-15.

Basu. (2008). Dynamic economic emission dispatch using non-dominated sorting genetic algorithm-ii. *Elect. Power Energy Syst, 30*, 140-149.

Basu, M. (2006). Particle swarm optimization based goal-attainment method for dynamic economic emission dispatch. *Electric Power Components and Systems, 34*(9), 1015–1025. doi:10.1080/15325000600596759

Basu, M. (2014). Teaching learning based optimization algorithm for multi-area economic dispatch. *Energy, 68*, 21–28. doi:10.1016/j.energy.2014.02.064

Bechert, T. E., & Chen, N. (1977). Area automatic generation control by multi-pass dynamic programming. *IEEE Transactions on Power Apparatus and Systems, 96*(5), 1460–1468. doi:10.1109/T-PAS.1977.32474

Begovi'c, M. M., & Phadke, A. G. (1992). Control of voltage stability using sensitivity analysis. *IEEE Transactions on Power Systems, 7*(1), 114–123. doi:10.1109/59.141694

Beiraghi, M., & Ranjbar, A. M. (2013). Online voltage security assessment based on wide area measurements. *IEEE Transactions on Power Delivery, 28*(2), 989–997. doi:10.1109/TPWRD.2013.2247426

Benavides, J. (n.d.). *Precios, inversión y economía política de la energía eléctrica.* Academic Press.

Benevent, E. (n.d.). *IEEE 802.15.4 -2006 standard.* Retrieved September 13, 2018, from https://www.unirc.it/documentazione/materiale_didattico/599_2009_192_7229.pdf

Bevrani, H., & Hiyama, T. (2009). On load–frequency regulation with time delays: Design and real-time implementation. *IEEE Transactions on Power Systems, 24*(1), 292–300.

Bhattacharya, A., & Chattopadhyay, P. K. (2010). Solving economic emission load dispatch problems using hybrid differential evolution. *Electrical Power and Energy Systems, 11*(2). doi.org/10.1016/j.asoc.2010.09.008

Bhatti, T. S., Al-Ademi, A. A. F., & Bansal, N. K. (1997). Load Frequency Control of Isolated Wind Diesel Hybrid Power Systems. *Energy Conservation and Mgmr, 38*(9), 829–837. doi:10.1016/S0196-8904(96)00108-2

Bhesdadiya, Parmar, Trivedi, & Jangir, Bhoye, & Jangir. (2016). Optimal Active and Reactive Power Dispatch Problem Solution using Whale Optimization Algorithm. *Indian Journal of Science and Technology, 9*(1), 1–6. doi:10.17485/ijst/2016/v9iS1/101941

Bhongade, S., Tyagi, B., & Gupta, H. O. (2011). Effect of SMES unit in load following contract in a restructured power system. *International Journal of Engineering Science and Technology, 3*(1), 255–264. doi:10.4314/ijest.v3i1.67652

Bijwe, P. R., & Kelapure, S. M. (2003). Nondivergent fast decoupled power flow methods. *IEEE Transactions on Power Systems, 18*(2), 633–688. doi:10.1109/TPWRS.2003.810905

Bijwe, P., Kothari, D., Nanda, J., & Lingamurthy, K. (1986). Optimal voltage control using a constant sensitivity matrix. *Electric Power Systems Research, 11*(3), 195–203. doi:10.1016/0378-7796(86)90034-9

Billinton, R., & Allan, R. N. (1996). *Reliability Evaluation of Power Systems* (2nd ed.). Plenum Press. doi:10.1007/978-1-4899-1860-4

Bishop, J., Axon, C., Bonilla, D., Tran, M., Banister, D., & McCulloch, M. (2013). Evaluating the impact of V2G services on the degradation of batteries in PHEV and EV. *Applied Energy, 111*, 206–218. doi:10.1016/j.apenergy.2013.04.094

Biswas, R. N., Mitra, S. K. & Naskar, M. K. (2014). A robust mobile anchor based localization system for wireless sensor networks using smart antenna. *International Journal of Ad-Hoc and Ubiquitous Computing, 15*(1/2/3), 23-37.

Bizon, N. (2013). Energy harvesting from the PV Hybrid Power Source. *Energy*, *52*, 297–307. doi:10.1016/j.energy.2013.02.006

Blanco-Rodriguez, P., Fernández-Serantes, L., Otero-Pazos, A., Calvo-Rolle, J., & de Cos Juez, F. (2017). Radon mitigation approach in a laboratory measurement room. *Sensors (Basel)*, *17*(5), 1090. doi:10.339017051090

Boukerche, A., Oliveira, H. A. B. F., Nakamura, E. F., & Loureiro, A. A. F. (2007). Localization systems for wireless sensor networks. *IEEE Wireless Communications*, *14*(6), 6–12. doi:10.1109/MWC.2007.4407221

Boukerche, A., Oliveira, H. A. B. F., Nakamura, E. F., & Loureiro, A. A. F. (2008). Secure localization algorithms for wireless sensor networks. *IEEE Communications Magazine*, *46*(4), 96–101. doi:10.1109/MCOM.2008.4481347

Bouktir, T., Herbadji, O., & Slimani, L. (2013). BiogeographyBased Optimization Approach for Solving Optimal Power Flow Problem. *International Journal of Hybrid Information Technology*, *6*(5), 183–196. doi:10.14257/ijhit.2013.6.5.17

Ca˜nizares, C. A., De Souza, A. C., & Quintana, V. H. (1996). Comparison of performance indices for detection of proximity to voltage collapse. *IEEE Transactions on Power Systems*, *11*(3), 1441–1450. doi:10.1109/59.535685

Ca˜nizares, C., & Faur, Z. T. (1999). Analysis of svc and tcsc controllers in voltage collapse. *IEEE Transactions on Power Systems*, *14*(1), 158–165. doi:10.1109/59.744508

Caballero, F., Sauma, E., & Yanine, F. (2013). Business optimal design of a grid-connected hybrid PV (photovoltaic)-wind energy system without energy storage for an Easter Island's block. *Energy*, *61*, 248–261. doi:10.1016/j.energy.2013.08.030

Cabral, C. V. T., Filho, D. O., Diniz, A. S. A. C., Martins, J. H., Toledo, O. M., & Neto, L. V. B. M. (2010). A stochastic method for stand-alone photovoltaic system sizing. *Solar Energy*, *84*(9), 1628–1636. doi:10.1016/j.solener.2010.06.006

Calovic, M. (1972). Linear regulator design for a load and frequency control. *IEEE Transactions on Power Apparatus and Systems*, *91*(6), 2271–2285. doi:10.1109/TPAS.1972.293383

Calvo-Rolle, J. L., Quintián, H., Corchado, E., & Ferreiro-Garcia, R. (2013). Intelligent model to obtain current extinction angle for a single phase half wave controlled rectifier with resistive and inductive load. In *Soft Computing Models in Industrial and Environmental Applications* (pp. 249–256). Springer.

Calvo-Rolle, J., Corchado, E., Quintian-Pardo, H., García, R., Román, J., & Hernández, P. (2012). A novel hybrid intelligent classifier to obtain the controller tuning parameters for temperature control. *Hybrid Artificial Intelligent Systems*, 677–689.

Canizares, C. A. (2002). Voltage stability assessment: concepts, practices and tools. *Power System Stability Subcommittee Special Publication IEEE/PES, Final Document* (thunderbox.uwaterloo. ca/˜ claudio/claudio. html# VSWG). IEEE.

Canizares, C. A., & Alvarado, F. L. (1993). Point of collapse and continuation methods for large ac/dc systems. *IEEE Transactions on Power Systems*, *8*(1), 1–8. doi:10.1109/59.221241

Carson, W. (1994). *Taylor, Power System Voltage Stability*. McGraw-Hill.

Casteleiro-Roca, J. L., Fernández-Serantes, L. A., Calvo-Rolle, J. L., Machón-González, I., Crespo-Ramos, M. J., & López-Garcia, H. (2013). Study of the effect of a geothermal heat exchanger over the ground. In *2013 International Conference on New Concepts in Smart Cities: Fostering Public and Private Alliances (SmartMILE)* (pp. 1–6). Academic Press. 10.1109/SmartMILE.2013.6708197

Casteleiro-Roca, J.-L., Gómez-González, J. F., Calvo-Rolle, J. L., Jove, E., Quintián, H., Martín, J. F. A., … Méndez-Perez, J. A. (2018). Prediction of the energy demand of a hotel using an artificial intelligence-based model. Lecture Notes in Computer Science, 10870. doi:10.1007/978-3-319-92639-1_49

Casteleiro-Roca, J.-L., Pérez, J. A. M., Piñón-Pazos, A. J., Calvo-Rolle, J. L., & Corchado, E. (2019). Intelligent Model for Electromyogram (EMG) Signal Prediction During Anesthesia. *Journal of Multiple-Valued Logic & Soft Computing, 32.*

Casteleiro-Roca, J.-L., Barragán, A. J., Segura, F., Calvo-Rolle, J. L., & Andújar, J. M. (2019). Fuel Cell Output Current Prediction with a Hybrid Intelligent System. *Complexity*, 2019.

Casteleiro-Roca, J.-L., Gómez-González, J. F., Calvo-Rolle, J. L., Jove, E., Quintián, H., Gonzalez Diaz, B., & Mendez Perez, J. A. (2019). Short-Term Energy Demand Forecast in Hotels Using Hybrid Intelligent Modeling. *Sensors (Basel), 19*(11), 2485. doi:10.339019112485

Casteleiro-Roca, J.-L., Jove, E., Gonzalez-Cava, J. M., Méndez Pérez, J. A., Calvo-Rolle, J. L., & Blanco Alvarez, F. (2018). Hybrid model for the ANI index prediction using Remifentanil drug and EMG signal. *Neural Computing & Applications*. doi:10.100700521-018-3605-z

Casteleiro-Roca, J.-L., Jove, E., Sánchez-Lasheras, F., Méndez-Pérez, J.-A., Calvo-Rolle, J.-L., & De Cos Juez, F. J. (2017). Power Cell SOC Modelling for Intelligent Virtual Sensor Implementation. *Journal of Sensors, 2017*, 1–10. doi:10.1155/2017/9640546

Casteleiro-Roca, J., Quintián, H., Calvo-Rolle, J., Corchado, E., & del Carmen Meizoso-López, M. (2013). Intelligent Model to Obtain Initial and Final Conduction Angle of a Diode in a Half Wave Rectifier with a Capacitor Filter. In *International Joint Conference SOCO'13-CISIS'13-ICEUTE'13*. Springer International Publishing.

Chademo Association. (2019). *CHAdeMO releases the latest version of the protocol enabling up to 400KW*. Available at: https://www.chademo.com/chademo-releases-the-latest-version-of-the-protocol-enabling-up-to-400kw

Chaiyatham, Ngamroo, Pothiya, & Vachirasricirikul. (2009). Design of Optimal Fuzzy Logic-PID Controller using Bee Colony Optimization for Frequency Control in An Isolated Wind-Diesel System. *Proceedings of IEEE Transmission & Distribution Conference & Exposition: Asia and Pacific,* 1-4.

Chakravorty, M., & Das, D. (2001). Voltage stability analysis of radial distribution networks. *International Journal of Electrical Power & Energy Systems, 23*(2), 129–135. doi:10.1016/S0142-0615(00)00040-5

Chappa & Thakur. (2018, July). A Fast Online Voltage Instability Detection in Power Transmission System Using Wide-Area Measurements. *Iranian Journal of Science and Technology, Transactions of Electrical Engineering.*

Chaturvedi, D. K., Satsangi, P. S., & Kalra, P. K. (1999). Load frequency control: A generalized neural network approach. *Electrical Power & Energy Systems, 21*(6), 405–415. doi:10.1016/S0142-0615(99)00010-1

Chebbo, A., Irving, M., & Sterling, M. (1992). Voltage collapse proximity indicator: Behavior and implications. *IEE Proceedings. Generation, Transmission and Distribution, 139*(3), 241–252. doi:10.1049/ip-c.1992.0037

Chen, Y.-L. (1996). Weak bus-oriented optimal multi-objective VAR planning. *IEEE Transactions on Power Systems, 11*(4), 1885–1890. doi:10.1109/59.544659

Chen, Y.-L., Chang, C.-W., & Liu, C.-C. (1995). Efficient methods for identifying weak nodes in electrical power networks. *IEE Proceedings. Generation, Transmission and Distribution, 142*(3), 317–322. doi:10.1049/ip-gtd:19951728

Chen, Y., Yang, J., Trappe, W., & Martin, R. P. (2010). Detecting and localizing identity-based attacks in wireless and sensor networks. *IEEE Transactions on Vehicular Technology, 59*(5), 2418–2434. doi:10.1109/TVT.2010.2044904

CIGRE Task Force 38.02.11 Report, W. (1994). *Indices predicting voltage collapse including dynamic phenomena*. CIGRE.

Collares-Pereira, M., & Rabl, A. (1979). Derivation of method for predicting long term average energy delivery of non-concentrating and concentrating solar collectors. *Solar Energy, 22*, 155–170. doi:10.1016/0038-092X(79)90100-2

Corchado, E., & Baruque, B. (2012). WeVoS-ViSOM: An ensemble summarization algorithm for enhanced data visualization. *Neurocomputing*, *75*(1), 171–184. doi:10.1016/j.neucom.2011.01.027

Corsi, S., & Taranto, G. N. (2008). A real-time voltage instability identification algorithm based on local phasor measurements. *IEEE Transactions on Power Systems*, *23*(3), 1271–1279. doi:10.1109/TPWRS.2008.922586

Crisan, O., & Liu, M. (1994). Voltage collapse prediction using an improved sensitivity approach. *Electric Power Systems Research*, *28*(3), 181–190. doi:10.1016/0378-7796(94)90032-9

Crossbow. (n.d.). *Data sheet of Crossbow MICAz mote*. Retrieved September 13, 2018, from http://edge.rit.edu/edge/P08208/public/Controls_Files/MICaZ-DataSheet.pdf

Cross, N., Herman, R., & Gaunt, C. T. (2006). *Investigating the usefulness of beta pdf to describe parameters in reliability analyses. In Proceedings of Probabilistic Methods Applied to Power Systems*. PMAPS.

Crow, M. L. (2002). *Computational methods for electric power systems*. CRC Press. doi:10.1201/9781420058284

Cutsem, T. V. (2000). Voltage instability: Phenomena, countermeasures, and analysis methods. *Proceedings of the IEEE*, *88*(2), 208–227. doi:10.1109/5.823999

Da Costa, G. (2002). Modified newton method for reactive dispatching. *International Journal of Electrical Power & Energy Systems*, *24*(10), 815–819. doi:10.1016/S0142-0615(02)00013-3

de Souza, A. Z. (2000). Discussion on some voltage collapse indices. *Electric Power Systems Research*, *53*(1), 53–58. doi:10.1016/S0378-7796(99)00042-5

De Souza, A. Z., Canizares, C. A., & Quintana, V. H. (1997). New techniques to speed up voltage collapse computations using tangent vectors. *IEEE Transactions on Power Systems*, *12*(3), 1380–1387. doi:10.1109/59.630485

Deeb, N., & Shahidehpour, S. (1988). An efficient technique for reactive power dispatch using a revised linear programming approach. *Electric Power Systems Research*, *15*(2), 121–134. doi:10.1016/0378-7796(88)90016-8

Deeb, N., & Shahidehpour, S. (1990). Linear reactive power optimization in a large power network using the decomposition approach. *IEEE Transactions on Power Systems*, *5*(2), 428–438. doi:10.1109/59.54549

Del Valle, Y., Harley, R. G., Hernandez, J. C., Mohagheghi, S., & Venayagamoorthy, G. K. (2008). Particle swarm optimization: Basic concepts, variants and applications in power systems. *IEEE Transactions on Evolutionary Computation*, *12*(2), 171–195. doi:10.1109/TEVC.2007.896686

Delbecq, A. L., Van de Ven, A. H., & Gustafson, D. H. (1975). *Group techniques for program planning: A guide to nominal group and Delphi processes* Glenview, IL: Scott, Foresman.

Department of Minerals and Energy - Republic of South Africa. (2003, November). *White paper on renewable energy.*

Dev, A., & Sarkar, M. K. (2019). Robust Higher Order Observer Based Non-linear Super Twisting Load Frequency Control for Multi Area Power Systems via Sliding Mode. *International Journal of Control, Automation, and Systems*, *17*(7), 1814–1825. doi:10.100712555-018-0529-4

Dev, A., Sarkar, M. K., Asthana, P., & Narzary, D. (2019). Event-triggered adaptive integral higher order sliding mode control for load frequency problems in multi area power systems. *Iranian Journal of Science and Technology. Transaction of Electrical Engineering*, *43*(1), 137–152. doi:10.100740998-018-0078-0

Diao, R., Sun, K., Vittal, V., Keefe, R. J., Richardson, M. R., Bhatt, N., ... Sarawgi, S. K. (2009). Decision tree based online voltage security assessment using PMU measurements. *IEEE Transactions on Power Systems*, *24*(2), 832–839. doi:10.1109/TPWRS.2009.2016528

Dipayan Guha, P. K. (2016). Load frequency control of interconnected power system using grey wolf optimization. *Swarm and Evolutionary Computation*, *27*(2), 97–115. doi:10.1016/j.swevo.2015.10.004

Dobson, I., & Lu, L. (1992). Computing an optimum direction in control space to avoid saddle node bifurcation and voltage collapse in electric power systems. *IEEE Transactions on Automatic Control*, *37*(10), 1616–1620. doi:10.1109/9.256397

Dobson, I., & Lu, L. (1993). New methods for computing a closest saddle node bifurcation and worst case load power margin for voltage collapse. *IEEE Transactions on Power Systems*, *8*(3), 905–913. doi:10.1109/59.260912

Dommel, H. W., & Tinney, W. F. (1968). Optimal power flow solutions. *IEEE Transactions on Power Apparatus and Systems*, *PAS-87*(10), 1866–1876. doi:10.1109/TPAS.1968.292150

Dong, L., & Severance, F. L. (2007). Position estimation with moving beacons in wireless sensor networks. In *Proceedings of International Conference on Wireless Communications and Networking* (pp. 2317–2321). Kowloon, Hong Kong: IEEE. 10.1109/WCNC.2007.433

Doty, S., & Turner, W. C. (2004). *Energy management handbook*. CRC Press.

Duman, S., Güvenç, U., Sönmez, Y., & Yörükeren, N. (2012). Optimal power flow using gravitational search algorithm. *Energy Conversion and Management*, *59*, 86–95. doi:10.1016/j.enconman.2012.02.024

Dursun, E., & Kilic, O. (2012). Comparative evaluation of different power management strategies of a stand-alone PV/Wind/PEMFC hybrid power system. *International Journal of Electrical Power & Energy Systems*, *34*(1), 81–89. doi:10.1016/j.ijepes.2011.08.025

Dwivedi, S. K. (n.d.). *Load Modelling In Optimal Power Flow Studies*. Retrieved from http://ethesis@nitrkl.ac.in

Dzobo, O., Gaunt, C. T., & Herman, R. (2011). Investigating the use of probability distribution functions in reliability-worth analysis of electric power systems. *International Journal of Electrical Power & Energy Systems*, *37*(1), 110–116. doi:10.1016/j.ijepes.2011.12.013

Dzobo, O., Gaunt, C. T., & Herman, R. (2012a). *Customer interruption cost for composite reliability analysis, Proceedings of Probabilistic Methods Applied to Power Systems*. PMAPS.

Dzobo, O., Gaunt, C. T., & Herman, R. (2012b). Reliability worth assessment of electricity consumers: A South African case study. *Journal of Energy in Southern Africa*, *23*(3), 31–39. doi:10.17159/2413-3051/2012/v23i3a3170

Dzobo, O., & Xia, X. (2017). Optimal operation of smart multi-energy hub systems incorporating energy hub coordination and demand response strategy. *Journal of Renewable and Sustainable Energy*, *9*(4), 045501. doi:10.1063/1.4993046

Edimu, M., Gaunt, C. T., & Herman, R. (2011). Applying Probability Distribution Functions to Model System Failures Due to Adverse Weather. *17th Power Systems Computation Conference*.

Egido Cortés, I., Rouco Rodríguez, L., Alonso Perez, J., Porras Muñoz, E., & Ruiz Mendoza, J. L. (n.d.). Impacto en regulación frecuencia-potencia de los cambios de programa en escalón de las unidades generadoras. *Revista de La Asociación de Ingenieros Del ICAI*.

Elaiw, Xia, & Shehata. (n.d.). *Minimization of fuel costs and gaseous emissions of electric power generation by model predictive control*. Math Prob Eng Article ID 906958.

Elaiw, A. M., Xia, X., & Shehata, A. M. (2012). Application of model predictive control to optimal dynamic dispatch of generation with emission limitations. *Electric Power Systems Research*, *84*(1), 31–44. doi:10.1016/j.epsr.2011.09.024

El-Dib, El-Metwally, Hosam, Osrnan, & Youssef. (2004). Load flow solution using hybrid particle swarm optimization. *ICEEC 2004,International conference electrical electronic and computer engineering*, 742–746.

El-Dib, El-Metwally, Hosam, Osrnan, & Youssef. (2006). Maximum loadability of power systems using hybrid particle swarm optimization. *Electric Power Systems Research*, *76*, 485–492.

Elgerd, O. I. (1971). *Electrical Energy systems theory: an Introduction* (2nd ed.). McGraw Hill.

Elgerd, O. I., & Fosha, C. E. (1970). Optimum megawatt-frequency control of diverse-area electric energy systems. *IEEE Transactions on Power Apparatus and Systems*, *PAS-89*(4), 556–563. doi:10.1109/TPAS.1970.292602

Elgerd, O. l. (1982). *Electic energy systems theory-an introduction*. TMH.

El-Keib, A., & Ma, X. (1995). Application of artificial neural networks in voltage stability assessment. *IEEE Transactions on Power Systems*, *10*(4), 1890–1896. doi:10.1109/59.476054

Energía y sociedad. (n.d.a). *Energía y sociedad*. Author.

Energía y sociedad. (n.d.b). *Energías renovables: tecnología, economía, evolución e integración en el sistema eléctrico*. Author.

Erol-Kantarci, M., & Mouftah, H. (2011). Wireless sensor networks for cost efficient residential energy management in the smart grid. *IEEE Transactions on Smart Grid*, *1*, 320–325.

European Commission. (2003). EUR 20871 – Renewable energy technologies and Kyoto protocol mechanisms. Author.

Fabra, N. (2007). Ajustes regulatorios en el sector eléctrico español, La electricidad: mercado, inversiones y garantía de suministro. *ENEconomía Industrial*, 364.

Fahrioglu, M., & Alvarado, F. L. (2000). Designing incentive compatible contracts for effective demand management. *IEEE Transactions on Power Systems*, *15*(4), 1255–1260. doi:10.1109/59.898098

Fahrioglu, M., & Alvarado, F. L. (2001). Using utility information to calibrate customer demand management behaviour models. *IEEE Transactions on Power Systems*, *450*(16), 317–322. doi:10.1109/59.918305

Farmani, F., Parvizimosaed, M., Monsef, H., & Rahimi-Kian, A. (2018). A conceptual model of a smart energy management system for a residential building equipped with CCHP system. *International Journal of Electrical Power & Energy Systems*, *95*, 523–536. doi:10.1016/j.ijepes.2017.09.016

Farrokhabadi, M., Ca˜nizares, C., & Bhattacharya, K. (2016). Unit Commitment For Isolated Microgrids Considering Frequency Control. *IEEE Transactions on Smart Grid*, *9*(4), 3270–3280. doi:10.1109/TSG.2016.2629982

Fasugba, M., & Krein, P. (2011). Cost benefits and vehicle-to-grid regulation services of unidirectional charging of electric vehicles. *2011 IEEE Energy Conversion Congress And Exposition*. doi:10.1109/ecce.2011.6063856

Feliachi, A. (1987). Load frequency control using reduced order models and local observers. *International Journal of Electrical Energy Systems*, *7*(2), 72–75.

FERC. (2005). *Principles for efficient and reliable reactive power supply and consumption*. FERC Staff Reports, USA, Docket No. AD05-1-000.

Ferguson, C. D. (2011). *Nuclear Energy: What Everyone Needs to Know*. Oxford University Press.

Fernández Serantes, L. A. (2014). *Deseño e implementación dun emulador de fontes de corrente continua a partir dun convertedor DC/DC*. Academic Press.

Fernandez-Serantes, L. A., Berger, H., Stocksreiter, W., & Weis, G. (2016). Ultra-High Frequent Switching with GaN-HEMTs using the Coss-Capacitances as non-dissipative Snubbers. In *PCIM Europe 2016; International Exhibition and Conference for Power Electronics, Intelligent Motion, Renewable Energy and Energy Management* (pp. 1–8). Academic Press.

Fernández-Serantes, L. A., Estrada Vázquez, R., Casteleiro-Roca, J., Calvo-Rolle, J., & Corchado, E. (2014). Hybrid Intelligent Model to Predict the SOC of a LFP Power Cell Type. In M. Polycarpou, A. P. L. F. de Carvalho, J.-S. Pan, M. Wozniak, H. Quintian, & E. Corchado (Eds.), *Hybrid Artificial Intelligence Systems*. Springer International Publishing. doi:10.1007/978-3-319-07617-1_49

Find and Compare Cars. (2018). Retrieved from https://www.fueleconomy.gov/feg/findacar.shtml

Flatabo, N., Ognedal, R., & Carlson, T. (1990). Voltage stability condition in a power transmission system calculated by sensitivity methods. *IEEE Transactions on Power Systems*, *5*(4), 1286–1293. doi:10.1109/59.99379

Fosha, C. E., & Elgerd, O. I. (1970). The megawatt frequency control problem: A new approach via optimal control theory. *IEEE Transactions on Power Apparatus and Systems*, *89*(4), 563–577. doi:10.1109/TPAS.1970.292603

Fukuyama, Y., & Yoshida, H. (2001). A particle swarm optimization for reactive power and voltage control in electrical power systems. *Proceedings of 2001 congress on evolutionary computation, 1*, 87–93. 10.1109/CEC.2001.934375

Fulwani, D., Bandyopadhyay, B., & Fridman, L. (2011). Non-linear sliding surface: Towards high performance robust control. *IET Control Theory & Applications*, *6*(2), 235–242. doi:10.1049/iet-cta.2010.0727

GaleanoGonzález, D. A., & Botero Castro, H. A. (n.d.). Representación mediante grafcet del accionamiento de generadores sincrónicos en una central hidroeléctrica modernizada. *Dyna, Revista de La Facultad de Minas - Universidad de Colombia*.

Galiana, F. (1984). Bound estimates of the severity of line outages in power system contingency analysis and ranking. *IEEE Transactions on Power Apparatus and Systems*, *PAS-103*(9), 2612–2624. doi:10.1109/TPAS.1984.318430

Gan, D., Qu, Z., & Wu, X. (2003). Loadability of power systems with steady-state and dynamic securityconstraints. *International Journal of Electrical Power & Energy Systems*, *25*(2), 91–96. doi:10.1016/S0142-0615(02)00030-3

Gao, B., Morison, G. K., & Kundur, P. (1992). Voltage stability evaluation using modal analysis. *IEEE Transactions on Power Systems*, *7*(4), 1529–1542. doi:10.1109/59.207377

Garcia Iglesias, D., Jove Perez, E., Penyafiel Verdu, P., Valverde, I., Gonzalez Vasserot, M., & Rubin Lopez, J. M. (2017). P1034 Frequency domain analysis of surface ECG enables distinction of patients with brugada syndrome. *EP Europace, 19*(3), iii222--iii223.

Gezici, S. (2008). A survey on wireless position estimation. *Wireless Personal Communications*, *44*(3), 263–282. doi:10.100711277-007-9375-z

Ghoshal, S. P. (2003). Multi area frequency and tie line power flow control with fuzzy logic based integral gain scheduling. *Journal of Institute of Engineers*, *84*, 135–141.

Gilboy, J. (2019). *BMW, Porsche Reveal Prototype EV Fast Charger That Gives 62 Miles of Range in 3 Minutes*. Retrieved from http://www.thedrive.com/tech/25468/bmw-porsche-reveal-prototype-ev-fast-charger-that-gives-62-miles-of-range-in-3-minutes

Glavic, M., & Cutsem, T. V. (2009). Wide area detection of voltage instability from synchronized phasor measurements. Part II: Simulation results. *IEEE Transactions on Power Systems*, *24*(3), 1417–1425. doi:10.1109/TPWRS.2009.2023272

Glavic, M., Novosel, D., Heredia, E., Kosterev, D., Salazar, A., Ashrafi, F. H., & Donnelly, M. (2012). See it fast to keep calm: Real time voltage control under stressed conditions. *IEEE Power & Energy Magazine*, *10*(4), 43–55. doi:10.1109/MPE.2012.2196332

Goharrizi, A. Y., & Asghari, R. (2007). A novel line stability index (NLSI) for voltage stability assessment of power systems. *Proceedings of the International Conference on Power Systems.*

Gonzalez-Cava, J. M., Arnay, R., Pérez, J. A. M., León, A., Martin, M., Jove-Perez, E., … de Cos Juez, F. J. (2017). A machine learning based system for analgesic drug delivery. In *International Joint Conference SOCO17-CISIS17-ICEUTE17 León, Spain, September 6--8, 2017, Proceeding* (pp. 461–470). Academic Press.

Gonzalez-Cava, J. M., Reboso, J. A., Casteleiro-Roca, J. L., Calvo-Rolle, J. L., & Méndez Pérez, J. A. (2018). A novel fuzzy algorithm to introduce new variables in the drug supply decision-making process in medicine. *Complexity.*

Guenther, C., Schott, B., Hennings, W., Waldowski, P., & Danzer, M. (2013). Model-based investigation of electric vehicle battery aging by means of vehicle-to-grid scenario simulations. *Journal of Power Sources*, *239*, 604–610. doi:10.1016/j.jpowsour.2013.02.041

Guha, Roy, & Banerjee. (2016). Load frequency control of large scale power system using quasi-oppositional grey wolf optimization algorithm. *Engineering Science and Technology, an International Journal, 19*(1), 1693–1713.

Hassan, H. A., & Zellagui, M. (2017). Application of Grey Wolf Optimizer Algorithm for Optimal Power Flow Of Two-Terminal HVDC Transmission System. *Advances in Electrical And Electronic Engineering - Power Engineering And Electrical Engineering*, *15*(5), 701–712.

Hcbson, E. (1980). Network constrained reactive power control using linear programming. *IEEE Transactions on Power Apparatus and Systems*, *PAS-99*(3), 868–877. doi:10.1109/TPAS.1980.319715

He, F., Wu, D., Yin, Y., & Guan, Y. (2013). Optimal deployment of public charging stations for plug-in hybrid electric vehicles. *Transportation Research Part B: Methodological*, *47*, 87–101. doi:10.1016/j.trb.2012.09.007

Hegazy, Y. G., Salama, M. M. A., & Chikhani, A. Y. (2003). Adequacy assessment of distributed generation systems using Monte Carlo simulations. *IEEE Transactions on Power Systems*, *18*(1), 48–52. doi:10.1109/TPWRS.2002.807044

Heinzelman, W. B., Chandrakasan, A. P., & Balakrishnan, H. (2002). An application-specific protocol architecture for wireless microsensor networks. *IEEE Transactions on Wireless Communications*, *1*(4), 660–670. doi:10.1109/TWC.2002.804190

Hersey, P., & Blanchard, K. H. (1969). *Management of organizational behavior: Utilizing human resources.* Academic Press.

He, T., Kolluri, S., Mandal, S., Galvan, F., & Rasigoufard, P. (2004). Identification of weak locations in bulk transmission systems using voltage stability margin index. In *Power engineering society general meeting, 2004* (pp. 1814–1819). IEEE. doi:10.1109/PES.2004.1373193

Hong, Y.-Y., & Gau, C.-H. (1994). Voltage stability indicator for identification of the weakest bus/area in power systems. *IEE Proceedings. Generation, Transmission and Distribution*, *141*(4), 305–309. doi:10.1049/ip-gtd:19949985

Hove, T. (2000). A method for predicting long-term average performance of photovoltaic systems. *Renewable Energy*, *21*(2), 207–229. doi:10.1016/S0960-1481(99)00131-7

Hsu, Y. Y. & Chan, W. C. (1984). Optimal variable structure controller for load-frequency control of interconnected hydro-thermal power systems. *Electrical Power & Energy Systems*, *6*(4), 221–229.

Hsu, Y.-Y., & Kuo, H.-C. (1992). Fuzzy-set based contingency ranking. *IEEE Transactions on Power Systems*, *7*(3), 1189–1196. doi:10.1109/59.207333

Huang, S., Hodge, B., Taheripour, F., Pekny, J., Reklaitis, G., & Tyner, W. (2011). The effects of electricity pricing on PHEV competitiveness. *Energy Policy*, *39*(3), 1552–1561. doi:10.1016/j.enpol.2010.12.029

Huggins, R. A. (2010). *Energy Storage*. Springer. doi:10.1007/978-1-4419-1024-0

Iguchi-Cartigny, J., Ruiz, P. M., Simplot-Ryl, D., Stojmenovic, I., & Yago, C. M. (2009). Localized minimum-energy broadcasting for wireless multihop networks with directional antennas. *IEEE Transactions on Computers*, *58*(1), 120–131. doi:10.1109/TC.2008.125

Indulkar, C. S., & Raj, B. (1995). Application of fuzzy controller to automatic generation control. *Electric Machines and Power Systems*, *23*(2), 209–220. doi:10.1080/07313569508955618

International Energy Agency. (2011). *Technology Roadmaps - Electric and plug-in hybrid electric vehicles.* Retrieved from http://www.ieahev.org/assets/1/7/EV_PHEV_Roadmap.pdf

Irisarri, G. D., Mokhtari, S., Tong, J., & Wang, X. (1997). Maximum loadability of power systems Using interior point non-linear optimization methods. *IEEE Transactions on Power Systems*, *12*(1), 162–172. doi:10.1109/59.574936

Jeddiand, B., & Vahidinasab, V. (2014). A modified harmony search method for environmental/economic load dispatch of real-world power systems. *Energy Conversion and Management*, *78*, 66175.

Jiang, L., Yao, W., Wu, Q. H., Wen, J. Y., & Cheng, S. J. (2012). Delay- dependent stability for load frequency control with constant and time varying delays. *IEEE Transactions on Power Systems*, *27*(2), 932–941. doi:10.1109/TPWRS.2011.2172821

Jin, X., & Chen, X. Y. (2012). Study on the SMES Application Solutions for Smart Grid. *Physics Procedia*, *36*, 902–907. doi:10.1016/j.phpro.2012.06.227

Jove, E., Aláiz-Moretón, H., Casteleiro-Roca, J., Corchado, E., & Calvo-Rolle, J. (2014). Modeling of Bicomponent Mixing System Used in the Manufacture of Wind Generator Blades. In E. Corchado, J. Lozano, H. Quintián, & H. Yin (Eds.), *Intelligent Data Engineering and Automated Learning-IDEAL*. Springer International Publishing.

Jove, E., Alaiz-Moretón, H., Garcia-Rodriguez, I., Benavides-Cuellar, C., Casteleiro-Roca, J. L., & Calvo-Rolle, J. L. (2017). PID-ITS: an intelligent tutoring system for PID tuning learning process. In *International Joint Conference SOCO17-CISIS17-ICEUTE17 León, Spain, September 6--8, 2017, Proceeding* (pp. 726–735). Academic Press.

Jove, E., Casteleiro-Roca, J.-L., Quintián, H., Méndez-Pérez, J. A., & Calvo-Rolle, J. L. (2019). Outlier Generation and Anomaly Detection Based on Intelligent One-Class Techniques over a Bicomponent Mixing System. In *International Workshop on Soft Computing Models in Industrial and Environmental Applications* (pp. 399–410). Academic Press.

Jove, E., Casteleiro-Roca, J.-L., Quintián, H., Méndez-Pérez, J. A., & Calvo-Rolle, J. L. (n.d.). A fault detection system based on unsupervised techniques for industrial control loops. *Expert Systems*, e12395.

Jove, E., Gonzalez-Cava, J. M., Casteleiro-Roca, J.-L., Quintián, H., Méndez-Pérez, J. A., & Calvo-Rolle, J. L. (2019). Anomaly Detection on Patients Undergoing General Anesthesia. In *International Joint Conference: 12th International Conference on Computational Intelligence in Security for Information Systems (CISIS 2019) and 10th International Conference on EUropean Transnational Education (ICEUTE 2019)* (pp. 141–152). Academic Press.

Jove, E., Gonzalez-Cava, J. M., Casteleiro-Roca, J.-L., Quintián, H., Méndez-Pérez, J. A., Calvo-Rolle, J. L., … Reboso, J. (2018). Remifentanil dose prediction for patients during general anesthesia. Lecture Notes in Computer Science, 10870. doi:10.1007/978-3-319-92639-1_45

Jove, E., Blanco-Rodríguez, P., Casteleiro-Roca, J. L., Moreno-Arboleda, J., López-Vázquez, J. A., de Cos Juez, F. J., & Calvo-Rolle, J. L. (2018). *Attempts prediction by missing data imputation in engineering degree* (Vol. 649). Advances in Intelligent Systems and Computing. doi:10.1007/978-3-319-67180-2_16

Jove, E., Casteleiro-Roca, J.-L., Quintián, H., Méndez-Pérez, J. A., & Calvo-Rolle, J. L. (2018). A New Approach for System Malfunctioning over an Industrial System Control Loop Based on Unsupervised Techniques. In M. Graña, J. M. López-Guede, O. Etxaniz, Á. Herrero, J. A. Sáez, H. Quintián, & E. Corchado (Eds.), *International Joint Conference SOCO'18-CISIS'18-ICEUTE'18* (pp. 415–425). Cham: Springer International Publishing.

Jove, E., Gonzalez-Cava, J. M., Casteleiro-Roca, J. L., Pérez, J. A. M., Calvo-Rolle, J. L., & de Cos Juez, F. J. (2018). *An intelligent model to predict ANI in patients undergoing general anesthesia* (Vol. 649). Advances in Intelligent Systems and Computing. doi:10.1007/978-3-319-67180-2_48

Jove, E., Gonzalez-Cava, J. M., Casteleiro-Roca, J.-L., Méndez-Pérez, J.-A., Antonio Reboso-Morales, J., Javier Pérez-Castelo, F., ... Luis Calvo-Rolle, J. (2018). Modelling the hypnotic patient response in general anaesthesia using intelligent models. *Logic Journal of the IGPL*, *27*(2), 189–201. doi:10.1093/jigpal/jzy032

Jove, E., López, J. A. V., Fernández-Ibáñez, I., Casteleiro-Roca, J. L., & Calvo-Rolle, J. L. (2018). Hybrid intelligent system topredict the individual academic performance of engineering students. *International Journal of Engineering Education*, *34*(3), 895–904.

Juang, C. F., & Lu, C. F. (2002). Power system load frequency control with fuzzy gain scheduling designed by new genetic algorithms. IEEE International Conference on Fuzzy Systems, 1, 64–68.

Kanellos, F. D., Tsouchnikas, A. I., & Hatziargyriou, N. D. (2005). Micro-Grid Simulation during Grid-Connected andIslanded Modes of Operation. *Proceedings of International Conference on Power Systems Transients (IPST'05).*

Karysinos, M. E., Nikolaidis, V. C., & Vorunas, C. D. (2005). Loadability limits and emergency Counter measures against voltage collapse. *Proceeding of the 15th Power system computation conference*, 22–26.

Kessel, P., & Glavitsch, H. (1986). Estimating the voltage stability of a power system. *IEEE Transactions on Power Delivery*, *PWRD-1*(3), 346–354. doi:10.1109/TPWRD.1986.4308013

Khajavi, P., Monsef, H., & Abniki, H. (2010). Load profile reformation through demand response programs using smart grid. *Proc. Of the international symposium on modern electric power systems.*

Khamis, Ghani, Kim, Kamarudin, & Aras. (2017). Voltage and Frequency Control of Microgrid Systems with Demand Response. *International Journal of Simulation: Systems. Science & Technology*, *25*(1), 1–6.

Khodabakhshian, A., & Edrisi, M. (2008). A new robust PID load frequency controller. *Control Engineering Practice*, *16*(9), 1069–1080. doi:10.1016/j.conengprac.2007.12.003

Kim, E., & Kim, K. (2010). Distance estimation with weighted least squares for mobile beacon-based localization in wireless sensor networks. *IEEE Signal Processing Letters*, *17*(6), 559–562. doi:10.1109/LSP.2010.2047169

Kirschen, D., & Strbac, G. (1999). Tracing active and reactive power between generators and loads using real and imaginary currents. *IEEE Transactions on Power Systems*, *14*(4), 1312–1319. doi:10.1109/59.801890

Kumano, T., Yokoyama, A., & Sekine, Y. (1994). Fast monitoring and optimal preventive control of voltage instability. *International Journal of Electrical Power & Energy Systems*, *16*(2), 117–125. doi:10.1016/0142-0615(94)90048-5

Kundur. (1994). Power System Stability and Control. New York: McGraw-Hill.

Kundur, P. (1994). *Power System Stability and Control.* New York: McGraw-Hill.

Kundur, P., Balu, N. J., & Lauby, M. G. (1994). *Power system stability and control* (Vol. 7). McGraw-hill New York.

Kundur, P., Paserba, J., Ajjarapu, V., Anderson, G., Bose, A., Canizares, C., ... Vittal, V. (2004). Definition and classification of power system stability- IEEE joint task force on stability terms and definitions. *IEEE Transactions on Power Systems, 19*(3), 1387–1401. doi:10.1109/TPWRS.2004.825981

Kundur, P., Paserba, J., Vittal, V., & Anderson, G. (2006). Closure of definition and classification of power system stability. *IEEE Transactions on Power Systems, 21*(1), 446. doi:10.1109/TPWRS.2005.861952

Kwatny, H. G., Kalnitsky, K. C., & Bhatt, A. (1975). An optimal tracking approach to load frequency control. *IEEE Transactions on Power Apparatus and Systems, 94*(5), 1635–1643. doi:10.1109/T-PAS.1975.32006

Lachs, W. R., & Sutanto, D. (1992). Voltage instability in interconnected power systems: A simulation approach. *IEEE Transactions on Power Systems, 7*(2), 753–761. doi:10.1109/59.141782

Lazarov, Notton, Zarkov, & Bochev. (2005). Hybrid Power Systems with Renewable Energy Sources – Types, Structures, Trends For Research And Development. *Proceedings of Eleventh International Conference On Electrical Machines, Drives And Power Systems.*

Lee, B. H., & Lee, K. Y. (1991). A study on voltage collapse mechanism in electricpower systems. *IEEE Transactions on Power Systems, 6*(3), 966–973. doi:10.1109/59.119236

Lee, D. J., & Wang, L. (2008). Small-signal stability analysis of an autonomous hybrid renewable energy power generation/energy storage system part I: Time-domain simulations. *IEEE Transactions on Energy Conversion, 23*(1), 311–320. doi:10.1109/TEC.2007.914309

Lee, S., Kim, E., Kim, C., & Kim, K. (2009). Localization with a mobile beacon based on geometric constraints in wireless sensor networks. *IEEE Transactions on Wireless Communications, 8*(12), 5801–5805. doi:10.1109/TWC.2009.12.090319

Li, H., & Song, Y. (2002). Identification of weak bus bars in large scale power system. In *International conference on Power system technology (Powercon)* (*Vol. 3*, pp.1700–1704). Academic Press. 10.1109/ICPST.2002.1067823

Li, A., Law, M. Y., & Wong, K. P. (1997). Development of constrained genetic algorithm load flow method. *IEE Proceedings. Generation, Transmission and Distribution, 144*(2), 91–99. doi:10.1049/ip-gtd:19970847

Liao, K., & Xu, Y. (2017). A robust load frequency control scheme for power systems based on second order sliding mode and disturbance observer. *IEEE Transactions on Industrial Informatics, 14*(7), 3076–3086. doi:10.1109/TII.2017.2771487

Liaw, C. M. (1994). Design of a reduced-order adaptive LFC for an interconnected hydrothermal power system. *International Journal of Control, 60*(6), 1051–1063. doi:10.1080/00207179408921510

Liberti, J. C., & Rappaport, T. S. (1999). *Smart antennas for wireless communications: IS-95 and third generation CDMA applications*. Prentice Hall.

Li, J., Liu, X., & Su, X. (2018). Sliding mode observer-based load frequency control of multi area power systems under delayed inputs attack. *Chinese Control and Decision Conference*. 10.1109/CCDC.2018.8407768

Lim, K. Y., Wang, Y., & Zhou, R. (1996). Robust decentralized load-frequency control of multi-area power systems. *IEE Proceedings. Generation, Transmission and Distribution, 143*(5), 377–386. doi:10.1049/ip-gtd:19960452

Lin, X., & Lei, Y. (2017). *Coordinated Control Strategies for SMES-Battery Hybrid Energy Storage Systems* (Vol. 5). IEE Access.

List of Electric Vehicles. (2018). (2018). Retrieved from https://evrater.com/evs

Liu, J. H., & Chu, C. C. (2014). Wide area measurement based voltage stability indicators by modified coupled single port models. *IEEE Transactions on Power Systems*, *29*(2), 756–764. doi:10.1109/TPWRS.2013.2284475

Lof, P.-A., Andersson, G., & Hill, D. (1993). Voltage stability indices for stressed power systems. *IEEE Transactions on Power Systems*, *8*(1), 326–335. doi:10.1109/59.221224

Lof, P.-A., Smed, T., Andersson, G., & Hill, D. (1992). Fast calculation of a voltage stability index. *IEEE Transactions on Power Systems*, *7*(1), 54–64. doi:10.1109/59.141687

Lomi, A., & Thukaram, D. (2012). Optimum reactive power dispatch for alleviation of voltage deviations. *TELKOMNIKA*, *10*(2), 257–264. doi:10.12928/telkomnika.v10i2.793

Long, W., Jiao, J., Liang, X., & Tang, M. (2018). An exploration-enhanced grey wolf optimizer to solve high-dimensional numerical optimization. *Engineering Applications of Artificial Intelligence*, *68*(2), 63–80. doi:10.1016/j.engappai.2017.10.024

López García, H., & Machón González, I. (2004). Self-organizing map and clustering for wastewater treatment monitoring. *Engineering Applications of Artificial Intelligence*, *17*(3), 215–225. doi:10.1016/j.engappai.2004.03.004

Lucia López, J., & Meneu Ferrer, V. (2005). Los precios en los mercados reestructurados de electricidad: algunas lecciones básicas para la negociación derivada. *CICE Nº 69.*

M.J Bradley and Associates LLC. (2013). Electric Vehicle Grid Integration in the U.S., Europe, and China. *International Council on Clean Transportation.* Retrieved from https://www.theicct.org/sites/default/files/publications/EVpolicies_final_July11.pdf

Machowski, J., Bialek, J. W. & Bumby. (2008). *Power System Dynamics.* John Wiley and Sons.

MacKay, D. J. C. (2009). *Sustainable Energy: Without the Hot Air.* Uit Cambridge Limited.

Maiti, Acharya, Chakraborty, Konar, & Janarthana. (2008). Tuning PID and PIλDδ Controllers using the Integral Time Absolute Error Criterion. *Proceedings of 4th IEEE International Conference on Information and Automation for Sustainability, 516.*

Malette, M., & Venkataramanan, G. (2010). *Financial incentives to encourage demand response participation by plug-in hybrid electric vehicle owners. In IEEE Energy Conversation Congress and Exposition* (pp. 4278–4284). ECCE.

Mamandur, K., & Chenoweth, R. (1981). Optimal control of reactive power flow for improvements in voltage profiles and for real power loss minimization. *IEEE Transactions on Power Apparatus and Systems*, (7): 3185–3194. doi:10.1109/TPAS.1981.316646

Mangoli, M. K., Lee, K. Y., & Park, Y. M. (1993). Optimal real and reactive power control using linear programming. *Electric Power Systems Research*, *26*(1), 1–10. doi:10.1016/0378-7796(93)90063-K

Manuel Vilar-Martinez, X., Aurelio Montero-Sousa, J., Luis Calvo-Rolle, J., & Luis Casteleiro-Roca, J. (2014). Expert system development to assist on the verification of" TACAN" system performance. *Dyna (Bilbao)*, *89*(1), 112–121.

Mary, R. (2008). Optimum cost of generation for Maximum loadability limit of power system using hybrid particle swarm optimization. *International Journal of Electrical Power & Energy Systems*, *30*(8), 486–490. doi:10.1016/j.ijepes.2008.04.001

Mashburn, W. H. (1989). *Managing energy resources in times of dynamic change.* Academic Press.

MathWorks. (2019). *Fuzzy logic toolbox.* Retrieved September 13, 2018, from https://www.mathworks.com/help/pdf_doc/fuzzy/fuzzy.pdf

Ma, Z., Callaway, D., & Hiskens, I. (2013). Decentralized Charging Control of Large Populations of Plug-in Electric Vehicles. *IEEE Transactions on Control Systems Technology*, *21*(1), 67–78. doi:10.1109/TCST.2011.2174059

Mendel, J. M. (1995). Fuzzy logic systems for engineering: A tutorial. *Proceedings of the IEEE*, *83*(3), 345–377. doi:10.1109/5.364485

Menezes, T., da Silva, L. C., Affonso, C., & Da Costa, V. F. (2004). Mvar management on the pre-dispatch problem for improving voltage stability margin. *IEE Proceedings. Generation, Transmission and Distribution*, *151*(6), 665–672. doi:10.1049/ip-gtd:20040802

Mirjalili,S. (2015).Moth-flame optimization algorithm: A novel nature-inspired heuristic paradigm. *Knowledge-Based System*, 89, 228-49.doi.org/10.1016/j.knosys.2015.07.006.

Mirjalili, S., Mirjalili, S. M., & Lewis, A. (2014). Grey Wolf Optimizer. *Advances in Engineering Software*, *69*(1), 46–61. doi:10.1016/j.advengsoft.2013.12.007

Mi, Y., Fu, Y., Li, D., Wang, C., Loh, P. C., & Weng, P. (2016). The sliding mode load frequency control for hybrid power system based on disturbance observer. *International Journal of Electrical Power & Energy Systems*, *74*, 446–452. doi:10.1016/j.ijepes.2015.07.014

Mi, Y., Fu, Y., Wang, C., & Wang, P. (2013). Decentralized Sliding Mode Load Frequency Control for Multi-Area Power Systems. *IEEE Transactions on Power Systems*, *28*(4), 4301–4309. doi:10.1109/TPWRS.2013.2277131

Mi, Y., Hua, X., Liu, Y., Fu, Y., Wang, C., Wang, P., & Loh, P. C. (2017). Sliding mode load frequency control for multi area time-delay power system with wind power integration. *IET Generation, Transmission & Distribution*, *11*(18), 4644–4643. doi:10.1049/iet-gtd.2017.0600

Modarresi, J., Gholipour, E., & Khodabakhshian, A. (2016). A comprehensive review of the voltage stability indices. *Renewable & Sustainable Energy Reviews*, *63*, 1–12. doi:10.1016/j.rser.2016.05.010

Moger, T. (2016). *Reactive power planning and operation of power systems with wind farms for voltage stability improvement* (PhD dissertation). Department of Electrical Engineering, Indian Institute of Science, Bangalore, India.

Moger, T., & Dhadbanjan, T. (2013a). An improved approach for evaluation of reactive power sources contribution to reactive load and loss. *Annual India Conference (INDICON)*. 10.1109/INDCON.2013.6725961

Moger, T., & Dhadbanjan, T. (2013b). Reactive power loss allocation to consumers in a deregulated power system. *Fifth International Conference on Power and Energy Systems (ICPS)*.

Moger, T., & Dhadbanjan, T. (2015). A novel index for identification of weak nodes for reactive compensation to improve voltage stability. *IET Generation, Transmission & Distribution*, *9*(14), 1826–1834. doi:10.1049/iet-gtd.2015.0054

Moger, T., & Dhadbanjan, T. (2017). Evaluation of Reactive Power Support and Loss Allocation in a Pool Based Competitive Electricity Market. *International Journal of Emerging Electric Power Systems*, *18*(2), 1–19. doi:10.1515/ijeeps-2016-0201

Moger, T., Johnson, T., & Dhadbanjan, T. (2018). Significance of Reactive Power Loss and Its Application to System Voltage Stability. In *IEEMA Engineer Infinite Conference (eTechNxT)*. IEEE. 10.1109/ETECHNXT.2018.8385343

Moghavvemi, M., & Faruque, M. O. (2001). Technique for assessment of voltage stability in illconditioned radial distribution network. *IEEE Power Engineering Review*, *21*(1), 58–60. doi:10.1109/39.893345

Moghavvemi, M., & Faruque, O. (1998). Real-time contingency evaluation and ranking technique. *IEE Proceedings. Generation, Transmission and Distribution*, *145*(5), 517–524. doi:10.1049/ip-gtd:19982179

Moghavvemi, M., & Omar, F. M. (1998). Technique for contingency monitoring and voltage collapse prediction. *IEE Proceedings. Generation, Transmission and Distribution*, *145*(6), 634–640. doi:10.1049/ip-gtd:19982355

Mohamed, A., Jasmon, G. B., & Yusoff, S. (1989). A static voltage collapse indicator using line stability factors. *Journal of Information Technology*, *7*, 73–85.

Mohamed, F. A., & Koivo, H. N. (2010). System modelling and online optimal management of MicroGrid using Mesh Adaptive Direct Search. *International Journal of Electrical Power & Energy Systems*, *32*(5), 398–407. doi:10.1016/j.ijepes.2009.11.003

Mohammadi, H., & Behghani, M. (2015). PMU based voltage securing assessment of power systems exploiting principal component analysis and decision trees. *International Journal of Electrical Power & Energy Systems*, *64*, 655–663. doi:10.1016/j.ijepes.2014.07.077

Momoh, J. A., El-Hawary, M., & Adapa, R. (1999). A review of selected optimal power flow literature to 1993.PART I: Nonlinear and quadratic programming approaches. *IEEE Transactions on Power Systems*, *14*(1), 96–104.

Mondal, S., Gokul, T. V., & Mahanta, C. (2012). Chattering Free Sliding Mode Controller for Mismatched Uncertain System. *6th International Conference on Industrial and Information Systems*. 10.1109/ICIInfS.2012.6304790

Mondal, S., & Mahanta, C. (2011). Nonlinear Sliding Surface based Second Order Sliding Mode Controller for Uncertain Linear Systems. *Communications in Nonlinear Science and Numerical Simulation, Elsevier*, *16*(9), 3760–3769. doi:10.1016/j.cnsns.2010.12.020

Montero Sousa, J. A., Casteleiro Roca, J. L., & Calvo Rolle, J. L. (2017a). El sector eléctrico desde su inicio hasta la segunda guerra mundial. *Dyna (Bilbao)*, *92*(1), 43–47. doi:10.6036/7947

Montero-Sousa, J. A., Fernåndez-Serantes, L. A., Casteleiro-Roca, J.-L., Vilar-Martinez, X. M., & Calvo-Rolle, J. L. (2017). Energy Management Strategies to Improve Electrical Networks Using Storage Systems. In Renewable and Alternative Energy: Concepts, Methodologies, Tools, and Applications (pp. 1500–1514). IGI Global.

Montero-Sousa, J.-A., Fernandez-Serantes, L.-A., Casteleiro-Roca, J.-L., Vilar-Martnez, X.-M., & Calvo-Rolle, J.-L. (2017). *Energy storage management for generation-distribution facilities*. Federacion Asociaciones Ingenieros Industriales Espana Alameda De Mazarredo.

Moorthi, V. R., & Aggarwal, R. P. (1972). Suboptimal and near optimal control of a load frequency control system. *Proceedings of the Institution of Electrical Engineers*, *119*(11), 1653–1660. doi:10.1049/piee.1972.0329

Moradi, M. H., & Abedini, M. (2012). A combination of genetic algorithm and particle swarm optimization for optimal DG location and sizing in distribution systems. *International Journal of Electrical Power & Energy Systems*, *34*(1), 66–74. doi:10.1016/j.ijepes.2011.08.023

Morais, H., Faria, P., & Vale, Z. (2014). Demand response design and use based on network locational marginal prices. *International Journal of Electrical Power & Energy Systems*, *61*, 18091. doi:10.1016/j.ijepes.2014.03.024

Musio, M., Lombardi, P., & Damiano, A. (2010). Vehicles to grid (V2G) concept applied to a Virtual Power Plant structure. *The XIX International Conference on Electrical Machines - ICEM 2010*. 10.1109/ICELMACH.2010.5608261

Musirin, I., & Rahman, T. K. A. (2002). Estimating maximum loadability for weak bus identification using FVSI. *IEEE Power Engineering Review*, *22*(11), 50–52. doi:10.1109/MPER.2002.1045568

Mwasilu, F., Justo, J., Kim, E., Do, T., & Jung, J. (2014). Electric vehicles and smart grid interaction: A review on vehicle to grid and renewable energy sources integration. *Renewable & Sustainable Energy Reviews*, *34*, 501–516. doi:10.1016/j.rser.2014.03.031

Niknam, T., Golestaneh, F., & Malekpour, A. (2012). Probabilistic energy and operation management of a microgrid containing wind/photovoltaic/fuel cell generation and energy storage devices based on point estimate method and self-adaptive gravitational search algorithm. *Energy*, *43*(1), 427–437. doi:10.1016/j.energy.2012.03.064

Nwulu, N. I., & Xia, X. (2015). Implementing a model predictive control strategy on the dynamic economic emission dispatch problem with game theory based demand response programs. *Energy*, *91*, 404419. doi:10.1016/j.energy.2015.08.042

Nwulu, N. I., & Xia, X. (2015). Multi-objective dynamic economic emission dispatch of electric power generation integrated with game theory based demand response programs. *Energy Conversion and Management*, *89*, 963–974. doi:10.1016/j.enconman.2014.11.001

Nwulu, N. I., Xia, X., & Zhang, J. (2013). Determining the optimal incentive and number of retrofits for a demand response program in South Africa. *Proc. Of the 5th international conference on applied energy.*

Ordacgi Filho, J. M. (2010). *Brazilian blackout 2009: Blackout watch.* Protection, Automation and Control World.

Osorio, G. J., Lujano-Rojas, J. M., Matias, J. C. O., & Catal, J. P. S. (2015). A probabilistic approach to solve the economic dispatch problem with intermittent renewable energy sources. *Energy*, *82*, 949–959. doi:10.1016/j.energy.2015.01.104

Ou, C.-H. (2011). A localization scheme for wireless sensor networks using mobile anchors with directional antennas. *IEEE Sensors Journal*, *11*(7), 1607–1616. doi:10.1109/JSEN.2010.2102748

Overbye, T. J., & DeMarco, C. L. (1991). Improved techniques for power system voltage stability assessment using energy methods. *IEEE Transactions on Power Systems*, *6*(4), 1446–1452. doi:10.1109/59.116988

Padhy, Panda, & Mahapatra. (2017). A modified GWO technique based cascade PI-PD controller for AGC of power systems in presence of Plug in Electric Vehicles. *Engineering Science and Technology, an International Journal, 20*(2), 427-442.

Padillo, F., Luna, J. M., Cano, A., & Ventura, S. (2016). A data structure to speed-up machine learning algorithms on massive datasets. In *International Conference on Hybrid Artificial Intelligence Systems* (pp. 365–376). Academic Press. 10.1007/978-3-319-32034-2_31

Pal, M. K. (1992, February). Voltage Stability Conditions Considering Load Characteristics. *IEEE Transactions on Power Systems*, *7*(1), 243–249. doi:10.1109/59.141710

Pan, C. T., & Liaw, C. M. (1989). An adaptive controller for power system and load frequency control. *IEEE Transactions on Power Systems*, *4*(1), 122–1228. doi:10.1109/59.32469

Pandey, Irfaan, & Bhardwaj. (2014). Automatic Generation Control of Multi-Area Power System with Wind Power Generation. *International Journal of Science. Engineering and Technology*, *2*(6), 1364–1369.

Pandey, S. K., Mohanty, S. R., & Kishor, N. (2013). A literature survey on load–frequency control for conventional and distribution generation power systems. *Elsevier Renewable and Sustainable Energy Reviews*, *25*, 318–334. doi:10.1016/j.rser.2013.04.029

Papaefthymiou, G., Schavemaker, P. H., van der Sluis, L., Kling, W. L., Kurowicka, D., & Cooke, R. M. (2006). Integration of stochastic generation in power systems. *International Journal of Electrical Power & Energy Systems*, *28*(9), 655–667. doi:10.1016/j.ijepes.2006.03.004

Parise, G., Martirano, L., & Kermani, M. (2017). Designing a Power Control Strategy in a Microgrid Using PID / Fuzzy Controller Based on Battery Energy Storage. *Proceedings of 2017 IEEE International Conference on Environment and Electrical Engineering.*

Parks, K., Denholm, P., & Markel, T. (2007). Costs and Emissions Associated with Plug-In Hybrid Electric Vehicle Charging in the Xcel Energy Colorado Service Territory. *National Renewable Energy Laboratory*. Retrieved from https://www.nrel.gov/docs/fy07osti/41410.pdf

Parvania, M., & Fotuhi-Firuzabad, M. (2010). Demand response scheduling by stochastic SCUC. *IEEE Transactions on Smart Grid*, *1*(1), 8998. doi:10.1109/TSG.2010.2046430

Patwari, N., Ash, J. N., Kyperountas, S., Hero, A. O., Moses, R. L., & Correal, N. S. (2005). Locating the nodes: Cooperative localization in wireless sensor networks. *IEEE Signal Processing Magazine*, *22*(4), 54–69. doi:10.1109/MSP.2005.1458287

Pazzi, R. W. N., & Boukerche, A. (2008). Mobile data collector strategy for delay-sensitive applications over wireless sensor network. *Computer Communications*, *31*(5), 1028–1039. doi:10.1016/j.comcom.2007.12.024

Peterson, S., Apt, J., & Whitacre, J. (2010). Lithium-ion battery cell degradation resulting from realistic vehicle and vehicle-to-grid utilization. *Journal of Power Sources*, *195*(8), 2385–2392. doi:10.1016/j.jpowsour.2009.10.010

Pieltain Fernandez, L., Gomez San Roman, T., Cossent, R., Mateo Domingo, C., & Frias, P. (2011). Assessment of the Impact of Plug-in Electric Vehicles on Distribution Networks. *IEEE Transactions on Power Systems*, *26*(1), 206–213. doi:10.1109/TPWRS.2010.2049133

Po-Hung, C., & Hong-Chan, C. (1995). Large-scale economic dispatch by genetic algorithm. *IEEE Transactions on Power Systems*, *12*(2), 128–149.

Prasad, S., Purwar, S., & Kishor, N. (2017). Non-linear sliding mode load frequency control in multi area power system. *Control Engineering Practice*, *61*, 81–92. doi:10.1016/j.conengprac.2017.02.001

Prasad, S., Purwar, S., & Kishor, N. (2018). Load frequency regulation using observer based nonlinear sliding mode control. *Electrical Power and Energy Systems*, *104*, 178–193. doi:10.1016/j.ijepes.2018.06.035

Prathibha, M., & Bhavani, M. (2014). Automatic Generation Control in Restructured Power System with Wind Integrated System. *Proceedings of International Conference on Innovations in Engineering and Technology (ICIET'14), Engineering and Technology*.

Project Group Turkey. (2015). *Report on Blackout in Turkey on 31st March 2015*. European Network of Transmission System Operators for Electricity.

Pudjianto, D., Ahmed, S., & Strbac, G. (2002). Allocation of VAR support using LP and NLP based optimal power flows. *IEE Proceedings. Generation, Transmission and Distribution*, *149*(4), 377–383. doi:10.1049/ip-gtd:20020200

Qiu, J., & Shahidehpour, S. (1987). A new approach for minimizing power losses and improving voltage profile. *IEEE Transactions on Power Systems*, *2*(2), 287–295. doi:10.1109/TPWRS.1987.4335121

Quintián, H., Casteleiro-Roca, J.-L., Perez-Castelo, F. J., Calvo-Rolle, J. L., & Corchado, E. (2016). Hybrid intelligent model for fault detection of a lithium iron phosphate power cell used in electric vehicles. In *International Conference on Hybrid Artificial Intelligence Systems* (pp. 751–762). Academic Press. 10.1007/978-3-319-32034-2_63

Quintián, H., Calvo-Rolle, J. L., & Corchado, E. (2014). A Hybrid Regression System Based on Local Models for Solar Energy Prediction. *Informatica*, *25*(2), 265–282. doi:10.15388/Informatica.2014.14

Quintián, H., & Corchado, E. (2017). Beta scale invariant map. *Engineering Applications of Artificial Intelligence*, *59*, 218–235. doi:10.1016/j.engappai.2017.01.002

Rahimian, M. A., & Tavazoei, M. S. (2013). Improving integral square error performance with implementable fractional-order PI controllers. *Optimal Control Applications And Methods Wiley Online Library*, *35*(3), 303–313. doi:10.1002/oca.2069

Rahman, M. Z., & Kleeman, L. (2009). Paired measurement localization: A robust approach for wireless localization. *IEEE Transactions on Mobile Computing*, *8*(8), 1087–1102. doi:10.1109/TMC.2008.173

Ramakrishna, K. S. S., Sharma, P., & Bhatti, T. (2010). Automatic generation control of interconnected power systems with diverse sources of power generation. *International Journal of Engineering Science and Technology*, *2*(5), 51–65. doi:10.4314/ijest.v2i5.60102

Ramakrishnan, S., & Thyagarajan, T. (2012). Fuzzy logic-based transmission power control algorithm for energy efficient MAC protocol in wireless sensor networks. *International Journal of Communication Networks and Distributed Systems*, *9*(3/4), 247–265. doi:10.1504/IJCNDS.2012.048873

Rampurkar, V., Pentayya, P., Mangalvedekar, H. A., & Kazi, F. (2016). Cascading Failure Analysis for Indian Power Grid. *IEEE Transactions on Smart Grid*, *7*(4), 1951–1960. doi:10.1109/TSG.2016.2530679

Rao & Mishra. (2014). Comparative study of P, PI and PID controller for speed control of VSI-fed induction motor. *International Journal of Engineering Development and Research*, *2*(2), 2740–2744.

Rao, K. R. (2011). *Energy and Power Generation Handbook: Established and Emerging Technologies*. American Society of Mechanical Engineers. doi:10.1115/1.859551

Ray, G., Prasad, A. N., & Prasad, G. D. (1999). A new approach to the design of robust load-frequency controller for large scale power systems. *Electric Power Systems Research*, *51*(1), 13–22. doi:10.1016/S0378-7796(98)00125-4

Ray, G., Sitansu, D., & Bhattacharyya, T. K. (2004). Multi-area load Frequency control of power systems: A decentralized variable structure Approach. *Electric Power Components and Systems*, *33*(3), 315–331. doi:10.1080/15325000590474672

Rerkpreedapong, D., Hasanovic, A., & Feliachi, A. (2003). Robust load frequency control using genetic algorithms and linear matrix inequalities. *IEEE Transactions on Power Systems*, *18*(2), 855–861. doi:10.1109/TPWRS.2003.811005

Richardson, D. (2013). Electric vehicles and the electric grid: A review of modeling approaches, Impacts, and renewable energy integration. *Renewable & Sustainable Energy Reviews*, *19*, 247–254. doi:10.1016/j.rser.2012.11.042

Roy, R., & Kailath, T. (1989). ESPRIT-estimation of signal parameters via rotational invariance techniques. *IEEE Transactions on Acoustics, Speech, and Signal Processing*, *37*(7), 984–995. doi:10.1109/29.32276

Sabah Nimma, K., Al-Falahi, M. D. A., Nguyen, H. D., Jayasinghe, S. D. G., & Mahmoud, T. S. (2018). Grey Wolf Optimization-Based Optimum Energy-Management and Battery-Sizing Method for Grid-Connected Microgrids. *Energies*, *11*(4), 847. doi:10.3390/en11040847

Sajan, K. S., Kumar, V., & Tyagi, B. (2015). Genetic algorithm based support vector machine for online voltage stability monitoring. *International Journal of Electrical Power & Energy Systems*, *73*, 200–208. doi:10.1016/j.ijepes.2015.05.002

Salmerón, J. (2010). *Las renovables en el mercado de la electricidad: perspectivas de futuro*. Academic Press.

Sargolzaei, A., Yen, K. K., & Abdelghani, M. N. (2016). Preventing time delay switch attacks on load frequency control in distributed power systems. *IEEE Transactions on Smart Grid*, *7*(2), 1176–1185.

Sarkar, M. K., Dev, A., Asthana, P., & Narzary, D. (2018). Chattering free robust adaptive integral higher order sliding mode control for load frequency problems in multi area power systems. *IET Control Theory & Applications*, *12*(9), 1216–1227. doi:10.1049/iet-cta.2017.0735

Sato, H. (2002). Computation of power system loadability limits. *IEEE Conference on Transmission and Distribution, 3*(2), 1707–1011. 10.1109/TDC.2002.1177711

Saxena, S., MacDonald, J., & Moura, S. (2015). Charging ahead on the transition to electric vehicles with standard 120 V wall outlets. *Applied Energy, 157*, 720–728. doi:10.1016/j.apenergy.2015.05.005

Setlhaolo, D., Xia, X., & Zhang, J. (2014). Optimal scheduling of household appliances for demand response. *Electric Power Systems Research, 116*, 248. doi:10.1016/j.epsr.2014.04.012

Shafie-Khah, M., & Siano, P. (2018). A stochastic home energy management system considering satisfaction cost and response fatigue. *IEEE Transactions on Industrial Informatics, 14*(2), 629–638. doi:10.1109/TII.2017.2728803

Shayanfar, H. A., Shayeghi, H., & Younesi, A. (2015). Optimal PID Controller Design Using Krill Herd Algorithm for Frequency Stabilizing in an Isolated Wind-Diesel System. *Proceedings of Int'l Conf. Artificial Intelligence.*

Shen, J., Jiang, C., & Li, B. (2015). Controllable Load Management Approaches in Smart Grids. *Energies, 8*(1), 11187–11202. doi:10.3390/en81011187

Sichitiu, M. L., & Ramadurai, V. (2004). Localization of wireless sensor networks with a mobile beacon. In *Proceedings of International Conference on Mobile Ad-hoc and Sensor Systems* (pp. 174-183). Fort Lauderdale, FL: IEEE. 10.1109/MAHSS.2004.1392104

Singh, S., Joshi, H., Chanana, S., & Verma, R. K. (2014). Impact of superconducting magnetic energy storage on the frequency stability of an isolated hybrid power system. *Proceedings of International Conference on Computing for sustainable global development,* 141-145.

Singh, S., Singh, M., Chanana, S., & Semwal, S. (2015). Frequency regulation of isolated hybrid Wind/Diesel, power generation with fuel cell system. In Lecture Notes in Electrical Engineering, (pp. 853–862). Springer Verlag Publication.

Singh, G. K. (2013). Solar power generation by PV (photovoltaic) technology: A review. *Energy, 53*, 1–13. doi:10.1016/j.energy.2013.02.057

Singh, G., Singh, S., Raju, G., & Thukaram, D. (1996). Methodology for alleviation of voltage excursions in large power systems. *International Journal of Electrical Power & Energy Systems, 18*(3), 167–173. doi:10.1016/0142-0615(95)00018-6

Singh, J., Singh, S., & Srivastava, S. (2007). An approach for optimal placement of static VAR compensators based on reactive power spot price. *IEEE Transactions on Power Systems, 22*(4), 2021–2029. doi:10.1109/TPWRS.2007.907374

Singh, S., Singh, M., Chanana, S., & Raveendhra, D. (2013). Operation and control of a hybrid wind-diesel-battery energy system connected to micro-grid. *Proceedings of International conference on control and automation, robotics and embedded system, 1*(6). 10.1109/CARE.2013.6733758

Singh, S., Singh, S. K., Chanana, S., & Singh, Y. P. (2014). Frequency regulation of an isolated hybrid power system with a battery energy storage system. In *Proceedings of Power and Energy Systems: towards Sustainable Energy (PES-TSE), Amrita Vishwa Vidyapeetham Bengaluru Campus*. Bengaluru, India: Amrita School of Engineering. doi:10.1109/PESTSE.2014.6805286

Singh, S., Verma, R. K., Shakya, A. K., & Pratap Singh, S. (2017). Frequency Regulation of Microgrid Connected Hybrid Power System with SMES. *Technology and Economics Smart Grids Sustain Energy, 2*(1), 2–13. doi:10.100740866-017-0028-3

Sinha, N., Chakraborti, R., & Chattopadhyay, P. (2003). Evolutionary programming techniques for economic load dispatch. *IEEE Transactions on Evolutionary Computation, 7*(1), 83–94. doi:10.1109/TEVC.2002.806788

Sinha, S. K., Patel, R. N., & Prasad, R. (2010). Application of GA and PSO tuned fuzzy controller for AGC of three area thermal-hydropower system. *International Journal of Computer Theory and Engineering*, *2*(2), 1793–8201.

Slochanal & Shunmugalatha. (2008). Maximum loadability of a powersystem using multi agent-based hybrid particle swarm optimization. *Electric Power Components and Systems*, *36*(6), 575–86. doi.org/10.1080/15325000701801579

Society of Automotive Engineers (2012), Electric vehicle and plugin hybrid electric vehicle conductive charge coupler. *SAE Standard J, 1772*.

Sode-Yome, A., & Mithulananthan, N. (2004). Comparison of shunt capacitor, svc and STATCOM in static voltage stability margin enhancement. *International Journal of Electrical Engineering Education*, *41*(2), 158–171. doi:10.7227/IJEEE.41.2.7

Solar Energy Services for Professionals. (n.d.). Retrieved from http://www.soda-is.com/eng/services/service_invoke/gui.php?xml_descript=hc3v5_invoke_hour_demo.xml#parameters

Song, M., Alvehag, K., Widen, J., & Parisio, A. (2014). Estimating the impacts of demand response by simulating household behaviours under price and CO2 signals. *Electric Power Systems Research*, *111*, 103–114. doi:10.1016/j.epsr.2014.02.016

Sortomme, E., & El-Sharkawi, M. (2012). Optimal Scheduling of Vehicle-to-Grid Energy and Ancillary Services. *IEEE Transactions on Smart Grid*, *3*(1), 351–359. doi:10.1109/TSG.2011.2164099

Sousa, J. A. M., Roca, J. L. C., & Rolle, J. L. C. (2017b). Evolución del sector eléctrico tras la Segunda Guerra Mundial. *Dyna (Bilbao)*, *92*(3), 280–284.

Srivastava, S., Velayutham, A., Agrawal, K., & Bakshi, A. (2012). *Report of the enquiry committee on grid disturbance in northern region on 30th July 2012 and in northern, eastern & northeastern region on 31st July 2012*. Academic Press.

Ssu, K.-F., Ou, C.-H., & Jiau, H. C. (2005). Localization with mobile anchor points in wireless sensor networks. *IEEE Transactions on Vehicular Technology*, *54*(3), 1187–1197. doi:10.1109/TVT.2005.844642

Stagg, G. W., El-Abiad, A. H., & El-Abiad, A. (1968). *Computer methods in power system analysis* (Vol. 9). McGraw-Hill New York.

Sun, Y., Wang, Y., Wei, Z., Sun, G., & Wu, X. (2018). Robust H_∞ load frequency control of multi area power system with time delay: A sliding mode control approach. *IEEE/CAA Journal of Automatica Sinic*, *5*(2), 610-617.

Sun, Y., Bollen, M. H. J., & Ault, G. W. (2006). Probabilistic reliability evaluation of distribution systems with DER and Microgrids. *Proceeding of International Conference on Probabilistic Methods Applied to Power Systems*, 1-8. 10.1109/PMAPS.2006.360214

Talaq, J. H., El-Hawary, F., & El-Hawary, M. E. (1994). A summary of environmental/economic dispatch algorithms. *IEEE Transactions on Power Systems*, *9*(3), 150816. doi:10.1109/59.336110

Tamura, Y., Mori, H., & Iwamoto, S. (1983). Relationship between voltage instability and multiple load flow solutions in electric power systems. *IEEE Transactions on Power Apparatus and Systems*, *PAS-102*(5), 1115–1125. doi:10.1109/TPAS.1983.318052

Tan, W., Chang, S., & Zhou, R. (2017). Load frequency control of power system with nonlinearities. *IET Generation, Transmission & Distribution*, *11*(17), 4307–4313. doi:10.1049/iet-gtd.2017.0599

Tan, X. A., Li, Q. B., & Wang, H. A. (2013). Advances and trends of energy storage technology in Microgrid. *International Journal of Electrical Power & Energy Systems*, *44*(1), 179–191. doi:10.1016/j.ijepes.2012.07.015

Taylor & Erickson. (2005). Recording and analyzing the July 2 cascading outage Western USA power system. *IEEE Computer Applications in Power*, *10*(1), 26–30.

Taylor, C. W. (1994). *Power system voltage stability*. McGraw-Hill Companies.

Tazvinga, H., & Dzobo, O. (2018, April). Hybrid energy systems for rural communities in Zimbabwe. *EAI International Conference for Research, Innovation and Development for Africa*, 480-492. 10.4108/eai.20-6-2017.2270759

Tazvinga, H., Xia, X., & Zhang, J. (2013). Minimum cost solution of photovoltaic–diesel–battery hybrid power systems for remote consumers. *Solar Energy*, *96*, 292–299. doi:10.1016/j.solener.2013.07.030

Teeparthi, K., & Vinod Kumar, D. M. (2016). Grey wolf optimization algorithm based dynamic security constrained optimal power flow. *Proceedings of National Power System Conference 2016*. 10.1109/NPSC.2016.7858858

Thomas, D., Deblecker, O., & Ioakimidis, C. S. (2018). Optimal operation of an energy management system for a grid-connected smart building considering photovoltaics' uncertainty and stochastic electric vehicles' driving schedule. *Applied Energy*, *210*, 1188–1206. doi:10.1016/j.apenergy.2017.07.035

Thukaram, D., Jenkins, L., Khincha, H., Yesuratnam, G., & Kumar, B. R. (2004).Monitoring the effects of on-load tap changing transformers on voltage stability. In *International conference on Power system technology* (*Vol. 1*, pp. 419–424). Academic Press.

Thukaram, D., Jenkins, L., & Visakha, K. (2006). Optimum allocation of reactive power for voltage stability improvement in ac-dc power systems. *IEE Proceedings. Generation, Transmission and Distribution*, *153*(2), 237–246. doi:10.1049/ip-gtd:20045210

Thukaram, D., & Lomi, A. (2000). Selection of static VAR compensator location and size for system voltage stability improvement. *Electric Power Systems Research*, *54*(2), 139–150. doi:10.1016/S0378-7796(99)00082-6

Thukaram, D., & Parthasarathy, K. (1996). Optimal reactive power dispatch algorithm for voltage stability improvement. *International Journal of Electrical Power & Energy Systems*, *18*(7), 461–468. doi:10.1016/0142-0615(96)00004-X

Thukaram, D., Parthasarathy, K., & Prior, D. (1984). Improved algorithm for optimum reactive power allocation. *International Journal of Electrical Power & Energy Systems*, *6*(2), 72–74. doi:10.1016/0142-0615(84)90011-5

Thukaram, D., & Yesuratnam, G. (2006). Comparison of optimum reactive power schedule with different objectives using lp technique. *International Journal of Emerging Electric PowerSystems*, *7*(3), 1–31.

Thukaram, D., & Yesuratnam, G. (2008). Optimal reactive power dispatch in a large power system with ac-dc and facts controllers. *IET Generation, Transmission & Distribution*, *2*(1), 71–81. doi:10.1049/iet-gtd:20070163

Tina, G., & Gagliano, S. (2011). Probabilistic modelling of hybrid solar/wind power system with solar tracking system. *Renewable Energy*, *36*(6), 1719–1727. doi:10.1016/j.renene.2010.12.001

Tiranuchit, A., & Thomas, R. (1988). A posturing strategy against voltage instabilities in electric power systems. *IEEE Transactions on Power Systems*, *3*(1), 87–93. doi:10.1109/59.43177

Tomsovic, K. (1992). A fuzzy linear programming approach to the reactive power/voltage control problem. *IEEE Transactions on Power Systems*, *7*(1), 287–293. doi:10.1109/59.141716

Tsay, T. S. (2011). Load frequency control of interconnected power system with governor backlash nonlinearities. *International Journal of Electrical Power & Energy Systems*, *33*(9), 1542–1549. doi:10.1016/j.ijepes.2011.06.005

Tungadio, D. H., & Bansal, R. C. (2017). Active power reserve estimation of two interconnected Microgrids. *Energy Procedia*, *105*, 3909–3914. doi:10.1016/j.egypro.2017.03.807

Uehara, T., Ludin, G. A., Datta, M., & Senjyu, T. (2015). Frequency Control using Real-Time Pricing for Isolated Power Systems. *Proceedings of 9th International Conference on Power Electronics-ECCE Asia*. 10.1109/ICPE.2015.7167929

Urdaneta, A. J., Gomez, J. F., Sorrentino, E., Flores, L., & Diaz, R. (1999). A hybrid genetic algorithm for optimal reactive power planning based upon successive linear programming. *IEEE Transactions on Power Systems*, *14*(4), 1292–1298. doi:10.1109/59.801887

Utray Fabra, J. (n.d.). *El mercado de la electricidad: Algunas reflexiones y algunas propuestas*. Academic Press.

Van Cutsem. (1998). *Costas Vournas*. Voltage Stability of Electric Power Systems.

Van Cutsem, T. (1991). A method to compute reactive power margins with respect to voltage collapse. *IEEE Transactions on Power Systems*, *6*(1), 145–156. doi:10.1109/59.131057

Vega Vega, R., Quintián, H., Calvo-Rolle, J. L., Herrero, Á., & Corchado, E. (2018). Gaining deep knowledge of Android malware families through dimensionality reduction techniques. *Logic Journal of the IGPL*, *27*(2), 160–176. doi:10.1093/jigpal/jzy030

Velusami, S., & Ramar, K. (1997). Design of observer-based decentralized load-frequency controllers for interconnected power systems. *International Journal of Power and Energy Systems*, *17*(2), 152–160.

Venikov, V., Stroev, V., Idelchick, V., & Tarasov, V. (1975). Estimation of electrical power system steady-state stability in load flow calculations. *IEEE Transactions on Power Apparatus and Systems*, *94*(3), 1034–1041. doi:10.1109/T-PAS.1975.31937

Venkatesh, B., Arunagiri, A., & Gooi, H. (2003). Unified opf method for maximizing voltage stability margin using successive fuzzy LP. *Electric Power Systems Research*, *64*(2), 119–128. doi:10.1016/S0378-7796(02)00152-9

Venkatesh, P., Gnanadas, R., & Padhy, N. P. (2003). Comparison and application of evolutionary programming techniques to combined economic emission dispatch with line flow constrained. *IEEE Transactions on Power Systems*, *18*(2), 688–697. doi:10.1109/TPWRS.2003.811008

Visakha, K., Thukaram, D., & Jenkins, L. (2004a). Application of upfc for system security improvement under normal and network contingencies. *Electric Power Systems Research*, *70*(1), 46–55. doi:10.1016/j.epsr.2003.11.011

Visakha, K., Thukaram, D., & Jenkins, L. (2004b). An approach for real power scheduling to improve system stability margins under normal and network contingencies. *Electric Power Systems Research*, *71*(2), 109–117. doi:10.1016/j.epsr.2004.01.005

Vivekanandan, V., & Wong, V. W. S. (2007). Concentric anchor beacon localization algorithm for wireless sensor networks. *IEEE Transactions on Vehicular Technology*, *56*(5), 2733–2744. doi:10.1109/TVT.2007.899962

Wang, C., & Xiao, L. (2007). Sensor localization under limited measurement capabilities. *IEEE Network*, *21*(3), 16–23. doi:10.1109/MNET.2007.364254

Wang, F., Zhou, L., Ren, H., Liu, X., Talari, S., Shafie-khah, M., & Catalão, J. P. (2018). Multi-Objective Optimization Model of Source–Load–Storage Synergetic Dispatch for a Building Energy Management System Based on TOU Price Demand Response. *IEEE Transactions on Industry Applications*, *54*(2), 1017–1028. doi:10.1109/TIA.2017.2781639

Wang, W., Srinivasan, V., Wang, B., & Chua, K.-C. (2008). Coverage for target localization in wireless sensor networks. *IEEE Transactions on Wireless Communications*, *7*(2), 667–676. doi:10.1109/TWC.2008.060611

Wang, Y., Wang, C., Lin, F., Li, W., Wang, L. Y., & Zhao, J. (2013). Incorporating generator equivalent model into voltage stability analysis. *IEEE Transactions on Power Systems*, *28*(4), 4857–4866. doi:10.1109/TPWRS.2013.2273501

Winkler, H. (2005). Renewable energy policy in South Africa: Policy options for renewable electricity. *Energy Policy*, *33*(1), 27–38. doi:10.1016/S0301-4215(03)00195-2

Wong, K., & Wong, Y. (1994). Genetic and genetic/simulated-annealing approaches to economic dispatch. *IEEE Proc Generation Transmission Distribution, 141*(5), 507–13. 10.1049/ip-gtd:19941354

Wu, X., Cho, J., D'Aurio, B. J., & Lee, S. (2007). Mobility-assisted relocation for self-deployment in wireless sensor networks. *IEICE Transactions on Communications*, *E90-B*(8), 2056–2069. doi:10.1093/ietcom/e90-b.8.2056

Xiao, B., Chen, H., & Zhou, S. (2008). Distributed localization using a moving beacon in wireless sensor networks. *IEEE Transactions on Parallel and Distributed Systems*, *19*(5), 587–600. doi:10.1109/TPDS.2007.70773

Xia, X., & Elaiw, A. M. (2010). Optimal dynamic economic dispatch of generation: A review. *Electric Power Systems Research*, *80*(8), 975–986. doi:10.1016/j.epsr.2009.12.012

Yamashita, K., & Taniguchi, T. (1986). Optimal observer design for load frequency control. *International Journal of Electrical Power & Energy Systems*, *8*(2), 93–100. doi:10.1016/0142-0615(86)90003-7

Yang, H. X., Lu, L., & Burnett, J. (2003). Weather data and probability analysis of hybrid photovoltaic-wind power generation systems in Hong Kong. *Renewable Energy*, *28*(11), 1813–1824. doi:10.1016/S0960-1481(03)00015-6

Yesuratnam, G., & Thukaram, D. (2007). Optimum reactive power dispatch and identification of critical on-load tap changing (OLTC) transformers. *Electric Power Components and Systems*, *35*(6), 655–674. doi:10.1080/15325000601139641

Yilmaz, M., & Krein, P. (2013). Review of Battery Charger Topologies, Charging Power Levels, and Infrastructure for Plug-In Electric and Hybrid Vehicles. *IEEE Transactions on Power Electronics*, *28*(5), 2151–2169. doi:10.1109/TPEL.2012.2212917

Yoon, J. H., Bladick, R., & Novoselac, A. (2014). Demand response for residential buildings based on dynamic price of electricity. *Energy and Building*, *80*, 53141. doi:10.1016/j.enbuild.2014.05.002

Zargar, M. Y., Mufti, M. D., & Ahmed Lone, S. (2017). Voltage and frequency control of a hybrid Wind-Diesel system using SVC and predictively controlled SMES. *Proceedings of 6th International Conference on Computer Applications in Electrical Engineering-Recent Advances (CERA)*.

Zeinalzadeh, A., Mohammadi, Y., & Moradi, M. H. (2015). Optimal multi objective placement and sizing of multiple DGs and shunt capacitor banks simultaneously considering load uncertainty via MOPSO approach. *International Journal of Electrical Power & Energy Systems*, *67*, 336–349. doi:10.1016/j.ijepes.2014.12.010

Zhang, B., & Yu, F. (2008). An energy efficient localization algorithm for wireless sensor networks using a mobile anchor node. In *Proceedings of International Conference on Information and Automation* (pp. 215-219). Zhangjiajie, China: IEEE.

Zhang, C. K., Jiang, L., Wu, Q. H., He, Y., & Wu, M. (2013). Delay dependent robust load frequency control for time delay power systems. *IEEE Transactions on Power Systems*, *28*(3), 2192–2201. doi:10.1109/TPWRS.2012.2228281

Zhang, W., Li, F., & Tolbert, L. M. (2007). Review of reactive power planning: Objectives, constraints, and algorithms. *IEEE Transactions on Power Systems*, *22*(4), 2177–2186. doi:10.1109/TPWRS.2007.907452

Zhang, W., & Tolbert, L. M. (2005). Survey of reactive power planning methods. In *Power engineering society general meeting* (pp. 1430–1440). IEEE.

Zhang, Y., & Ren, Z. (2005). Optimal reactive power dispatch considering costs of adjusting the control devices. *IEEE Transactions on Power Systems*, *20*(3), 1349–1356. doi:10.1109/TPWRS.2005.851920

Zheng, K., Wang, H., Li, H., Xiang, W., Lei, L., Qiao, J., & Shen, X. S. (2017). Energy- efficient localization and tracking of mobile devices in wireless sensor networks. *IEEE Transactions on Vehicular Technology*, *66*(3), 2714–2726. doi:10.1109/TVT.2016.2584104

Zhenjie, X., & Changjia, C. (2006). A localization scheme with mobile beacon for wireless sensor networks. In *Proceedings of 6th International Conference on ITS Telecommunications* (pp. 1017-1020). Chengdu, China: IEEE. 10.1109/ITST.2006.288725

Zhou, X., Wang, P., & Gao, Z. (2018). ADMM-Based Decentralized Charging Control of Plug-In Electric Vehicles with Coupling Constraints in Distribution Networks. In *37th Chinese Control Conference (CCC)* (pp. 2512-2517). Wuhan, China: IEEE. 10.23919/ChiCC.2018.8483082

Zhou, X., Xiao, Y. C., & Jian, X. J. (2011). Development of SMES Technology and Its Applications in Power Grid. *Proceedings of IEEE International Conference on Applied Superconductivity and Electromagnetic Devices, 260-269.*

Zhu, J., & Xiong, X. (2003). Optimal reactive power control using modified interior point method. *Electric Power Systems Research*, *66*(2), 187–192. doi:10.1016/S0378-7796(03)00078-6

Zobian, A., & Ilic, M. D. (1996). A steady state voltage monitoring and control algorithm using localized least square minimization of load voltage deviations. *IEEE Transactions on Power Systems*, *11*(2), 929–938. doi:10.1109/59.496177

Zooghby, A. E. (2005). *Smart antenna engineering*. Boston: Artech House.

Zou, Y., & Chakrabarty, K. (2004). Sensor deployment and target localization in distributed sensor networks. *ACM Transactions on Embedded Computing Systems*, *3*(1), 61–91. doi:10.1145/972627.972631

About the Contributors

Juan Aurelio was born in Ferrol (A Coruña), Spain, in 1965. He received Degree in Law from UNED in 2000, Degree in Geography and History from the University of Santiago de Compostela in 1990 and Social Graduate from the University of A Coruña in 1994. Since 2005 he is a professor-tutor at the National University of Distance Education (UNED). Currently, I do my doctoral thesis in the Official Doctoral Program in Energy and Marine Propulsion, with the title Study of energy storage based strategies for the optimization of generation and consumption. I am the author of several articles in journals with impact factor in JCR and has presented several press releases in national congresses.

Kehinde Awodele is a senior lecturer in the Department of Electrical Engineering at the University of Cape Town.

Sandeep Bhongade (1974) received M.E. degree in Electrical Engineering from V.J.T.I. Mumbai, Mumbai University (India) in 2003 and Ph. D degree from IIT Roorkee in 2012. Presently, he is a faculty member in the Electrical Engineering Department at Shri G.S Institute of Technology & Science, Indore (M.P)-India. His research interests include Smart microgrids, Power System Restructuring, Power System Operations and Control, Energy management system, Distributed Generation, Renewable Energy Sources.

Rathindra Nath Biswas received his M.E degree in Electronics and Tele-Communication Engineering from Jadavpur University, India in 2008. Currently, he is associated as lecturer in Electronics and Tele-Communication Engineering at Acharya Jagadish Chandra Bose Polytechnic, India since 2008.

Jose Luis Calvo-Rolle was born in A Coruna, Spain, in 1974. He received the M.S. and Ph.D. degrees in Industrial Engineering from the University of Leon, Leon, Spain, in 2004, and 2007, respectively. He is Associate Professor of Automatic Control and the head of Industrial Engineering Department, Faculty of Engineering, University of A Coruna, Spain. His main research interests have been centered in applying expert system technology to the diagnosis and control systems and in intelligent training systems for control engineering, optimization and education.

José Luis Casteleiro-Roca received the B.S. from University of Coruna in 2003, the M.S. in Industrial Engineering from the University of Leon in 2012, and now he is a Ph.D student in the University of Coruna. He has been Technical Engineer in the Spanish Navy since 2004 (in the Arsenal from Ferrol); he worked in Missiles workshop for 11 years, performing tests and certifying the SM-1 and SM-2 missiles from the Spanish Navy. Now, he works in the fire weapon control workshop, performing the maintenance and repairing different weapons in the F-100 type vessels. Since 2014, he is also part of the teaching and research staff of the UDC as a part-time associate professor in the Industrial Engineering Department.

Hemanthakuamar Chappa received M.Tech degree from Maulana Azad National Institute of Technology(MANIT), Bhopal, India. He is currently working towards Ph.D degree in Department of Electical Engineering, MANIT, Bhopal. He also got Commonwealth Split-site fellowship to study in Durham University, UK. His research area includes power system stability, synchrophasor application to power systems and wide area monitoring of power systems.

Ark Dev received the M.Tech degree in Electrical Engineering from National Institute of Technology Manipur, India in 2017. He is currently pursuing Ph.D in Electrical Engineering from National Institute of Technology Manipur, India. His research interest is in the field of control systems, nonlinear control, sliding mode controller and observer design.

Thukaram Dhadbanjan received the B.E. degree in electrical engineering from Osmania University, Hyderabad, India, in 1974, the M.Tech. degree in integrated power systems from Nagpur University, Nagpur, India, in 1976, and the Ph.D. degree from the Indian Institute of Science, Bangalore, India, in 1986. Since 1976, he has been with the Indian Institute of Science, Bangalore, India, as a Research Fellow and Member of Faculty in various positions, where, since 2002, he has been a Professor with the Department of Electrical Engineering. His research interests include computer-aided power system analysis, reactive power optimization, voltage stability, and AI applications in power systems. He is a senior member of IEEE (USA).

Oliver Dzobo is a Senior Lecturer in the Department of Electrical & Electronic Science Engineering at the University of Johannesburg, South Africa, since November 2016. Prior to joining University of Johannesburg, I have been involved in various projects including the national energy efficiency audit project in Zimbabwe. Current research topics I am pursuing involves power system control and the implementation of new energy system optimization algorithms in smart grids that will provide the electricity user and the power utility with the capability to manage risk, pollutants emission, and energy efficiency directly, along with reducing costs, the possibility of taking financial positions where available, long term and inter-period Conditional Value-at risk (CVaR) constraints.

Luis Alfonso Fernandez-Serantes was born in Viveiro, Spain, in 1992. He received the BSc. in Electrical Engineering from the University of A Coruna in 2014, Spain, and the MSc. in Engineering from the FH Joanneum University in Applied Sciences in 2016, Austria. After a research experience in Infineon Technologies AG doing the master thesis, he joined the FH Joanneum as Power Electronics Researcher in the Institute of Electronic Engineering, in Austria. His research has been focused in power electronic systems.

Tomás González-Ayuso earned a Bachelor's degree in chemistry in 1982 from the Complutense University of Madrid. 1981 – 1986. Predoctoral work at the Institute of Catalysis and Petrochemistry (ICP-CSIC). Study of surfacial and textural properties of natural silicates and its application as catalytic supports. Preparation and test of different types of supported metal catalysts (nickel, platinum, rhodium).

1986 – 2003. Chemical engineer in Técnicas Reunidas SA, Tecnología Grupo INI, Tecnología y Gestión de la Innovación, Fusinco SL, and Olea Madrid SL., Hydrometallurgy, electrochemistry and fuel cells. Collaboration in several projects of development of molten carbonate fuel cells (Picon project: Development of molten carbonates fuel cell technology. Activities of research transfer of technology of molten carbonates fuel cells from International Fuel Cells to Spanish program of molten carbonate fuel cells. Collaboration with the Institute of Catalysis and the Ciemat in the development of components for molten carbonate fuel cells. Collaboration in the construction and operation of 500 kW Molten carbonate fuel cell plant (S Agustín de Guadalix). Assembling and test of molten carbonate fuel cells from watts to 5 kW. 2004 – 2008. Researcher in the Ciemat. Participation in projects of Unit of Fuel cells and systems integration. Development and fabrication of components for molten carbonates fuel cells. Assembling and test of molten carbonate fuel cells from watts to 5 kW. Fuel cells integration and applications. Doctoral thesis in 2007 "Nickel-Cobalt cathodes for molten carbonates fuel cells" presented in the Autónoma University of Madrid. 2009 – Today. Head of Group of systems integration (2009). Head of Unit of Fuel cells and System Integration of CIEMAT (2012). Participation on projects related to fuel cells applications to integrate with renewable energy and development of applications to integrated systems. Integration of PEMFC with photovoltaic panels, electrolyzers, hydrogen storage as hydrides, batteries. Control software. Integration of PEMFC with a motorized chair. Integration of PEMFC into an uninterrupted power system to increase its autonomy. Design and construction of test stations for fuel cells testing. Design and construction of a multifuel processor (GN, biogas, bioethanol) to supply a hydrogen rich flow to fuel cells. Program of utilization and optimization of energy resources of the Community of Madrid through the validation of the fuel cells technologies, PEMFC and SOFC – ENERCAM (S-0505/ENE/000304, 2006-2009) and Energy diversification through generation systems based on fuel cells, DIVERCEL-CM Ref: (S2009/Jan-1475, 2010-2014).

Esteban Jove-Pérez received the Master's Degree in Industrial Engineering from the University of Leon, Leon, Spain, in 2014. After two years working in the automotive industry, he joined the University of A Coruna, Spain, where he is Professor of Power Electronics in the Faculty of Engineering since 2016. He is a Ph.D student in the University of La Laguna and his research has been focused on using intelligent techniques to model nonlinear systems.

Tushar Kumar is currently pursuing their Ph.D from Maulana Azad National Institute of Technology Bhopal. Their area of interests include Electricity markets, renewable Energy, and Smart grids.

M. Maheswari is currently working as Professor in the Department of Electrical and Electronics Engineering in Nalla Malla Reddy Engineering College, Hyderabad.She completed her B.E. Electrical and Electronics Engineering degree in K.S.Rangasamy College of TechnologyThiruchengode with first class with distinction, M.E. Power Systems in Government College of Technology, Coimbatore and with first class with distinction and Ph.D. at Anna University, Chennai in the year 2015. She has 18 years of teaching experience and various administrative positions. She published her research work in 15 various international journals and 14 International Conferences. She is also a reviewer of leading journals.

Swarup Kumar Mitra received his B.Tech Degree in Electronics and Tele-Communication Engineering from Kalyani University India in 2000. and has achieved his M.Tech in VLSI Design and Microelectronics Technology from Jadavpur University, India, 2007. He has more than ten years of teaching experience He is attached as Associate Professor in Department of ECE MCKV Institute of Engineering, Liluah, Howrah India and awarded with Ph.D. (Engg.) in "Studies in Data gathering Schemes in Wireless Sensor Networks" from Jadavpur University. His present area of research in Wireless Sensor Network and its architecture, etc.

Tukaram Moger received B.E. degree in electrical and electronics engineering from Karnatak University, Dharwad, Karnataka, India, in 2001, M.Tech. degree in electrical engineering from Indian Institute of Technology Kanpur (IITK), UP, India, in 2005, and Ph.D. degree in electrical engineering from Indian Institute of Science (IISc) Bengaluru, India, in 2016. He has been associated with academic institutions as a faculty member and currently working as assistant professor in the Department of Electrical and Electronics Engineering, National Institute of Technology Karnataka (NITK), Surathkal, Mangaluru, India. His research interests include grid Integration of renewable energy, solar photovoltaic systems, microgrid, reactive power and voltage control, power system operation and planning, power system deregulation. He is a senior member of IEEE (USA), IEEE Power & Energy Society (PES), member of IEEE Eta-Kappa Nu (Mu Xi Chapter of IISc), IET (UK), CIGRE, Institution of Engineers (India), and life member of Indian Society for Technical Education (ISTE), System Society of India (SSI) and Soft Computing Research Society (SCRS) of India. He also holds Chartered Engineer (India) certificate.

Suvabrata Mukherjee was born in 1985 in Asansol, West Bengal, India. He received his B-Tech degree in Electrical Engineering from Asansol Engineering College, Asansol, Burdwan, India in 2007; M-Tech degree from Asansol Engineering College, Asansol, India in 2013. Presently he is working as Assistant Professor in the department of Electrical and Electronics Engineering, NSHM knowledge campus, Durgapur, India. His field of research interest includes Load flow analysis, Optimal Power flow, Power system state estimation, Evolutionary computing techniques.

Mrinal Kanti Naskar received his B.Tech. (Hons) and M.Tech from E&ECE Department, IIT Kharagpur, India in 1987 and 1989 respectively. He served as a faculty member at NIT, Jamshedpur and NIT, Durgapur during 1991–1996 and 1996–1999 respectively. Currently, he is a Professor in the Department of Electronics and Tele-Communication Engineering, Jadavpur University, Kolkata, India. His research interests include Computer Networks and Embedded Computing.

Héctor Quintián was born in A Coruna, Spain. He received the M.S. degree in Industrial Engineering from the University of Leon, Leon, Spain, in 2010, and Ph.D. degree in Computer Science from University of Salamanca in 2017. He is Assistance Professor of Automatic Control and the head of Industrial Engineering Department, Faculty of Engineering, University of A Coruna, Spain. His main research interests have been centered in neural networks, unsupervised learning and control systems and in intelligent training systems for control engineering, optimization and education.

Mahiraj Singh Rawat received the B. Tech. in Electrical and Electronics Engineering form U.P. Technical University in 2007 and M. Tech. in Power System from National Institute of Technology, Hamirpur, in 2011. He is currently working as an Assistant Professor in Electrical Engineering Department at National Insititute of Technology, Uttarakhand, India. He is also pursuing Ph. D. in Electrical Engineering from National Institute of Technology, Kurukshetra, Haryana, India. His current research is focused on voltage stability, renewable energy integration and distributed generation.

Provas Kumar Roy was born in 1973 at Mejia, Bankura, West Bengal, India. He Received the B.E Degree in Electrical Engineering from R. E. College, Durgapur, Burdwan, India in 1997; M.E Degree in Electrical Machine from Jadavpur University, Kolkata, India in 2001 and Ph.D from NIT Durgapur in 2011. Presently he is working as Professor in the department of Electrical Engineering, Kalyani Government Engineering College, Kalyani, West Bengal, India. His field of research interest includes Economic Load Dispatch, Optimal Power flow, FACTS, Unit Commitment, Automatic Generation Control, Power System Stabilizer, Radial Distribution System, State Estimation and Evolutionary computing techniques.

Mrinal Kanti Sarkar completed his Ph.D in Electrical Engineering from National Institute of Technology Durgapur, India in the year 2015. Currently, he is an Assistant Professor and was former Head of the Department of Electrical Engineering at National Institute of Technology Manipur, India. His research interest is in the field of magnetic levitation systems, DC-DC converter control, sliding mode controller and observer design.

Yanxia Sun is an associate professor at University of Johannesburg. Her research is focussed on particle swarm optimization, artificial intelligence, advanced control theory and nonlinear dynamics. This goes hand in hand with other problems like the application of optimization algorithms to control an unmanned aerial vehicle, applied the neural networks to the optimization algorithms.

Shelly Vadhera received the B. Eng. in electrical engineering from Thapar University in 1994, and the M.Tech. and Ph. D. from National Institute of Technology, Kurukshetra in 1998 and 2013 respectively. She is currently working as an Associate Professor in the Department of Electrical Engineering at National Institute of Technology, Kurukshetra, Haryana, India. Her current research is focused on artificial intelligence techniques, renewable energy application in power system.

Suma Veeraganti is working as an Assistant Professor in Electrical and Electronics Engineering at Malla Reddy Enginnering College. She has 6 years of Teaching experience. She completed M.Tech in Power Systems at KL University, Vijayawada, Andhra Pradesh. She had completed B.Tech in EEE at V.R.Siddhartha Engineering College, Vijayawada, Andhra Pradesh. Her Research interests are Power system Deregulation, power system stability.

Ritu Verma is doing her Master's in Power Electronics from the Department of Electrical Engineering, at Shri G.S Institute of Technology & Science, Indore (M.P)-India. Her area of interest includes Micro-Grid, Control system application in different areas, Power system.

Xose Manuel Vilar Martinez was born in Mugardos, Spain, in 1963. He received the B.S. degree in Industrial Engineering from the University of Coruña, Coruña, Spain, in 1998. Also he received the M.S. degree in Industrial Engineering from the University of Leon, Leon, Spain, in 2012 and the M.S. degree in Marine Engineering from the University of Coruña 2014. He works in Indra Sistemas S.A. since 2000, and he was partial time professional teacher at the Industrial Engineering Department, Faculty of Engineering, University of A Coruna. His main research interests have been focused in applying expert systems technology to the diagnosis and control systems.

Index

www.ingramcontent.com/pod-product-compliance
Lightning Source LLC
LaVergne TN
LVHW062117300926
849161LV00010B/2482